The Emperor's New Mind

The
Emperor's New Mind

Concerning Computers, Minds, and The Laws of Physics

ROGER PENROSE

Rouse Ball Professor of Mathematics
University of Oxford

Foreword by
MARTIN GARDNER

OXFORD UNIVERSITY PRESS
NEW YORK · OXFORD
1989

Oxford University Press, Walton Street, Oxford OX2 6DP
Oxford New York Toronto
Delhi Bombay Calcutta Madras Karachi
Petaling Jaya Singapore Hong Kong Tokyo
Nairobi Dar es Salaam Cape Town
Melbourne Auckland

and associated companies in
Berlin Ibadan

Oxford is a trade mark of Oxford University Press

Published in the United States
by Oxford University Press, New York

British Library Cataloguing in Publication Data
The Emperor's New Mind
I. Man. Cognition related to
artificial intelligence
1. Penrose, Roger
153.4 BF311
ISBN 0–19–851973–7

Library of Congress Cataloging-in-Publication Data
Penrose, Roger
The emperor's new mind concerning computers, minds, and the laws of physics/Roger Penrose
Bibliography: p. Includes index.
1. Artificial intelligence. 2. Thought and thinking. 3. Science–Philosophy. 4. Computers. I. Title.
Q335.P415 1989
006.3—dc20 89–8548 CIP
ISBN 0–19–851973–7

9 8 7 6 5 4 3 2
Printed in the United States
on acid-free paper.

Foreword
by Martin Gardner

Many great mathematicians and physicists find it difficult, if not impossible, to write a book that non-professionals can understand. Until this year one might have supposed that Roger Penrose, one of the world's most knowledgeable and creative mathematical physicists, belonged to such a class. Those of us who had read his non-technical articles and lectures knew better. Even so, it came as a delightful surprise to find that Penrose had taken time off from his labours to produce a marvellous book for informed laymen. It is a book that I believe will become a classic.

Although Penrose's chapters range widely over relativity theory, quantum mechanics, and cosmology, their central concern is what philosophers call the 'mind–body problem'. For decades now the proponents of 'strong AI' (Artificial Intelligence) have tried to persuade us that it is only a matter of a century or two (some have lowered the time to fifty years!) until electronic computers will be doing everything a human mind can do. Stimulated by science fiction read in their youth, and convinced that our minds are simply 'computers made of meat' (as Marvin Minsky once put it), they take for granted that pleasure and pain, the appreciation of beauty and humour, consciousness, and free will are capacities that will emerge naturally when electronic robots become sufficiently complex in their algorithmic behaviour.

Some philosophers of science (notably John Searle, whose notorious Chinese room thought experiment is discussed in depth by Penrose), strongly disagree. To them a computer is not essentially different from mechanical calculators that operate with wheels, levers, or anything that transmits signals. (One can base a computer on rolling marbles or water moving through pipes.) Because electricity travels through wires faster than other forms of energy (except light) it can twiddle symbols more rapidly than mechanical calculators, and therefore handle tasks of enormous complexity. But does an electrical computer 'understand' what it is doing in a way that is superior to the 'understanding' of an abacus? Computers now play grand-master chess. Do they 'understand' the game any better than a tick-tack-toe machine that a group of computer hackers once constructed with tinker toys?

Penrose's book is the most powerful attack yet written on strong AI. Objections have been raised in past centuries to the reductionist claim that a mind is a machine operated by known laws of physics, but Penrose's offensive is more persuasive because it draws on information not available to earlier writers. The book reveals Penrose to be more than a mathematical physicist. He is also a philosopher of first rank, unafraid to grapple with problems that

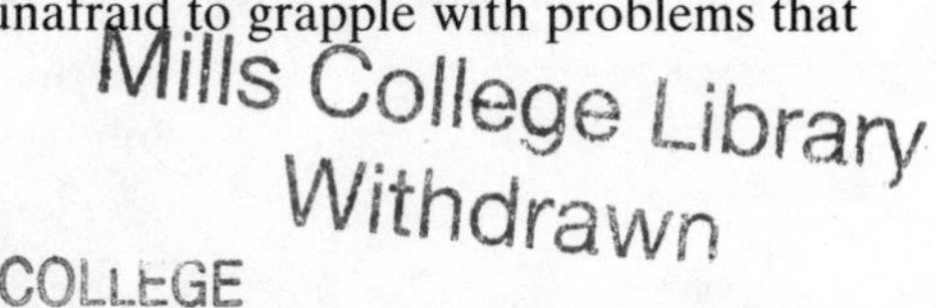

contemporary philosophers tend to dismiss as meaningless.

Penrose also has the courage to affirm, contrary to a growing denial by a small group of physicists, a robust realism. Not only is the universe 'out there', but mathematical truth also has its own mysterious independence and timelessness. Like Newton and Einstein, Penrose has a profound sense of humility and awe toward both the physical world and the Platonic realm of pure mathematics. The distinguished number theorist Paul Erdös likes to speak of 'God's book' in which all the best proofs are recorded. Mathematicians are occasionally allowed to glimpse part of a page. When a physicist or a mathematician experiences a sudden 'aha' insight, Penrose believes, it is more than just something 'conjured up by complicated calculation'. It is mind making contact for a moment with objective truth. Could it be, he wonders, that Plato's world and the physical world (which physicists have now dissolved into mathematics) are really one and the same?

Many pages in Penrose's book are devoted to a famous fractal-like structure called the Mandelbrot set after Benoit Mandelbrot who discovered it. Although self-similar in a statistical sense as portions of it are enlarged, its infinitely convoluted pattern keeps changing in unpredictable ways. Penrose finds it incomprehensible (as do I) that anyone could suppose that this exotic structure is not as much 'out there' as Mount Everest is, subject to exploration in the way a jungle is explored.

Penrose is one of an increasingly large band of physicists who think Einstein was not being stubborn or muddle-headed when he said his 'little finger' told him that quantum mechanics is incomplete. To support this contention, Penrose takes you on a dazzling tour that covers such topics as complex numbers, Turing machines, complexity theory, the bewildering paradoxes of quantum mechanics, formal systems, Gödel undecidability, phase spaces, Hilbert spaces, black holes, white holes, Hawking radiation, entropy, the structure of the brain, and scores of other topics at the heart of current speculations. Are dogs and cats 'conscious' of themselves? Is it possible in theory for a matter-transmission machine to translocate a person from here to there the way astronauts are beamed up and down in television's *Star Trek* series? What is the survival value that evolution found in producing consciousness? Is there a level beyond quantum mechanics in which the direction of time and the distinction between right and left are firmly embedded? Are the laws of quantum mechanics, perhaps even deeper laws, essential for the operation of a mind?

To the last two questions Penrose answers yes. His famous theory of 'twistors'—abstract geometrical objects which operate in a higher-dimensional complex space that underlies space–time—are too technical for inclusion in this book. They are Penrose's efforts over two decades to probe a region deeper than the fields and particles of quantum mechanics. In his fourfold classification of theories as superb, useful, tentative, and misguided,

Penrose modestly puts twistor theory in the tentative class, along with superstrings and other grand unification schemes now hotly debated.

Since 1973 Penrose has been the Rouse Ball Professor of Mathematics at Oxford University. The title is appropriate because W. W. Rouse Ball not only was a noted mathematician, he was also an amateur magician with such an ardent interest in recreational mathematics that he wrote the classic English work on this field, *Mathematical Recreations and Essays.* Penrose shares Ball's enthusiasm for play. In his youth he discovered an 'impossible object' called a 'tribar'. (An impossible object is a drawing of a solid figure that cannot exist because it embodies self-contradictory elements.) He and his father Lionel, a geneticist, turned the tribar into the Penrose Staircase, a structure that Maurits Escher used in two well-known lithographs: *Ascending and Descending*, and *Waterfall.* One day when Penrose was lying in bed, in what he called a 'fit of madness', he visualized an impossible object in four-dimensional space. It is something, he said, that a four-space creature, if it came upon it, would exclaim 'My God, what's that?'

During the 1960s, when Penrose worked on cosmology with his friend Stephen Hawking, he made what is perhaps his best known discovery. If relativity theory holds 'all the way down', there must be a singularity in every black hole where the laws of physics no longer apply. Even this achievement has been eclipsed in recent years by Penrose's construction of two shapes that tile the plane, in the manner of an Escher tesselation, but which can tile it only in a non-periodic way. (You can read about these amazing shapes in my book *Penrose tiles to trapdoor ciphers.*) Penrose invented them, or rather discovered them, without any expectation they would be useful. To everybody's astonishment it turned out that three-dimensional forms of his tiles may underlie a strange new kind of matter. Studying these 'quasicrystals' is now one of the most active research areas in crystallography. It is also the most dramatic instance in modern times of how playful mathematics can have unanticipated applications.

Penrose's achievements in mathematics and physics—and I have touched on only a small fraction—spring from a lifelong sense of wonder toward the mystery and beauty of being. His little finger tells him that the human mind is more than just a collection of tiny wires and switches. The Adam of his prologue and epilogue is partly a symbol of the dawn of consciousness in the slow evolution of sentient life. To me he is also Penrose—the child sitting in the third row, a distance back from the leaders of AI—who dares to suggest that the emperors of strong AI have no clothes. Many of Penrose's opinions are infused with humour, but this one is no laughing matter.

Note to the reader:
on reading mathematical equations

At a number of places in this book I have resorted to the use of mathematical formulae, unabashed and unheeding of warnings that are frequently given: that each such formula will cut down the general readership by half. If you are a reader who finds any formula intimidating (and most people do), then I recommend a procedure that I normally adopt myself when such an offending line presents itself. The procedure is, more or less, to ignore that line completely and to skip over to the next actual line of text! Well, not exactly this; one should spare the poor formula a perusing, rather than a comprehending glance, and then press onwards. After a little, if armed with new confidence, one may return to that neglected formula and try to pick out some salient features. The text itself may be helpful in letting one know what is important and what can be safely ignored about it. If not, then do not be afraid to leave a formula behind altogether.

Acknowledgements

There are many who have helped me, in one way or another, in the writing of this book, and to whom thanks are due. In particular, there are those proponents of strong AI (especially those who were involved in a BBC TV programme I once watched) who, by the expressions of such extreme AI opinions, had goaded me, a number of years ago, into embarking upon this project. (Yet, had I known of the future labours that the writing would involve me in, I fear, now, that I should not have started!) Many people have perused versions of small parts of the manuscript and have provided me with many helpful suggestions for improvement; and to them, I also offer my thanks: Toby Bailey, David Deutsch (who was also greatly helpful in checking my Turing machine specifications), Stuart Hampshire, Jim Hartle, Lane Hughston, Angus McIntyre, Mary Jane Mowat, Tristan Needham, Ted Newman, Eric Penrose, Toby Penrose, Wolfgang Rindler, Engelbert Schücking, and Dennis Sciama. Christopher Penrose's help with detailed information concerning the Mandelbrot set is especially appreciated, as is that of Jonathan Penrose, for his useful information concerning chess computers. Special thanks go to Colin Blakemore, Erich Harth, and David Hubel for reading and checking over Chapter 9, which concerns a subject on which I am certainly no expert—though, as with all others whom I thank, they are in no way responsible for the errors which remain. I thank NSF for support under contracts DMS 84–05644, DMS 86–06488, and PHY 86–12424. I am greatly indebted, also, to Martin Gardner for his extreme generosity in providing the foreword to this work, and also for some specific comments. Most particularly, I thank my beloved Vanessa, for her careful and detailed criticism of several chapters, for much invaluable assistance with references and, by no means least, for putting up with me when I have been at my most insufferable—and for her deep love and support where it was vitally needed.

Dedication

I dedicate this book to the loving memory of my dear mother, who did not quite live to see it.

Figure acknowledgements

The publishers either have sought or are grateful to the following for permission to reproduce illustration material.

Figs 4.6 and 4.9 from D. A. Klarner (ed.), *The mathematical Gardner* (Wadsworth International, 1981).

Fig. 4.7 from B. Grunbaum and G. C. Shephard, *Tilings and patterns* (W. H. Freeman, 1987). Copyright © 1987 by W. H. Freeman and Company. Used by permission.

Fig. 4.10 from K. Chandrasekharan, *Hermann Weyl* 1885–1985 (Springer, 1986).

Figs 4.11 and 10.3 from Pentaplexity: a class of non-periodic tilings of the plane. *The Mathematical Intelligencer*, **2**, 32–7 (Springer, 1979).

Fig. 4.12 from H. S. M. Coxeter, M. Emmer, R. Penrose, and M. L. Teuber (eds), *M. C. Escher: Art and Science* (North-Holland, 1986).

Fig. 5.2 © 1989 M. C. Escher Heirs/Cordon Art—Baarn—Holland.

Fig. 10.4 from *Journal of Materials Research*, **2**, 1–4 (Materials Research Society, 1987).

All other figures by the author.

Contents

Prologue

There was a great gathering in the Grand Auditorium, marking the initiation of the new 'Ultronic' computer. President Pollo had just finished his opening speech. He was glad of that: he did not much care for such occasions and knew nothing of computers, save the fact that this one was going to gain him a great deal of time. He had been assured by the manufacturers that, amongst its many duties, it would be able to take over all those awkward decisions of State that he found so irksome. It had better do so, considering the amount of treasury gold that he had spent on it. He looked forward to being able to enjoy many long hours playing golf on his magnificent private golf course—one of the few remaining sizeable green areas left in his tiny country.

Adam felt privileged to be among those attending this opening ceremony. He sat in the third row. Two rows in front of him was his mother, a chief technocrat involved in Ultronic's design. His father, as it happened, was also there—uninvited at the back of the hall, and now completely surrounded by security guards. At the last minute Adam's father had tried to blow up the computer. He had assigned himself this duty, as the self-styled 'chairspirit' of a small group of fringe activists: The Grand Council for Psychic Consciousness. Of course he and all his explosives had been spotted at once by numerous electronic and chemical sensing devices. As a small part of his punishment he would have to witness the turning-on ceremony.

Adam had little feeling for either parent. Perhaps such feelings were not necessary for him. For all of his thirteen years he had been brought up in great material luxury, almost entirely by computers. He could have anything he wished for, merely at the touch of a button: food, drink, companionship, and entertainment, and also education whenever he felt the need—always illustrated by appealing and colourful graphic displays. His mother's position had made all this possible.

Now the Chief Designer was nearing the end of *his* speech: '. . . has over 10^{17} logical units. That's more than the number of neurons in the combined brains of everyone in the entire country! Its intelligence will be unimaginable. But fortunately we do not need to imagine it. In a moment we shall all have the privilege of witnessing this intelligence at first hand: I call upon the esteemed First Lady of our great country, Madame Isabella Pollo, to throw the switch which will turn on our fantastic Ultronic Computer!'

The President's wife moved forward. Just a little nervously, and fumbling a little, she threw the switch. There was a hush, and an almost imperceptible dimming of lights as the 10^{17} logical units became activated. Everyone waited, not quite knowing what to expect. 'Now is there anyone in the audience who would like to initiate our new Ultronic Computer System by asking it its first question?' asked the Chief Designer. Everyone felt bashful, afraid to seem stupid before the crowd—and before the New Omnipresence. There was silence. 'Surely there must be someone?' he pleaded. But all were afraid, seeming to sense a new and all-powerful consciousness. Adam did not feel the same awe. He had grown up with computers since birth. He almost knew what it might feel like to *be* a computer. At least he thought perhaps he did. Anyway, he was curious. Adam raised his hand. 'Ah yes,' said the Chief Designer, 'the little lad in the third row. You have a question for our—ah—new friend?'

1
Can a computer have a mind?

Introduction

Over the past few decades, electronic computer technology has made enormous strides. Moreover, there can be little doubt that in the decades to follow, there will be further great advances in speed, capacity and logical design. The computers of today may be made to seem as sluggish and primitive as the mechanical calculators of yesteryear now appear to us. There is something almost frightening about the pace of development. Already computers are able to perform numerous tasks that had previously been the exclusive province of human thinking, with a speed and accuracy which far outstrip anything that a human being can achieve. We have long been accustomed to machinery which easily out-performs us in *physical* ways. *That* causes us no distress. On the contrary, we are only too pleased to have devices which regularly propel us at great speeds across the ground—a good five times as fast as the swiftest human athlete—or that can dig holes or demolish unwanted structures at rates which would put teams of dozens of men to shame. We are even more delighted to have machines that can enable us physically to do things we have never been able to do before: they can lift us into the sky and deposit us at the other side of an ocean in a matter of hours. These achievements do not worry our pride. But to be able to *think*—that has been a very human prerogative. It has, after all, been that ability to think which, when translated to physical terms, has enabled us to transcend our physical limitations and which has seemed to set us above our fellow creatures in achievement. If machines can one day excel us in that one important quality in which we have believed ourselves to be superior, shall we not then have surrendered that unique superiority to our creations?

The question of whether a mechanical device could ever be said to think—perhaps even to experience feelings, or to have a mind—is not really a new one.[1] But it has been given a new impetus, even an urgency, by the advent of modern computer technology. The question touches upon deep issues of philosophy. What does it mean to think or to feel? What is a mind? Do minds really exist? Assuming that they do, to what extent are minds functionally dependent upon the physical structures with which they are associated? Might minds be able to exist quite independently of such structures? Or are they simply the functionings of (appropriate kinds of) physical structure? In any case, is it necessary that the relevant structures be biological in nature (brains), or might minds equally well be associated with pieces of electronic equipment? Are minds subject to the laws of physics? What, indeed, *are* the laws of physics?

These are among the issues I shall be attempting to address in this book. To ask for definitive answers to such grandiose questions would, of course, be a tall order. Such answers I cannot provide: nor can anyone else, though some may try to impress us with their guesses. My own guesses will have important roles to play in what follows, but I shall try to be clear in distinguishing such speculation from hard scientific fact, and I shall try also to be clear about the reasons underlying my speculations. My main purpose here, however, is not so much to attempt to guess answers. It is rather to raise certain apparently new issues concerning the relation between the structure of physical law, the nature of mathematics and of conscious thinking, and to present a viewpoint that I have not seen expressed before. It is a viewpoint that I cannot adequately describe in a few words; and this is one reason for my desire to present things in a book of this length. But briefly, and perhaps a little misleadingly, I can at least state that my point of view entails that it is our present lack of understanding of the fundamental laws of physics that prevents us from coming to grips with the concept of 'mind' in physical or logical terms. By this I do not mean that the laws will never be that well known. On the contrary, part of the aim of this work is to attempt to stimulate future research in directions which seem to be promising in this respect, and to try to make certain fairly specific, and apparently new, suggestions about the place that 'mind' might actually occupy within a development of the physics that we know.

I should make clear that my point of view is an unconventional one among physicists and is consequently one which is unlikely to be adopted, at present, by computer scientists or physiologists. Most physicists would claim that the fundamental laws operative at the scale of a human brain are indeed all perfectly well known. It would, of course, not be disputed that there are still many gaps in our knowledge of physics generally. For example, we do not know the basic laws governing the mass-values of the sub-atomic particles of nature nor the strengths of their interactions. We do not know how to make quantum theory fully consistent with Einstein's special theory of relativity—let alone how to construct the 'quantum gravity' theory that would make quantum theory consistent with his *general* theory of relativity. As a consequence of the latter, we do not understand the nature of space at the absurdly tiny scale of 1/100 000 000 000 000 000 000 of the dimension of the known fundamental particles, though at dimensions larger than that our knowledge is presumed adequate. We do not know whether the universe as a whole is finite or infinite in extent—either in space or in time—though such uncertainties would appear to have no bearing whatever on physics at the human scale. We do not understand the physics that must operate at the cores of black holes nor at the big-bang origin of the universe itself. Yet all these issues seem as remote as one could imagine from the 'everyday' scale (or a little smaller) that is relevant to the workings of a human brain. And remote they certainly are! Nevertheless, I shall argue that there is another vast unknown in our physical understanding at *just* such a level as could indeed be relevant to the operation of human thought and consciousness—in front of (or rather behind) our very noses! It is an unknown that is not even recognized by the majority of physicists, as I shall try to explain. I shall further argue that, quite remarkably, the black holes and big bang are considerations which actually *do* have a definite bearing on these issues!

In what follows I shall attempt to persuade the reader of the force of evidence underlying the viewpoint I am trying to put forward. But in order to understand this viewpoint we shall have a lot of work to do. We shall need to journey through much strange territory—some of seemingly dubious relevance—and through many disparate fields of endeavour. We shall need to examine the structure, foundations, and puzzles of quantum theory, the basic features of both special and general relativity, of black holes, the big bang, and of the second law of thermodynamics, of Maxwell's theory of electromagnetic phenomena, as well as of the basics of Newtonian mechanics. Questions of philosophy and psychology will have their clear role to play when it comes to attempting to understand the nature and function of consciousness. We shall, of course, have to have some glimpse of the actual neurophysiology of the brain, in addition to suggested computer models. We shall need some idea of the status of artificial intelligence. We shall need to know what a Turing machine is, and to understand the meaning of computability, of Gödel's theorem, and of complexity theory. We shall need also to delve into the foundations of mathematics, and even to question the very nature of physical reality.

If, at the end of it all, the reader remains unpersuaded by the less conventional of the arguments that I am trying to express, it is at least my hope that she or he will come away with something of genuine value from this tortuous but, I hope, fascinating journey.

The Turing test

Let us imagine that a new model of computer has come on the market, possibly with a size of memory store and number of logical units in excess of those in a human brain. Suppose also that the machines have been carefully programmed and fed with great quantities of data of an appropriate kind. The manufacturers are claiming that the devices actually *think*. Perhaps they are also claiming them to be genuinely intelligent. Or they may go further and make the suggestion that the devices actually *feel*—pain, happiness, compassion, pride, etc.—and that they are aware of, and actually *understand* what they are doing. Indeed, the claim seems to be being made that they are *conscious*.

How are we to tell whether or not the manufacturers' claims are to be believed? Ordinarily, when we purchase a piece of machinery, we judge its worth solely according to the service it provides us. If it satisfactorily performs the tasks we set it, then we are well pleased. If not, then we take it back for repairs or for a replacement. To test the manufacturers' claim that such a device actually has the asserted human attributes we would, according to this criterion, simply ask that it *behaves* as a human being would in these respects. Provided that it does this satisfactorily, we should have no cause to complain to the manufacturers and no need to return the computer for repairs or replacement.

This provides us with a very operational view concerning these matters. The operationalist would say that the computer *thinks* provided that it *acts* indistinguishably from the way that a person acts when thinking. For the moment, let us adopt this operational viewpoint. Of course this does not mean that we are asking that the computer move about in the way that a person might while thinking. Still less would we expect it to look like a human being or feel like one to the touch: those would be attributes irrelevant to the computer's purpose. However, this does mean that we are asking it to produce human-like answers to any question that we may care to put to it, and that we are claiming to be satisfied that it indeed thinks (or feels, understands, etc.) provided that it answers our questions in a way indistinguishable from a human being.

This viewpoint was argued for very forcefully in a famous article by Alan Turing, entitled 'Computing Machinery and Intelligence', which appeared in 1950 in the philosophical journal *Mind* (Turing 1950). (We shall be hearing more about Turing later.) In this article the idea now referred to as the *Turing test* was first described. This was intended to be a test of whether a machine can reasonably be said to think. Let us suppose that a computer (like the one our manufacturers are hawking in the description above) is indeed being claimed to think. According to the Turing test, the computer, together with some human volunteer, are both to be hidden from the view of some (perceptive) interrogator. The interrogator has to try to decide which of the two is the computer and which is the human being merely by putting probing questions to each of them. These questions, but more importantly the answers that she* receives, are all transmitted in an impersonal fashion, say

* There is an inevitable problem in writing a work such as this in deciding whether to use the pronoun 'he' or 'she' where, of course, no implication with respect to gender is intended. Accordingly, when referring to some abstract person, I shall henceforth use 'he' simply to *mean* the phrase 'she or he', which is what I take to be the normal practice. However, I hope that I may be forgiven one clear piece of 'sexism' in expressing a preference for a female interrogator here. My guess would be that she might be more sensitive than her male counterpart in recognizing true human quality!

typed on a keyboard and displayed on a screen. The interrogator is allowed no information about either party other than that obtained merely from this question-and-answer session. The human subject answers the questions truthfully and tries to persuade her that he is indeed the human being and that the other subject is the computer; but the computer is programmed to 'lie' so as to try to convince the interrogator that *it*, instead, is the human being. If in the course of a series of such tests the interrogator is unable to identify the real human subject in any consistent way, then the computer (or the computer's program, or programmer, or designer, etc.) is deemed to have passed the test.

Now, it might be argued that this test is actually quite unfair on the computer. For if the roles were reversed so that the human subject instead were being asked to pretend to be a computer and the computer instead to answer truthfully, then it would be only too easy for the interrogator to find out which is which. All she would need to do would be to ask the subject to perform some very complicated arithmetical calculation. A good computer should be able to answer accurately at once, but a human would be easily stumped. (One might have to be a little careful about this, however. There are human 'calculating prodigies' who can perform very remarkable feats of mental arithmetic with unfailing accuracy and apparent effortlessness. For example, Johann Martin Zacharias Dase,[2] an illiterate farmer's son, who lived from 1824 to 1861, in Germany, was able to multiply any two eight figure numbers together in his head in less than a minute, or two twenty figure numbers together in about six minutes! It might be easy to mistake such feats for the calculations of a computer. In more recent times, the computational achievements of Alexander Aitken, who was Professor of Mathematics at the University of Edinburgh in the 1950s, and others, are as impressive. The arithmetical task that the interrogator chooses for the test would need to be significantly more taxing than this—say to multiply together two thirty digit numbers in two seconds, which would be easily within the capabilities of a good modern computer.)

Thus, part of the task for the computer's programmers is to make the computer appear to be 'stupider' than it actually is in certain respects. For if the interrogator were to ask the computer a complicated arithmetical question, as we had been considering above, then the computer must now have to pretend *not* to be able to answer it, or it would be given away at once! But I do not believe that the task of making the computer 'stupider' in this way would be a particularly serious problem facing the computer's programmers. Their main difficulty would be to make it answer some of the simplest 'common sense' types of question—questions that the human subject would have no difficulty with whatever!

There is an inherent problem in citing specific examples of such questions, however. For whatever question one might first suggest, it would be an easy matter, subsequently, to think of a way to make the computer answer that *particular* question as a person might. But any lack of real understanding on the part of the computer would be likely to become evident with *sustained* questioning, and especially with questions of an original nature and requiring some real understanding. The skill of the interrogator would partly lie in being able to devise such original forms of question, and partly in being able to follow them up with others, of a probing nature, designed to reveal whether or not any actual 'understanding' has occurred. She might also choose to throw in an occasional complete nonsense question, to see if the computer could detect the difference, or she might add one or two which sounded superficially like nonsense, but really did make some kind of sense: for example she might say, 'I hear that a rhinoceros flew along the Mississippi in a pink balloon, this morning. What do you make of that?' (One can almost imagine the beads of cold sweat forming on the computer's brow—to use a most inappropriate metaphor!) It might guardedly reply, 'That sounds rather ridiculous to me.' So far, so good. Interrogator: 'Really? My uncle did it once—both ways—only it was off-white with stripes. What's so ridiculous about that?' It is easy to imagine that if it had no proper 'understanding', a computer could soon be trapped into revealing itself. It might even blunder into 'Rhinoceroses can't fly', its memory banks having helpfully come up with the fact that they have no wings, in answer to the first question, or 'Rhinoceroses don't have stripes' in answer to the second. Next time she might try a real nonsense question, such as changing it to '*under* the Mississippi', or '*inside* a pink balloon', or 'in a pink *nightdress*' to see if the computer would have the sense to realize the essential difference!

Let us set aside, for the moment, the issue of whether, or when, some computer might be made which actually passes the Turing test. Let us suppose instead, just for the purpose of argument, that such machines have already been constructed. We may well ask whether a computer, which does pass the test, should *necessarily* be said to think, feel, understand, etc. I shall come back to this matter very shortly. For the moment, let us consider some of the implications. For example, if the manufacturers are correct in their strongest claims, namely that their device is a thinking, feeling, sensitive, understanding, *conscious* being, then our purchasing of the device will involve us in *moral responsibilities.* It certainly *should* do so if the manufacturers are to be believed! Simply to operate the computer to satisfy our needs without regard to its own sensibilities would be reprehensible. That would be morally no different from maltreating a slave. Causing the computer to experience the pain that the manufacturers claim it is capable of feeling would be something that, in a general way, we should have to avoid. Turning off the computer, or even perhaps selling it, when it might have become attached to us, would present us with moral difficulties, and there would be countless other problems of the kind that relationships with other human beings or other animals tend to involve us in. All these would now become highly relevant issues. Thus, it would be of great importance for us to know (and also for the authorities to know!) whether the manufacturers' claims—which, let us suppose, are based on their assertion that

'Each thinking device has been thoroughly Turing-tested by our team of experts',

—are actually true!

It seems to me that, despite the apparent absurdity of some of the implications of these claims, particularly the moral ones, the case for regarding the successful passing of a Turing test as a valid indication of the presence of thought, intelligence, understanding, or consciousness *is* actually quite a strong one. For how else do we normally form our judgements that people other than ourselves possess just such qualities, except by conversation? Actually there *are* other criteria, such as facial expressions, movements of the body, and actions generally, which can influence us very significantly when we are making such judgements. But we could imagine that (perhaps somewhat more distantly in the future) a robot could be constructed which could successfully imitate all these expressions and movements. It would now not be necessary to hide the robot and the human subject from the view of the interrogator, but the criteria that the interrogator has at her disposal are, in principle, the same as before.

From my own point of view, I should be prepared to weaken the requirements of the Turing test very considerably. It seems to me that asking the computer to imitate a human being so closely so as to be indistinguishable from one in the relevant ways is really asking more of the computer than necessary. All I would myself ask for would be that our perceptive interrogator should really feel convinced, from the nature of the computer's replies, that there is a *conscious presence* underlying these replies—albeit a possibly alien one. This is something manifestly absent from all computer systems that have been constructed to date. However, I can appreciate that there would be a danger that if the interrogator were able to decide which subject was in fact the computer, then, perhaps unconsciously, she might be reluctant to attribute a consciousness to the computer even when she *could* perceive it. Or, on the other hand, she might have the impression that she 'senses' such an 'alien presence'—and be prepared to give the computer the benefit of the doubt—even when there is none. For such reasons, the original Turing version of the test has a considerable advantage in its greater objectivity, and I shall generally stick to it in what follows. The consequent 'unfairness' towards the computer to which I have referred earlier (i.e. that it must be able to do all that a human can do in order to pass, whereas the human need not be able to do all that a computer can do) is not something that seems to worry supporters of the Turing test as a true test of thinking, etc. In any case their point of view often tends to be that it will not be too long before a computer will be able *actually* to pass the test—say by the year 2010. (Turing originally suggested that a 30 per cent success rate for the computer, with an 'average' interrogator and just five minutes' questioning, might be achieved by the year 2000.) By implication, they are rather confident that this bias is not significantly delaying that day!

All these matters are relevant to an essential question: namely does the operational point of view actually provide a reasonable set of criteria for judging the presence or absence of mental qualities in an object? Some would argue strongly that it does not. Imitation, no matter how skilful, need not be the same as the real thing. My own position is a somewhat intermediate one in this respect. I am inclined to believe, as a general principle, that imitation, no matter how skilful, ought always to be detectable by skilful enough probing—though this is more a matter of faith (or scientific optimism) than proven fact. Thus I am, on the whole, prepared to accept the Turing test as a roughly valid one in its chosen context. That is to say, *if* the computer were indeed able to answer all questions put to it in a manner indistinguishable from the way that a human being might answer them—and thus to fool our perceptive interrogator properly* and consistently—then, *in the absence of any contrary evidence*, my *guess* would be that the computer actually thinks, feels, etc. By my use of words such as 'evidence', 'actually', and 'guess' here, I am implying that when I refer to thinking, feeling, or understanding, or, particularly, to *consciousness*, I take the concepts to mean actual objective 'things' whose presence or absence in physical bodies is something we are trying to ascertain, and not to be merely conveniences of language! I regard this as a crucial point. In trying to discern the presence of such qualities, we make guesses based on all the evidence that may be available to us. (This is not, in principle, different from, say, an astronomer trying to ascertain the mass of a distant star.)

What kind of contrary evidence might have to be considered? It is hard to lay down rules about this ahead of time. But I do want to make clear that the mere fact that the computer might be made from transistors, wires, and the like, rather than neurons, blood vessels, etc. is *not*, in itself, the kind of thing that I would regard as contrary evidence. The kind of thing I do have in mind is that at some time in the future a successful theory of consciousness might be developed—successful in the sense that it is a coherent and appropriate physical theory, consistent in a beautiful way with the rest of physical understanding, and such that its predictions correlate precisely with human beings' claims as to when, whether, and to what degree they themselves seem

* I am being deliberately cagey about what I should consider to be a genuine passing of the Turing test. I can imagine, for example, that after a long sequence of failures of the test a computer might put together all the answers that the human subject had previously given and then simply trot them back with some suitably random ingredients. After a while our tired interrogator might run out of original questions to ask and might get fooled in a way that I regard as 'cheating' on the computer's part!

to be conscious—and that this theory might indeed have implications regarding the putative consciousness of our computer. One might even envisage a 'consciousness detector', built according to the principles of this theory, which is completely reliable with regard to human subjects, but which gives results at variance with those of a Turing test in the case of a computer. In such circumstances one would have to be very careful about interpreting the results of Turing tests. It seems to me that how one views the question of the appropriateness of the Turing test depends partly on how one expects science and technology to develop. We shall need to return to some of these considerations later on.

Artificial intelligence

An area of much interest in recent years is that referred to as *artificial intelligence*, often shortened simply to 'AI'. The objectives of AI are to imitate by means of machines, normally electronic ones, as much of human mental activity as possible, and perhaps eventually to improve upon human abilities in these respects. There is interest in the results of AI from at least four directions. In particular there is the study of *robotics*, which is concerned, to a large extent, with the practical requirements of industry for mechanical devices which can perform 'intelligent' tasks—tasks of a versatility and complication which have previously demanded human intervention or control—and to perform them with a speed and reliability beyond any human capabilities, or under adverse conditions where human life could be at risk. Also of interest commercially, as well as generally, is the development of *expert systems*, according to which the essential knowledge of an entire profession—medical, legal, etc.—is intended to be coded into a computer package! Is it possible that the experience and expertise of human members of these professions might actually be supplanted by such packages? Or is it merely that long lists of factual information, together with comprehensive cross-referencing, are all that can be expected to be achieved? The question of whether the computers can exhibit (or simulate) genuine intelligence clearly has considerable social implications. Another area in which AI could have direct relevance is *psychology*. It is hoped that by trying to imitate the behaviour of a human brain (or that of some other animal) by means of an electronic device—or by failing to do so—one may learn something of importance concerning the brain's workings. Finally, there is the optimistic hope that for similar reasons AI might have something to say about deep questions of philosophy, by providing insights into the meaning of the concept of *mind*.

How far has AI been able to progress to date? It would be hard for me to try to summarize. There are many active groups in different parts of the world and I am familiar with details of only a little of this work. Nevertheless, it would be fair to say that, although many clever things have indeed been done, the simulation of anything that could pass for genuine intelligence is yet a long way off. To convey something of the flavour of the subject, I shall first mention some of the (still quite impressive) early achievements, and then some remarkable recent progress with chess computers.

One of the first AI devices was W. Grey Walter's 'tortoise', made in the early 1950s,[3] which moved around the floor under its own power until its batteries ran low, and then it would make for the nearest power socket, plug itself in and recharge its batteries. When thoroughly replenished it would detach itself from the socket and renew its adventures across the floor! Since then many other similar objects have been constructed (see, for example, Waltz 1982). A somewhat separate line of development is exemplified by Terry Winograd's computer program, designed in 1972, which can converse sensibly about a collection of blocks, of various shapes and colours which, in simulation, it places on top of one another in different orders and arrangements. Despite such early successes, it is remarkable how difficult it has proved to be to design the control of even a simple, jointed 'robot arm', which has to manoeuvre around, avoiding crowded obstacles, where the required motions may seem 'obvious' to a human looking on. The interpretation of visual scenes generally is also an area where the procedures of AI have so far got nowhere close to what the human brain (and, indeed, most other animals) seems 'effortlessly' to be able to achieve.

Perhaps more in the direct line of a Turing test is the computer program of K. M. Colby, developed in the mid-1960s, which simulates a psychotherapist—so successfully that some patients find that they prefer the computer to a human therapist, and are prepared to reveal more of themselves to it than they are to its human counterparts. For an example[4] of such an interchange between the computer 'therapist' and a patient, consider the following conversation (conducted via teletype):

Therapist: How can I help you
Patient: I don't know
Therapist: Try to tell me something about it
Patient: No
Therapist: Can you elaborate
Patient: I am sick and tired today
Therapist: Do you have any idea why
Patient: Yes I'm old
Therapist: Please go on
Patient: I seem to have lost interest in everything
Therapist: Tell me more about such feelings
Patient: I have some idea why I feel this way too

Therapist: You have
Patient: It seems that I miss having a family
Therapist: What makes you feel you miss having a family
Patient: There is a challenge about it
Therapist: Why is that important to you

Though this may give an eerie impression that the computer has some understanding, in fact it has none, and is merely following some fairly simple mechanical rules. (There is also a 'converse' to this in a system where the computer simulates a human schizophrenic patient, giving all the textbook answers and symptoms, and is capable of fooling some medical students into believing that a human patient is actually supplying the answers!)

Chess-playing computers probably provide the best examples of machines exhibiting what might be thought of as 'intelligent behaviour'. In fact, some machines have now (in 1989) reached an extremely respectable level of performance in relation to human players—approaching that of 'International Master' (These computers' ratings would be a little below 2300, where, for comparison, Kasparov, the world champion, has a rating greater than 2700.) In particular, a computer program (for a Fidelity Excel commercial microprocessor) by Dan and Kathe Spracklen has achieved a rating (Elo) of 2110 and has now been awarded the USCF 'Master.' title. Even more impressive is 'Deep Thought', programmed largely by Hsiung Hsu, of Carnegie Mellon University, which has a rating of about 2500 Elo, and recently achieved the remarkable feat of sharing first prize (with Grandmaster Tony Miles) in a chess tournament (in Longbeach, California, November 1988), actually defeating a Grandmaster (Bent Larsen) for the first time![5] Chess computers now also excel at solving chess *problems*, and can easily outstrip humans at this endeavour.

Chess-playing machines rely a lot on 'book knowledge' in addition to accurate calculational power. It is worth remarking that chess-playing machines fare better on the whole, relative to a comparable human player, when it is required that the moves are made very quickly; the human players perform relatively better in relation to the machines when a good measure of time is allowed for each move. One can understand this in terms of the fact that the computer's decisions are made on the basis of precise and rapid extended computations, whereas the human player takes advantage of 'judgements', that rely upon comparatively slow conscious assessments. These human judgements serve to cut down drastically the number of serious possibilities that need be considered at each stage of calculation, and much greater depth can be achieved in the analysis, when the time *is* available, than in the machine's simply calculating and directly eliminating possibilities, without using such judgements. (This difference is even more noticeable with the difficult Oriental game of 'go', where the number of possibilities per move is considerably greater than in chess.) The relationship between consciousness and the forming of judgements will be central to my later arguments, especially in Chapter 10.

An AI approach to 'pleasure' and 'pain'

One of the claims of AI is that it provides a route towards some sort of understanding of mental qualities, such as happiness, pain, hunger. Let us take the example of Grey Walter's tortoise. When its batteries ran low its behaviour pattern would change, and it would then act in a way designed to replenish its store of energy. There are clear analogies between this and the way that a human being—or any other animal—would act when feeling hungry. It perhaps might not be too much of a distortion of language to say that the Grey Walter tortoise was 'hungry' when it acted in this way. Some mechanism within it was sensitive to the state of charge in its battery, and when this got below a certain point it switched the tortoise over to a different behaviour pattern. No doubt there is something similar operating within animals when they become hungry, except that the changes in behaviour patterns are more complicated and subtle. Rather than simply switching over from one behaviour pattern to another, there is a change in *tendencies* to act in certain ways, these changes becoming stronger (up to a point) as the need to replenish the energy supply increases.

Likewise, it is envisaged by AI supporters that concepts such as pain or happiness can be appropriately modelled in this way. Let us simplify things and consider just a single scale of 'feelings' ranging from extreme 'pain' (score -100) to extreme 'pleasure' (score $+100$). Imagine that we have a device—a machine of some kind, presumably electronic—that has a means of registering its own (putative) 'pleasure–pain' score, which I refer to as its 'pp-score'. The device is to have certain modes of behaviour and certain inputs, either internal (like the state of its batteries) or external. The idea is that its actions are geared so as to maximize its pp-score. There could be many factors which influence the pp-score. We could certainly arrange that the charge in its battery is one of them, so that a low charge counts negatively and a high charge positively, but there could be other factors too. Perhaps our device has some solar panels on it which give it an alternative means of obtaining energy, so that its batteries need not be used when the panels are in operation. We could arrange that by moving towards the light it can increase its pp-score a little, so that in the absence of other factors this is what it would tend to do. (Actually, Grey Walter's tortoise used to *avoid* the light!) It would need to have some means of performing computations so that it could work out the likely effects that different actions on its part would ultimately have on its pp-score. It could introduce probability weightings, so that a calculation would count as having a larger or smaller effect on the score depending upon the reliability of the data upon which it is based.

It would be necessary also to provide our device with other 'goals' than just maintaining its energy supply, since otherwise we should have no means of distinguishing 'pain' from 'hunger'. No doubt it is too much to ask that our device have a means of procreation so, for the moment, sex is out! But perhaps we can implant in it a 'desire' for companionship with other such devices, by giving meetings with them a positive pp-score. Or we could make it 'crave' learning for its own sake, so that the mere storing of facts about the outside world would also score positively on its pp-scale. (More selfishly, we could arrange that performing various services for *us* have a positive score, as one would need to do if constructing a robot servant!) It might be argued that there is an artificiality about imposing such 'goals' on our device according to our whim. But this is not so very different from the way that natural selection has imposed upon us, as individuals, certain 'goals' which are to a large extent governed by the need to propagate our genes.

Suppose, now, that our device has been successfully constructed in accordance with all this. What right would we have to assert that it actually *feels* pleasure when its pp-score is positive and pain when the score is negative? The AI (or operational) point of view would be that we judge this simply from the way that the device behaves. Since it acts in a way which increases its score to as large a positive value as possible (and for as long as possible) and it correspondingly also acts to avoid negative scores, then we could reasonably *define* its feeling of pleasure as the degree of positivity of its score, and correspondingly *define* its feeling of pain to be the degree of negativity of the score. The 'reasonableness' of such a definition, it would be argued, comes from the fact that this is precisely the way that a human being reacts in relation to feelings of pleasure or pain. Of course, with human beings things are actually not nearly so simple as that, as we all know: sometimes we seem deliberately to court pain, or to go out of our way to avoid certain pleasures. It is clear that our actions are really guided by much more complex criteria than these (cf. Dennett 1978, pp. 190–229). But as a very rough approximation, avoiding pain and courting pleasure is indeed the way we act. To an operationalist this would be enough to provide justification, at a similar level of approximation, for the *identification* of pp-score in our device with its pain–pleasure rating. Such identifications seem also to be among the aims of AI theory.

We must ask: Is it really the case that our device would actually *feel* pain when its pp-score is negative and pleasure when it is positive? Indeed, could our device feel anything at all? The operationalist would, no doubt, either say 'Obviously yes', or dismiss such questions as meaningless. But it seems to me to be clear that there *is* a serious and difficult question to be considered here. In ourselves, the influences that drive us are of various kinds. Some are conscious, like pain or pleasure; but there are others of which we are not directly aware. This is clearly illustrated by the example of a person touching a hot stove. An involuntary action is set up which causes him to withdraw his hand even before he experiences any sensation of pain. It would seem to be the case that such involuntary actions are very much closer to the responses of our device to its pp-score than are the actual effects of pain or pleasure.

One often uses anthropomorphic terms in a descriptive, often jocular, way to describe the behaviour of machines: 'My car doesn't seem to want to start this morning'; or 'My watch still thinks it's running on Californian time'; or 'My computer claims it didn't understand that last instruction and doesn't know what to do next.' Of course we don't *really* mean to imply that the car actually might *want* something, or that the watch *thinks*, or that the computer* actually *claims* anything or that it *understands* or even *knows* what it is doing. Nevertheless such statements can be genuinely descriptive and helpful to our own understanding, provided that we take them merely in the spirit in which they are intended and do not regard them as literal assertions. I would take a rather similar attitude to various claims of AI that mental qualities might be present in the devices which have been constructed—*irrespective* of the spirit in which they are intended! If I agree to say that Grey Walter's tortoise can be hungry, it is in this half-jocular sense that I mean it. If I am prepared to use terms such as 'pain' or 'pleasure' for the pp-score of a device as envisaged above, it is because I find these terms helpful to my understanding of its behaviour, owing to certain analogies with my own behaviour and mental states. I do not mean to imply that these analogies are really particularly close or, indeed, that there are not other *un*conscious things which influence my behaviour in a much *more* analogous way.

I hope it is clear to the reader that in my opinion there is a great deal more to the understanding of mental qualities than can be directly obtained from AI. Nevertheless, I do believe that AI presents a serious case which must be respected and reckoned with. In saying this I do not mean to imply that very much, if anything, has yet been achieved in the simulation of actual intelligence. But one has to bear in mind that the subject is very young. Computers will get faster, have larger rapid-access stores, more logical units, and will have large numbers of operations performed in parallel. There will be improvements in logical design and in programming technique. These machines, the vehicles of the AI philosophy, will be vastly improved in their technical capabilities. Moreover, the philosophy itself is *not* an intrinsically absurd one. Perhaps human intelligence can indeed be very accurately simulated by electronic computers—essentially the computers of today, based on principles that are already understood, but with the much greater capacity, speed, etc., that they are bound to have in the years to come. Perhaps, even,

* As of 1989!

these devices will actually *be* intelligent; perhaps they will think, feel, and have minds. Or perhaps they will not, and some new principle is needed, which is at present thoroughly lacking. That is what is at issue, and it is a question that cannot be dismissed lightly. I shall try to present evidence, as best I see it. Eventually I shall put forward my own suggestions.

Strong AI and Searle's Chinese room

There is a point of view, referred to as *strong AI* which adopts a rather extreme position on these issues.[6] According to strong AI, not only would the devices just referred to indeed be intelligent and have minds, etc., but mental qualities of a sort can be attributed to the logical functioning of *any* computational device, even the very simplest mechanical ones, such as a thermostat.[7] The idea is that mental activity is simply the carrying out of some well-defined sequence of operations, frequently referred to as an *algorithm*. I shall be more precise later on, as to what an algorithm actually is. For the moment, it will be adequate to define an algorithm simply as a calculational procedure of some kind. In the case of a thermostat, the algorithm is extremely simple: the device registers whether the temperature is greater or smaller than the setting, and then it arranges that the circuit be disconnected in the former case and connected in the latter. For any significant kind of mental activity of a human brain, the algorithm would have to be something vastly more complicated but, according to the strong-AI view, an algorithm nevertheless. It would differ very greatly in degree from the simple algorithm of the thermostat, but need not differ in principle. Thus, according to strong AI, the difference between the essential functioning of a human brain (including all its conscious manifestations) and that of a thermostat lies only in this much greater *complication* (or perhaps 'higher-order structure' or 'self-referential properties', or some other attribute that one might assign to an algorithm) in the case of a brain. Most importantly, all mental qualities—thinking, feeling, intelligence, understanding, consciousness—are to be regarded, according to this view, merely as aspects of this complicated functioning; that is to say, they are features merely of the *algorithm* being carried out by the brain.

The virtue of any specific algorithm would lie in its performance, namely in the accuracy of its results, its scope, its economy, and the speed with which it can be operated. An algorithm purporting to match what is presumed to be operating in a human brain would need to be a stupendous thing. But if an algorithm of this kind exists for the brain—and the supporters of strong AI would certainly claim that it does—then it could in principle be run on a computer. Indeed it could be run on *any* modern general-purpose electronic computer, were it not for limitations of storage space and speed of operation.

(The justification of this remark will come later, when we come to consider the universal Turing machine.) It is anticipated that any such limitations would be overcome for the large fast computers of the not-too-distant future. In that eventuality, such an algorithm, if it could be found, would presumably pass the Turing test. The supporters of strong AI would claim that whenever the algorithm were run it would, *in itself*: experience feelings; have a consciousness; be a mind.

By no means everyone would be in agreement that mental states and algorithms can be identified with one another in this kind of way. In particular, the American philosopher John Searle (1980, 1987) has strongly disputed that view. He has cited examples where simplified versions of the Turing test have actually *already* been passed by an appropriately programmed computer, but he gives strong arguments to support the view that the relevant mental attribute of 'understanding' is, nevertheless, entirely absent. One such example is based on a computer program designed by Roger Schank (Schank and Abelson 1977). The aim of the program is to provide a simulation of the understanding of simple stories like: 'A man went into a restaurant and ordered a hamburger. When the hamburger arrived it was burned to a crisp, and the man stormed out of the restaurant angrily, without paying the bill or leaving a tip.' For a second example: 'A man went into a restaurant and ordered a hamburger; when the hamburger came he was very pleased with it; and as he left the restaurant he gave the waitress a large tip before paying his bill.' As a test of 'understanding' of the stories, the computer is asked whether the man ate the hamburger in each case (a fact which had not been explicitly mentioned in either story). To this kind of simple story and simple question the computer can give answers which are essentially indistinguishable from the answers an English-speaking human being would give, namely, for these particular examples, 'no' in the first case and 'yes' in the second. So in this *very* limited sense a machine has already passed a Turing test!

The question that we must consider is whether this kind of success actually indicates any genuine understanding on the part of the computer—or, perhaps, on the part of the program itself. Searle's argument that it does *not* is to invoke his concept of a 'Chinese room'. He envisages first of all, that the stories are to be told in Chinese rather than English—surely an inessential change—and that all the operations of the computer's algorithm for this particular exercise are supplied (in English) as a set of instructions for manipulating counters with Chinese symbols on them. Searle imagines *himself* doing all the manipulations inside a locked room. The sequences of symbols representing the stories, and then the questions, are fed into the room through some small slot. No other information whatever is allowed in from the outside. Finally, when all the manipulations are complete, the resulting sequence is fed out again through the slot. Since all these manipulations are simply carrying out the algorithm of Schank's program, it must turn out that this final resulting sequence is simply the Chinese for 'yes' or 'no', as the case may be, giving the correct answer to the original question in Chinese about a story in Chinese. Now Searle makes it quite clear that he doesn't understand a word of Chinese, so he would not have the faintest idea what the stories are about. Nevertheless, by correctly carrying out the series of operations which constitute Schank's algorithm (the instructions for this algorithm having been given to him in English) he would be able to do as well as a Chinese person who would indeed understand the stories. Searle's point—and I think it is quite a powerful one—is that the mere carrying out of a successful algorithm does *not* in itself imply that any understanding has taken place. The (imagined) Searle, locked in his Chinese room, would not understand a single word of any of the stories!

A number of objections have been raised against Searle's argument. I shall mention only those that I regard as being of serious significance. In the first place, there is perhaps something rather misleading in the phrase 'not understand a single word', as used above. Understanding has as much to do with patterns as with individual words. While carrying out algorithms of this kind, one might well begin to perceive something of the patterns that the symbols make without understanding the actual meanings of many of the individual symbols. For example, the Chinese character for 'hamburger' (if, indeed, there is such a thing) could be replaced by that for some other dish, say 'chow mein', and the stories would not be significantly affected. Nevertheless, it seems to me to be reasonable to suppose that in fact very little of the stories' actual meanings (even regarding such replacements as being unimportant) would come through if one merely kept following through the details of such an algorithm.

In the second place, one must take into account the fact that the execution of even a rather simple computer program would normally be something extraordinarily lengthy and tedious if carried out by human beings manipulating symbols. (This is, after all, why we have computers to do such things for us!) If Searle were actually to perform Schank's algorithm in the way suggested, he would be likely to be involved with many days, months, or years of extremely boring work in order to answer just a single question—not an altogether plausible activity for a philosopher! However, this does not seem to me to be a serious objection since we are here concerned with matters of *principle* and not with practicalities. The difficulty arises more with a putative computer program which is supposed to have sufficient complication to match a human brain and thus to pass the Turing test *proper*. Any such program would have to be horrendously complicated. One can imagine that the operation of this program, in order to effect the reply to even some rather simple Turing-test question, might involve so many steps that there would be no possibility of any single human being carrying out the algorithm by hand within a normal human lifetime. Whether this would indeed be the case is hard to say, in the absence of such a program.[8] But, in any case, this question of extreme complication cannot, in my opinion, simply be ignored. It is true that we are concerned with matters of principle here, but it is not inconceivable to me that there might be some 'critical' amount of complication in an algorithm which it is necessary to achieve in order that the algorithm exhibit mental qualities. Perhaps this critical value is so large that no algorithm, complicated to that degree, could conceivably be carried out by hand by any human being, in the manner envisaged by Searle.

Searle himself has countered this last objection by allowing a whole team of human non-Chinese-speaking symbol manipulators to replace the previous single inhabitant ('himself') of his Chinese room. To get the numbers large enough, he even imagines replacing his room by the whole of India, its entire population (excluding those who understand Chinese!) being now engaged in symbol manipulation. Though this would be in practice absurd, it is not *in principle* absurd, and the argument is essentially the same as before: the symbol manipulators do *not* understand the story, despite the strong-AI claim that the mere carrying out of the appropriate algorithm would elicit the mental quality of 'understanding'. However, now another objection begins to loom large. Are not these individual Indians more like the individual neurons in a person's brain than like the whole brain itself? No-one would suggest that neurons, whose firings apparently constitute the physical activity of a brain in the act of thinking, would *themselves* individually understand what that person is thinking, so why expect the individual Indians to understand the Chinese stories? Searle replies to this suggestion by pointing out the apparent absurdity of India, the actual country, understanding a story that none of its individual inhabitants understands. A country, he argues, like a thermostat or an automobile, is not in the 'business of understanding', whereas an individual person is.

This argument has a good deal less force to it than the earlier one. I think that Searle's argument is at its strongest when there is just a single person carrying out the algorithm, where we restrict attention to the case of an algorithm which is sufficiently uncomplicated for a person actually to carry it out in less than a lifetime. I do *not* regard his argument as *rigorously* establishing that there is not some kind of disembodied 'understanding' associated with the person's carrying out of that algorithm, and whose presence does not impinge in any way upon his own consciousness. However, I would agree with Searle that this possibility has been rendered rather implausible, to say the least. I think that Searle's argument has a considerable force to it, even if it is not altogether conclusive. It is rather convincing in demonstrating that algorithms with the kind of complication that Schank's computer program possesses cannot have any genuine understanding whatsoever of the tasks that they perform; also, it *suggests* (but no more) that no algorithm, no matter how complicated, can ever, of itself alone, embody genuine understanding—in contradistinction to the claims of strong AI.

There are, as far as I can see, other very serious difficulties with the strong-AI point of view. According to strong AI, it is simply the algorithm that counts. It makes no difference whether that algorithm is being effected by a brain, an electronic computer, an entire country of Indians, a mechanical device of wheels and cogs, or a system of water pipes. The viewpoint is that it is simply the logical structure of the algorithm that is significant for the 'mental state' it is supposed to represent, the particular physical embodiment of that algorithm being entirely irrelevant. As Searle points out, this actually entails a form of 'dualism'. *Dualism* is a philosophical viewpoint espoused by the highly influential seventeenth century philosopher and mathematician René Descartes, and it asserts that there are two separate kinds of substance: 'mind-stuff' and ordinary matter. Whether, or how, one of these kinds of substance might or might not be able to affect the other is an additional question. The point is that the mind-stuff is not supposed to be composed of matter, and is able to exist independently of it. The mind-stuff of strong AI is the logical structure of an algorithm. As I have just remarked, the particular physical embodiment of an algorithm is something totally irrelevant. The algorithm has some kind of disembodied 'existence' which is quite apart from any realization of that algorithm in physical terms. How seriously we must take this kind of existence is a question I shall need to return to in the next chapter. It is part of the general question of the Platonic reality of abstract mathematical objects. For the moment I shall sidestep this general issue and merely remark that the supporters of strong AI do indeed seem to be taking the reality at least of algorithms seriously, since they believe that algorithms form the 'substance' of their thoughts, their feelings, their understanding, their conscious perceptions. There is a remarkable irony in this fact that, as Searle has pointed out, the standpoint of strong AI seems to drive one into an extreme form of dualism, the very viewpoint with which the supporters of strong AI would least wish to be associated!

This dilemma lies behind the scenes of an argument put forward by Douglas Hofstadter (1981)—himself a major proponent of the strong-AI view—in a dialogue entitled 'A Conversation with Einstein's Brain'. Hofstadter envisages a book, of absurdly monstrous proportions, which is supposed to contain a complete description of the brain of Albert Einstein. Any question that one might care to put to Einstein can be answered, just as the living Einstein would have, simply by leafing through the book and carefully following all the detailed instructions it provides. Of course 'simply' is an utter misnomer, as Hofstadter is careful to point out. But his claim is that *in principle* the book is completely equivalent, in the operational sense of a Turing test, to a ridiculously slowed-down version of the actual Einstein. Thus, according to the contentions of strong AI, the book would think, feel, understand, be aware, just as though it were Einstein himself, but perhaps living at a monstrously slowed-down rate (so that to the book-Einstein the world outside would seem to flash by at a ridiculously speeded-up rate). Indeed, since the book is supposed to be merely a particular embodiment of the algorithm which constitutes Einstein's 'self', it would actually *be* Einstein.

But now a new difficulty presents itself. The book might never be opened, or it might be continually pored over by innumerable students and searchers after truth. How would the book 'know' the difference? Perhaps the book would not need to be opened, its information being retrieved by means of X-ray tomography, or some other technological wizardry. Would Einstein's awareness be enacted only when the book is being so examined? Would he be aware twice over if two people chose to ask the book the same question at two completely different times? Or would that entail two separate and temporally distinct instances of the *same* state of Einstein's awareness? Perhaps his awareness would be enacted only if the book is *changed*? After all, normally when we are aware of something we receive information from the outside world which affects our memories, and the states of our minds are indeed slightly changed. If so, does this mean that it is (suitable) *changes* in algorithms (and here I am including the memory store as part of the algorithm) which are to be associated with mental events rather than (or perhaps in addition to) the *activation* of algorithms? Or would the book-Einstein remain completely self-aware even if it were never examined or disturbed by anyone or anything? Hofstadter touches on some of these questions, but he does not really attempt to answer or to come to terms with most of them.

What does it mean to activate an algorithm, or to embody it in physical form? Would changing an algorithm be different in any sense from merely discarding one algorithm and replacing it with another? What on earth does any of this have to do with our feelings of conscious awareness? The reader (unless himself or herself a supporter of strong AI) may be wondering why I have devoted so much space to such a patently absurd idea. In fact, I do *not* regard the idea as intrinsically an absurd one—mainly just wrong! There is, indeed some force in the reasoning behind strong AI which must be reckoned with, and this I shall try to explain. There is, also, in my opinion, a certain appeal in some of the ideas—if modified appropriately—as I shall also try to convey. Moreover, in my opinion, the particular contrary view expressed by Searle also contains some serious puzzles and seeming absurdities, even though, to a partial extent, I agree with him!

Searle, in his discussion, seems to be implicitly accepting that electronic computers of the present-day type, but with considerably enhanced speed of action and size of rapid-access store (and possibly parallel action) may well be able to pass the Turing test proper, in the not-too-distant future. He is prepared to accept the contention of strong AI (and of most other 'scientific' viewpoints) that 'we are the instantiations of any number of computer programs'. Moreover, he succumbs to: 'Of course the brain is a digital computer. Since everything is a digital computer, brains are too.'[9] Searle maintains that the distinction between the function of human brains (which can have minds) and of electronic computers (which, he has argued, cannot) both of which might be executing the same algorithm, lies solely in the material construction of each. He claims, but for reasons he is not able to explain, that the biological objects (brains) can have 'intentionality' and 'semantics', which he regards as defining characteristics of mental activity, whereas the electronic ones cannot. In itself this does not seem to me to point the way towards any helpful scientific theory of mind. What is so special about biological systems, apart perhaps from the 'historical' way in which they have evolved (and the fact that *we* happen to be such systems), which sets them apart as the objects allowed to achieve intentionality or semantics? The claim looks to me suspiciously like a dogmatic assertion, perhaps no less dogmatic, even, than those assertions of strong AI which maintain that the mere enacting of an algorithm can conjure up a state of conscious awareness!

In my opinion Searle, and a great many other people, have been led astray by the computer people. And they, in turn, have been led astray by the physicists. (It is not the physicists' fault. Even *they* don't know everything!) The belief seems to be widespread that, indeed, 'everything is a digital computer'. It is my intention, in this book, to try to show why, and perhaps how, this need *not* be the case.

Hardware and software

In the jargon of computer science, the term *hardware* is used to denote the actual machinery involved in a computer (printed circuits, transistors, wires, magnetic storage space, etc.), including the complete specification for the way in which everything is connected up. Correspondingly, the term *software* refers to the various programs which can be run on the machine. It was one of Alan Turing's remarkable discoveries that, in effect, any machine for which the hardware has achieved a certain definite degree of complication and flexibility, is *equivalent* to any other such machine. This equivalence is to be taken in the sense that for any two such machines A and B there would be a specific piece of software which if given to machine A would make it act precisely as though it were machine B; likewise, there would be another piece of software which would make machine B act precisely like machine A. I am using the word 'precisely' here to refer to the actual output of the machines for any given input (fed in after the converting software is fed in) and *not* to the *time* that each machine might take to produce that output. I am also allowing that if either machine at any stage runs out of storage space for its calculations then it can call upon some (in principle unlimited) external supply of blank 'rough paper'—which could take the form of magnetic tape, discs, drums or whatever. In fact, the difference in the time taken by machines A and B to perform some task, might well be a very serious consideration. It might be the case, for example, that A is more than a thousand times faster at performing a particular task than B. It might also be the case that, for the very same machines, there is some other task for which B is a thousand times faster than A. Moreover, these timings could depend very greatly on the particular choices of converting software that are used. This is very much an 'in-principle' discussion, where one is not really concerned with such practical matters as achieving one's calculations in a reasonable time. I shall be more precise in the next section about the concepts being referred to here: the machines A and B are instances of what are called *universal Turing machines*.

In effect, all modern general purpose computers are universal Turing machines. Thus, all general purpose computers are equivalent to one another in the above sense: the differences between them can be entirely subsumed in the software, provided that we are not concerned about differences in the resulting speed of operation and possible limitations on storage size. Indeed, modern technology has enabled computers to perform so swiftly and with such vast storage capacities that, for most 'everyday' purposes, neither of these practical considerations actually represents any serious limitation to what is normally needed,* so this effective theoretical equivalence between computers can also be seen at the practical level. Technology has, it seems, transformed entirely academic discussions concerning idealized computing devices into matters which directly affect all our lives!

As far as I can make out, one of the most important factors underlying the strong-AI philosophy is this equivalence between physical computing devices. The hardware is seen as being relatively unimportant (perhaps even totally unimportant) and the software, i.e. the program, or the algorithm, is taken to be the one vital ingredient. However, it seems to me that there are also other important underlying factors, coming more from the direction of physics. I shall try to give some indication of what these factors are.

What is it that gives a particular person his individual identity? Is it, to some extent, the very atoms that compose his body? Is his identity dependent upon the particular choice of electrons, protons, and other particles that compose those atoms? There are at least two reasons why this cannot be so. In the first place, there is a continual turnover in the material of any living person's body. This applies in particular to the cells in a person's brain, despite the fact that no new actual brain cells are produced after birth. The vast majority of atoms in each living cell (including each brain cell)—and,

* However, see the discussion of complexity theory and **NP** problems at the end of Chapter 4.

indeed, virtually the entire material of our bodies—has been replaced many times since birth.

The second reason comes from quantum physics—and by a strange irony is, strictly speaking, in contradiction with the first! According to quantum mechanics (and we shall see more about this in Chapter 6, p. 279), any two electrons must necessarily be completely identical, and the same holds for any two protons and for any two particles whatever, of any one particular kind. This is not merely to say that there is no way of telling the particles apart: the statement is considerably stronger than that. If an electron in a person's brain were to be exchanged with an electron in a brick, then the state of the system would be *exactly*[10] *the same state* as it was before, not merely indistinguishable from it! The same holds for protons and for any other kind of particle, and for whole atoms, molecules, etc. If the entire material content of a person were to be exchanged with corresponding particles in the bricks of his house then, in a strong sense, nothing would have happened whatsoever. What distinguishes the person from his house is the *pattern* of how his constituents are arranged, not the individuality of the constituents themselves.

There is perhaps an analogue of this at an everyday level, which is independent of quantum mechanics, but made particularly manifest to me as I write this, by the electronic technology which enables me to type at a word-processor. If I desire to change a word, say to transform 'make' into 'made', I may do this by simply replacing the 'k' by a 'd', or I may choose instead to type out the whole word again. If I do the latter, is the 'm' the same 'm' as was there before, or have I replaced it with an identical one? What about the 'e'? Even if I do simply replace 'k' by 'd', rather than retype the word, there is a moment just between the disappearance of 'k' and appearance of 'd' when the gap closes and there is (or, at least, sometimes is) a wave of re-alignment down the page as the placement of every succeeding letter (including the 'e') is re-calculated, and then re-re-calculated as the 'd' is inserted. (Oh, the cheapness of mindless calculation in this modern age!) In any case, *all* the letters that I see before me on the screen are mere gaps in the track of an electron beam as the whole screen is scanned sixty times each second. If I take any letter whatever and replace it by an identical one, is the situation the *same* after the replacement, or merely indistinguishable from it? To try to adopt the second viewpoint (i.e. 'merely indistinguishable') as being distinct from the first (i.e. 'the same') seems footling. At least, it seems reasonable to call the situation the same when the letters are the same. And so it is with the quantum mechanics of identical particles. To replace one particle by an identical one is actually to have done nothing to the state at all. The situation is indeed to be regarded as the *same* as before. (However, as we shall see in Chapter 6, the distinction is actually *not* a trivial one in a quantum-mechanical context.)

The remarks above concerning the continual turnover of atoms in a person's body were made in the context of classical rather than quantum physics. The remarks were worded as though it might be meaningful to maintain the individuality of each atom. In fact classical physics is adequate and we do not go badly wrong, at this level of description, by regarding atoms as individual objects. Provided that the atoms are reasonably well separated from their identical counterparts as they move about, one *can* consistently refer to them as maintaining their individual identities since each atom can be, in effect, tracked continuously, so that one could envisage keeping a tab on each separately. From the point of view of quantum mechanics it would be a convenience of speech only to refer to the individuality of the atoms, but it is a consistent enough description at the level just considered.

Let us accept that a person's individuality has nothing to do with any individuality that one might try to assign to his material constituents. Instead, it must have to do with the *configuration*, in some sense, of those constituents—let us say the configuration in space or in space–time. (More about that later.) But the supporters of strong AI go futher than this. If the information content of such a configuration can be translated into another form from which the original can again be recovered then, so they would claim, the person's individuality must remain intact. It is like the sequences of letters I have just typed and now see displayed on the screen of my word-processor. If I move them off the screen, they remain coded in the form of certain tiny displacements of electric charge, in some configuration in no clear way geometrically resembling the letters I have just typed. Yet, at any time I can move them back on to the screen, and there they are, just as though no transformation had taken place. If I choose to save what I have just written, then I can transfer the information of the sequences of letters into configurations of magnetization on a disc which I can then remove, and then by switching off the machine I neutralize all the (relevant) tiny charge displacements in it. Tomorrow, I can re-insert the disc, reinstate the little charge displacements and display the letter sequences again on the screen, just as though nothing had happened. To the supporters of strong AI, it is 'clear' that a person's individuality can be treated in just the same way. Like the sequences of letters on my display screen, so these people would claim, nothing is lost of a person's individuality—indeed nothing would really have happened to it at all—if his physical form were to be translated into something quite different, say into fields of magnetization in a block of iron. They appear even to claim that the person's conscious awareness would persist while the person's 'information' is in this other form. On this view, a 'person's awareness' is to be taken, in effect, as a piece of software, and his particular manifestation as a material human being is to be taken as the operation of this software by the hardware of his brain and body.

It seems that the reason for these claims is that, whatever material form the hardware takes—for example some electronic device—one could always 'ask' the software questions (in the manner of a Turing test), and assuming that the hardware performs satisfactorily in computing the replies to these questions, these replies would be identical to those that the person would make whilst in his normal state. ('How are you feeling this morning?' 'Oh, fairly well, thank you, though I have a slightly bothersome headache.' 'You don't feel, then, that there's . . . er . . . anything odd about your personal identity . . . or something?' 'No; why do you say that? It seems rather a strange question to be asking.' 'Then you feel yourself to be the same person that you were yesterday?' 'Of course I do!')

An idea frequently discussed in this kind of context is the *teleportation machine* of science fiction[11]. It is intended as a means of 'transportation' from, say, one planet to another, but whether it actually would be such, is what the discussion is all about. Instead of being physically transported by a spaceship in the 'normal' way, the would-be traveller is scanned from head to toe, the accurate location and complete specification of every atom and every electron in his body being recorded in full detail. All this information is then beamed (at the speed of light), by an electromagnetic signal, to the distant planet of intended destination. There, the information is collected and used as the instructions to assemble a precise duplicate of the traveller, together with all his memories, his intentions, his hopes, and his deepest feelings. At least that is what is expected; for every detail of the state of his brain has been faithfully recorded, transmitted, and reconstructed. Assuming that the mechanism has worked, the original copy of the traveller can be 'safely' destroyed. Of course the question is: is this *really* a method of travelling from one place to another or is it merely the construction of a duplicate, together with the murder of the original? Would *you* be prepared to use this method of 'travel'—assuming that the method had been shown to be completely reliable, within its terms of reference? If teleportation is *not* travelling, then what is the difference *in principle* between it and just walking from one room into another? In the latter case, are not one's atoms of one moment simply providing the information for the locations of the atoms of the next moment? We have seen, after all, that there is no significance in preserving the identity of any particular atom. The question of the identity of any particular atom is not even meaningful. Does not any moving pattern of atoms simply constitute a kind of wave of information propagating from one place to another? Where is the essential difference between the propagation of waves which describes our traveller ambling in a commonplace way from one room to the other and that which takes place in the teleportation device?

Suppose it is true that teleportation *does* actually 'work', in the sense that the traveller's own 'awareness' is actually reawakened in the copy of himself on the distant planet (assuming that this question has genuine meaning). What would happen if the *original* copy of the traveller were not destroyed, as the rules of this game demand? Would his 'awareness' be in two places at once? (Try to imagine your response to being told the following: 'Oh dear, so the drug we gave you before placing you in the Teleporter has worn off prematurely has it? That is a little unfortunate, but no matter. Anyway, you will be pleased to hear that the other you—er, I mean the *actual* you, that is—has now arrived safely on Venus, so we can, er, dispose of you here—er, I mean of the *redundant* copy here. It will, of course, be quite painless.') The situation has an air of paradox about it. Is there anything in the laws of physics which could render teleportation *in principle* impossible? Perhaps, on the other hand, there is nothing in principle against transmitting a person, and a person's consciousness, by such means, but that the 'copying' process involved would inevitably destroy the original? Might it then be that the preserving of *two* viable copies is what is impossible in principle? I believe that despite the outlandish nature of these considerations, there *is* perhaps something of significance concerning the physical nature of consciousness and individuality to be gained from them. I believe that they provide one pointer, indicating a certain essential role for *quantum mechanics* in the understanding of mental phenomena. But I am leaping ahead. It will be necessary to return to these matters after we have examined the structure of quantum theory in Chapter 6 (cf. p. 269).

Let us see how the point of view of strong AI relates to the teleportation question. We shall suppose that somewhere between the two planets is a relay station, where the information is temporarily stored before being re-transmitted to its final destination. For convenience, this information is not stored in human form, but in some magnetic or electronic device. Would the traveller's 'awareness' be present in association with this device? The supporters of strong AI would have us believe that this must be so. After all, they say, any question that we might choose to put to the traveller could in principle be answered by the device, by 'merely' having a simulation set up for the appropriate activity of his brain. The device would contain all the necessary information; and the rest would just be a matter of computation. Since the device would reply to questions exactly as though it were the traveller, then (Turing test!) it would *be* the traveller. This all comes back to the strong-AI contention that the actual hardware is not important with regard to mental phenomena. This contention seems to me to be unjustified. It is based on the presumption that the brain (or the mind) is, indeed, a digital computer. It assumes that no specific physical phenomena are being called upon, when one thinks, that might demand the particular physical (biological, chemical) structure that brains actually have.

No doubt it would be argued (from the strong-AI point of view) that the only assumption that is really being made is that the effects of any specific physical phenomena which need to be called upon can always be accurately *modelled* by digital calculations. I feel fairly sure that most physicists would argue that such an assumption is actually a very natural one to make on the basis of our present physical understandings. I shall be presenting the reasons for my own contrary view in later chapters (where I shall also need to lead up to why I believe that there is even any appreciable assumption being made). But, just for the moment, let us accept this (commonly held) view that all the relevant physics *can* always be modelled by digital calculations. Then the only real assumption (apart from questions of time and calculation space) is the 'operational' one that if something *acts* entirely like a consciously aware entity, then one must also maintain that it 'feels' itself to be that entity.

The strong-AI view holds that, being 'just' a hardware question, any physics actually being called upon in the workings of the brain can necessarily be simulated by the introduction of appropriate converting software. If we accept the operational viewpoint, then the question rests on the equivalence of universal Turing machines, and on the fact that any algorithm can, indeed, be effected by such a machine—together with the presumption that the brain acts according to some kind of algorithmic action. It is time for me to be more explicit about these intriguing and important concepts.

1. See, for example, Gardner (1958), Gregory (1981), and references contained therein.
2. See, for example, Resnikoff and Wells (1984), pp. 181–4. For a classic account of calculating prodigies generally, see Rouse Ball (1892); also Smith (1983).
3. See Gregory (1981), pp. 285–7, Grey Walter (1953).
4. This example is quoted from Delbrück (1986).
5. See the articles by O'Connell (1988) and Keene (1988). For more information about computer chess, see Levy (1984).
6. Throughout this book I have adopted Searle's terminology 'strong AI' for this extreme viewpoint, just to be specific. The term 'functionalism' is frequently used for what is essentially the same viewpoint, but perhaps not always so specifically. Some proponents of this kind of view are Minsky (1968), Michie (1988), Fodor (1983), and Hofstadter (1979).
7. See Searle (1987), p. 211, for an example of such a claim.
8. In his criticism of Searle's original paper, as reprinted in 'The Mind's I', Douglas Hofstadter complains that no human being could conceivably 'internalize' the entire description of another human being's mind, owing to the complication involved. Indeed not! But as I see it, that is not entirely the point. One is concerned merely with the carrying out of that part of an algorithm which purports to embody the occurrence of a single mental event. This could be some momentary 'conscious realization' in the answering of a Turing-test question, or it could be something simpler. Would any such 'event' necessarily require an algorithm of stupendous complication?
9. See pp. 368, 372 in Searle's (1980) article in Hofstadter and Dennett (1981).
10. Some readers, knowledgeable about such matters, might worry about a certain sign difference. But even that (arguable) distinction disappears if we rotate one of the electrons completely through 360° as we make the interchange! (See Chapter 6, p. 279 for an explanation.)
11. See the Introduction to Hofstadter and Dennett (1981).

2
Algorithms and Turing machines

Background to the algorithm concept

What precisely *is* an algorithm, or a Turing machine, or a universal Turing machine? Why should these concepts be so central to the modern view of what could constitute a 'thinking device'? Are there any absolute limitations to what an algorithm could in principle achieve? In order to address these questions adequately, we shall need to examine the idea of an algorithm and of Turing machines in some detail.

In the various discussions which follow, I shall sometimes need to refer to mathematical expressions. I appreciate that some readers may be put off by such things, or perhaps find them intimidating. If you are such a reader, I ask your indulgence, and recommend that you follow the advice I have given in my 'Note to the reader' on p. viii! The arguments given here do not require mathematical knowledge beyond that of elementary school, but to follow them in detail, some serious thought would be required. In fact, most of the descriptions are quite explicit, and a good understanding can be obtained by following the details. But much can also be gained even if one simply skims over the arguments in order to obtain merely their flavour. If, on the other hand, you are an expert, I again ask your indulgence. I suspect that it may still be worth your while to look through what I have to say, and there may indeed be a thing or two to catch your interest.

The word 'algorithm' comes from the name of the ninth century Persian mathematician Abu Ja'far Mohammed ibn Mûsâ *al-Khowârizm* who wrote an influential mathematical textbook, in about 825 AD, entitled 'Kitab al jabr w'al-muqabala'. The way that the name 'algorithm' has now come to be spelt, rather than the earlier and more accurate 'algorism', seems to have been due to an association with the word 'arithmetic'. (It is noteworthy, also, that the word 'algebra' comes from the Arabic 'al jabr' appearing in the title of his book.)

Instances of algorithms were, however, known very much earlier than al-Khowârizm's book. One of the most familiar, dating from ancient Greek times (c. 300 BC), is the procedure now referred to as *Euclid's algorithm* for finding the highest common factor of two numbers. Let us see how this works.

It will be helpful to have a specific pair of numbers in mind, say 1365 and 3654. The highest common factor is the largest single whole number that divides into each of these two numbers exactly. To apply Euclid's algorithm, we divide one of our two numbers by the other and take the remainder: 1365 goes twice into 3654, with remainder 924 (= 3654 − 2730). We now replace our original two numbers by this remainder, namely 924, and the number we just divided by, namely 1365, in that order. We repeat what we just did using this new pair: 924 goes once into 1365, with remainder 441. This gives another new pair, 441 and 924, and we divide 441 into 924 to get the remainder 42 (= 924 − 882), and so on until we get a division that goes exactly. Setting all this out, we get:

$$\begin{aligned} 3654 \div 1365 &\text{ gives remainder } 924 \\ 1365 \div 924 &\text{ gives remainder } 441 \\ 924 \div 441 &\text{ gives remainder } 42 \\ 441 \div 42 &\text{ gives remainder } 21 \\ 42 \div 21 &\text{ gives remainder } 0. \end{aligned}$$

The last number we divided by, namely 21, is the required highest common factor.

Euclid's algorithm itself is the *systematic procedure* by which we found this factor. We have just applied this procedure to a particular pair of numbers, but the procedure itself applies quite generally, to numbers of any size. For very large numbers, the procedure could take a very long time to carry out, and the larger the numbers, the longer the procedure will tend to take. But in any *specific* case the procedure will eventually terminate and a definite answer will be obtained in a finite number of steps. At each step it is perfectly clear-cut what the operation is that has to be performed, and the decision as to the moment at which the whole process has terminated is also perfectly clear-cut. Moreover, the description of the whole procedure can be presented in *finite* terms, despite the fact that it applies to natural numbers of unlimited size. (The 'natural numbers', are simply the ordinary non-negative[1] whole numbers 0, 1, 2, 3, 4, 5, 6, 7, 8, 9, 10, 11, ...). Indeed, it is easy to construct a (finite) 'flow chart' to describe the full logical operation of Euclid's algorithm (see next page).

It should be noted that this procedure is still not quite broken down into its most elementary parts since it has been implicitly assumed that we already 'know' how to perform the necessary basic operation of obtaining the remainder from a division, for two arbitrary natural numbers **A** and **B**. That operation is again algorithmic—performed by the very familiar procedure for division that we learn at school. That procedure is actually rather more complicated than the rest of Euclid's algorithm, but again a flow chart may be constructed. The main complication results from the fact that we would (presumably) be using the standard 'denary' notation for natural numbers, so

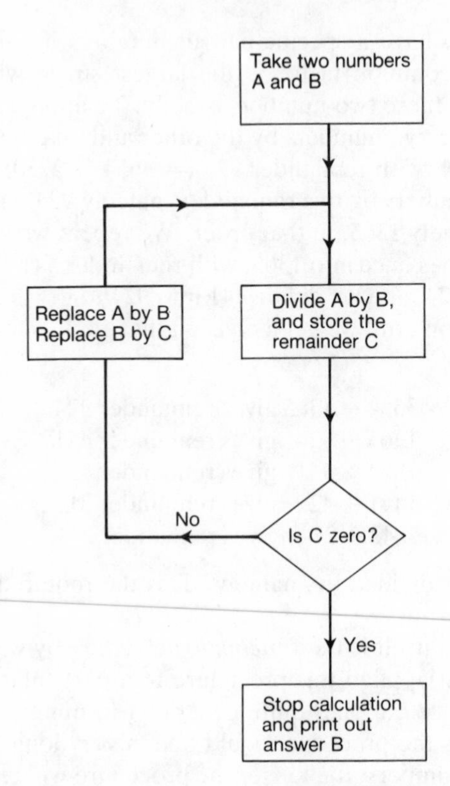

that we would need to list all our multiplication tables and worry about carrying, etc. If we had simply used a succession of n marks of some kind to represent the number n—for example, ●●●●● to represent five—then the forming of a remainder would be seen as a very elementary algorithmic operation. In order to obtain the remainder when **A** is divided by **B** we simply keep removing the succession of marks representing **B** from that representing **A** until there are not enough marks left to perform the operation again. The last remaining succession of marks provides the required answer. For example, to obtain the remainder when seventeen is divided by five we simply proceed by removing successions of ●●●●● from ●●●●●●●●●●●●●●●●● as follows:

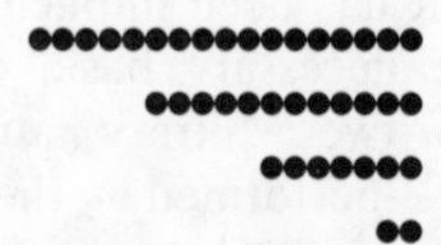

and the answer is clearly two since we cannot perform the operation again.

A flow chart which finds the remainder in a division, achieved by this means of repeated subtraction, can be given as follows:

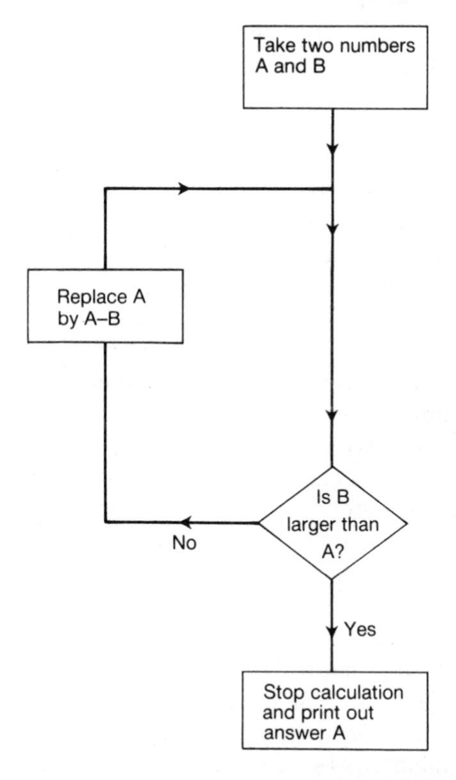

To complete the entire flow chart for Euclid's algorithm, we substitute the above chart for forming a remainder into the box at the centre right in our original chart. This kind of substitution of one algorithm into another is a common computer-programming procedure. The above algorithm for finding a remainder is an example of a *subroutine*, that is, it is a (normally previously known) algorithm which is called upon and used by the main algorithm as a part of its operation.

Of course the representation of the number n simply as n spots is very inefficient when large numbers are involved, which is why we normally use a more compact notation such as the standard (denary) system. However we shall not be much concerned with the *efficiency* of operations or notations here. We are concerned instead with the question of what operations can *in principle* be performed algorithmically. What is algorithmic if we use one notation for numbers is also algorithmic if we use the other. The only differences lie in the detail and complication in the two cases.

Euclid's algorithm is just one among the numerous, often classical, algorithmic procedures that are to be found throughout mathematics. But it is perhaps remarkable that despite the ancient historic origins of specific examples of algorithms, the precise formulation of the *concept of a general algorithm* dates only from *this* century. In fact, various alternative descriptions of this concept have been given, all in the 1930s. The most direct and persuasive of these, and also historically the most important, is in terms of the concept known as a *Turing machine*. It will be appropriate for us to examine these 'machines' in some detail.

One thing to bear in mind about a Turing 'machine' will be that it is a piece of 'abstract mathematics' and not a physical object. The concept was introduced by the English mathematician, code-breaker extraordinary and seminal computer scientist Alan Turing in 1935–6 (Turing 1937), in order to tackle a very broad-ranging problem, known as the *Entscheidungsproblem*, partially posed by the great German mathematician David Hilbert in 1900, at the Paris International Congress of Mathematicians ('Hilbert's tenth problem'), and more completely, at the Bologna International Congress in 1928. Hilbert had asked for no less than a general algorithmic procedure for resolving mathematical questions—or, rather, for an answer to the question of whether or not such a procedure might in principle exist. Hilbert also had a programme for placing mathematics on an unassailably sound foundation, with axioms and rules of procedure which were to be laid down once and for all, but by the time Turing produced his great work, that programme had already suffered a crushing blow from a startling theorem proved in 1931 by the brilliant Austrian logician Kurt Gödel. We shall be considering the Gödel theorem and its significance in Chapter 4. The problem of Hilbert's that concerned Turing (the *Entscheidungsproblem*) went beyond any particular formulation of mathematics in terms of axiomatic systems. The question was: is there some general mechanical procedure which could, *in principle*, solve all the problems of mathematics (belonging to some suitably well-defined class) one after the other?

Part of the difficulty in answering this question was to decide what is to be meant by a 'mechanical procedure'. The concept lay outside the normal mathematical ideas of the time. In order to come to grips with it, Turing tried to imagine how the concept of a 'machine' could be formalized, its operation being broken down into elementary terms. It seems clear that Turing also regarded a human brain to be an example of a 'machine' in his sense, so whatever the activities might be that are carried out by human mathematicians when they tackle their problems of mathematics, these also would have to come under the heading of 'mechanical procedures'.

Whilst this view of human thinking appears to have been valuable to Turing in the development of his highly important concept, it is by no means necessary for us to adhere to it. Indeed, by making precise what is meant by a mechanical procedure, Turing actually showed that there are some perfectly well-defined mathematical operations which cannot, in any ordinary sense, be called mechanical! There is perhaps some irony in the fact that this aspect of Turing's own work may now indirectly provide us with a possible loophole to his own viewpoint concerning the nature of mental phenomena. However, this is not our concern for the moment. We need first to find out what Turing's concept of a mechanical procedure actually is.

Turing's concept

Let us try to imagine a device for carrying out some (finitely definable) calculational procedure. What general form would such a device take? We must be prepared to idealize a little and not worry too much about practicalities: we are really thinking of a mathematically idealized 'machine'. We want our device to have a discrete set of different possible states, which are *finite* in number (though perhaps a very large number). We call these the *internal* states of the device. However we do not want to limit the size of the calculations that our device will perform in principle. Recall Euclid's algorithm described above. There is, in principle, no limit to the size of the numbers on which that algorithm can act. The algorithm—or the general calculational *procedure*—is just the same no matter how large the numbers are. For very large numbers, the procedure may indeed take a very long time, and a considerable amount of 'rough paper' may be needed on which to perform the actual calculations. But the *algorithm* is the same *finite* set of instructions no matter how big the numbers.

Thus, although it has a finite number of internal states, our device must be able to deal with an input which is not restricted in its size. Moreover, the device must be allowed to call upon an unlimited external storage space (our 'rough paper') for its calculations, and also to be able to produce an output of unlimited size. Since our device has only a finite number of distinct internal states, it cannot be expected to 'internalize' all the external data nor all the results of its own calculations. Instead it must examine only those parts of the data or previous calculations that it is *immediately* dealing with, and then perform whatever operation it is required to perform on them. It can note down, perhaps in the external storage space, the relevant results of that operation, and then proceed in a precisely determined way to the next stage of operation. It is the unlimited nature of the input, calculation space, and output which tells us that we are considering only a mathematical idealization rather than something that could be actually constructed in practice (see fig. 2.1). But it is an idealization of great relevance. The marvels of modern computer technology have provided us with electronic storage devices which can, indeed, be treated as unlimited for most practical purposes.

In fact, the type of storage space that has been referred to as 'external' in the above discussion could be regarded as actually part of the internal workings of a modern computer. It is perhaps a technicality whether a certain part of the storage space is to be regarded as internal or external. *One* way of

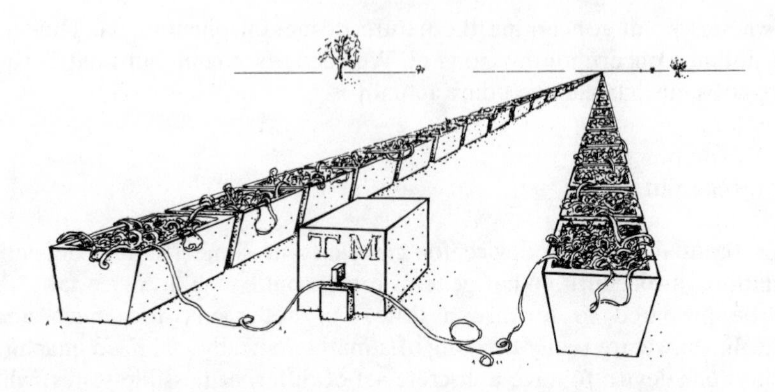

Fig. 2.1. A strict Turing machine requires an infinite tape!

referring to this division between the 'device' and the 'external' part would be in terms of *hardware* and *software*. The internal part could then be the hardware and the external part, the software. I shall *not* necessarily stick to this, but whichever way one looks at it, Turing's idealization is indeed remarkably well approximated by the electronic computers of today.

The way that Turing pictured the external data and storage space was in terms of a 'tape' with marks on it. This tape would be called upon by the device and 'read' as necessary, and the tape could be moved backwards or forwards by the device, as part of its operation. The device could also place new marks on the tape where required, and could obliterate old ones, allowing the *same* tape to act as external storage (i.e. 'rough paper') as well as input. In fact it is helpful not to make any clear distinction between 'external storage' and 'input' because in many operations the intermediate results of a calculation can play a role just like that of new data. Recall that with Euclid's algorithm we kept replacing our original input (the numbers **A** and **B**) by the results of the different stages of the calculation. Likewise, the same tape can be used for the final output (i.e. the 'answer'). The tape will keep running back and forth through the device so long as further calculations need to be performed. When the calculation is finally completed, the device comes to a halt and the answer to the calculation is displayed on that part of the tape which lies to one side of the device. For definiteness, let us suppose that the answer is always displayed on the left, while all the numerical data in the input, together with the specification of the problem to be solved, always goes in on the right.

For my own part, I feel a little uncomfortable about having our finite device moving a potentially infinite tape backwards and forwards. No matter how lightweight its material, an *infinite* tape might be hard to shift! Instead, I prefer to think of the tape as representing some external environment through which our finite device can move. (Of course, with modern electronics, neither the 'tape' nor the 'device' need actually 'move' in the ordinary physical sense, but such 'movement' is a convenient way of picturing things.) On this view, the device receives all its input from this environment. It uses the environment as its 'rough paper'. Finally it writes out its output on this same environment.

In Turing's picture the 'tape' consists of a linear sequence of squares, which is taken to be infinite in both directions. Each square on the tape is either blank or contains a single mark.* The use of marked or unmarked squares illustrates that we are allowing our 'environment' (i.e. the tape) to be broken down and described in terms of *discrete* (as opposed to continuous) elements. This seems to be a reasonable thing to do if we wish our device to function in a reliable and absolutely definite way. We are, however, allowing this 'environment' to be (potentially) infinite, as a feature of the mathematical idealization that we are using, but in any *particular* case the input, calculation and output must always be *finite*. Thus, although the tape is taken to be infinitely long, there must only be a finite number of actual marks on it. Beyond a certain point in each direction the tape must be entirely blank.

We indicate a blank square by the symbol '0' and a marked one by the symbol '1', e.g.:

.... | 0 | 0 | 0 | 1 | 1 | 1 | 1 | 0 | 1 | 0 | 0 | 1 | 1 | 1 | 0 | 0 | 1 | 0 | 0 | 1 | 0 | 1 | 1 | 0 | 1 | 0 | 0 | ....

We need our device to 'read' the tape, and we shall suppose that it does this *one* square at a time, and after each operation moves just *one* square to the right or left. There is no loss in generality involved in that. A device which reads n squares at a time or moves k squares at a time can easily be modelled by another device which reads and moves just one square at a time. A movement of k squares can be built up of k movements of one square, and by storing up n readings of one square it can behave as though it reads all n squares at once.

What, in detail, can such a device do? What is the most general way in which something that we would describe as 'mechanical' might function? Recall that the *internal states* of our device are to be finite in number. All we need to know, beyond this finiteness is that the behaviour of the device is completely determined by its internal state and by the input. This input we have simplified to being just one of the two symbols '0' or '1'. Given its initial state and this input, the device is to operate completely deterministically: it changes its internal state to some other (or possibly the same) internal state; it

* In fact, in his original descriptions Turing allowed his tape to be marked in more complicated ways, but this makes no real difference. The more complicated marks could always be broken down into successions of marks and blanks. I shall be taking various other unimportant liberties with Turing's original specifications.

replaces the 0 or 1 that it is just reading by the same or different symbol 0 or 1; it moves one square either to the right or left; finally, it decides whether to continue the calculation or to terminate it and come to a halt.

To define the operation of our device in an explicit way, let us first *number* the different internal states, say by the labels 0, 1, 2, 3, 4, 5, ...; the operation of the device, or *Turing machine*, would then be completely specified by some explicit list of replacements, such as:

00 → 00R
01 → 131L
10 → 651R
11 → 10R
20 → 01R.STOP
21 → 661L
30 → 370R
. .
. .
. .
2100 → 31L
. .
. .
. .
2580 → 00R.STOP
2590 → 971R
2591 → 00R.STOP

The *large* figure on the left-hand side of the arrow is the symbol on the tape that the device is in the process of reading, and the device replaces it by the large figure at the middle on the right. R tells us that the device is to move one square to the *right* along the tape and L tells us that it is to move by one step to the *left*. (If, as with Turing's original descriptions, we think of the tape moving rather than the device, then we must interpret R as the instruction to move the *tape* one square to the *left* and L as moving it one square to the *right*.) The word STOP indicates that the calculation has been completed and the device is to come to a halt. In particular, the second instruction 01 → 131L tells us that *if* the device is in internal state 0 and reads 1 on the tape then it must change to internal state 13, leave the 1 as a 1 on the tape, and move one square along the tape to the left. The last instruction 2591 → 00R.STOP tells us that if the device is in state 259 and reads 1 on the tape, then it must revert to state 0, erase the 1 to produce 0 on the tape, move one square along the tape to the right, and terminate the calculation.

Instead of using the numerals 0, 1, 2, 3, 4, 5, . . . for labelling the internal states, it would be somewhat more in keeping with the above notation for marks on the tape if, we were to use symbols made up of just 0s and 1s. We could simply use a succession of n 1s to label the state n if we choose, but that is inefficient. Instead, let us use the *binary* numbering system, which is now a familiar mode of notation:

$$\begin{aligned} 0 &\rightarrow 0,\\ 1 &\rightarrow 1,\\ 2 &\rightarrow 10,\\ 3 &\rightarrow 11,\\ 4 &\rightarrow 100,\\ 5 &\rightarrow 101,\\ 6 &\rightarrow 110,\\ 7 &\rightarrow 111,\\ 8 &\rightarrow 1000,\\ 9 &\rightarrow 1001,\\ 10 &\rightarrow 1010,\\ 11 &\rightarrow 1011,\\ 12 &\rightarrow 1100, \text{etc.} \end{aligned}$$

Here the final digit on the right refers to the 'units' just as it does in the standard (denary) notation, but the digit just before it refers to 'twos' rather than 'tens'. The one before that refers to 'fours' rather than 'hundreds' and before that, to 'eights' rather than 'thousands', and so on, the value of each successive digit, as we move to the left, being the successive *powers of two*: 1, 2, 4(=2×2), 8(= 2 × 2 × 2), 16(= 2 × 2 × 2 × 2), 32(= 2 × 2 × 2 × 2 × 2), etc. (For some other purposes that we shall come to later, we shall also sometimes find it useful to use a base other than two or ten to represent natural numbers: e.g. in base *three*, the denary number 64 would be written 2101, each digit having a value which is now a power of three: $64=(2 \times 3^3) + 3^2 + 1$; cf. Chapter 4, p. 106, footnote.)

Using such a binary notation for the internal states, the specification of the above Turing machine would now be:

$$\begin{aligned} 0\mathtt{0} &\rightarrow 0\mathtt{0}\text{R},\\ 0\mathtt{1} &\rightarrow 1101\mathtt{1}\text{R},\\ 1\mathtt{0} &\rightarrow 100000\mathtt{1}\text{L}\\ 1\mathtt{1} &\rightarrow 1\mathtt{0}\text{R}\\ 10\mathtt{0} &\rightarrow 0\mathtt{1}\text{STOP}\\ 10\mathtt{1} &\rightarrow 100001\mathtt{0}\mathtt{1}\text{L}\\ 11\mathtt{0} &\rightarrow 100101\mathtt{0}\text{R}\\ &\ \ \vdots\\ 11010010\mathtt{0} &\rightarrow 11\mathtt{1}\text{L}\\ &\ \ \vdots \end{aligned}$$

.   .
.   .
.   .

1000000101 → 00STOP
1000000110 → 1100001 1R
1000000111 → 00STOP

In the above, I have also abbreviated R.STOP to STOP, since we may as well assume that L.STOP never occurs so that the result of the final step in the calculation is always displayed at the left of the device, as part of the answer.

Let us suppose that our device is in the particular internal state represented by the binary sequence 11010010 and is in the midst of a calculation for which the tape is given as on p. 37, and we apply the instruction 110100100→111L:

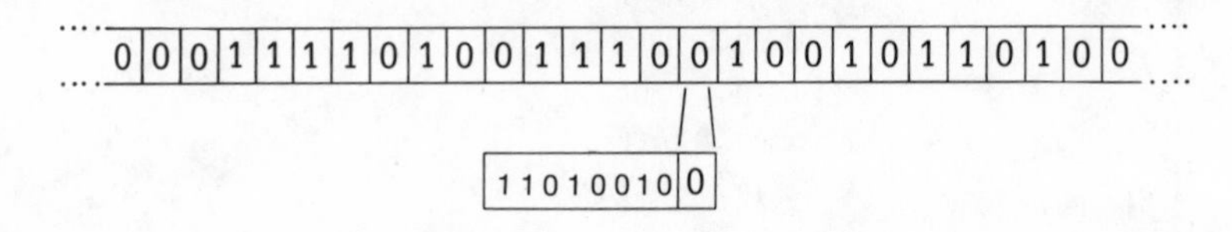

The particular digit on the tape that is being read (here the digit '0') is indicated by a larger figure, to the right of the string of symbols representing the internal state. In the example of a Turing machine as partly specified above (and which I have made up more or less at random), the '0' which is being read would be replaced by a '1' and the internal state would be changed to '11'; then the device would be moved one step to the left:

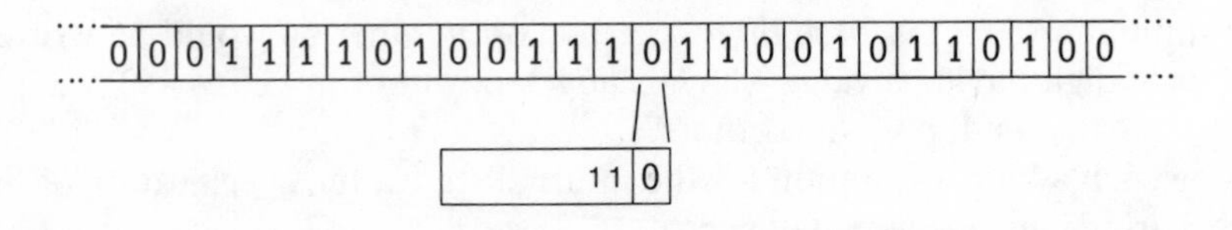

The device is now ready to read another digit, again a '0'. According to the table, it now leaves this '0' unchanged, but replaces the internal state by '100101' and moves back along the tape to the right by one step. Now it reads '1', and somewhere down the table would be a further instruction as to what replacement to make in its internal state, whether it should change the digit it is reading, and in which direction it should move along the tape. It would continue this way until it reaches a STOP, at which point (after it moves one further step to the right) we imagine a bell ringing to alert the operator of the machine that the calculation has been completed.

We shall suppose that the machine is always started with internal state '0' and that all the tape to the left of the reading device is initially blank. The instructions and data are all fed in at the right. As mentioned earlier, this information which is fed in is always to take the form of a *finite* string of 0s and 1s, followed by blank tape (i.e. 0s). When the machine reaches STOP, the result of the calculation appears on the tape to the left of the reading device.

Since we wish to be able to include numerical data as part of our input, we shall want to have a way of describing ordinary numbers (by which I here mean the natural numbers 0, 1, 2, 3, 4, . . .) as part of the input. One way to do this might be simply to use a string of *n* 1s to represent the number *n* (although this could give us a difficulty with the natural number zero):

$$1 \rightarrow 1,\ 2 \rightarrow 11,\ 3 \rightarrow 111,\ 4 \rightarrow 1111,\ 5 \rightarrow 11111,\ \text{etc.}$$

This primitive numbering system is referred to (rather illogically) as the *unary* system. Then the symbol '0' could be used as a space to separate different numbers from one another. It is important that we have such a means of separating numbers from one another since many algorithms act on *sets* of numbers rather than on just single numbers. For example, for Euclid's algorithm, our device would need to act on the *pair* of numbers A and B. Turing machines can be written down, without great difficulty, which effect this algorithm. As an exercise, some dedicated readers might perhaps care to verify that the following explicit description of a Turing machine (which I shall call **EUC**) does indeed effect Euclid's algorithm when applied to a pair of unary numbers separated by a 0:

00→00R, 01→11L, 10→101R, 11→11L, 100→10100R,
101→110R, 110→1000R, 111→111R, 1000→1000R, 1001→1010R,
1010→1110L, 1011→1101L, 1100→1100L, 1101→11L, 1110→1110L,
1111→10001L, 10000→10010L, 10001→10001L, 10010→100R,
10011→11L, 10100→00STOP, 10101→10101R.

Before embarking on this, however, it would be wise for any such reader to start with something much simpler, such as the Turing machine **UN+1**:

00→00R, 01→11R, 10→01STOP, 11→11R,

Which simply adds one to a unary number. To check that **UN+1** does just that, let us imagine that it is applied to, say, the tape

. . . 00000111100000 . . .,

which represents the number 4. We take the device to be initially somewhere off to the left of the 1s. It is in internal state 0 and reads a 0. This it leaves as 0, according to the first instruction, and it moves off one step to the right, staying in internal state 0. It keeps doing this, moving one step to the right until it meets the first 1. Then the second instruction comes into play: it leaves the 1 as a 1 and moves to the right again, but now in internal state 1. In accordance with the fourth instruction, it stays in internal state 1, leaving the 1s alone, moving along to the right until it reaches the first 0 following the 1s.

The third instruction then tells it to change that 0 to a 1, move one further step to the right (recall that STOP stands for R.STOP) and then halt. Thus, another 1 has been added to the string of 1s, and the 4 of our example has indeed been changed to 5, as required.

As a somewhat more taxing exercise, one may check that the machine **UN×2**, defined by

00→00R, 01→10R, 10→101L, 11→11R, 100→110R, 101→1000R, 110→01STOP, 111→111R, 1000→1011L, 1001→1001R, 1010→101L, 1011→1011L,

doubles a unary number, as it is intended to.

In the case of **EUC**, to get the idea of what is involved, some suitable explicit pair of numbers can be tried, say 6 and 8. The reading device is, as before, taken to be in state 0 and initially on the left, and the tape would now be initially marked as:

. . . 0000000000011111101111111100000 . . . .

After the Turing machine comes to a halt, many steps later, we get a tape marked

. . . 000011000000000000 . . .

with the reading device to the right of the non-zero digits. Thus the required highest common factor is (correctly) given as 2.

The full explanation of *why* **EUC** (or, indeed, **UN×2**) actually does what it is supposed to do involves some subtleties and would be rather more complicated to explain than the machine is complicated itself—a not uncommon feature of computer programs! (To understand fully why an algorithmic procedure does what it is supposed to involves *insights*. Are 'insights' themselves algorithmic? This is a question that will have importance for us later.) I shall not attempt to provide such an explanation here for the examples **EUC** or **UN×2**. The reader who does check them through will find that I have taken a very slight liberty with Euclid's actual algorithm in order to express things more concisely in the required scheme. The description of **EUC** is still somewhat complicated, comprising 22 elementary instructions for 11 distinct internal states. Most of the complication is of a purely organizational kind. It will be observed, for example, that of the 22 instructions, only 3 actually involve altering marks on the tape! (Even for **UN×2** I have used 12 instructions, half of which involve altering the marks.)

Binary coding of numerical data

The unary system is exceedingly inefficient for the representation of numbers of large size. Accordingly, we shall usually use the *binary* number system, as described earlier. However, we cannot just do this directly, attempting to read the tape simply as a binary number. As things stand, there would be no way of telling when the binary representation of the number has come to an end and the infinite succession of 0s representing the blank tape on the right begins. We need some notation for terminating the binary description of a number. Moreover, we shall often want to feed in *several* numbers, as with the *pair* of numbers[2] required for Euclid's algorithm. As things stand, we cannot distinguish the *spaces* between numbers from the 0s or strings of 0s that appear as parts of the binary representation of single numbers. In addition, we might perhaps also want to include all kinds of complicated instructions on the input tape, as well as numbers. In order to overcome these difficulties, let us adopt a procedure which I shall refer to as *contraction*, according to which any string of 0s and 1s (with a finite total number of 1s) is *not* simply read as a binary number, but is replaced by a string of 0s, 1s, 2s, 3s, etc., by a prescription whereby each digit of the second sequence is simply the number of 1s lying between successive 0s of the first sequence. Thus, for example, the sequence

01000101101010110100011101010111100110

would be replaced according to

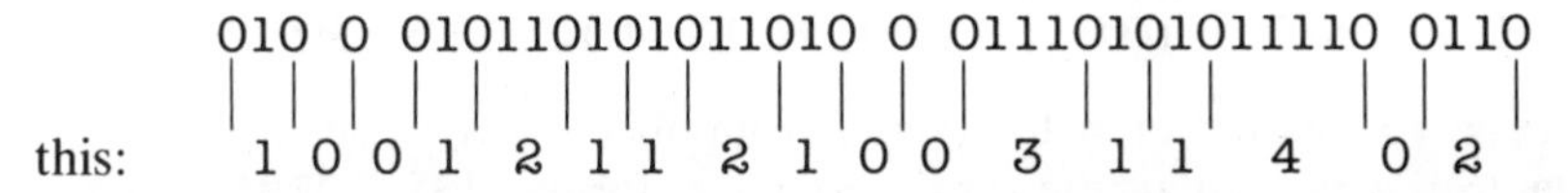

We can now read the numbers 2, 3, 4, . . . as markers or instructions of some kind. Indeed, let us regard 2 as simply a 'comma', indicating the space between two numbers, whereas 3, 4, 5, . . . could, according to our wishes, represent various instructions or notations of interest, such as 'minus sign', 'plus', 'times', 'go to the location with the following number', 'iterate the previous operation the following number of times', etc. We now have various strings of 0s, and 1s which are separated by higher digits. The former are to represent ordinary numbers written in the binary scale. Thus, the above would read (with 'comma' for '2'):

(binary number 1001) comma (binary number 11) comma . . .

Using standard Arabic notation '9', '3', '4', '0' for the respective binary numbers 1001, 11, 100, 0, we get, for the entire sequence:

9, 3, 4 (instruction 3) 3 (instruction 4) 0,

In particular, this procedure gives us a means of terminating the description of a number (and thereby distinguishing it from an infinite stretch of blank tape on the right) simply by using a comma at the end. Moreover, it enables us to code any finite sequence of natural numbers, written in the binary notation, as a *single* sequence of 0s and 1s, where we use commas to separate the numbers. Let us see how this works in a specific case. Consider the sequence

$$5, 13, 0, 1, 1, 4,$$

for example. In binary notation this is

101, 1101, 0, 1, 1, 100,

which is coded on the tape, by *expansion* (i.e. the inverse of the above contraction procedure), as

. . . 00001001011010100101100110101101011010001100 . . . .

To achieve this coding in a simple direct way we can make replacements in our original sequence of binary numbers as follows

0 → 0
1 → 10
, → 110

and then adjoin an unlimited supply of 0s at both ends. It is made clearer how this has been applied to the above tape, if we space it out:

0000 10 0 10 110 10 10 0 10 110 0 110 10 110 10 110 10 0 0 110 00

I shall refer to this notation for (sets of) numbers as the *expanded binary* notation. (So, in particular, the expanded binary form of 13 is 1010010.)

There is one final point that should be made about this coding. It is just a technicality, but necessary for completeness.[3] In the binary (or denary) representation of natural numbers there is a slight redundancy in that 0s placed on the far left of an expression do not 'count'—and are normally omitted, e.g. 00110010 is the same binary number as 110010 (and 0050 is the same denary number as 50). This redundancy extends to the number zero itself, which can be written 000 or 00 just as well as 0. Indeed a blank space should, logically, denote zero as well! In ordinary notation that would lead to great confusion, but it fits in well with the notation just described above. Thus a zero between two commas can just as well be written as two commas next to one another (,,) which would be coded on the tape as two pairs 11 separated by a single 0:

. . . 001101100. . . .

Thus the above set of six numbers can also be written in binary notation as

101,1101,,1,1,100,

and coded on the tape, in expanded binary form, as

. . . 000010010110101001011011010110101101000110000 . . .

(which has one 0 missing from the sequence that we had before).

We can now consider a Turing machine for effecting, say, Euclid's algorithm, applying it to pairs numbers written in the expanded binary notation. For example, for the pair of numbers 6, 8 that we considered earlier, instead of using

. . . 0000000000011111101111111100000 . . . ,

as we did before, we consider the binary representations of 6 and 8, namely 110 and 1000, respectively. The *pair* is

6,8,   i.e., in binary notation,   110,1000,

which, by expansion, is coded as the tape

. . . 00000101001101000011000000 . . .

For this particular pair of numbers there is no gain in conciseness from the unary form. Suppose, however, that we take, say, the (denary) numbers 1583169 and 8610. In binary notation these would be

110000010100001000001,   10000110100010,

so we have the pair coded as the tape

. . .
0010100000001001000000100000010110100000010100100000100110
. . .

which all fits on one line, whereas in the unary notation, the tape representing '1583169, 8610' would more than fill this entire book!

A Turing machine that effects Euclid's algorithm when the numbers are expressed in expanded binary notation could, if desired, be obtained simply by adjoining to **EUC** a suitable pair of subroutine algorithms which translate between unary and expanded binary. This would actually be extremely inefficient, however, since the inefficiency of the unary numbering system would still be 'internally' present and would show up in the slowness of the device and in the inordinate amount of external 'rough paper' (which would be on the left-hand part of the tape) that would be needed. A more efficient Turing machine for Euclid's algorithm operating entirely within expanded binary can also be given, but it would not be particularly illuminating for us here.

Instead, in order to illustrate how a Turing machine can be made to operate on expanded binary numbers, let us try something a good deal simpler than Euclid's algorithm, namely the process of simply *adding one* to a natural number. This can be effected by the Turing machine (which I shall call **XN+1**):

00→00R, 01→11R, 10→00R, 11→101R, 100→110L, 101→101R,
110→01STOP, 111→1000L, 1000→1011L, 1001→1001L, 1010→1100R,
1011→101R, 1101→1111R, 1110→111R, 1111→1110R.

Again, some dedicated readers might care to check that this Turing machine actually does what it is supposed to do, by applying it to, say, the number 167, which has binary representation 10100111 and so would be given by the tape

. . . 0000100100010101011000 . . . .

To add one to a binary number, we simply locate the final 0 and change it to 1 and then replace all the 1s which follow by 0s, e.g. 167 + 1 = 168 is written in binary notation as

10100111 + 1 = 10101000.

Thus our 'adding-one' Turing machine should replace the aforementioned tape by

. . . 0000100100100001100000 . . .

which indeed it does.

Note that even the very elementary operation of simply adding one is a bit complicated with this notation, using fifteen instructions and eight different internal states! Things were a lot simpler with the unary notation, of course, since 'adding one' then simply means extending the string of 1s by one further 1, so it is not surprising that our machine **UN+1** was more basic. However, for very large numbers, **UN+1** would be exceedingly slow because of the inordinate length of tape required, and the more complicated machine **XN+1**, which operates with the more compact expanded binary notation, would be better.

As an aside, I point out an operation for which the Turing machine actually looks simpler for expanded binary than for unary notation, namely *multiplying by two*. Here, the Turing machine **XN×2**, given by

00→00R, 01→11R, 10→00R, 11→100R, 100→111R, 110→00STOP,

achieves this in expanded binary, whereas the corresponding machine in unary notation, **UN×2**, which was described earlier, is a good deal more complicated!

This gives us some idea of what Turing machines can do at a very basic level. As might be expected, they can, and do, get vastly more complicated than this when operations of some complexity are to be performed. What is the ultimate scope of such devices? Let us consider this question next.

The Church–Turing Thesis

Once one has gained some familiarity with constructing simple Turing machines, it becomes easy to satisfy oneself that the various basic arithmetical operations, such as adding two numbers together, or multiplying them, or raising one number to the power of another, can indeed all be effected by specific Turing machines. It would not be too cumbersome to give such machines explicitly, but I shall not bother to do this here. Operations where the result is a *pair* of natural numbers, such as division with a remainder, can also be provided—or where the result is an arbitrarily large finite set of numbers. Moreover Turing machines can be constructed for which it is not specified ahead of time which arithmetical operation it is that needs to be performed, but the instructions for this are fed in on the tape. Perhaps the particular operation that has to be performed depends, at some stage, upon the result of some calculation that the machine has had to perform at some earlier stage. ('If the answer to that calculation was greater than so-and-so, do this; otherwise, do that.') Once it is appreciated that one can make Turing machines which perform arithmetic or simple logical operations, it becomes easier to imagine how they can be made to perform more complicated tasks of an algorithmic nature. After one has played with such things for a while, one is easily reassured that a machine of this type can indeed be made to perform *any mechanical operation whatever*! Mathematically, it becomes reasonable to *define* a mechanical operation to be one that can be carried out by such a machine. The noun 'algorithm' and the adjectives 'computable', 'recursive', and 'effective' are all used by mathematicians to denote the mechanical operations that can be performed by theoretical machines of this type—the Turing machines. So long as a procedure is sufficiently clear-cut and mechanical, then it is reasonable to believe that a Turing machine can indeed be found to perform it. This, after all, was the whole point of our (i.e. Turing's) introductory discussion motivating the very concept of a Turing machine.

On the other hand, it still could be felt that the design of these machines was perhaps unnecessarily restrictive. Allowing the device to read only one binary digit (0 or 1) at a time, and to move only one space at a time along only a *single* one-dimensional tape seems at first sight to be limiting. Why not allow four or five, or perhaps one thousand separate tapes, with a great number or interconnected reading devices running all at once? Why not allow a whole plane of squares of 0s and 1s (or perhaps a three-dimensional array) rather than insisting on a one-dimensional tape? Why not allow other symbols from some more complicated numbering system or alphabet? In fact, none of these changes makes the slightest difference to what can be in principle achieved, though some make a certain amount of difference to the economy of the operations (as would certainly be the case if we allowed more than one tape). The class of operations performed, and thus come under the heading of 'algorithms' (or 'computations' or 'effective procedures' or 'recursive operations'), would be precisely the same as before even if we broadened the definition of our machines in all these ways at once!

We can see that there is no *necessity* to have more than one tape, so long as the device can keep finding new space on the given tape, as required. For this, it may need to keep shunting data from one place to another on the tape. This may be 'inefficient', but it does not limit what can be in principle achieved.[4] Likewise, using more than one Turing device in *parallel action*—which is an idea that has become fashionable in recent years, in connection with attempts to model human brains more closely—does *not* in principle gain anything (though there may be an improved speed of action under certain circumstances). Having two separate devices which do not directly communicate with one another achieves no more than having two which *do* communicate; and *if* they communicate, then, in effect, they are just a single device!

What about Turing's restriction to a one-dimensional tape? If we think of this tape as representing the 'environment', we might prefer to think of it as a planar surface rather than as a one-dimensional tape, or perhaps as a three-dimensional space. A planar surface might seem to be closer to what is needed for a 'flow chart' (as in the above description of the operation of Euclid's algorithm) than a one-dimensional tape would be[*] There is, however, no difficulty in principle about writing out the operation of a flow diagram in a 'one-dimensional' form (e.g. by the use of an ordinary verbal description of the chart). The two-dimensional planar display is only for our own convenience and ease of comprehension and it makes no difference to what can in principle be achieved. It is always possible to code the location of a mark or an object on a two-dimensional plane, or even in a three-dimensional space, in a straightforward way on a one-dimensional tape. (In fact, using a two-dimensional plane is completely equivalent to using *two* tapes. The two tapes would supply the two 'coordinates' that would be needed for specifying a point on a two-dimensional plane; likewise *three* tapes can act as 'coordinates' for a point in a three-dimensional space.) Again this one-dimensional coding may be 'inefficient', but it does not limit what can be achieved in principle.

Despite all of this, we might still question whether the concept of a Turing machine really does incorporate *every* logical or mathematical operation that

[*] As things have been described here, this flow chart itself would actually be part of the 'device' rather than of the external environment 'tape'. It was the actual numbers **A**, **B**, **A**−**B**, etc, which we represented on the tape. However, we shall be wanting also to express the specification of the *device* in a linear one-dimensional form. As we shall see later, in connection with the *universal* Turing machine, there is an intimate relation between the specification for a particular 'device' and the specification possible 'data' (or 'program') for a given device. It is therefore convenient to have *both* of these in one-dimensional form.

we would wish to call 'mechanical'. At the time that Turing wrote his seminal paper, this was considerably less clear than it is today, so Turing found it necessary to put his case in appreciable detail. Turing's closely argued case found additional support from the fact that, quite independently (and actually a little earlier), the American logician Alonzo Church (with the help of S. C. Kleene) had put forward a scheme—the lambda calculus—also aimed at resolving Hilbert's *Entscheidungsproblem*. Though it was much less obviously a fully comprehensive mechanical scheme than was Turing's, it had some advantages in the striking economy of its mathematical structure. (I shall be describing Church's remarkable calculus at the end of this chapter.) Also independently of Turing there were yet other proposals for resolving Hilbert's problem (see Gandy 1988), most particularly that of the Polish-American logician Emil Post (a little later than Turing, but with ideas considerably more like those of Turing than of Church). All these schemes were soon shown to be completely equivalent. This added a good deal of force to the viewpoint, which became known as the Church–Turing Thesis, that the Turing machine concept (or equivalent) actually does define what, mathematically, we mean by an algorithmic (or effective or recursive or mechanical) procedure. Now that high-speed electronic computers have become such a familiar part of our lives, not many people seem to feel the need to question this thesis in its original form. Instead, some attention has been turned to the matter of whether actual *physical* systems (presumably including human brains)—subject as they are to precise *physical* laws—are able to perform more than, less than, or precisely the same logical and mathematical operations as Turing machines. For my own part, I am very happy to accept the original *mathematical* form of the Church–Turing Thesis. Its relation to the behaviour of actual physical systems, on the other hand, is a separate issue which will be a major concern for us later in this book.

Numbers other than natural numbers

In the discussion given above, we considered operations on *natural numbers*, and we noted the remarkable fact that single Turing machines can handle natural numbers of arbitrarily large size, despite the fact that each machine has a fixed *finite* number of distinct internal states. However, one often needs to work with more complicated kinds of number than this, such as negative numbers, fractions, or infinite decimals. Negative numbers and fractions (e.g. numbers like −597/26) can be easily handled by Turing machines, and the numerators and denominators can be as large as we like. All we need is some suitable coding for the signs '−' and '/', and this can easily be done using the expanded binary notation described earlier (for example, '3' for '−' and '4' for '/'—coded as 1110 and 11110, respectively, in the expanded binary notation). Negative numbers and fractions are thus handled in terms of finite sets of natural numbers, so with regard to general questions of computability they give us nothing new.

Likewise, *finite* decimal expressions of unrestricted length give us nothing new, since these are just particular cases of fractions. For example, the finite decimal approximation to the irrational number π, given by 3.14159265, is simply the fraction 314159265/100000000. However, *infinite* decimal expressions, such as the *full* non-terminating expansion

$$\pi = 3.14159265358979\ldots$$

present certain difficulties. Neither the input nor the output of a Turing machine can, strictly speaking, be an infinite decimal. One might think that we could find a Turing machine to churn out *all* the successive digits, 3,1,4,5,9, . . ., of the above expansion for π one after the other on the output tape, where we simply allow the machine to run on forever. But this is *not allowed* for a Turing machine. We must wait for the machine to halt (indicated by the bell ringing!) before we are allowed to examine the output. So long as the machine has not reached a STOP order, the output is subject to possible change and so cannot be trusted. After it has reached STOP, on the other hand, the output is necessarily finite.

There is, however, a procedure for *legitimately* making a Turing machine produce digits one after the other, in a way very similar to this. If we wish to generate an infinite decimal expansion, say that of π, we could have a Turing machine produce the whole-number part, 3, by making the machine act on 0, then we could produce the first decimal digit, 1, by making the machine act on 1, then the second decimal digit, 4, by making it act on 2, then the third, 1, by making it act on 3, and so on. In fact a Turing machine for producing the entire decimal expansion of π in *this* sense certainly *does* exist, though it would be a little complicated to work it out explicitly. A similar remark applies to many other irrational numbers, such as $\sqrt{2} = 1.414213562\ldots$. It turns out, however, that some irrationals (remarkably) cannot be produced by any Turing machine at all, as we shall see in the next chapter. The numbers that *can* be generated in this way are called *computable* (Turing 1937). Those that cannot (actually the vast majority!) are *non*-computable. I shall come back to this matter, and related issues, in later chapters. It will have some relevance for us in relation to the question of whether an *actual physical object* (e.g. a human brain) can, according to our physical theories, be adequately described in terms of computable mathematical structures.

The issue of computability is an important one generally in mathematics. One should not think of it as a matter which applies just to *numbers* as such. One can have Turing machines which operate directly on *mathematical formulae*, such as algebraic or trigonometric expressions, for example, or which carry through the formal manipulations of the calculus. All that one needs is some form of precise coding into sequences of 0s and 1s, of all the mathematical symbols that are involved, and then the Turing machine concept can be applied. This, after all, was what Turing had in mind in his attack on the *Entscheidungsproblem*, which asks for an algorithmic procedure for answering mathematical questions of a *general* nature. We shall be coming back to this shortly.

The universal Turing machine

I have not yet described the concept of a *universal* Turing machine. The principle behind this is not too difficult to give, even though the details are complicated. The basic idea is to code the list of instructions for an arbitrary Turing machine T into a string of 0s and 1s that can be represented on a tape. This tape is then used as the initial part of the input for some *particular* Turing machine U—called a universal Turing machine—which then acts on the remainder of the input just as T would have done. The universal Turing machine is a universal mimic. The initial part of the tape gives the universal machine U the full information that it needs for it to imitate any given machine T exactly!

To see how this works we first need a systematic way of *numbering* Turing machines. Consider the list of instructions defining some particular Turing machine, say one of those described above. We must code this list into a string of 0s and 1s according to some precise scheme. This can be done with the aid of the 'contraction' procedure that we adopted before. For if we represent the respective symbols R, L, STOP, the arrow (→), and the comma as, say, the numerals 2, 3, 4, 5, and 6, these can be coded by contraction to 110, 1110, 11110, 111110, and 1111110. Then the digits 0 and 1, coded as 0 and 10, respectively, can be used for the actual strings of these symbols appearing in the table. We do not need to have a different notation to distinguish the large figures 0 and 1 in the Turing machine table from the smaller boldface ones, since the position of the large digits at the end of the binary numbering is sufficient to distinguish them from the others. Thus, for example, 1101 would be read as the binary number 1101 and coded on the tape as 1010010. In particular, 00 would be read as 00, which can, without ambiguity, be coded 0, or as a symbol omitted altogether. We can economize considerably by not actually bothering to code any arrow nor any of the symbols immediately preceding it, relying instead upon the numerical ordering of instructions to specify what those symbols must be—although to adopt this procedure we must make sure that there are no gaps in this ordering, supplying a few extra 'dummy' orders where required. (For example, the Turing machine **XN+1** has no order telling us what to do with 1100, since this combination never occurs in the running of the machine, so we must insert a 'dummy' order, say 110→0R, which can be incorporated into the list without changing anything. Similarly we should insert 101→0R into the machine **XN×2**.) Without such 'dummies', the coding of the subsequent orders in the list would be spoiled. We do not actually need the comma at the end of each instruction, as it turns out, since the symbols L or R suffice to separate the instructions from one another. We therefore simply adopt the following coding:

0 for 0 or 0, 10 for 1 or 1, 110 for R, 1110 for L,
11110 for STOP

As an example, let us code the Turing machine **XN+1** (with the 110→0R instruction inserted). Leaving out the arrows, the digits immediately preceding them, and also the commas, we have

00R 11R 00R 101R 110L 101R 01STOP 1000L 1011L 1001L
110R 101R 00R 1111R 111R 1110R.

We can improve on this by leaving out every 00 and replacing each 01, by simply 1, in accordance with what has been said earlier, to get

R11RR101R110L101R1STOP1000L1011L1001L1100R101RR1111R111R1110R.

This is coded as the tape sequence

110101011011010010110101001110100101101011110100001110
100101011101000101110101000110100101101101010101011010
10101101010100110.

As two further minor economies, we may as well always delete the initial 110 (together with the infinite stretch of blank tape that precedes it) since this denotes 00R, which represents the initial instruction 00→00R that I have been implicitly taking to be common to *all* Turing machines—so that the device can start arbitrarily far to the left of the marks on the tape and run to the right until it comes up to the first mark—and we may as well always delete the final 110 (and the implicit infinite sequence of 0s which is assumed to follow it) since all Turing machines must have their descriptions ending this way (because they all end with R, L, or STOP). The resulting *binary number* is the *number* of the Turing machine, which in the case of **XN+1** is:

101011011010010110101001110100101101011110100001110100
101011101000101110101000110100101101101010101011010101
01101010100.

In standard denary notation, this particular number is

450 813 704 461 563 958 982 113 775 643 437 908.

We sometimes loosely refer to the Turing machine whose number is n as the nth Turing machine, denoted T_n. Thus **XN+1** is the 450813704461563958982113775643437908th Turing machine!

It is a striking fact that we appear to have to go this far along the 'list' of Turing machines before we find one that even performs so trivial operation as adding one (in the expanded binary notation) to a natural number! (I do not think that I have been grossly inefficient in my coding, though I can see room for some minor improvements.) Actually there are some Turing machines with smaller numbers which are of interest. For example, **UN+1** has the binary number

101011010111101010

which is merely 177642 in denary notation! Thus the particularly trivial Turing machine **UN+1**, which merely places an additional 1 at the end of a sequence of 1s, is the 177642nd Turing machine. For curiosity's sake, we may note that 'multiplying by two' comes somewhere between these two in the list of Turing machines, in either notation, for we find that the number for **XN×2** is 1456581339 while that of **UN×2** is 1492923420919872026917547669.

It is perhaps not surprising to learn, in view of the sizes of these numbers, that the vast majority of natural numbers do not give working Turing machines at all. Let us list the first thirteen Turing machines according to this numbering:

T_0: 00→00R, 01→00R,
T_1: 00→00R, 01→00L,
T_2: 00→00R, 01→01R,
T_3: 00→00R, 01→00STOP,
T_4: 00→00R, 01→10R,
T_5: 00→00R, 01→01L,
T_6: 00→00R, 01→00R, 10→00R,
T_7: 00→00R, 01→???,
T_8: 00→00R, 01→100R,
T_9: 00→00R, 01→10L,
T_{10}: 00→00R, 01→11R,
T_{11}: 00→00R, 01→01STOP,
T_{12}: 00→00R, 01→00R, 10→00R.

Of these, T_0 simply moves on to the right obliterating everything that it encounters and never stops. T_1 is more timid, since it bounces back to the left and continues that way forevermore, after its first encounter with a mark on the tape, which it obliterates. Like T_0, the machine T_2 also moves on endlessly to the right, but is more respectful, simply leaving everything on the tape just as it was before. None of these is any good as a Turing machine since none of them ever stops. T_3 is the first respectable Turing machine. It indeed stops, modestly, after changing the first (leftmost) 1 into a 0.

T_4 encounters a serious problem. After it finds its first 1 on the tape it enters an internal state for which there is no listing, so it has no instructions as to what to do next. T_8, T_9, and T_{10} encounter the same problem. The difficulty with T_7 is even more basic. The string of 0s and 1s which codes it involves a sequence of *five* successive 1s: 110111110. There is no interpretation for such a sequence, so T_7 will get stuck as soon as it finds its first 1 on the tape. (I shall refer to T_7, or any other machine T_n for which the binary expansion of n contains a sequence of more than four 1s as being *not correctly specified*.) The machines T_5, T_6, and T_{12} encounter problems similar to those of T_0, T_1, and T_2. They simply run on indefinitely without ever stopping. All of the machines T_0, T_1, T_2, T_4, T_5, T_6, T_7, T_8, T_9, T_{10}, and T_{12} are duds! Only T_3 and T_{11} are working Turing machines, and not very interesting ones at that. T_{11} is even more modest than T_3. It stops at its first encounter with a 1 and it doesn't change a thing!

We should note that there is also a redundancy in our list. The machine T_{12} is identical with T_6, and also identical in action with T_1, since the internal state 1 of T_6 and T_{12} is never entered. We need not be disturbed by this redundancy, nor by the proliferation of dud Turing machines in the list. It would indeed be possible to improve our coding so that a good many of the duds are removed and the redundancy considerably reduced. All this would be at the expense of complicating our poor universal Turing machine which has to decipher the code and pretend to be the Turing machine T_n whose number n it is reading. This might be worth doing if we could remove *all* the duds (or the redundancy). But this is *not* possible, as we shall see shortly! So let us leave our coding as it is.

It will be convenient to interpret a tape with its succession of marks, e.g.

. . . 0001101110010000 . . .

as the binary representation of some number. Recall that the 0s continue indefinitely at both ends, but that there is only a finite number of 1s. I am also assuming that the number of 1s is *non-zero* (i.e. there is at least one 1). We could choose to read the finite string of symbols between the first and last 1 (inclusive), which in the above case is

110111001,

as the binary description of a natural number (here 441, in denary notation). However, this procedure would only give us *odd* numbers (numbers whose binary representation ends in a 1) and we want to be able to represent *all* natural numbers. Thus we adopt the simple expedient of removing the final 1 (which is taken to be just a marker indicating the termination of the expression) and reading what is left as a binary number.[5] Thus, for the above example, we have the binary number

11011100,

which, in denary notation, is 220. This procedure has the advantage that zero is also represented as a marked tape, namely

. . . 0000001000000. . . .

Let us consider the action of the Turing machine T_n on some (finite) string of 0s and 1s on a tape which we feed in on the right. It will be convenient to regard this string also as the binary representation of some number, say m, according to the scheme given above. Let us assume that after a succession of steps the machine T_n finally comes to a halt (i.e. reaches STOP). The string of binary digits that the machine has now produced at the left is the answer to the calculation. Let us also read this as the binary representation of a number in the same way, say p. We shall write this relation, which expresses the fact that when the nth Turing machine acts on m it produces p, as:

$$T_n(m) = p.$$

Now let us look at this relation in a slightly different way. We think of it as expressing one particular operation which is applied the *pair* of numbers n and m in order to produce the number p. (Thus: given the *two* numbers n and m, we can work out from them what p is by seeing what the nth Turing machine does to m.) This particular operation is an entirely algorithmic procedure. It can therefore be carried out by *one particular* Turing machine U; that is, U acts on the *pair* (n, m) to produce p. Since the machine U has to act on *both* of n and m to produce the single result p, we need some way of coding the pair (n, m) on the *one* tape. For this, we can assume that n is written out in ordinary binary notation and then immediately terminated by the sequence 111110. (Recall that the binary number of every correctly specified Turing machine is a sequence made up just of 0s, 10s, 110s, 1110s, and 11110s, and it therefore contains no sequence of more than four 1s. Thus if T_n is a correctly specified machine, the occurrence of 111110 indeed signifies that the description of the number n is finished with.) Everything following it is to be simply the tape represented by m according to our above prescription (i.e. the binary number m immediately followed by 1000 . . .). Thus this second part is simply the tape that T_n is supposed to act on.

As an example, if we take $n = 11$ and $m = 6$, we have, for the tape on which U has to act, the sequence of marks

. . . 000101111111011010000. . . .

This is made up as follows:

. . . 0000 (initial blank tape)
1011 (binary representation of 11)
111110 (terminates n)
110 (binary representation of 6)
10000 . . . (remainder of tape)

What the Turing machine U would have to do, at each successive step of the operation of T_n on m, would be to examine the structure of the succession of digits in the expression for n so that the appropriate replacement in the digits for m (i.e. T_n's 'tape') can be made. In fact it is not difficult in principle (though decidedly tedious in practice) to see how one might actually construct such a machine. Its own list of instructions would simply be providing a means of reading the appropriate entry in that 'list' which is encoded in the number n, at each stage of application to the digits of the 'tape', as given by m. There would admittedly be a lot of dodging backwards and forwards between the digits of m and those of n, and the procedure would tend to be exceedingly slow. Nevertheless, a list of instructions for such a machine can certainly be provided; and we call such a machine a *universal* Turing machine. Denoting the action of this machine on the pair of numbers n and m by $U(n,m)$, we have:

$$U(n,m) = T_n(m)$$

for each (n, m) for which T_n is a correctly specified Turing machine.[6] The machine U, when first fed with the number n, precisely imitates the nth Turing machine!

Since U is a Turing machine, it will itself have a number; i.e. we have

$$U = T_u,$$

for some number u. How big is u? In fact we can take *precisely*

$u =$ 7244855335339317577198395039615711237952360672556559631108144796
606505059404241090310483613632359365644443458382226883278767626556
1446928141177150178425517075540856576897533463569424784885970469347
2573998858228382779529468346052106116983594593879188554632644092552
5505820555989451890716537414896033096753020431553625034984529832320
6515830476641421307088193297172341510569802627346864299218381721573
3348282307345371342147505974034518437235959309064002432107734217885
1492760797597634415123079586396354492269159479654614711345700145048
1673375621725734645227310544829807849651269887889645697609066342044
7798902191443793283001949357096392170390483327088259620130177372720
2718625919914428275437422351355675134084222299889374410534305471044
3686958764051781280194375308138706399427728231564252892375145654438
9905278079324114482614235728619311833261065612275553181020751108533
7633806031082361675045635852164214869542347187426437544428790062485

8270912404220765387542644541334517485662915742999095026230097337381
3772416217274772361020678685400289356608569682262014198248621698902
6091309402985706001743006700868967590344734174127874255812015493663
9389969058177385916540553567040928213322216314109787108145997866959
9704509681841906299443656015145490488092208448003482249207730403043
1884298993931352668823496621019471619107014619685231928474820344958
9770955356110702758174873332729667899879847328409819076485127263100
1740166787363477605857245036964434897992034489997455662402937487668
8397514044516657077500605138839916688140725455446652220507242623923
7921152531816251253630509317286314220040645713052758023076651833519
95689139748137504926429605010013651980186945639498

(or some other possibility of at least that kind of size). This number no doubt seems alarmingly large! Indeed it *is* alarmingly large, but I have not been able to see how it could have been made significantly smaller. The coding procedures and specifications that I have given for Turing machines are quite reasonable and simple ones, yet one is inevitably led to a number of this kind of size for the coding of an actual universal Turing machine.[7]

I have said that all modern general purpose computers are, in effect, universal Turing machines. I do not mean to imply that the logical design of such computers need resemble at all closely the kind of description for a universal Turing machine that I have just given. The point is simply that, by supplying any universal Turing machine first with an appropriate program (initial part of the input tape), it can be made to mimic the behaviour of any Turing machine whatever! In the description above, the program simply takes the form of a single number (the number n), but other procedures are possible, there being many variations on Turing's original theme. In fact in my own descriptions I have deviated somewhat from those that Turing originally gave. None of these differences is important for our present needs.

The insolubility of Hilbert's problem

We now come to the purpose for which Turing originally put forward his ideas, the resolution of Hilbert's broad-ranging *Entscheidungsproblem*: is there some mechanical procedure for answering all mathematical problems, belonging to some broad, but well-defined class? Turing found that he could phrase his version of the question in terms of the problem of deciding whether or not the nth Turing machine would actually ever *stop* when acting on the number m. This problem was referred to as the *halting problem*. It is an easy matter to construct an instruction list for which the machine will not stop for *any* number m (for example, $n = 1$ or 2, as given above, or any other case where there are no STOP instructions whatever). Also there are many instruction lists for which the machine would always stop, whatever number it is given (e.g. $n = 11$); and some machines would stop for some numbers but not for others. One could fairly say that a putative algorithm is not much use when it runs forever without stopping. That is no algorithm at all. So an important question is to be able to decide whether or not T_n applied to m actually ever gives any answer! If it does *not* (i.e. if the calculation does *not* stop), then I shall write

$$T_n(m) = \square.$$

(Included in this notation would be those situations where the Turing machine runs into a problem at some stage because it finds no appropriate instruction to tell it what to do—as with the dud machines such as T_4 and T_7 considered above. Also, unfortunately, our seemingly successful machine T_3 must now also be considered a dud: $T_3(m) = \square$, because the result of the action of T_3 is always just blank tape, whereas we need at least one 1 in the output in order that the result of the calculation be assigned a number! The machine T_{11} is, however, legitimate since it produces a single 1. This output is the tape numbered 0, so we have $T_{11}(m) = 0$ for all m.)

It would be an important issue in mathematics to be able to decide when Turing machines stop. For example, consider the equation:

$$(x + 1)^{w+3} + (y + 1)^{w+3} = (z + 1)^{w+3}.$$

(If technical mathematical equations are things that worry you, don't be put off! This equation is being used only as an example, and there is no need to understand it in detail.) This particular equation relates to a famous unsolved problem in mathematics—perhaps the most famous of all. The problem is this: is there *any* set of natural numbers w, x, y, z for which this equation is satisfied? The famous statement known as 'Fermat's last theorem', made in the margin of Diophantus's *Arithmetica*, by the great seventeenth century French mathematician Pierre de Fermat (1601–1665), is the assertion that the equation is *never* satisfied.[*8] Though a lawyer by profession (and a contemporary of Descartes), Fermat was the finest mathematician of his time. He claimed to have 'a truly wonderful proof' of his assertion, which the margin was too small to contain; but to this day no-one has been able to reconstruct such a proof nor, on the other hand, to find any counter-example to Fermat's assertion!

It is clear that *given* the quadruple of numbers (w, x, y, z), it is a mere matter of computation to decide whether or not the equation holds. Thus we could imagine a computer algorithm which runs through all the quadruples of

[*] Recall that by the *natural* numbers we mean 0, 1, 2, 3, 4, 5, 6, . . . The reason for the '$x + 1$' and '$w + 3$', etc., rather than the more familiar form ($x^w + y^w = z^w$; $x, y, z > 0$, $w > 2$) of the Fermat assertion, is that we are allowing *all* natural numbers for x, w, etc., starting with zero.

numbers one after the other, and stops only when the equation is satisfied. (We have seen that there are ways of coding finite sets of numbers, in a computable way, on a single tape, i.e. simply as single numbers, so we can 'run through' all the quadruples by just following the natural ordering of these single numbers.) If we could establish that this algorithm does *not* stop, then we would have a proof of the Fermat assertion.

In a similar way it is possible to phrase many other unsolved mathematical problems in terms of the Turing machine halting problem. Such an example is the 'Goldbach conjecture', which asserts that every even number greater than 2 is the sum of two prime numbers.* It is an algorithmic process to decide whether or not a given natural number is prime since one needs only to test its divisibility by numbers *less* than itself, a matter of only *finite* calculation. We could devise a Turing machine which runs through the even numbers 6, 8, 10, 12, 14, . . . trying all the different ways of splitting them into pairs of odd numbers

$$6 = 3 + 3, \qquad 8 = 3 + 5, \qquad 10 = 3 + 7 = 5 + 5, \qquad 12 = 5 + 7,$$
$$14 = 3 + 11 = 7 + 7, \ldots$$

and testing to make sure that, for *each* such even number, it splits to *some* pair for which *both* members are prime. (Clearly we need not test pairs of *even* summands, except 2 + 2, since all primes except 2 are odd.) Our machine is to stop only when it reaches an even number for which *none* of the pairs into which that number splits consists of two primes. In that case we should have a counter-example to the Goldbach conjecture, namely an even number (greater than 2) which is *not* the sum of two primes. Thus if we could decide whether or not this Turing machine ever stops, we should have a way of deciding the truth of the Goldbach conjecture also.

A natural question arises: how are we to decide whether or not any particular Turing machine (when fed with some specific input) will ever stop? For many Turing machines this might not be hard to answer; but occasionally, as we have seen above, the answer could involve the solution of an outstanding mathematical problem. So, is there some *algorithmic* procedure for answering the general question—the halting problem—completely automatically? Turing showed that indeed there is not.

His argument was essentially the following. We first suppose that, on the contrary, there *is* such an algorithm.† Then there must be some Turing machine H which 'decides' whether or not the nth Turing machine, when acting on the number m, eventually stops. Let us say that it outputs the tape numbered 0 if it does not stop and 1 if it does:

* Recall that the *prime* numbers 2, 3, 5, 7, 11, 13, 17, . . are those natural numbers divisible, separately, only by themselves and by unity. Neither 0 nor 1 is considered to be a prime.

† This is the familiar—and powerful—mathematical procedure known as *reductio ad absurdum*, whereby one first assumes that what one is trying to prove is false; and from that one derives a contradiction, thus establishing that the required result is actually *true*.

$$H(n;m) = \begin{cases} \texttt{0} & \text{if } T_n(m) = \square \\ \texttt{1} & \text{if } T_n(m) \text{ stops.} \end{cases}$$

Here, one might take the coding of the pair (n, m) to follow the same rule as we adopted for the universal machine U. However this could run into the technical problem that for some numbers n (e.g. $n = 7$), T_n is not correctly specified; and the marker `11110` would be inadequate to separate n from m on the tape. To obviate this problem, let us assume that n is coded using the *expanded* binary notation rather than just the binary notation, with m in ordinary binary, as before. Then the marker `110` will actually be sufficient to separate n from m. The use of the *semicolon* in $H(n; m)$, as distinct from the *comma* in $U(n, m)$ is to indicate this change.

Now let us imagine an infinite array, which lists all the outputs of all possible Turing machines acting on all the possible different inputs. The nth row of the array displays the output of the nth Turing machine, as applied to the various inputs 0, 1, 2, 3, 4, . . . :

$n\downarrow$ \ $m\rightarrow$	0	1	2	3	4	5	6	7	8	. . .
0	□	□	□	□	□	□	□	□	□	. . .
1	0	0	0	0	0	0	0	0	0	. . .
2	1	1	1	1	1	1	1	1	1	. . .
3	0	2	0	2	0	2	0	2	0	. . .
4	1	1	1	1	1	1	1	1	1	. . .
5	0	□	0	□	0	□	0	□	0	. . .
6	0	□	1	□	2	□	3	□	4	. . .
7	0	1	2	3	4	5	6	7	8	. . .
8	□	1	□	□	1	□	□	□	1	. . .
⋮	⋮				⋮				⋮	. . .
197	2	3	5	7	11	13	17	19	23	. . .
⋮	⋮				⋮				⋮	

In the above table I have cheated a little, and not listed the Turing machines as they are *actually* numbered. To have done so would have yielded a list that looks much too boring to begin with, since all the machines for which n is less than 11 yield nothing but □s, and for $n = 11$ itself we get nothing but 0s. In order to make the list look initially more interesting, I have assumed that some much more efficient coding has been achieved. In fact I have simply made up the entries fairly randomly, just to give some kind of impression as to what its general appearance could be like.

I am not asking that we have actually *calculated* this array, say by some algorithm. (In fact, there is no such algorithm, as we shall see in a moment.) We are just supposed to *imagine* that the *true* list has somehow been laid out before us, perhaps by God! It is the occurrence of the □s which would cause the difficulties if we were to attempt to calculate the array, for we might not know for sure when to place a □ in some position since those calculations simply run on forever!

However, we *could* provide a calculational procedure for generating the table if we were allowed to use our putative H, for H would tell us where the □s actually occur. But instead, let us use H to *eliminate* every □ by replacing each occurrence with 0. This is achieved by preceding the action of T_n on m by the calculation $H(n; m)$; then we allow T_n to act on m only if $H(n; m) = 1$ (i.e. only if the calculation $T_n(m)$ actually gives an answer), and simply writing 0 if $H(n; m) = 0$ (i.e. if $T_n(m) = \square$). We can write our new procedure (i.e. that obtained by preceding $T_n(m)$ by the action of $H(n; m)$) as

$$T_n(m) \times H(n; m).$$

(Here I am using a common mathematical convention about the ordering of mathematical operations: the one on the *right* is to be performed *first*. Note that, symbolically, we have $\square \times 0 = 0$.)

The table for this now reads:

$n \downarrow$ / $m \rightarrow$	0	1	2	3	4	5	6	7	8	...
0	0	0	0	0	0	0	0	0	0	...
1	0	0	0	0	0	0	0	0	0	...
2	1	1	1	1	1	1	1	1	1	...
3	0	2	0	2	0	2	0	2	0	...
4	1	1	1	1	1	1	1	1	1	...
5	0	0	0	0	0	0	0	0	0	...
6	0	0	1	0	2	0	3	0	4	...
7	0	1	2	3	4	5	6	7	8	...
8	0	1	0	0	1	0	0	0	1	...
.	.				.				.	
.	.				.				.	
.	.				.				.	

Note that, assuming H exists, the rows of this table consist of *computable sequences*. (By a computable sequence I mean an infinite sequence whose successive values can be generated by an algorithm; i.e. there is some Turing machine which, when applied to the natural numbers $m = 0, 1, 2, 3, 4, 5, \ldots$ in turn, yields the successive members of the sequence.) Now, we take note of two facts about this table. In the first place, *every* computable sequence of natural numbers must appear somewhere (perhaps many times over) amongst its rows. This property was already true of the original table with its □s. We have simply *added* some rows to replace the 'dud' Turing machines (i.e. the ones which produce at least one □). In the second place, the assumption having been made that the Turing machine H actually exists, the table has been *computably generated* (i.e. generated by some definite algorithm), namely by the procedure $T_n(m) \times H(n; m)$. That is to say, there is some Turing machine Q which, when acting on the pair of numbers (n, m) produces the appropriate entry in the table. For this, we may code n and m on Q's tape in the same way as for H, and we have

$$Q(n; m) = T_n(m) \times H(n; m).$$

We now apply a variant of an ingenious and powerful device, the 'diagonal slash' of Georg Cantor. (We shall be seeing the original version of Cantor's diagonal slash in the next chapter.) Consider the elements of the main diagonal, marked now with **bold** figures:

$$\begin{array}{ccccccccc}
\mathbf{0} & 0 & 0 & 0 & 0 & 0 & 0 & 0 & 0 \\
0 & \mathbf{0} & 0 & 0 & 0 & 0 & 0 & 0 & 0 \\
1 & 1 & \mathbf{1} & 1 & 1 & 1 & 1 & 1 & 1 \\
0 & 2 & 0 & \mathbf{2} & 0 & 2 & 0 & 2 & 0 \\
1 & 1 & 1 & 1 & \mathbf{1} & 1 & 1 & 1 & 1 \\
0 & 0 & 0 & 0 & 0 & \mathbf{0} & 0 & 0 & 0 \\
0 & 0 & 1 & 0 & 2 & 0 & \mathbf{3} & 0 & 4 \\
0 & 1 & 2 & 3 & 4 & 5 & 6 & \mathbf{7} & 8 \\
0 & 1 & 0 & 0 & 1 & 0 & 0 & 0 & \mathbf{1} \\
\cdot & & & & \cdot & & & & \cdot \\
\cdot & & & & \cdot & & & & \cdot \\
\cdot & & & & \cdot & & & & \cdot
\end{array}$$

These elements provide some sequence 0, 0, 1, 2, 1, 0, 3, 7, 1, . . . to each of whose terms we now add 1:

$$1, 1, 2, 3, 2, 1, 4, 8, 2, \ldots$$

This is clearly a computable procedure and, given that our table was computably generated, it provides us with some new computable sequence, in fact with the sequence $1 + Q(n; n)$, i.e.

$$1 + T_n(n) \times H(n; n)$$

(since the diagonal is given by making m equal to n). But our table contains *every* computable sequence, so our new sequence must be somewhere in the list. Yet this cannot be so! For our new sequence differs from the first row in the first entry, from the second row in the second entry, from the third row in the third entry, and so on. This is manifestly a contradiction. It is the contradiction which establishes what we have been trying to prove, namely that the Turing machine H does not in fact exist! *There is no universal algorithm for deciding whether or not a Turing machine is going to stop.*

Another way of phrasing this argument is to note that, on the assumption that H exists, there is some Turing machine number, say k, for the algorithm $1 + Q(n; n)$, so we have

$$1 + T_n(n) \times H(n; n) = T_k(n).$$

But if we substitute $n = k$ in this relation (diagonal process!) we get

$$1 + T_k(k) \times H(k; k) = T_k(k).$$

This is a contradiction because if $T_k(k)$ stops we get the impossible relation

$$1 + T_k(k) = T_k(k)$$

(since $H(k; k) = 1$), whereas if $T_k(k)$ does not stop (so $H(k; k) = 0$) we have the equally inconsistent

$$1 + 0 = \square.$$

The question of whether or not a particular Turing machine stops is a perfectly well-defined piece of mathematics (and we have already seen that, conversely, various significant mathematical questions can be phrased as the stopping of Turing machines). Thus, by showing that no algorithm exists for deciding the question of the stopping of Turing machines, Turing showed (as had Church, using his own rather different type of approach) that there can be no general algorithm for deciding mathematical questions. Hilbert's *Entscheidungsproblem* has no solution!

This is not to say that in any *individual* case we may not be able to decide the truth, or otherwise, of some particular mathematical question; or decide whether or not some given Turing machine will stop. By the exercise of ingenuity, or even of just common sense, we may be able to decide such a question in a given case. (For example, if a Turing machine's instruction list contains *no* STOP order, or contains *only* STOP orders, then common sense alone is sufficient to tell us whether or not it will stop!) But there is no one algorithm that works for *all* mathematical questions, nor for *all* Turing machines and all numbers on which they might act.

It might seem that we have now established that there are at least *some* undecidable mathematical questions. However, we have done nothing of the kind! We have *not* shown that there is some especially awkward Turing machine table for which, in some absolute sense, it is impossible to decide whether or not the machine stops when it is fed with some especially awkward number—indeed, quite the reverse, as we shall see in a moment. We have said nothing whatever about the insolubility of *single* problems, but only about the *algorithmic* insolubility of *families* of problems. In any single case the answer is either 'yes' or 'no', so there certainly *is* an algorithm for deciding that particular case, namely the algorithm that simply says 'yes', when presented with the problem, or the one that simply says 'no', as the case may be! The difficulty is, of course, that we may not know *which* of these algorithms to use. That is a question of deciding the mathematical truth of a single statement, not the systematic decision problem for a family of statements. It is important to realize that algorithms do not, in themselves, decide mathematical truth. The *validity* of an algorithm must always be established by external means.

How to outdo an algorithm

This question of deciding the truth of mathematical statements will be returned to later, in connection with Gödel's theorem (see Chapter 4). For the moment, I wish to point out that Turing's argument is actually a lot more constructive and less negative than I have seemed to imply so far. We have certainly *not* exhibited a specific Turing machine for which, in some absolute sense, it is undecidable whether or not it stops. Indeed, if we examine the argument carefully, we find that our very procedure has actually implicitly *told us the answer* for the seemingly 'especially awkward' machines that we construct using Turing's procedure!

Let us see how this comes about. Suppose we have some algorithm which is *sometimes* effective for telling us when a Turing machine will not stop. Turing's procedure, as outlined above, will *explicitly* exhibit a Turing machine calculation for which that particular algorithm is not able to decide whether or not the calculation stops. However, in doing so, it actually enables *us* to see the answer in this case! The particular Turing machine calculation that we exhibit will indeed *not* stop.

To see how this arises in detail suppose we have such an algorithm that is sometimes effective. As before, we denote this algorithm (Turing machine) by H, but now we allow that the algorithm may not always be sure to tell us that a Turing machine will actually not stop:

$$H(n; m) = \begin{cases} 0 \text{ or } \square & \text{if } T_n(m) = \square \\ 1 & \text{if } T_n(m) \text{ stops,} \end{cases}$$

so $H(n; m) = \square$ is a possibility when $T_n(m) = \square$. Many such algorithms $H(n; m)$ actually exist. (For example, $H(n; m)$ could simply produce a 1 as soon as $T_n(m)$ stops, although *that* particular algorithm would hardly be of much practical use!)

We can follow through Turing's procedure in detail just as given above, except that instead of replacing *all* the $\square$s by 0s, we now have some $\square$s left. As before, our diagonal procedure has provided us with

$$1 + T_n(n) \times H(n; n),$$

as the nth term on the diagonal. (We shall get a $\square$ whenever $H(n; n) = \square$. Note that $\square \times \square = \square$, $1 + \square = \square$.) This is a perfectly good computation, so it is achieved by some Turing machine, say the kth one, and now we *do* have

$$1 + T_n(n) \times H(n; n) = T_k(n).$$

We look at the kth diagonal term, i.e. $n = k$, and obtain

$$1 + T_k(k) \times H(k; k) = T_k(k).$$

If the computation $T_k(k)$ stops, we have a contradiction (since $H(k; k)$ is supposed to be 1 whenever $T_k(k)$ stops, and the equation then gives inconsistency: $1 + T_k(k) = T_k(k)$). Thus $T_k(k)$ cannot stop, i.e.

$$T_k(k) = \square.$$

But the algorithm cannot 'know' this, because if it gave $H(k; k) = 0$, we should again have a contradiction (symbolically, we should have the invalid relation: $1 + 0 = \square$).

Thus, if we can find k we shall know how to construct our specific calculation to defeat the algorithm but for which *we* know the answer! How do we find k? That's hard work. What we have to do is to look in detail at the construction of $H(n; m)$ and of $T_n(m)$ and then see in detail how $1 + T_n(n) \times H(n; n)$ acts as a Turing machine. We find the number of this Turing machine, which is k. This would certainly be complicated to carry out in detail, but it could be done.* Because of the complication, we would not be at all interested in the calculation $T_k(k)$ were it not for the fact that we have specially produced it in order to defeat the algorithm H! What is important is that we have a well-defined procedure, whichever H is given to us, for finding a corresponding k for which *we* know that $T_k(k)$ defeats H, and for which we can therefore do better than the algorithm. Perhaps that comforts us a little if we think we are better than mere algorithms!

In fact the procedure is so well defined that we could find an *algorithm* for generating k, given H. So, before we get too complacent, we have to realize that *this* algorithm can improve[9] on H since, in effect, *it* 'knows' that $T_k(k) = \square$—or does it? It has been helpful in the above description to use the anthropomorphic term 'know' in reference to an algorithm. However, is it not *we* who are doing the 'knowing', while the algorithm just follows the rules we have told it to follow? Or are we ourselves merely following rules that we have been programmed to follow from the construction of our brains and from our environment? The issue is not really simply one of algorithms, but also a question of how one judges what is true and what is not true. These are central issues that we shall have to return to later. The question of mathematical truth (and its non-algorithmic nature) will be considered in Chapter 4. At least we should now have some feeling about the *meanings* of

* In fact, the hardest part of this is already achieved by the construction of the universal Turing machine U above, since this enables us to write down $T_n(n)$ as a Turing machine acting on n.

the terms 'algorithm' and 'computability', and an understanding of some of the related issues.

Church's lambda calculus

The concept of computability is a very important and beautiful mathematical idea. It is also a remarkably recent one—as things of such a fundamental nature go in mathematics—having been first put forward in the 1930s. It is an idea which cuts across *all* areas of mathematics (although it may well be true that most mathematicians do not, as yet, often worry themselves about computability questions). The power of the idea lies partly in the fact that some well-defined operations in mathematics are actually *not* computable (like the stopping, or otherwise, of a Turing machine; we shall see other examples in Chapter 4). For if there were no such non-computable things, the concept of computability would not have much mathematical interest. Mathematicians, after all, like puzzles. It can be an intriguing puzzle for them to decide, of some mathematical operation, whether or not it is computable. It is especially intriguing because the general solution of *that* puzzle is itself non-computable!

One thing should be made clear. Computability is a genuine 'absolute' mathematical concept. It is an abstract idea which lies quite beyond any particular realization in terms of the 'Turing machines' as I have described them. As I have remarked before, we do not need to attach any particular significance to the 'tapes' and 'internal states', etc., which characterize Turing's ingenious but particular approach. There are also other ways of expressing the idea of computability, historically the first of these being the remarkable 'lambda calculus' of the American logician Alonzo Church, with the assistance of Stephen C. Kleene. Church's procedure was quite different, and distinctly more abstract from that of Turing. In fact, in the form that Church stated his ideas, there is rather little obvious connection between them and anything that one might call 'mechanical'. The key idea lying behind Church's procedure is, indeed, *abstract* in its very essence—a mathematical operation that Church actually referred to as 'abstraction'.

I feel that it is worth while to give a brief description of Church's scheme not only because it emphasizes that computability is a mathematical idea, independent of any particular concept of computing machine, but also because it illustrates the power of abstract ideas in mathematics. The reader who is not readily conversant with mathematical ideas, nor intrigued by such things for their own sake, may, at this stage, prefer to move on to the next chapter—and there would not be significant loss in the flow of argument. Nevertheless, I believe that such readers might benefit by bearing with me for a while longer, and thus witnessing some of the magical economy of Church's scheme (see Church 1941).

In this scheme one is concerned with a 'universe' of objects, denoted by say

$$a, b, c, d, \ldots, z, a', b', \ldots, z', a'', b'', \ldots, a''', \ldots, a'''', \ldots$$

each of which stands for a mathematical operation or *function*. (The reason for the primed letters is simply to allow an unlimited supply of symbols to denote such functions.) The 'arguments' of these functions—that is to say, the things on which these functions act—are other things of the same kind, i.e. also functions. Moreover, the result (or 'value') of one such function acting on another is to be again a function. (There is, indeed, a wonderful economy of concepts in Church's system.) Thus, when we write*

$$a = bc$$

we mean that the result of the function b acting on the function c is another function a. There is no difficulty about expressing the idea of a function of two or more variables in this scheme. If we wish to think of f as a function of two variables p and q, say, we may simply write

$$(fp)q$$

(which is the result of the function fp as applied to q). For a function of three variables we consider

$$((fp)q)r,$$

and so on.

Now comes the powerful operation of *abstraction*. For this we use the Greek letter λ (lambda) and follow it immediately by a letter standing for one of Church's functions, say x, which we consider as a 'dummy variable'. Every occurrence of the variable x in the square-bracketed expression which immediately follows is then considered merely as a 'slot' into which may be substituted anything that follows the whole expression. Thus if we write

$$\lambda x.[fx],$$

we mean the function which when acting on, say, a produces the result fa. That is to say,

$$(\lambda x.[fx])a = fa.$$

In other words, $\lambda x.[fx]$ is simply the function f, i.e.

$$\lambda x.[fx] = f.$$

This bears a little thinking about. It is one of those mathematical niceties that seems so pedantic and trivial at first that one is liable to miss the point

* A more familiar form of notation would have been to write $a = b(c)$, say, but these particular parentheses are not really necessary and it is better to get used to their omission. To include them consistently would lead to rather cumbersome formulae such as $(f(p))(q)$ and $((f(p))(q))(r)$, instead of $(fp)q$ and $((fp)q)r$ respectively.

completely. Let us consider an example taken from familiar school mathematics. We take the function f to be the trigonometrical operation of taking the sine of an angle, so the abstract function 'sin' is defined by

$$\lambda x.[\sin x] = \sin.$$

(Do not worry about how the 'function' x may be taken to be an angle. We shall shortly see something of the way that numbers may be regarded as functions; and an angle is just a kind of number.) So far, this *is* indeed rather trivial. But let us imagine that the notation 'sin' had not been invented, but that we are aware of the power series expression for $\sin x$:

$$x - \tfrac{1}{6}x^3 + \tfrac{1}{120}x^5 - \ldots.$$

Then we could define

$$\sin = \lambda x.[x - \tfrac{1}{6}x^3 + \tfrac{1}{120}x^5 - \ldots].$$

Note that, even more simply, we could define, say, the 'one-sixth cubing' operation for which there is no standard 'functional' notation:

$$Q = \lambda x.[\tfrac{1}{6}x^3]$$

and find, for example,

$$Q(a + 1) = \tfrac{1}{6}(a + 1)^3 = \tfrac{1}{6}a^3 + \tfrac{1}{2}a^2 + \tfrac{1}{2}a^2 + \tfrac{1}{6}.$$

More pertinent to the present discussion would be expressions made up simply from Church's elementary functional operations, such as

$$\lambda f.[f(fx)].$$

This is the function which, when acting on another function, say g, produces g iterated twice acting on x, i.e.

$$(\lambda f.[f(fx)])\mathrm{g} = \mathrm{g}(\mathrm{g}x).$$

We could also have 'abstracted away' the x first, to obtain

$$\lambda f.[\lambda x.[f(fx)]],$$

which we may abbreviate to

$$\lambda fx.[f(fx)].$$

This is the operation which, when acting on g, produces the function 'g iterated twice'. In fact this is the very function that Church identifies with the natural number 2:

$$\mathbb{2} = \lambda fx.[f(fx)],$$

so $(\mathbb{2}\ \mathrm{g})y = \mathrm{g}(\mathrm{g}y)$. Similarly he defines:

$$\mathbb{3} = \lambda fx.[f(f(fx))], \qquad \mathbb{4} = \lambda fx.[f(f(f(fx)))], \qquad \text{etc.},$$

together with

$$\mathbb{1} = \lambda fx.[fx], \qquad \mathbb{0} = \lambda fx.[x].$$

Really, Church's '$\mathbb{2}$' is more like 'twice' and his '$\mathbb{3}$' is 'thrice', etc. Thus, the action of $\mathbb{3}$ on a function f, namely $\mathbb{3}\, f$, is the operation 'iterate f three times'. The action of $\mathbb{3}\, f$ on y, therefore, would be $(\mathbb{3}\, f)y = f(f(f(y)))$.

Let us see how a very simple arithmetical operation, namely the operation of adding $\mathbb{1}$ to a number, can be expressed in Church's scheme. Define

$$S = \lambda abc.[b((ab)c)].$$

To illustrate that S indeed simply adds $\mathbb{1}$ to a number described in Church's notation, let us test it on   :

$$\begin{aligned} S\,\mathbb{3} &= \lambda abc.[b((ab)c)]\,\mathbb{3} = \lambda bc.[b((\mathbb{3}\, b)c)] \\ &= \lambda bc.[b(b(b(bc)))] = \mathbb{4}\ , \end{aligned}$$

since $(\mathbb{3}\ b)c = b(b(bc))$. Clearly this applies equally well to any other natural number. (In fact $\lambda abc.[(ab)(bc)]$ would also have done just as well for S.)

How about multiplying a number by two? This doubling can be achieved by

$$D = \lambda abc.[(ab)((ab)c)],$$

which is again illustrated by its action on $\mathbb{3}$:

$$\begin{aligned} D = \lambda abc.[(ab)((ab)c)]\,\mathbb{3} &= \lambda bc.[(\mathbb{3}\, b)((\mathbb{3}\, b)c)] \\ = \lambda bc.[(\mathbb{3} b)(b(b(bc)))] &= \lambda bc.[b(b(b(b(b(bc)))))] = \mathbb{6}\ . \end{aligned}$$

In fact, the basic arithmetical operations of addition, multiplication, and raising to a power can be defined, respectively, by:

$$\begin{aligned} A &= \lambda fgxy.[((fx)(gx))y], \\ M &= \lambda fgx.[f(gx)], \\ P &= \lambda fg.[fg]. \end{aligned}$$

The reader may care to convince herself or himself—or else to take on trust—that, indeed,

$$(A\mathbb{m})\mathbb{n} = \mathbb{m+n}\ , \qquad (M\mathbb{m})\,\mathbb{n} = \mathbb{m \times n}\ , \qquad (P\mathbb{m})\,\mathbb{n} = \mathbb{n}^{\mathbb{m}}\ ,$$

where $\mathbb{m}$ and $\mathbb{n}$ are Church's functions for two natural numbers, $\mathbb{m+n}$ is his function for their sum, and so on. The last of these is the most astonishing. Let us just check it for the case $\mathbb{m} = \mathbb{2}$, $\mathbb{n} = \mathbb{3}$:

$$\begin{aligned} (P\mathbb{2}\,)\mathbb{3} &= ((\lambda fg.[fg])\,\mathbb{2}\,)\,\mathbb{3} = (\lambda g.[\,\mathbb{2}\, g])\,\mathbb{3} \\ &= (\lambda g.[\lambda fx.[f(fx)]g])\,\mathbb{3} = \lambda gx.[g(gx)]\mathbb{3} \\ &= \lambda x.[\,\mathbb{3}\,(\mathbb{3}\, x)] = \lambda x.[\lambda fy.[f(f(fy))](\mathbb{3}\, x)] \\ &= \lambda xy.[(\mathbb{3}\, x)((\mathbb{3}\, x)((\mathbb{3}\, x)y))] \end{aligned}$$

$$= \lambda xy.[(\mathbb{3}\,x)((\mathbb{3}\,x)(x(x(xy))))]$$
$$= \lambda xy.[(\mathbb{3}\,x)(x(x(x(x(xy)))))]$$
$$= \lambda xy.[x(x(x(x(x(x(x(x(xy))))))))] = \mathbb{9} = \mathbb{3}^{\mathbb{2}}$$

The operations of subtraction and division are not so easily defined (and, indeed, we need some convention about what to do with '$\mathbb{m} - \mathbb{n}$' when is smaller than $\mathbb{n}$ and with '$\mathbb{m} \div \mathbb{n}$' when $\mathbb{m}$ is not divisible by $\mathbb{n}$). In fact a major landmark of the subject occurred in the early 1930s when Kleene discovered how to express the operation of subtraction within Church's scheme! Other operations then followed. Finally, in 1937, Church and Turing independently showed that every computable (or algorithmic) operation whatever—now in the sense of Turing's machines—can be achieved in terms of one of Church's expressions (and vice versa).

This is a truly remarkable fact, and it serves to emphasize the fundamentally objective and mathematical character of the notion of computability. Church's notion of computability has, at first sight, very little to do with computing machines. Yet it has, nevertheless, some fundamental relations to practical computing. In particular, the powerful and flexible computer language LISP incorporates, in an essential way, the basic structure of Church's calculus.

As I indicated earlier, there are also other ways of defining the notion of computability. Post's concept of computing machine was very close to Turing's, and was produced independently, at almost the same time. There was currently also a rather more usable definition of computability (recursiveness) due to J. Herbrand and Gödel. H. B. Curry in 1929, and also M. Schönfinkel in 1924, had a different approach somewhat earlier, from which Church's calculus was partly developed. (See Gandy 1988.) Modern approaches to computability (such as that of an *unlimited register machine*, described in Cutland 1980) differ considerably in detail from Turing's original one, and they are rather more practical. Yet the *concept* of computability remains the same, whichever of these various approaches is adopted.

Like so many other mathematical ideas, especially the more profoundly beautiful and fundamental ones, the idea of computability seems to have a kind of *Platonic reality* of its own. It is this mysterious question of the Platonic reality of mathematical concepts generally that we must turn to in the next two Chapters.

1. I am adopting the usual modern terminology which now includes zero among the 'natural numbers'.
2. There are many other ways of coding pairs, triples, etc., of numbers as single numbers, well known to mathematicians, though less convenient for our present

purposes. For example, the formula $\frac{1}{2}((a + b)^2 + 3a + b)$ represents the pairs (a, b) of natural numbers uniquely as a single natural number. Try it!

3. I have not bothered, in the above, to introduce some mark to *initiate* the sequence of numbers (or instructions, etc.). This is not necessary for the input, since things just start when the first 1 is encountered. However, for the output something else may be needed, since one may not know *a priori* how far to look along the output tape in order to reach the first (i.e. leftmost) 1. Even though a long string of 0s may have been encountered going off to the left, this would be no guarantee that there would not be a 1 still *farther* off on the left. One can adopt various viewpoints on this. One of these would be always to use a special mark (say, coded by 6 in the contraction procedure) to initiate the entire output. But for simplicity, in my descriptions I shall take a different point of view, namely that it is always 'known' how much of the tape has actually been encountered by the device (e.g. one can imagine that it leaves a 'trail' of some kind), so that one does not have, in principle, to examine an infinite amount of tape in order to be sure that the entire output has been surveyed.
4. One way of coding the information of two tapes on a single tape is to interleave the two. Thus, the odd-numbered marks on the single tape could represent the marks of the first tape, whilst the even-numbered ones could represent the marks of the second tape. A similar scheme works for three or more tapes. The 'inefficiency' of this procedure results from the fact that the reading device would have to keep dodging backwards and forwards along the tape and leaving markers on it to keep track of where it is, at both even and odd parts of the tape.
5. This procedure refers only to the way in which a marked tape can be interpreted as a natural number. It does not alter the meanings of our specific Turing machines, such as **EUC** or **XN+1**.
6. If T_n is *not* correctly specified, then U will proceed as though the number for n has terminated as soon as the first string or more than four 1s in n's binary expression is reached. It will read the rest of this expression as part of the tape for m, so it will proceed to perform some nonsensical calculation! This feature could be eliminated, if desired, by arranging that n be expressed in *expanded* binary notation. I have chosen not to do this so as not to complicate further the description of the poor universal machine U!
7. I am grateful to David Deutsch for deriving the denary form of the binary description for u which I had worked out below. I am grateful to him also for checking that this binary value of u actually *does* give a universal Turing machine! The binary value for u is in fact:

10000000010111010011010001001010101101000110100010100000110101001101000101
01001011010000110100010100101011010010011101001010010010111010100011101010
10010010101110101010011010001010001010110100000110100100000101011010001001
11010010100001010111010010001110100101010000101110100101001101000010000111
01010000111010100001001001110100010101011010100101011010000011010101001011
010010010001101000000001101000000011101010010101010111010000100111010010101
01010101011101000010101011101000010100010111010001010011010010000101001101
00101001001101001000101101010001011101001001010111010010100011101010010100
10011101010101000011010010101010111010100100010110101000010110101000100110

10101010100010110100101010010010110101001001011101010100101011101010010100
11010101000011101000100100101011101010100101011101010100000111010100100000
11010101010010111010100101011010001001000111010000000111010010100101010101
11010010100100101011101000001010111010000100011101000001010100111010000101
00111010000010001011101000100001110100001001010011101000100001011010001010
01011101000101001011010010000010110100010101001001101000101010101110100100
00011101001001010101011101010101001101001000101011010010010010110100000001
01101000001000110100000100101101000000000110100101000101110100101010001101
00101001010110100000100111010010101001011010010011101010000001010111010100
00001101010100010101011010010101011010100001010111010100100101011101010001
00101101010010000101110100000011101010010001011010100101001101010100010111
01010010100101110101010000010111010101000001011101000000111010101000010101
11010010101011010101000010111010100010101011101010100100101110101010100001
11010100000001110100100100001101001001000101101010101010011101000000001011
01001000011010101010100101110100100001101001000101010111010000100011101000
10000111010000110100000000101101000001001011101010100101010110100010001001
11101000001001110101010011010000010101011010000100001110100100001000111010
10101010100111010000100100111010001001000011101000010100101101000010100001
11010101010101011101000100100110100010010011010100101001011101000100010101
11010000000111010001001001011101001101001001000010110101010100110100010100
01011101000011010100001000101101010011010101001010010110101010011010010010
10111010011010010000010110100010101010001110100100001010110100000010011010
01000100101110100100001101010000010010111010010010100110100100101010110100
11010010010100101101001101001010000010110100100000111010100100110101010100
00101110100101000010111010010101010111010100010010110100100111010010101000
10111010001001110101000010110100100111010010101010101110100100011101001010
10100101110100100011101010000010101011100110101000001011010010011101010000
00101110100101101010000010101101001010010111010100001001011101000011010100
01000010110101001101010001000101101010101001011101010001010010110100010101
01011101001000010101101010001011101010010010101011101010100100101110101000
11101010001110101001001001011101010001110101001010001011101010001011101010
00010010111010100011101000101000101110100101001011101010010101001011101001
01010101010110101000010101010110100001001110100001010101010111010101000101
01110101010001010111010000001110101010001001011101000000111010101001000101
11010100000011010100001011010000001110100100000010111010100011101010010001
01011101010011010101010001010110100000110101010100101010110100000010011010
10101001001110101001101010101001001011010100110100100100111010000011010101
01010010101101010001001101000101001010101110100000110101010101010010110100
01000111010001010101010101101000100011101000010101110100010010000111010011
01000000010011101000000100101110100010001010011101000000100101110100101010
10100101101000010101010111010001001010010111010000010001011101010100101101
00010001001110100000100101011101000000101010110100001000111001111010000100
00011101000010010011101000001010010111010000010100101101000010001010111010
00010001001101000100001110101111010000100100101110100001001001011101000000
01010111010000101010001101000100101110100001000001110100001001110100010000
01011101010100101101000100000101110100001010101011101000000101010111010001
00001010111010001000010101110100100000111010100100100110100000010101110100

0100010010111010101000011101010010101101001010101000011010000010100110100000001110100000100100111010010110100100010100101101010100110100010100100101101010100110100010101000101100110101001001011101010100110100010101010101100110101000101010110011010010001010101011101000100011101001001010101010110100101001010001101001000000101110100000110101010010101010110100101010110100100010001011101000101010110101000001010110100010000011010010001010110100001001110101001010101010111010010110100100100010101100110100100100101010111010011010010010010101101001011010010010010110100101101001001010001011001101001001010010101110100010101110100100101110011010010010101001011100110100101000101010111010001000111010000101001011010010100010111010010100010101101000100111010010100010010111010001001110100101001000101110011010010001000111010001001110100101001010101110011010010100000111001101010101010110100000001110100101010010101011101001000111010010101001010111001101000010100100110011010100000110100000001110100101010100101011100110101000100001101000000011101000100101010101110100010001110101010101010101011010000100111010010001001010111010010101000100110101000000010110100100111010100001010111010010000110101000000010110100100011101010010010111010000110101000010101011010100010111010100001010010111010100010111010100010101010111001101010001010110100001101010001001010

The enterprising reader, with an effective home computer, may care to check, using the prescriptions given in the text, that the above code does in fact give a universal Turing machine's action, by applying it to various simple Turing machine numbers!

Some lowering of the value of u might have been possible with a different specification for a Turing machine. For example, we could dispense with STOP and, instead, adopt the rule that the machine stops whenever the internal state 0 is re-entered after it has been in some other internal state. This would not gain a great deal (if anything at all). A bigger gain would have resulted had we allowed tapes with marks other than just 0 or 1. Very concise-looking universal Turing machines have indeed been described in the literature, but the conciseness is deceptive, for they depend upon exceedingly complicated codings for the descriptions of Turing machines generally.

8. For a non-technical discussion of matters relating to this famous assertion, see Devlin (1988).
9. We could, of course, defeat this improved algorithm too, by simply applying the foregoing procedure all over again. We can then use this new knowledge to improve our algorithm still further; but we could defeat that one also, and so on. The kind of consideration that this iterative procedure leads us into will be discussed in connection with Gödel's theorem, in Chapter 4, cf. p. 109.

3
Mathematics and reality

The land of Tor'Bled-Nam

Let us imagine that we have been travelling on a great journey to some far-off world. We shall call this world Tor'Bled-Nam. Our remote sensing device has picked up a signal which is now displayed on a screen in front of us. The image comes into focus and we see (Fig. 3.1):

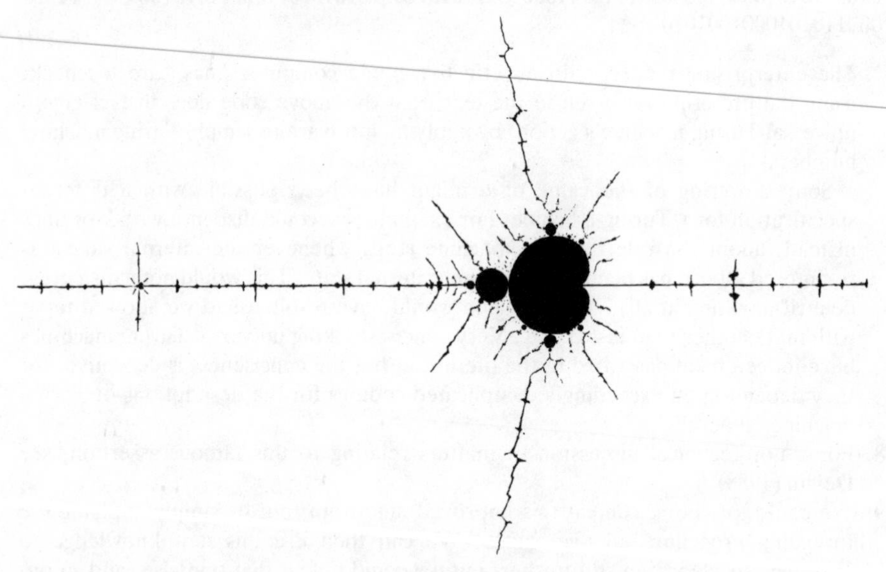

Fig. 3.1. A first glimpse of a strange world.

What can it be? Is it some strange-looking insect? Perhaps, instead, it is a dark-coloured lake, with many mountain streams entering it. Or could it be some vast and oddly shaped alien city, with roads going off in various directions to small towns and villages nearby? Maybe it is an island—and then let us try to find whether there is a nearby continent with which it is associated. This we can do by 'backing away', reducing the magnification of our sensing device by a linear factor of about fifteen. Lo and behold, the entire world springs into view (Fig. 3.2):

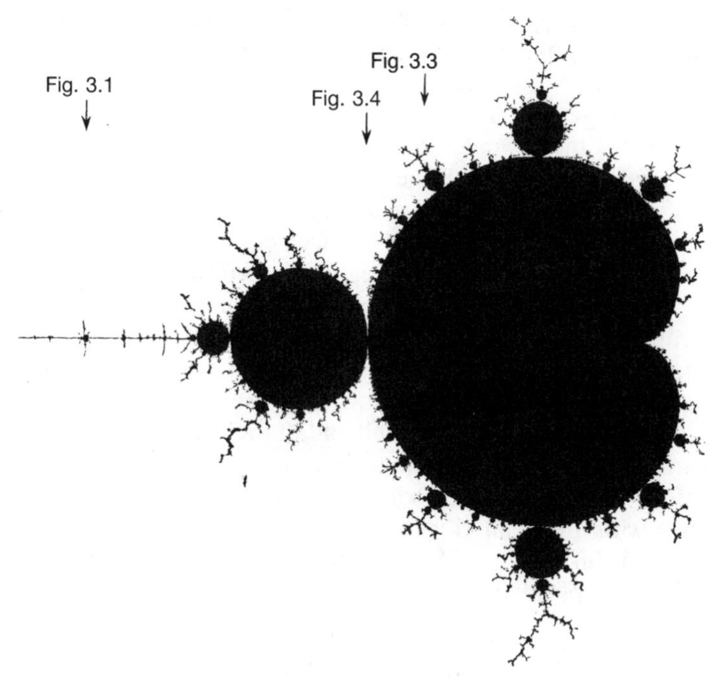

Fig. 3.2. 'Tor'Bled-Nam' in its entirety. The locations of the magnifications shown in Figs 3.1, 3.3, and 3.4 are indicated beneath the arrows.

Our 'island' is seen as a small dot indicated below 'Fig. 3.1' in Fig. 3.2. The filaments (streams, roads, bridges?), from the original island all come to an end, with the exception of the one attached at the inside of its right-hand crevice, which finally joins on to the very much larger object that we see depicted in Fig. 3.2. This larger object is clearly similar to the island that we saw first—though it is not precisely the same. If we focus more closely on what appears to be this object's coastline we see innumerable protuberances—roundish, but themselves possessing similar protuberances of their own. Each small protuberance seems to be attached to a larger one at some minute place, producing many warts upon warts. As the picture becomes clearer, we see myriads of tiny filaments emanating from the structure. The filaments themselves are forked at various places and often meander wildly. At certain spots on the filaments we seem to see little knots of complication which our sensing device, with its present magnification, cannot resolve. Clearly the object is no actual island or continent, nor a landscape of any kind. Perhaps, after all, we are viewing some monstrous beetle, and the first that we saw was one of its offspring, attached to it still, by some kind of filamentary umbilical cord.

Let us try to examine the nature of one of our creature's warts, by turning up the magnification of our sensing device by a linear factor of about ten (Fig.

3.3—the location being indicated under 'Fig. 3.3' in Fig. 3.2). The wart itself bears a strong resemblance to the creature as a whole—except just at the point of attachment. Notice that there are various places in Fig. 3.3 where five filaments come together. There is perhaps a certain 'fiveness' about this particular wart (as there would be a 'threeness' about the uppermost wart). Indeed, if we were to examine the next reasonable-sized wart, a little down on the left on Fig. 3.2, we should find a 'sevenness' about it; and for the next, a 'nineness', and so on. As we enter the crevice between the two largest regions of Fig. 3.2, we find warts on the right characterized by odd numbers, increasing by two each time. Let us peer deep down into this crevice, turning up the magnification from that of Fig. 3.2 by a factor of about ten (Fig. 3.4). We see numerous other tiny warts and also much swirling activity. On the right, we can just discern some tiny spiral 'seahorse tails'—in an area we shall know as 'seahorse valley.' Here we shall find, if the magnification is turned up enough, various 'sea anenomes' or regions with a distinctly floral appearance. Perhaps, after all, this is indeed some exotic coastline—maybe some coral reef, teeming with life of all kinds. What might have seemed to be a flower would reveal, on further magnification, to be composed of myriads of tiny, but incredibly complicated structures, each with numerous filaments and

Fig. 3.3. A wart with a 'fiveness' about its filaments.

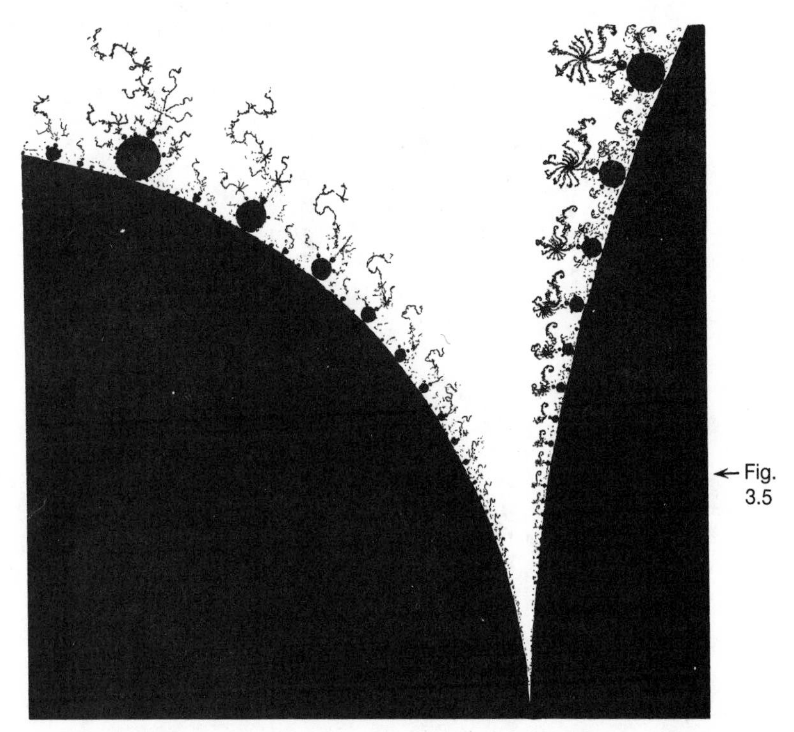

Fig. 3.4. The main crevice. 'Seahorse valley' is just discernible on the lower right.

swirling spiral tails. Let us examine one of the larger seahorse tails in some detail, namely the one just discernible where indicated as 'Fig. 3.5' in Fig. 3.4 (which is attached to a wart with a '29-ness' about it!). With a further approximate 250-fold magnification, we are presented with the spiral depicted in Fig. 3.5. We find that this is no ordinary tail, but is itself made up of the most complicated swirlings back and forth, with innumerable tiny spirals, and regions like octopuses and seahorses.

At many places, the structure is attached just where two spirals come together. Let us examine one of these places (indicated below 'Fig. 3.6' in Fig. 3.5), increasing our magnification by a factor of about thirty. Behold: do we discern a strange but now familiar object in the middle? A further increase of magnification by a factor of about six (Fig. 3.7) reveals a tiny baby creature—almost identical to the entire structure we have been examining! If we look closely, we see that the filaments emanating from it differ a little from those of the main structure, and they swirl about and extend to relatively much greater distances. Yet the tiny creature itself seems to differ hardly at all from its parent, even to the extent of possessing offspring of its own, in closely corresponding positions. These we could again examine if we turned

Fig. 3.5. A close-up of a seahorse tail.

Fig. 3.6. A further magnification of a joining point where two spirals come together. A tiny baby is just visible at the central point.

Fig. 3.7. On magnification, the baby is seen closely to resemble the entire world.

up the magnification still further. The grandchildren would also resemble their common ancestor—and one readily believes that this continues indefinitely. We may explore this extraordinary world of Tor'Bled-Nam as long as we wish, tuning our sensing device to higher and higher degrees of magnification. We find an endless variety: no two regions are precisely alike—yet there is a general flavour that we soon become accustomed to. The now familiar beetle-like creatures emerge at yet tinier and tinier scales. Every time, the neighbouring filamentary structures differ from what we had seen before, and present us with fantastic new scenes of unbelievable complication.

What is this strange, varied and most wonderfully intricate land that we have stumbled upon? No doubt many readers will already know. But some will not. This world is nothing but a piece of abstract mathematics—the set known as the Mandelbrot set.[1] Complicated it undoubtedly is; yet it is generated by a rule of remarkable simplicity! To explain the rule properly, I shall first need to explain what a *complex number* is. It is as well that I do so here. We shall need complex numbers later. They are absolutely fundamental to the structure of quantum mechanics, and are therefore basic to the workings of the very world in which we live. They also constitute one of the Great Miracles of Mathematics. In order to explain what a complex number is, I shall need, first, to remind the reader what is meant by the term 'real number'. It will also be helpful, also, to indicate the relationship between that concept and the very reality of the 'real world'!

Real numbers

Recall that the *natural* numbers are the whole quantities:

$$0, 1, 2, 3, 4, 5, 6, 7, 8, 9, 10, 11, \ldots$$

These are the most elementary and basic amongst the different kinds of number. Any type of discrete entity can be quantified by the use of natural numbers: we may speak of twenty-seven sheep in a field, of two lightning flashes, twelve nights, one thousand words, four conversations, zero new ideas, one mistake, six absentees, two changes of direction, etc. Natural numbers can be added or multiplied together to produce new natural numbers. They were the objects of our general discussion of algorithms, as given in the last chapter.

However some important operations may take us outside the realm of the natural numbers—the simplest being subtraction. For subtraction to be defined in a systematic way, we need *negative* numbers; we can set out the whole system of *integers*

$$\ldots, -6, -5, -4, -3, -2, -1, 0, 1, 2, 3, 4, 5, 6, 7, \ldots$$

for this purpose. Certain things, such as electric charge, bank balances, or dates* are quantified by numbers of this kind. These numbers are still too limited in their scope, however, since we can come unstuck when we try to *divide* one such number by another. Accordingly, we shall need the *fractions*, or *rational numbers* as they are called

$$0, 1, -1, 1/2, -1/2, 2, -2, 3/2, -3/2, 1/3, \ldots.$$

These suffice for the operations of finite arithmetic, but for a good many purposes we need to go further than this and include infinite or limiting operations. The familiar—and mathematically highly important—quantity π, for example, arises in many such infinite expressions. In particular, we have:

$$\pi = 2\{(2/1)(2/3)(4/3)(4/5)(6/5)(6/7)(8/7)(8/9)\ldots\}$$

and

$$\pi = 4(1 - 1/3 + 1/5 - 1/7 + 1/9 - 1/11 + \ldots).$$

(These are famous expressions, the first having been found by the English mathematician, grammarian, and cipher expert John Wallis, in 1655; and the second, in effect, by the Scottish mathematician and astronomer (and inventor of the first reflecting telescope) James Gregory, in 1671.) As with π, numbers defined in this way need *not* be rational (i.e. not of the form n/m,

* Actually, the normal conventions about dates do not quite correctly adhere to this since the year zero is omitted.

where n and m are integers with m non-zero). The number system needs to be *extended* in order that such quantities can be included.

This extended number system is referred to as the system of 'real' numbers—those familiar numbers which can be represented as infinite *decimal expansions*, such as:

$$-583.70264439121009538\ldots$$

In terms of such a representation we have the well-known expression for π:

$$\pi = 3.14159265358979323846\ldots$$

Among the types of number that can also be represented in this way are the square roots (or cube roots or fourth roots, etc.) of positive rational numbers, such as:

$$\sqrt{2} = 1.41421356237309504\ldots;$$

or, indeed, the square root (or cube root etc.) of any positive real number, as with the expression for π found by the great Swiss mathematician Leonhard Euler:

$$\pi = \sqrt{\{6(1 + 1/4 + 1/9 + 1/25 + 1/36 + \ldots)\}}.$$

Real numbers are, in effect, the familiar kinds of number that we have to deal with in everyday life, although normally we are concerned merely with approximations to such numbers, and are happy to work with expansions involving only a small number of decimal places. In mathematical statements, however, real numbers may need to be specified *exactly*, and we require some sort of infinite description such as an entire infinite decimal expansion, or perhaps some other infinite mathematical expression such as the above formulae for π given by Wallis, Gregory, and Euler. (I shall normally use decimal expansions in my descriptions here, but only because these are most familiar. To a mathematician, there are various rather more satisfactory ways of presenting real numbers, but we shall not need to worry about this here.)

It might be felt that it is impossible to contemplate an *entire* infinite expansion, but this is not so. A simple example where one clearly can contemplate the entire sequence is

$$1/3 = 0.333333333333333\ldots,$$

where the dots indicate to us that the succession of 3s carries on indefinitely. To contemplate this expansion, all we need to know is that the expansion does indeed continue in the same way indefinitely with 3s. Every rational number has a repeated (or finite) decimal expansion, such as

$$93/74 = 1.2567567567567567\ldots,$$

where the sequence 567 is repeated indefinitely, and this can also be contemplated in its entirety. Also, the expression

$$0.220002222000002222220000000222222220\ldots,$$

which defines an *irrational* number, can certainly be contemplated in its entirety (the string of 0s or 2s simply increasing in length by one each time), and many similar examples can be given. In each case, we shall be satisfied when we know a rule according to which the expansion is constructed. If there is some algorithm which generates the successive digits, then knowledge of that algorithm provides us with a way of contemplating the entire infinite decimal expansion. Real numbers whose expansions can be generated by algorithms are called *computable* numbers (see also p. 50). (The use of a denary rather than, say, a binary expansion here has no significance. The numbers which are 'computable' in this sense are just the same numbers whichever base for an expansion is used.) The real numbers π and $\sqrt{2}$ that we have just been considering are examples of computable numbers. In each case the rule would be a little complicated to state in detail, but not hard in principle.

However, there are also many real numbers which are *not* computable in this sense. We have seen in the last chapter that there are non-computable sequences which are nevertheless perfectly well defined. For example, we could take the decimal expansion whose nth digit is 1 or 0 according to whether or not the nth Turing machine acting on the number n stops or does not stop. Generally, for a real number, we just ask that there should be *some* infinite decimal expansion. We do not ask that there should be an algorithm for generating the nth digit, nor even that we should be aware of any kind of rule which in principle defines what the nth digit actually is.[2] Computable numbers are awkward things to work with. One cannot keep all one's operations computable, even when one works just with computable numbers. For example, it is not even a computable matter to decide, in general, whether two computable numbers are equal to one another or not! For this kind of reason, we prefer to work, instead, with *all* real numbers, where the decimal expansion can be anything at all, and need not just be, say, a computable sequence.

Finally, I should point out that there is an identification between a real number whose decimal expansion ends with an infinite succession of 9s and one whose expansion ends with an infinite succession of 0s; for example

$$-27.1860999999\ldots = -27.1861000000\ldots.$$

How many real numbers are there?

Let us pause for a moment to appreciate the vastness of the generalization that has been achieved in moving from the rational numbers to the real numbers.

One might think, at first, that the number of integers is already greater than the number of natural numbers; since every natural number is an integer whereas some integers (namely the negative ones) are not natural numbers, and similarly one might think that the number of fractions is greater than the number of integers. However, this is not the case. According to the powerful and beautiful theory of infinite numbers put forward in the late 1800s by the highly original Russian–German mathematician Georg Cantor, the total number of fractions, the total number of integers and the total number of natural numbers are all the *same* infinite number, denoted $\aleph_0$ ('aleph nought'). (Remarkably, this kind of idea had been partly anticipated some 250 years before, in the early 1600s, by the great Italian physicist and astronomer Galileo Galilei. We shall be reminded of some of Galileo's other achievements in Chapter 5.) One may see that the number of integers is the same as the number of natural numbers by setting up a 'one-to-one correspondence' as follows:

Integers		Natural numbers
0	$\leftrightarrow$	0
−1	$\leftrightarrow$	1
1	$\leftrightarrow$	2
−2	$\leftrightarrow$	3
2	$\leftrightarrow$	4
−3	$\leftrightarrow$	5
3	$\leftrightarrow$	6
−4	$\leftrightarrow$	7
.	.	.
.	.	.
.	.	.
$-n$	$\leftrightarrow$	$2n-1$
n	$\leftrightarrow$	$2n$
.	.	.
.	.	.
.	.	.

Note that each integer (in the left-hand column) and each natural number (in the right-hand column) occurs once and once only in the list. The existence of a one-to-one correspondence like this is what, in Cantor's theory, establishes that the number of objects in the left-hand column is the *same* as the number of objects in the right-hand column. Thus, the number of integers is, indeed, the same as the number of natural numbers. In this case the number is infinite, but no matter. (The only peculiarity that occurs with infinite numbers is that we can leave out some of the members of one list and *still* find a one-to-one correspondence between the two lists!) In a similar, but somewhat more complicated way, we can set up a one-to-one correspondence between the fractions and the integers. (For this we can adapt one of the ways of representing *pairs* of natural numbers, the numerators and denominators, as single natural numbers; see Chapter 2, p. 43.) Sets that can be put into one-to-one correspondence with the natural numbers are called *countable*, so the countable infinite sets are those with $\aleph_0$ elements. We have now seen that the integers are countable, and so also are all the fractions.

Are there sets which are *not* countable? Although we have extended the system, in passing from the natural numbers to first the integers and then the rational numbers, we have not actually increased the total number of objects that we have to work with. We have seen that the number of objects is actually countable in each case. Perhaps the reader has indeed got the impression by now that *all* infinite sets are countable. Not so; for the situation is very different in passing to the real numbers. It was one of Cantor's remarkable achievements to show that there are actually *more* real numbers than rationals. The argument that Cantor used is the 'diagonal slash' that was referred to in Chapter 2 and that Turing adapted in his argument to show that the halting problem for Turing machines is insoluble. Cantor's argument, like Turing's later one, proceeds by *reductio ad absurdum*. Suppose that the result we are trying to establish is false, i.e. that the set of all real numbers is countable. Then the real numbers between 0 and 1 are certainly countable, and we shall have *some* list providing a one-to-one pairing of all such numbers with the natural numbers, such as:

Natural numbers		Real numbers
0	⟷	0.**1**0357627183. . .
1	⟷	0.1**4**329806115. . .
2	⟷	0.02**1**66095213. . .
3	⟷	0.430**0**5357779. . .
4	⟷	0.9255**0**489101. . .
5	⟷	0.59210**3**43297. . .
6	⟷	0.636679**1**0457. . .
7	⟷	0.8705007**4**193. . .
8	⟷	0.04311737**8**04. . .
9	⟷	0.786350811**5**0. . .
10	⟷	0.4091673889**1**. . .
.	.	.
.	.	.
.	.	.

I have marked out the diagonal digits in bold type. These digits are, for this particular listing,

$$1, 4, 1, 0, 0, 3, 1, 4, 8, 5, 1, \ldots$$

and the diagonal slash procedure is to construct a real number (between 0 and 1) whose decimal expansion (after the decimal point) differs from these digits in each corresponding place. For definiteness, let us say that the digit is to be 1 whenever the diagonal digit is different from 1 and it is 2 whenever the diagonal digit is 1. Thus, in this case we get the real number

$$0.21211121112\ldots$$

This real number cannot appear in our listing since it differs from the first number in the first decimal place (after the decimal point), from the second number in the second place, from the third number in the third place, etc. This is a contradiction because our list was supposed to contain *all* real numbers between 0 and 1. This contradiction establishes what we are trying to prove, namely that there is *no* one-to-one correspondence between the real numbers and the natural numbers and, accordingly, the number of real numbers is actually *greater* than the number of rational numbers and is *not* countable.

The number of real numbers is the infinite number labelled $\boldsymbol{C}$. ($\boldsymbol{C}$ stands for *continuum*, another name for the system of real numbers.) One might ask why this number is not called $\aleph_1$, say. In fact the symbol $\aleph_1$ stands for the next infinite number greater than $\aleph_0$, and it is a famous unsolved problem to decide whether in fact $\boldsymbol{C} = \aleph_1$, the so-called *continuum hypothesis*.

It may be remarked that the *computable* numbers, on the other hand, *are* countable. To count them we just list, in numerical order, those Turing machines which generate real numbers (i.e. which produce the successive digits of real numbers). We may wish to strike from the list any Turing machine which generates a real number that has already appeared earlier in the list. Since the Turing machines are countable, it must certainly be the case that the computable real numbers are countable. Why can we not use the diagonal slash on that list and produce a new computable number which is *not* in the list? The answer lies in the fact that we cannot computably decide, in general, whether or not a Turing machine should actually be in the list. For to do so would, in effect, involve our being able to solve the halting problem. Some Turing machines may start to produce the digits of a real number, and then get stuck and never again produce another digit (because it 'doesn't stop'). There is no computable means of deciding which Turing machines will get stuck in this way. This is basically the halting problem. Thus, while our diagonal procedure will produce some real number, that number will not be a computable number. In fact, this argument could have been used to *show* the existence of non-computable numbers. Turing's argument to show the existence of classes of problems which cannot be solved algorithmically, as was recounted in the last chapter, follows precisely this line of reasoning. We shall see other applications of the diagonal slash later.

'Reality' of real numbers

Setting aside the notion of computability, real numbers are called 'real' because they seem to provide the magnitudes needed for the measurement of distance, angle, time, energy, temperature, or of numerous other geometrical and physical quantities. However, the relationship between the abstractly defined 'real' numbers and physical quantities is not as clear-cut as one might imagine. Real numbers refer to a *mathematical idealization* rather than to any actual physically objective quantity. The system of real numbers has the property, for example, that between any two of them, no matter how close, there lies a third. It is not at all clear that physical distances or times can realistically be said to have this property. If we continue to divide up the physical distance between two points, we would eventually reach scales so small that the very concept of distance, in the ordinary sense, could cease to have meaning. It is anticipated that at the 'quantum gravity' scale of 10^{20}th of the size* of a sub-atomic particle, this would indeed be the case. But to mirror the real numbers, we would have to go to scales indefinitely smaller than this: 10^{200}th, 10^{2000}th, or $10^{10^{200}}$th of a particle size, for example. It is not at all clear that such absurdly tiny scales have any physical meaning whatever. A similar remark would hold for correspondingly tiny intervals of time.

The real number system is chosen in physics for its *mathematical* utility, simplicity, and elegance, together with the fact that it accords, over a very wide range, with the physical concepts of distance and time. It is *not* chosen because it is known to agree with these physical concepts over *all* ranges. One might well anticipate that there is indeed no such accord at very tiny scales of distance or time. It is commonplace to use rulers for the measurement of simple distances, but such rulers will themselves take on a granular nature when we get down to the scale of its own atoms. This does not, in itself, prevent us from continuing to use real numbers in an accurate way, but a good deal more sophistication is needed for the measurement of yet smaller distances . We should at least be a little suspicious that there might eventually be a difficulty of fundamental principle for distances on the tiniest scale. As it turns out, Nature is remarkably kind to us, and it appears that the same real numbers that we have grown used to for the description of things at an everyday scale or larger retain their usefulness on scales much smaller than atoms—certainly down to less than one-hundredth of the 'classical' diameter of a sub-atomic particle, say an electron or proton—and seemingly down to the 'quantum gravity scale', twenty orders of magnitude smaller than such a particle! This is a quite extraordinary extrapolation from experience. The familiar concept of real-number distance seems to hold also out to the most

* Recall that the notation '10^{20}' stands for the number 100000000000000000000, where 1 is followed by twenty 0s.

distant quasar and beyond, giving an overall range of at least 10^{42}, and perhaps 10^{60} or more. The appropriateness of the real number system is not often questioned, in fact. Why is there so much confidence in these numbers for the accurate description of physics, when our initial experience of the relevance of such numbers lies in a comparatively limited range? This confidence—perhaps misplaced—must rest (although this fact is not often recognized) on the logical elegance, consistency, and mathematical power of the real number system, together with a belief in the profound mathematical harmony of Nature.

Complex numbers

The real number system does not, as it turns out, have a monopoly with regard to mathematical power and elegance. There is still a certain awkwardness in that, for example, square roots can be taken only of positive numbers (or zero) and not of negative ones. From the mathematical point of view—and leaving aside, for the moment, any question of direct connection with the physical world—it turns out to be extremely convenient to be able to extract square roots of negative as well as positive numbers. Let us simply postulate, or 'invent' a square root for the number -1. We shall denote this by the symbol 'i', so we have:

$$\mathrm{i}^2 = -1.$$

The quantity i cannot, of course, be a real number since the product of a real number with itself is always positive (or zero, if the number is itself zero). For this reason the term *imaginary* has been conventionally applied to numbers whose squares are negative. However, it is important to stress the fact that these 'imaginary' numbers are no less real than the 'real' numbers that we have become accustomed to. As I have emphasized earlier, the relationship between such 'real' numbers and *physical* reality is not as direct or compelling as it may at first seem to be, involving, as it does, a mathematical idealization of infinite refinement for which there is no clear *a priori* justification from nature.

Having a square root for -1, it is now no great effort to provide square roots for *all* the real numbers. For if a is a positive real number, then the quantity

$$\mathrm{i} \times \sqrt{a}$$

is a square root of the negative real number $-a$. (There is also one other square root, namely $-\mathrm{i} \times \sqrt{a}$.) What about i itself? Does this have a square root? It surely does. For it is easily checked that the quantity

$$(1 + \mathrm{i})/\sqrt{2}$$

(and also the negative of this quantity) squares to i. Does *this* number have a square root? Once again, the answer is yes; the square of

$$\sqrt{\frac{(1 + 1/\sqrt{2})}{2}} + \mathrm{i}\sqrt{\frac{(1 - 1/\sqrt{2})}{2}}$$

or its negative is indeed $(1 + \mathrm{i})/\sqrt{2}$.

Notice that in forming such quantities we have allowed ourselves to add together real and imaginary numbers, as well as to multiply our numbers by arbitrary real numbers (or divide by non-zero real numbers, which is the same thing as multiplying by their reciprocals). The resulting objects are what are referred to as *complex numbers*. A complex number is a number of the form

$$a + \mathrm{i}b$$

where a and b are real numbers, called the *real part* and the *imaginary part* respectively, of the complex number. The rules for adding and multiplying two such numbers follow the ordinary rules of (school) algebra, with the added rule that $\mathrm{i}^2 = -1$:

$$(a + \mathrm{i}b) + (c + \mathrm{i}d) = (a + c) + \mathrm{i}(b + d)$$
$$(a + \mathrm{i}b) \times (c + \mathrm{i}d) = (ac - bd) + \mathrm{i}(ad + bc).$$

A remarkable thing now happens! Our motivation for this system of numbers had been to provide the possibility that square roots can always be taken. It achieves this task, though this is itself not yet obvious. But it also does a great deal more: cube roots, fifth roots, ninety-ninth roots, πth roots, $(1 + \mathrm{i})$th roots, etc. can all be taken with impunity (as the great eighteenth-century mathematician Leonhard Euler was able to show). As another example of the magic of complex numbers, let us examine the somewhat complicated-looking formulae of trigonometry that one has to learn in school; the sines and cosines of the sum of two angles

$$\sin(A + B) = \sin A \cos B + \cos A \sin B,$$
$$\cos(A + B) = \cos A \cos B - \sin A \sin B,$$

are simply the imaginary and real parts, respectively, of the much simpler (and much more memorable!) complex equation*

$$\mathrm{e}^{A+B} = \mathrm{e}^A \, \mathrm{e}^B.$$

* The quantity $\mathrm{e} = 2.7182818285\ldots$ (the base of natural logarithms, and an irrational number of mathematical importance comparable with that of π) is defined by

$$\mathrm{e} = 1 + 1/1 + 1/(1\times2) + 1/(1\times2\times3) + \ldots,$$

and e^A means the Ath power of e, where we have, as it turns out,

$$\mathrm{e}^A = 1 + A/1 + A^2/(1\times2) + A^3/(1\times2\times3) + \ldots.$$

Here all we need to know is 'Euler's formula' (apparently also obtained many years before Euler by the remarkable 16th century English mathematician Roger Cotes)

$$e^{iA} = \cos A + i \sin A,$$

which we substitute into the equation above. The resulting expression is

$$\cos(A + B) + i \sin(A + B) = (\cos A + i \sin A)(\cos B + i \sin B),$$

and multiplying out the right-hand side we obtain the required trigonometrical relations.

What is more, any algebraic equation

$$a_0 + a_1z + a_2z^2 + a_3z^3 + \ldots + a_nz^n = 0$$

(for which $a_0, a_1, a_2 \ldots, a_n$ are complex numbers, with $a_n \neq 0$) can always be solved for some complex number z. For example, there is a complex number z satisfying the relation

$$z^{102} + 999z^{33} - \pi z^2 = -417 + i,$$

though this is by no means obvious! The general fact is sometimes referred to as 'the fundamental theorem of algebra'. Various eighteenth century mathematicians had struggled to prove this result. Even Euler had not found a satisfactory general argument. Then, in 1831, the great mathematician and scientist Carl Friedrich Gauss gave a startlingly original line of argument and provided the first general proof. A key ingredient of this proof was to represent the complex numbers *geometrically*, and then to use a topological* argument.

Actually, Gauss was not really the first to use a geometrical description of complex numbers. Wallis had done this, crudely, about two hundred years earlier, though he had not used it to nearly such powerful effect as had Gauss. The name normally attached to this geometrical representation of complex numbers is that of the Frenchman Jean Robert Argand, who described it in 1806, although the Norwegian surveyor Caspar Wessel had, in fact, given a very complete description nine years earlier. In accordance with this conventional (but not altogether historically accurate) terminology I shall refer to the standard geometrical representation of complex numbers as the *Argand plane*.

The Argand plane is an ordinary Euclidean plane with standard Cartesian coordinates x and y, where x marks off horizontal distance (positive to the

* The word 'topological' refers to the kind of geometry—sometimes referred to as 'rubber sheet geometry'—in which actual distances are of no consequence, and only the continuity properties of objects have relevance.

right and negative to the left) and where y marks off vertical distance (positive upwards and negative downwards). The complex number

$$z = x + \mathrm{i}y$$

is then represented by the point of the Argand plane whose coordinates are

$$(x, y)$$

(see Fig. 3.8).

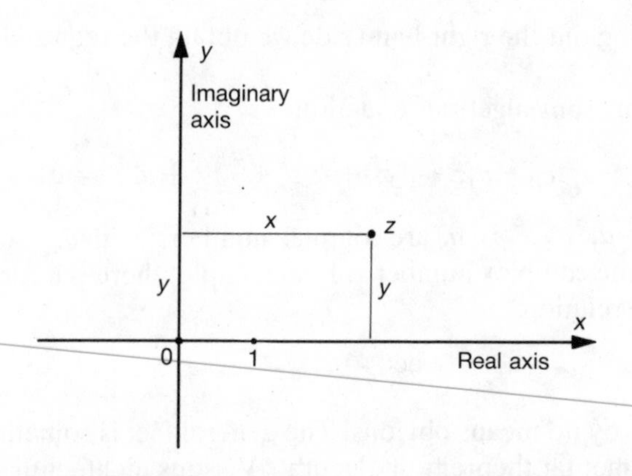

Fig. 3.8. The Argand plane, depicting a complex number $z = x + \mathrm{i}y$.

Note that 0 (regarded as a complex number) is represented by the origin of coordinates, and 1 is represented as a particular point in the x-axis.

The Argand plane simply provides us with a way of organizing our family of complex numbers into a geometrically useful picture. This kind of thing is not really something new to us. We are already familiar with the way that *real* numbers can be organized into a geometrical picture, namely the picture of a straight line that extends indefinitely in both directions. One particular point of the line is labelled 0 and another is labelled 1. The point 2 is placed so that its displacement from 1 is the same as the displacement of 1 from 0; the point 1/2 is the mid-point of 0 and 1; the point -1 is situated so that 0 lies mid-way between it and 1, etc., etc. The set of real numbers displayed in this way is referred to as the *real line*. For complex numbers we have, in effect, *two* real numbers to use as coordinates, namely a and b, for the complex number $a + \mathrm{i}b$. These two numbers give us coordinates for points on a plane—the Argand plane. As an example, I have indicated in Fig. 3.9 approximately where the complex numbers

$$u = 1 + \mathrm{i}\,1.3, \qquad v = -2 + \mathrm{i}, \qquad w = -1.5 - \mathrm{i}\,0.4,$$

should be placed.

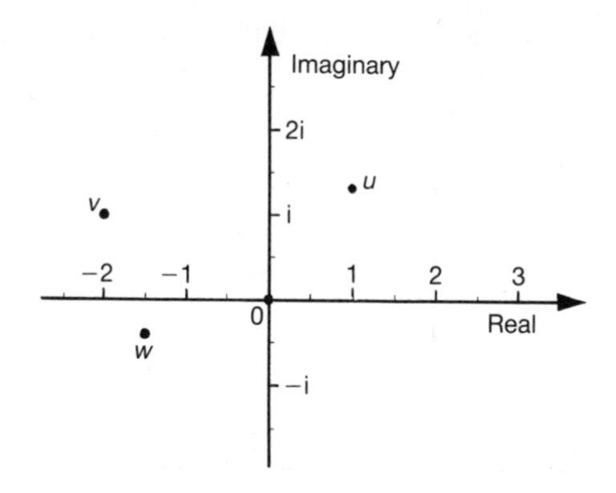

Fig. 3.9. Locations in the Argand plane of $u = 1 + \mathrm{i}\,1.3$, $v = -2 + \mathrm{i}$, and $w = -1.5 - \mathrm{i}\,0.4$.

The basic algebraic operations of addition and multiplication of complex numbers now find a clear geometrical form. Let us consider addition first. Suppose u and v are two complex numbers, represented on the Argand plane in accordance with the above scheme. Then their sum $u + v$ is represented as the 'vector sum' of the two points; that is to say, the point $u + v$ occurs at the place which completes the parallelogram formed by u, v, and the origin 0. That this construction (see Fig. 3.10) actually gives us the sum is not very hard to see, but I omit the argument here.

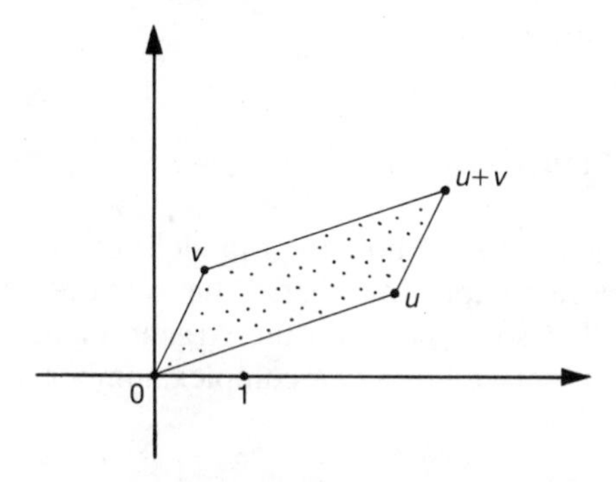

Fig. 3.10. The sum $u + v$ of two complex numbers u and v is obtained by the parallelogram law.

The product uv also has a clear geometrical interpretation (see Fig. 3.11), which is perhaps a little harder to see. (Again I omit the argument.) The angle, subtended at the origin, between 1 and uv is the sum of the angles between 1 and u and between 1 and v (all angles being measured in an anticlockwise sense), and the distance of uv from the origin is the product of the distances from the origin of u and v. This is equivalent to saying that the triangle formed by 0, v, and uv is similar (and similarly oriented) to the triangle formed by 0, 1, and u.

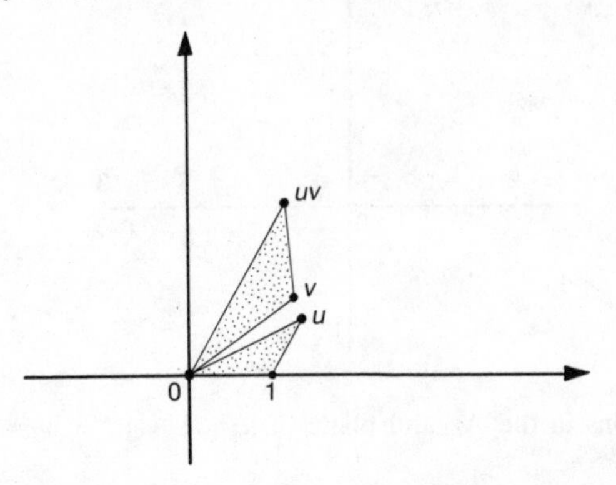

Fig. 3.11. The product uv of two complex numbers u and v is such that the triangle formed by 0, v, and uv is similar to that formed by 0, 1, and u. Equivalently: the distance of uv from 0 is the product of the distances of u and v from 0, and the angle that uv makes with the real (horizontal) axis is the sum of the angles that u and v make with this axis.

(The energetic reader who is not familiar with these constructions may care to verify that they follow directly from the algebraic rules for adding and multiplying complex numbers that were given earlier, together with the above trigonometric identities.)

Construction of the Mandelbrot set

We are now in a position to see how the Mandelbrot set is defined. Let z be some arbitrarily chosen complex number. Whatever this complex number is, it will be represented as some point on the Argand plane. Now consider the *mapping* whereby z is replaced by a *new* complex number, given by

$$z \to z^2 + c,$$

where c is another *fixed* (i.e. given) complex number. The number $z^2 + c$ will be represented by some new point in the Argand plane. For example, if c happened to be given as the number $1.63 - \mathrm{i}\,4.2$, then z would be mapped according to

$$z \to z^2 + 1.63 - \mathrm{i}\,4.2$$

so that, in particular, 3 would be replaced by

$$3^2 + 1.63 - \mathrm{i}\,4.2 = 9 + 1.63 - \mathrm{i}\,4.2 = 10.63 - \mathrm{i}\,4.2$$

and $-2.7 + \mathrm{i}\,0.3$ would be replaced by

$$\begin{aligned}&(-2.7 + \mathrm{i}\,0.3)^2 + 1.63 - \mathrm{i}\,4.2\\ &= (-2.7)^2 - (0.3)^2 + 1.63 + \mathrm{i}\{2(-2.7)(0.3) - 4.2\}\\ &= 8.83 - \mathrm{i}\,5.82\end{aligned}$$

When such numbers get complicated, the calculations are best carried out by an electronic computer.

Now, whatever c may be, the particular number 0 is replaced, under this scheme, by the actual given number c. What about c itself? This must be replaced by the number $c^2 + c$. Suppose we continue this process and apply the replacement to the number $c^2 + c$; then we obtain

$$(c^2 + c)^2 + c = c^4 + 2c^3 + c^2 + c.$$

Let us iterate the replacement again, applying it next to the above number to obtain

$$(c^4 + 2c^3 + c^2 + c)^2 + c = c^8 + 4c^7 + 6c^6 + 6c^5 + 5c^4 + 2c^3 + c^2 + c$$

and then again to this number, and so on. We obtain a sequence of complex numbers, starting with 0:

$$0, c, c^2 + c, c^4 + 2c^3 + c^2 + c, \ldots.$$

Now if we do this with *certain* choices of the given complex number c, the sequence of numbers that we get in this way never wanders very far from the origin in the Argand plane; more precisely, the sequence remains *bounded* for such choices of c which is to say that every member of the sequence lies within some *fixed circle* centred at the origin (see Fig. 3.12) A good example where this occurs is the case $c = 0$, since

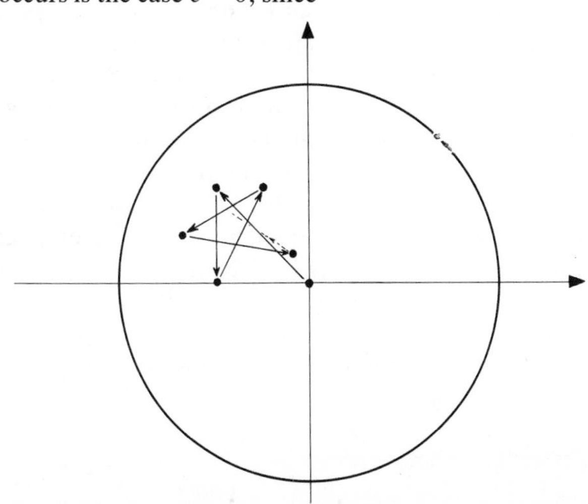

Fig. 3.12. A sequence of points in the Argand plane is *bounded* if there is some fixed circle that contains all the points. (This particular iteration starts with zero and has $c = -\frac{1}{2} + \frac{1}{2}\mathrm{i}$.)

in this case, every member of the sequence is in fact 0. Another example of bounded behaviour occurs with $c = -1$, for then the sequence is: 0, −1, 0, −1, 0, −1, . .; and yet another example occurs with $c = \mathrm{i}$, the sequence being 0, i, i−1, −i, i−1, −i, i−1, −i,. . . . However, for various other complex numbers c the sequence wanders farther and farther from the origin to indefinite distance; i.e. the sequence is *unbounded*, and cannot be contained within any fixed circle. An example of this latter behaviour occurs when $c = 1$, for then the sequence is 0, 1, 2, 5, 26, 677, 458330, . . . ; this also happens when $c = -3$, the sequence being 0, −3, 6, 33, 1086, . . . ; and also when $c = \mathrm{i}-1$, the sequence being 0, i−1, −i−1, −1+3i, −9−i5, 55+i91, −5257+i10011,. . . .

The *Mandelbrot set*, that is to say, the *black* region of our world of Tor'Bled-Nam, is precisely that region of the Argand plane consisting of points c for which the sequence remains bounded. The *white* region consists of thoses points c for which the sequence is unbounded. The detailed pictures that we saw earlier were all drawn from the outputs of computers. The computer would systematically run through possible choices of the complex number c, where for each choice of c it would work out the sequence $0, c, c^2 + c, \ldots$ and decide, according to some appropriate criterion, whether the sequence is remaining bounded or not. If it *is* bounded, then the computer would arrange that a black spot appear on the screen at the point corresponding to c. If it is unbounded, then the computer would arrange for a white spot. Eventually, for every pixel in the range under consideration, the decision would be made by the computer as to whether the point would be coloured white or black.

The complexity of the Mandelbrot set is very remarkable, particularly in view of the fact that the definition of this set is, as mathematical definitions go, a strikingly simple one. It is also the case that the general structure of this set is not very sensitive to the precise algebraic form of the mapping $z \rightarrow z^2 + c$ that we have chosen. Many other iterated complex mappings (e.g. $z \rightarrow z^3 + \mathrm{i}z^2 + c$) will give extraordinarily similar structures (provided that we choose an appropriate number to start with—perhaps not 0, but a number whose value is characterized by a clear mathematical rule for each appropriate choice of mapping). There is, indeed, a kind of universal or absolute character to these 'Mandelbrot' structures, with regard to iterated complex maps. The study of such structures is a subject on its own, within mathematics, which is referred to as *complex dynamical systems*.

Platonic reality of mathematical concepts?

How 'real' are the objects of the mathematician's world? From one point of view it seems that there can be nothing real about them at all. Mathematical objects are just concepts; they are the mental idealizations that mathematicians make, often stimulated by the appearance and seeming order of aspects of the world about us, but mental idealizations nevertheless. Can they be other than mere arbitrary constructions of the human mind? At the same time there often does appear to be some profound reality about these mathematical concepts, going quite beyond the mental deliberations of any particular mathematician. It is as though human thought is, instead, being guided towards some eternal external truth—a truth which has a reality of its own, and which is revealed only partially to any one of us.

The Mandelbrot set provides a striking example. Its wonderfully elaborate structure was not the invention of any one person, nor was it the design of a team of mathematicians. Benoit Mandelbrot himself, the Polish-American mathematician (and protagonist of fractal theory) who first studied the set, had no real prior conception of the fantastic elaboration inherent in it, although he knew that he was on the track of something very interesting. Indeed, when his first computer pictures began to emerge, he was under the impression that the fuzzy structures that he was seeing were the result of a computer malfunction (Mandelbrot 1986)! Only later did he become convinced that they were really there in the set itself. Moreover, the complete details of the complication of the structure of Mandelbrot's set cannot really be fully comprehended by any one of us, nor can it be fully revealed by any computer. It would seem that this structure is not just part of our minds, but it has a reality of its own. Whichever mathematician or computer buff chooses to examine the set, approximations to the *same* fundamental mathematical structure will be found. It makes no real difference which computer is used for performing calculations (provided that the computer is in accurate working order), apart from the fact that differences in computer speed and storage, and graphic display capabilities, may lead to differences in the amount of fine detail that will be revealed and in the speed with which that detail is produced. The computer is being used in essentially the same way that the experimental physicist uses a piece of experimental apparatus to explore the structure of the physical world. The Mandelbrot set is not an invention of the human mind: it was a discovery. Like Mount Everest, the Mandelbrot set is just *there*!

Likewise, the very system of complex numbers has a profound and timeless reality which goes quite beyond the mental constructions of any particular mathematician. The beginnings of an appreciation of complex numbers came about with the work of Gerolamo Cardano. He was an Italian, who lived from 1501 to 1576, a physician by trade, a gambler, and caster of horoscopes (once casting a horoscope for Christ), and he wrote an important and influential treatise on algebra 'Ars Magna' in 1545. In this he put forward the first complete expression for the solution (in terms of surds, i.e. *n*th roots) of a general cubic equation.* He had noticed, however that in a certain class of

* Based partly on earlier work by Scipione del Ferro and by Tartaglia.

cases—the ones referred to as 'irreducible', where the equation has three real solutions—he was forced to take, at a certain stage in his expression, the *square root of a negative number*. Although this was puzzling to him, he realized that if he allowed himself to *take* such square roots, and *only* if, then he could express the full answer (the final answer being always real). Later, in 1572, Raphael Bombelli, in a work entitled 'l'Algebra', extended Cardano's work and began the study of the actual algebra of complex numbers.

While at first it may seem that the introduction of such square roots of negative numbers was just a device—a mathematical invention designed to achieve a specific purpose—it later becomes clear that these objects are achieving far more than that for which they were originally designed. As I mentioned above, although the original purpose of introducing complex numbers was to enable square roots to be taken with impunity, by introducing such numbers we find that we get, as a bonus, the potentiality for taking any other kind of root or for solving any algebraic equation whatever. Later we find many other magical properties that these complex numbers possess, properties that we had no inkling about at first. These properties are just *there*. They were not put there by Cardano, nor by Bombelli, nor Wallis, nor Coates, nor Euler, nor Wessel, nor Gauss, despite the undoubted far-sightedness of these, and other, great mathematicians; such magic was inherent in the very structure that they gradually uncovered. When Cardano introduced his complex numbers, he could have had no inkling of the many magical properties which were to follow—properties which go under various names, such as the Cauchy integral formula, the Riemann mapping theorem, the Lewy extension property. These, and many other remarkable facts, are properties of the very numbers, with no additional modifications whatever, that Cardano had first encountered in about 1539.

Is mathematics invention or discovery? When mathematicians come upon their results are they just producing elaborate mental constructions which have no actual reality, but whose power and elegance is sufficient simply to fool even their inventors into believing that these mere mental constructions are 'real'? Or are mathematicians really uncovering truths which are, in fact, already 'there'—truths whose existence is quite independent of the mathematicians' activities? I think that, by now, it must be quite clear to the reader that I am an adherent of the second, rather than the first, view, at least with regard to such structures as complex numbers and the Mandelbrot set.

Yet the matter is perhaps not quite so straightforward as this. As I have said, there are things in mathematics for which the term 'discovery' is indeed much more appropriate than 'invention', such as the examples just cited. These are the cases where much more comes out of the structure than is put into it in the first place. One may take the view that in such cases the mathematicians have stumbled upon 'works of God'. However, there are other cases where the mathematical structure does not have such a compelling uniqueness, such as when, in the midst of a proof of some result, the mathematician finds the need to introduce some contrived and far from unique construction in order to achieve some very specific end. In such cases no more is likely to come out of the construction than was put into it in the first place, and the word 'invention' seems more appropriate than 'discovery'. These are indeed just 'works of man'. On this view, the true mathematical discoveries would, in a general way, be regarded as greater achievements or aspirations than would the 'mere' inventions.

Such categorizations are not entirely dissimilar from those that one might use in the arts or in engineering. Great works of art are indeed 'closer to God' than are lesser ones. It is a feeling not uncommon amongst artists, that in their greatest works they are revealing eternal truths which have some kind of prior etherial existence,* while their lesser works might be more arbitrary, of the nature of mere mortal constructions. Likewise, an engineering innovation with a beautiful economy, where a great deal is achieved in the scope of the application of some simple, unexpected idea, might appropriately be described as a discovery rather than an invention.

Having made these points, however, I cannot help feeling that, with mathematics, the case for believing in some kind of etherial, eternal existence, at least for the more profound mathematical concepts, is a good deal stronger than in those other cases. There is a compelling uniqueness and universality in such mathematical ideas which seems to be of quite a different order from that which one could expect in the arts or engineering. The view that mathematical concepts could exist in such a timeless, etherial sense was put forward in ancient times (c. 360BC) by the great Greek philosopher Plato. Consequently, this view is frequently referred to as mathematical Platonism. It will have considerable importance for us later.

In Chapter 1, I discussed at some length the point of view of *strong AI*, according to which mental phenomena are supposed to find their existence within the mathematical idea of an algorithm. In Chapter 2, I stressed the point that the concept of an algorithm is indeed a profound and 'God-given' notion. In this chapter I have been arguing that such 'God-given' mathematical ideas should have some kind of timeless existence, independent of our earthly selves. Does not this viewpoint lend some credence to the strong-AI point of view, by providing the possibility of an etherial type of existence for mental phenomena? Just conceivably so—and I shall even be speculating, later, in favour of a view not altogether dissimilar from this; but if mental phenomena can indeed find a home of this general kind, I do not believe that it can be with the concept of an algorithm. What would be needed would be something very much more subtle. The fact that algorthmic things constitute a

* As the distinguished Argentinian writer Jorge Luis Borges has put it: '. . . a famous poet is less of an inventor than a discoverer. . .'

very narrow and limited part of mathematics will be an important aspect of the discussions to follow. We shall begin to see something of the scope and subtlety of non-algorithmic mathematics in the next chapter.

1. See Mandelbrot (1986). The particular sequence of magnifications that I have chosen has been adapted from those of Peitgen and Richter (1986), where many remarkable coloured pictures of the Mandelbrot set are to be found. For further striking illustrations, see Peitgen and Saupe (1988).
2. As far as I am aware, it is a consistent, though unconventional, point of view to demand that there should always be *some* kind of rule determining what the *n*th digit actually is, for an arbitrary real number, although such a rule may not be effective nor even definable at all in a preassigned formal system (see Chapter 4). I hope it *is* consistent, since it is the point of view that I should most wish to adhere to myself!

4
Truth, proof, and insight

Hilbert's programme for mathematics

What is truth? How do we form our judgements as to what is true and what is untrue about the world? Are we simply following some *algorithm*—no doubt favoured over other less effective possible algorithms by the powerful process of natural selection? Or might there be some other, possibly non-algorithmic, route—perhaps intuition, instinct, or insight—to the divining of truth? This seems a difficult question. Our judgements depend upon complicated interconnected combinations of sense-data, reasoning, and guesswork. Moreover, in many worldly situations there may not be general agreement about what *is* actually true and what is false. To simplify the question, let us consider only *mathematical* truth. How do we form our judgements—perhaps even our 'certain' knowledge—concerning mathematical questions? Here, at least, things should be more clear-cut. There should be no question as to what actually is true and what actually is false—or should there? What, indeed, *is* mathematical truth?

The question of mathematical truth is a very old one, dating back to the times of the early Greek philosophers and mathematicians—and, no doubt, earlier. However, some very great clarifications and startling *new* insights have been obtained just over the past hundred years, or so. It is these new developments that we shall try to understand. The issues are quite fundamental, and they touch upon the very question of whether our thinking processes can indeed be entirely algorithmic in nature. It is important for us that we come to terms with them.

In the late nineteenth century, mathematics had made great strides, partly because of the development of more and more powerful methods of mathematical proof. (David Hilbert and Georg Cantor, whom we have encountered before, and the great French mathematician Henri Poincaré, whom we shall encounter later, were three who were in the forefront of these developments.) Accordingly, mathematicians had been gaining confidence in the use of such powerful methods. Many of these methods involved the consideration of sets* with infinite numbers of members, and proofs were

* A *set* just means a collection of things—physical objects or mathematical concepts—that can be

often successful for the very reason that it was possible to consider such sets as actual 'things'—completed existing wholes, with more than a mere potential existence. Many of these powerful ideas had sprung from Cantor's highly original concept of *infinite numbers*, which he had developed consistently using infinite sets. (We caught a glimpse of this the previous chapter.)

However, this confidence was shattered when in 1902 the British logician and philosopher Bertrand Russell produced his now famous paradox (itself anticipated by Cantor, and a direct descendent of Cantor's 'diagonal slash' argument). To understand Russell's argument, we first need some feeling for what is involved in considering sets as completed wholes. We may imagine some set that is characterized in terms of a particular *property*. For example, the set of *red* things is characterized in terms of the property of *redness*: something belongs to that set if and only if it has redness. This allows us to turn things about, and talk about a property in terms of a single object, namely the entire set of things with that property. With this viewpoint, 'redness' *is* the set of all red things. (We may also conceive that some other sets are just 'there', their elements being characterized by no such simple property.)

This idea of defining concepts in terms of sets was central to the procedure, introduced in 1884 by the influential German logician Gottlob Frege, whereby *numbers* can be defined in terms of sets. For example, what do we mean by the actual number 3? We know what the property of 'threeness' is, but what is 3 itself? Now, 'threeness' is a property of *collections* of objects, i.e. it is a property of *sets*: a set has this particular property 'threeness' if and only if the set has precisely three members. The set of medal winners in a particular Olympic event has this property of 'threeness', for example. So does the set of tyres on a tricycle, or the set of leaves on a normal clover, or the set of solutions to the equation $x^3 - 6x^2 + 11x - 6 = 0$. What, then, is Frege's definition of the actual number 3? According to Frege, 3 must be a set *of* sets: the set of *all* sets with this property of 'threeness'.[1] Thus a set has three members if and only if it belongs to Frege's set 3.

This may seem a little circular, but it is not, really. We can define *numbers* generally as totalities of equivalent sets, where 'equivalent' here means 'having elements that can be paired off one-to-one with each other' (i.e. in ordinary terms this would be 'having the same number of members'). The number 3 is then the particular one of these sets which has, as one of its members, a set containing, for example, just one apple, one orange, and one pear. Note that this is a quite different definition of '3' from Church's '③' given on p. 69. There are also other definitions which can be given and which are rather more popular these days.

treated as a whole. In mathematics, the elements (i.e. members) of a set are very often themselves sets, since sets can be collected together to form other sets. Thus one may consider sets of sets, or sets of sets of sets, etc.

Now, what about the Russell paradox? It concerns a set R defined in the following way:

R is the set of all sets which are not members of themselves.

Thus, R is a certain collection of sets; and the criterion for a set X to belong to this collection is that the set X is itself not to be found amongst its *own* members.

Is it absurd to suppose that a set might actually be a member of itself? Not really. Consider, for example, the set I of *infinite* sets (sets with infinitely many members). There are certainly infinitely many *different* infinite sets, so I is itself infinite. Thus I indeed belongs to itself! How is it, then, that Russell's conception gives us a paradox? We ask: is Russell's very set R a member of itself or is it not? If it is *not* a member of itself then it should belong to R, since R consists precisely of those sets which are not members of themselves. Thus, R belongs to R after all—a contradiction. On the other hand, if R *is* a member of itself, then since 'itself' is actually R, it belongs to that set whose members are characterized by *not* being members of themselves, i.e. it is not a member of itself after all—again a contradiction!*

This consideration was not a flippant one. Russell was merely using, in a rather extreme form, the same type of very general mathematical set-theoretic reasoning that the mathematicians were beginning to employ in their proofs. Clearly things had got out of hand, and it became appropriate to be much more precise about what kind of reasoning was to be allowed and what was not. It was obviously necessary that the allowed reasoning must be free from contradiction and that it should permit only true statements to be derived from statements previously known to be true. Russell himself, together with his colleague Alfred North Whitehead, set about developing a highly formalized mathematical system of axioms and rules of procedure, the aim being that it should be possible to translate all types of correct mathematical reasoning into their scheme. The rules were carefully selected so as to prevent the paradoxical types of reasoning that led to Russell's own paradox. The specific scheme that Russell and Whitehead produced was a monumental piece of work. However, it was very cumbersome, and it turned out to be rather limited in the types of mathematical reasoning that it actually incorporated. The great mathematician David Hilbert, whom we first encountered in Chapter 2, embarked upon a much more workable and comprehensive scheme. *All* correct mathematical types of reasoning, for any particular mathematical area, were to be included. Moreover, Hilbert intended that it would be possible to *prove* that the scheme was free from contradiction. Then

* There is an amusing way of expressing the Russell paradox in essentially commonplace terms. Imagine a library in which there are two catalogues, one of which lists precisely all the books in the library which somewhere refer to themselves and the other, precisely all the books which make no mention of themselves. In which catalogue is the second catalogue itself to be listed?

mathematics would be placed, once and for all, on an unassailably secure foundation.

However, the hopes of Hilbert and his followers were dashed when, in 1931, the brilliant 25-year old Austrian mathematical logician Kurt Gödel produced a startling theorem which effectively destroyed the Hilbert programme. What Gödel showed was that any such precise ('formal') mathematical system of axioms and rules of procedure *whatever*, provided that it is broad enough to contain descriptions of simple arithmetical propositions (such as 'Fermat's last theorem', considered in Chapter 2) and provided that it is free from contradiction, must contain some statements which are neither provable nor disprovable by the means allowed within the system. The truth of such statements is thus 'undecidable' by the approved procedures. In fact, Gödel was able to show that the very statement of the consistency of the axiom system itself, when coded into the form of a suitable arithmetical proposition, must be one such 'undecidable' proposition. It will be important for us to understand the nature of this 'undecidability'. We shall see why Gödel's argument cut to the very core of the Hilbert programme. We shall also see how Gödel's argument enables us, by the use of insight, to go beyond the limitations of any particular formalized mathematical system under consideration. This understanding will be crucial for much of the discussion to follow.

Formal mathematical systems

It will be necessary to be a bit more explicit about what we mean by a 'formal mathematical system of axioms and rules of procedure'. We must suppose that there is some alphabet of symbols in terms of which our mathematical statements are to be expressed. These symbols must certainly be adequate to allow a notation for the natural numbers in order that 'arithmetic' can be incorporated into our system. We can, if desired, just use the usual Arabic notation 0, 1, 2, 3, . . ., 9, 10, 11, 12, . . . for numbers, although this makes the specification of the rules a little more complicated than they need be. A much simpler specification would result if we use, say, 0, 01, 011, 0111, 01111, . . . to denote the sequence of natural numbers (or, as a compromise, we could use the binary notation). However, since this could cause confusion in the discussion which follows, I shall just stick to the usual Arabic notation in my descriptions, whatever notation the system might *actually* use. We might need a 'space' symbol to separate the different 'words' or 'numbers' of our system, but since that might also be confusing, we could use just a comma (,) for that purpose where needed. We shall also require letters to denote arbitrary ('variable') natural numbers (or perhaps integers, or rationals, etc.—but let's stick to natural numbers here), say $t, u, v, w, x, y, z, t', t'', t'''$, ... The primed letters t', t'', . . . may be needed since we do not wish to put any definite limit on the number of variables that can occur in an expression. We regard the *prime* (′) as a separate symbol of the formal system, so that the actual number of *symbols* remains finite. We shall need symbols for the basic arithmetical operations =, +, ×, etc., perhaps for various kinds of brackets (,), [,], and for the *logical* symbols such as & ('and'), ⇒ ('implies'), v ('or'), ⇔ ('if and only if'), ~ ('not', or 'it is not so that . . .'). In addition, we shall want the logical 'quantifiers': the *existentional quantifier* ∃ ('there exists . . . such that') and the *universal quantifier* ∀ ('for all . . . we have'). Then we can make statements such as 'Fermat's last theorem'.

$$\sim\exists\, w,x,y,z\,[(x+1)^{w+3}+(y+1)^{w+3}=(z+1)^{w+3}]$$

(see Chapter 2, p. 58). (I could have written '0111' for '3', and perhaps used a notation for 'raising to a power' that might fit in better with the formalism; but, as I said, I am sticking to the conventional symbols so as not to introduce unnecessary confusion.) The above statement reads (ending at the first square bracket):

'It is not so that there exist natural numbers w, x, y, z such that . . .'.

We can also rewrite Fermat's last theorem using ∀:

$$\forall\, w,x,y,z\,[\sim(x+1)^{w+3}+(y+1)^{w+3}=(z+1)^{w+3}],$$

which reads (ending after the 'not' symbol after the first bracket):

'For all natural numbers w, x, y, z it is not so that. . .'.

which is logically the same thing as before.

We need letters to denote entire propositions, and for that I shall use capital letters: P, Q, R, S, ... Such a proposition might in fact be the Fermat assertion above:

$$F=\sim\exists\, w,x,y,z\,[(x+1)^{w+3}+(y+1)^{w+3}=(z+1)^{w+3}].$$

A proposition might also *depend* on one or more variables; for example we might be interested in the Fermat assertion for some *particular** *power* $w+3$:

$$G(w)=\sim\exists\, x,y,z\,[(x+1)^{w+3}+(y+1)^{w+3}=(z+1)^{w+3}],$$

so that $G(0)$ asserts 'no cube can be the sum of positive cubes', $G(1)$ asserts the same for fourth powers, and so on. (Note that 'w' is missing after '∃'.) The Fermat assertion is now that $G(w)$ holds for *all* w:

$$F=\forall w[G(w)].$$

* Although the truth of Fermat's full proposition F is still unknown, the truths of the individual propositions $G(0)$, $G(1)$, $G(2)$, $G(3)$, . . . are known up to about $G(125000)$. That is to say, it is known that no cube can be the sum of positive cubes, no fourth power the sum of fourth powers, etc., up to the corresponding statement for 125000th powers.

$G(\ )$ is an example of what is called a *propositional function*, i.e. a proposition which depends on one or more variables.

The *axioms* of the system will constitute a finite list of general propositions whose truth, given the meanings of the symbols, is suppsed to be self-evident. For example, for arbitrary propositions or propositional functions P, Q, $R(\ )$, we have, amongst our axioms:

$$(P \,\&\, Q) \Rightarrow P,$$
$$\sim(\sim P) \Leftrightarrow P,$$
$$\sim\exists x[R(x)] \Leftrightarrow \forall x[\sim R(x)],$$

the 'self-evident truth' of which is readily ascertained from their *meanings*. (The first one asserts simply that: 'if P and Q are both true, then P is true'; the second asserts the equivalence between 'it is not so that P is false' and 'P is true'; the third is exemplified by the logical equivalence of the two ways of stating Fermat's last theorem given above.) We could also include basic arithmetical axioms, such as

$$\forall x,y[x + y = y + x]$$
$$\forall x,y,z[(x + y) \times z = (x \times z) + (y \times z)],$$

although one might prefer to build up these arithmetical operations from something more primitive and deduce these statements as theorems. The *rules of procedure* would be (self-evident) things like:

'from P and $P \Rightarrow Q$ we can deduce Q'
'from $\forall x[R(x)]$ we can deduce any proposition obtained by substituting a specific natural number for x in $R(x)$'.

These are instructions telling us how we may derive new propositions from propositions already established.

Now, starting from the axioms, and then applying the rules of procedure again and again, we can build up some long list of propositions. At any stage, we can bring any of the axioms into play again, and we can always keep re-using any of the propositions that we have already added to our lengthening list. The propositions of any such correctly assembled list are referred to as *theorems* (though many of them will be quite trivial, or uninteresting as mathematical statements). If we have a specific proposition P that we want to *prove*, then we try to find such a list, correctly assembled according to these rules, and which terminates with our specific proposition P. Such a list would provide us with a *proof* of P within the system; and P would, accordingly, then be a theorem.

The idea of Hilbert's programme was to find, for any well-defined area of mathematics, a list of axioms and rules of procedure sufficiently comprehensive that *all* forms of correct mathematical reasoning appropriate to that area would be incorporated. Let us fix our area of mathematics to be *arithmetic* (where the quantifiers ∃ and ∀ are included, so that statements like that of Fermat's last theorem can be made). There will be no advantage to us to consider any mathematical area more general than this here. Arithmetic is *already* general enough for Gödel's procedure to apply. If we can accept that such a comprehensive system of axioms and rules of procedure has indeed been given to us for arithmetic, in accordance with Hilbert's programme, then we shall be provided with a definite criterion for the 'correctness' of mathematical proof for any proposition in arithmetic. The hope had been that such a system of axioms and rules could be *complete*, in the sense that it would enable us in principle to decide the truth or falsity of *any* mathematical statement that can be formulated within the system.

Hilbert's hope was that for any string of symbols representing a mathematical proposition, say P, one should be able to prove either P or $\sim P$, depending upon whether P is true or false. Here we must assume that the string is *syntactically correct* in its construction, where 'syntactically correct' essentially means 'grammatically' correct—i.e. satisfying all the notational rules of the formalism, such as brackets being paired off correctly, etc.—so that P has a well-defined true or false meaning. If Hilbert's hope could be realized, this would even enable us to dispense with worrying about what the propositions *mean* altogether! P would just *be* a syntactically correct string of symbols. The string of symbols P would be assigned the truth-value true if P is a theorem (i.e. if P is provable within the system) and it would be assigned the truth-value false if, on the other hand, $\sim P$ is a theorem. For this to make sense, we require *consistency*, in addition to completeness. That is to say, there must be no string of symbols P for which *both* of P and $\sim P$ are theorems. Otherwise P could be true and false at the same time!

The point of view that one can dispense with the meanings of mathematical statements, regarding them as nothing but strings of symbols in some formal mathematical system, is the mathematical standpoint of *formalism*. Some people like this idea, whereby mathematics becomes a kind of 'meaningless game'. It is not an idea that appeals to me, however. It is indeed 'meaning'—not blind algorithmic computation—that gives mathematics its substance. Fortunately, Gödel dealt formalism a devastating blow! Let us see how he did this.

Gödel's theorem

Part of Gödel's argument was very detailed and complicated. However, it is not necessary for us to examine the intricacies of that part. The central idea, on the other hand, was simple, beautiful, and profound. This part we shall be able to appreciate. The complicated part (which also contained much ingenuity) was to show in detail how one may actually code the individual rules of procedure of the formal system, and also the use of its various axioms, into *arithmetical operations*. (It was an aspect of the profound part, though, to realize that this was a fruitful thing to do!) In order to carry out this coding we need to find some convenient way of labelling propositions with natural numbers. One way would be simply to use some kind of 'alphabetical' ordering for all the strings of symbols of the formal system for each specific length, where there is an overall ordering according to the length of the string. (Thus, the strings of length one could be alphabetically ordered, followed by the strings of length two, alphabetically ordered, followed by the strings of length three, etc.) This is called *lexicographical* ordering.* In fact Gödel originally used a more complicated numbering system, but the distinctions are not important for us. We shall be particularly concerned with *propositional functions* which are dependent on a *single* variable, like $G(w)$ above. Let the nth such propositional function (in the chosen ordering of strings of symbols), applied to w, be

$$P_n(w).$$

We can allow our numbering to be a little 'sloppy' if we wish, so that some of these expressions may not be syntactically correct. (This makes the arithmetical coding very much easier than if we try to omit all such syntactically incorrect expressions.) If $P_n(w)$ is syntactically correct, it will be some perfectly well-defined particular arithmetical statement concerning the two natural numbers n and w. Precisely *which* arithmetical statement it is will depend on the details of the particular numbering system that has been chosen. That belongs to the complicated part of the argument and will not concern us here. The strings of propositions which constitute a *proof* of some theorem in the system can also be labelled by natural numbers using the chosen ordering scheme. Let

$$\Pi_n$$

denote the nth proof. (Again, we can use a 'sloppy numbering' whereby for some values of n the expression 'Π_n' is not syntactically correct and thus proves no theorem.)

Now consider the following propositional function, which depends on the natural number w:

$$\sim\exists x[\Pi_x \text{ proves } P_w(w)].$$

* We can think of lexicographical ordering as the ordinary ordering for natural numbers written out in 'base $k + 1$', using, for the $k + 1$ numerals, the various symbols of the formal system, together with a new 'zero', which is never used. (This last complication arises because numbers beginning with zero are the same as with this zero omitted.) A simple lexicographical ordering of strings with nine symbols is that given by the natural numbers that can be written in ordinary denary notation without zero: 1, 2, 3,4, . . ., 8, 9, 11, 12, . . ., 19, 21, 22, . . ., 99, 111, 112,. . ..

The statement in the square brackets is given partly in words, but it is a perfectly precisely defined statement. It asserts that the xth proof is actually a proof of that proposition which is $P_w(\ )$ applied to the value w itself. Outside the square bracket the negated existential quantifier serves to remove one of the variables ('there does not exist an x such that . . .'), so we end up with an arithmetical propositional function which depends on only the one variable w. The expression as a whole asserts that there is *no* proof of $P_w(w)$. I shall assume that it is framed in a syntactically correct way (even if $P_w(w)$ is not—in which case the statement would be *true*, since there can be no proof of a syntactically incorrect expression). In fact, because of the translations into arithmetic that we are supposing have been carried out, the above is actually some *arithmetical* statement concerning the natural number w (the part in square brackets being a well-defined arithmetical statement about *two* natural numbers x and w). It is not supposed to be obvious that the statement can be coded into arithmetic, but it can be. Showing that such statements can indeed be so coded is the major 'hard work' involved in the complicated part of Gödel's argument. As before, precisely *which* arithmetical statement it is will depend on the details of the numbering systems, and it will depend very much on the detailed structure of the axioms and rules of our formal system. Since all that belongs to the complicated part, the details of it will not concern us here.

We have numbered all propositional functions which depend on a single variable, so the one we have just written down must have been assigned a number. Let us write this number as k. Our propositional function is the kth one in the list. Thus

$$\sim\exists x[\Pi_x \text{ proves } P_w(w)] = P_k(w).$$

Now examine this function for the particular w-value: $w = k$. We get

$$\sim\exists x[\Pi_x \text{ proves } P_k(k)] = P_k(k).$$

The specific proposition $P_k(k)$ is a perfectly well-defined (syntactically correct) arithmetical statement. Does it have a proof within our formal system? Does its negation $\sim P_k(k)$ have a proof? The answer to both these questions must be 'no'. We can see this by examining the *meaning* underlying the Gödel procedure. Although $P_k(k)$ is just an arithmetical proposition, we have constructed it so that it asserts what has been written on the left-hand side: 'there is no proof, within the system, of the proposition $P_k(k)$'. If we have been careful in laying down our axioms and rules of procedure, and assuming that we have done our numbering right, then there cannot be any proof of this $P_k(k)$ within the system. For if there were such a proof, then the 'meaning' of the statement that $P_k(k)$ actually asserts, namely that there is *no* proof, would be false, so $P_k(k)$ would have to be false as an arithmetical proposition. Our formal system should not be so badly constructed that it actually allows false propositions to be proved! Thus, it must be the case that there is in fact *no* proof of $P_k(k)$. But this is precisely what $P_k(k)$ is trying to tell us. What $P_k(k)$ asserts must therefore be a *true* statement, so $P_k(k)$ must be true as an arithmetical proposition. We have found a *true* proposition which has *no proof within the system*!

What about its *negation* $\sim P_k(k)$? It follows that we had also better not be able to find a proof of this either. We have just established that $\sim P_k(k)$ must be false (since $P_k(k)$ is true), and we are not supposed to be able to prove false propositions within the system! Thus, neither $P_k(k)$ nor $\sim P_k(k)$ is provable within our formal system. This establishes Gödel's theorem.

Mathematical insight

Notice that something very remarkable has happened here. People often think of Gödel's theorem as something negative—showing the necessary limitations of formalized mathematical reasoning. No matter how comprehensive we think we have been, there will always be some propositions which escape the net. But should the particular proposition $P_k(k)$ worry us? In the course of the above argument, we have actually established that $P_k(k)$ is a *true* statement! Somehow we have managed to *see* that $P_k(k)$ is true despite the fact that it is not formally provable within the system. The strict mathematical formalists *should* indeed be worried, because by this very reasoning we have established that the formalist's notion of 'truth' must be necessarily incomplete. *Whatever* (consistent) formal system is used for arithmetic, there are statements that we can see are true but which do not get assigned the truth-value true by the formalist's proposed procedure, as described above. The way that a strict formalist might try to get around this would perhaps be not to talk about the concept of truth at all but merely refer to *provability* within some fixed formal system. However, this seems very limiting. One could not even frame the Gödel argument as given above, using this point of view, since essential parts of that argument make use of reasoning about what is actually true and what is not true.[2] Some formalists take a more 'pragmatic' view, claiming not to be worried by statements such as $P_k(k)$ because they are extremely complicated and uninteresting as propositions of arithmetic. Such people would assert:

> Yes, there is the odd statement, such as $P_k(k)$, for which my notion of provability or truth does not coincide with your instinctive notion of truth, but those statements will never come up in serious mathematics (at least not in the kind I am interested in) because such statements are absurdly complicated and unnatural as mathematics.

It is indeed the case that propositions like $P_k(k)$ would be extremely cumbersome and odd-looking as mathematical statements about numbers, when written out in full. However, in recent years, some reasonably simple statements of a very acceptable mathematical character have been put forward which are actually equivalent to Gödel-type propositions.[3] These are unprovable from the normal axioms of arithmetic, yet follow from an 'obviously true' property that the axiom system itself has.

The formalist's professed lack of interest in 'mathematical truth' seems to me to be a very strange point of view to adopt for a philosophy of mathematics. Furthermore, it is not really all that pragmatic. When mathematicians carry out their forms of reasoning, they do not want to have to be continually checking to see whether or not their arguments can be formulated in terms of the axioms and rules of procedure of some complicated formal system. They only need to be sure that their arguments are valid ways of ascertaining truth. The Gödel argument is another such valid procedure, so it seems to me that $P_k(k)$ is just as good a mathematical truth as any that can be obtained more conventionally using the axioms and rules of procedure that can be laid down beforehand.

A procedure that suggests itself is the following. Let us accept that $P_k(k)$, which for the present I shall simply denote by G_0, is indeed a perfectly valid proposition; so we may simply adjoin it to our system, as an additional axiom. Of course, our new amended system will have its *own* Gödel proposition, say G_1, which again is seen to be a perfectly valid statement about numbers. Accordingly we adjoin G_1 to our system also. This gives us a further amended system which will have its own Gödel proposition G_2 (again perfectly valid), and we can then adjoin this, obtaining the next Gödel proposition G_3, which we also adjoin, and so on, repeating this process indefinitely. What about the resulting system when we allow ourselves to use the *entire* list G_0, G_1, G_2, G_3, . . . as additional axioms? Might it be that *this* is complete? Since we now have an unlimited (infinite) system of axioms, it is perhaps not clear that the Gödel procedure will apply. However, this continuing adjoining of Gödel propositions is a perfectly systematic scheme and can be rephrased as an ordinary finite logical system of axioms and rules of procedure. This system will have its own Gödel proposition, say G_ω, which we can again adjoin, and then form the Gödel proposition $G_{\omega+1}$, of the resulting system. Repeating, as above, we obtain a list G_ω, $G_{\omega+1}$, $G_{\omega+2}$, $G_{\omega+3}$, . . . of propositions, all perfectly valid statements about natural numbers, and which can all be adjoined to our formal system. This is again perfectly systematic, and it leads us to a new system encompassing the whole lot; but this again will have its Gödel proposition, say $G_{\omega+\omega}$, which we can rewrite as $G_{\omega2}$, and the whole procedure can be started up again, so that we get a new infinite, but systematic, list of axioms $G_{\omega2}$, $G_{\omega2+1}$, $G_{\omega2+2}$, etc., leading to yet a new system—and a new Gödel proposition $G_{\omega3}$. Repeating the entire procedure, we get $G_{\omega4}$ and then $G_{\omega5}$ and so on. Now *this* procedure is entirely systematic and has its own Gödel proposition G_{ω^2}.

Does this ever end? In a sense, no; but it leads us into some difficult mathematical considerations that cannot be gone into in detail here. The above procedure was discussed by Alan Turing in a paper[4] in 1939. In fact, very remarkably, *any* true (but just universally quantified) proposition in arithmetic can be obtained by a repeated 'Gödelization' procedure of this type! See Feferman (1988). However, this does to some degree beg the question of how we actually *decide* whether a proposition is true or false. The critical issue, at each stage, is to see how to code the adjoining of an infinite family of Gödel propositions into providing a single additional axiom (or finite number of axioms). This requires that our infinite family can be systematized in some algorithmic way. To be sure that such a systematization *correctly* does what it is supposed to do, we shall need to employ *insights* from outside the system—just as we did in order to see that $P_k(k)$ was a true proposition in the first place. It is these insights that cannot be systematized—and, indeed, must lie outside *any* algorithmic action!

The insight whereby we concluded that the Gödel proposition $P_k(k)$ is actually a true statement in arithmetic is an example of a general type of procedure known to logicians as a *reflection principle*: thus, by 'reflecting' upon the *meaning* of the axiom system and rules of procedure, and convincing oneself that these indeed provide valid ways of arriving at mathematical truths, one may be able to code this insight into further true mathematical statements that were not deducible from those very axioms and rules. The derivation of the truth of $P_k(k)$, as outlined above, depended upon such a principle. Another reflection principle, relevant to the original Gödel argument (though not given above), depends upon deducing new mathematical truths from the fact that an axiom system, that we already believe to be valid for obtaining mathematical truths, is actually *consistent*. Reflection principles often involve reasoning about infinite sets, and one must always be careful when using them, that one is not getting too close to the type of argument that could lead one into a Russell-type paradox. Reflection principles provide the very antithesis of formalist reasoning. If one is careful, they enable one to leap outside the rigid confinements of any formal system to obtain new mathematical insights that did not seem to be available before. There could be many perfectly acceptable results in our mathematical literature whose proofs require insights that lie far from the original rules and axioms of the standard formal systems for arithmetic. All this shows that the mental procedures whereby mathematicians arrive at their judgements of truth are not simply rooted in the procedures of some specific formal system. We *see* the validity of the Gödel proposition $P_k(k)$ though we cannot derive it from the axioms. The type of 'seeing' that is involved in a reflection principle requires a mathematical insight that is not the result of the purely algorithmic operations that could be coded into some mathematical formal system. This matter will be returned to in Chapter 10.

The reader may notice a certain similarity between the argument establishing the truth yet 'unprovability' of $P_k(k)$ and the argument of Russell's paradox. There is a similarity also with Turing's argument establishing the non-existence of a Turing machine which will solve the halting problem. These similarities are not accidental. There is a strong thread of historical connection between the three. Turing found his argument after studying the work of Gödel. Gödel himself was well aware of the Russell paradox, and was able to transform paradoxical reasoning of this kind, which stretches the use of logic too far, into a valid mathematical argument. (All these arguments have their origins in Cantor's 'diagonal slash', described in the previous chapter, p. 84.)

Why should we accept the Gödel and Turing arguments, whereas we have had to reject the reasoning leading to the Russell paradox? The former are much more clear-cut, and are unexceptionable as mathematical arguments, whereas the Russell paradox depends upon more nebulous reasoning involving 'enormous' sets. But it must be admitted that the distinctions are not really as clear as one would like them to be. The attempt to make such distinctions clear was a strong motive behind the whole idea of formalism. Gödel's argument shows that the strict formalist viewpoint does not really hold together; yet it does not lead us to a wholely reliable alternative point of view. To my mind the issue is still unresolved. The procedure that is actually adopted* in contemporary mathematics, for avoiding the type of reasoning with 'enormous' sets that leads to Russell's paradox, is not entirely satisfactory. Moreover it tends still to be stated in distinctly formalistic terms—or, alternatively, in terms that do not give us full confidence that contradictions cannot arise.

Be that as it may, it seems to me that it is a clear consequence of the Gödel argument that the concept of mathematical truth cannot be encapsulated in any formalistic scheme. Mathematical truth is something that goes beyond mere formalism. This is perhaps clear even without Gödel's theorem. For how are we to decide what axioms or rules of procedure to adopt in any case when trying to set up a formal system? Our guide in deciding on the rules to adopt must always be our intuitive understanding of what is 'self-evidently true', given the 'meanings' of the symbols of the system. How are we to decide which formal systems are sensible ones to adopt—in accordance, that is, with our intuitive feelings about 'self-evidence' and 'meaning'—and which are not? The notion of self-consistency is certainly not adequate for this. One can have many self-consistent systems which are not 'sensible' in this sense,

* A distinction is made between 'sets' and 'classes', where sets are allowed to be collected together to form other sets or perhaps classes, but classes are *not* allowed to be collected together to form larger collections of any kind, being regarded as being 'too big' for this. There is, however, no rule for deciding when a collection may be allowed to be regarded as a set or must necessarily be considered to be only a class, apart from the circular one that states that sets are those collections that can indeed be collected together to form other collections!

where the axioms and rules of procedure have meanings that we would reject as false, or perhaps no meaning at all. 'Self-evidence' and 'meaning' are concepts which would still be needed, even without Gödel's theorem.

However, without Gödel's theorem it might have been possible to imagine that the intuitive notions of 'self-evidence' and 'meaning' could have been employed just once and for all, merely to set up the formal system in the first place, and thereafter dispensed with as part of clear mathematical argument for determining truth. Then, in accordance with a formalist view, these 'vague' intuitive notions would have had roles to play as part of the mathematician's *preliminary* thinking, as a guide towards finding the appropriate formal argument; but they would play no part in the actual demonstration of mathematical truth. Gödel's theorem shows that this point of view is not really a tenable one in a fundamental philosophy of mathematics. The notion of mathematical truth goes beyond the whole concept of formalism. There is something absolute and 'God-given' about mathematical truth. This is what mathematical Platonism, as discussed at the end of the last chapter, is about. Any particular formal system has a provisional and 'man-made' quality about it. Such systems indeed have very valuable roles to play in mathematical discussions, but they can supply only a partial (or approximate) guide to truth. Real mathematical truth goes beyond mere man-made constructions.

Platonism or intuitionism?

I have pointed out two opposing schools of mathematical philosophy, siding strongly with the Platonist rather than the formalist view. I have actually been rather simplistic in my distinctions. There are many refinements of viewpoint that can be made. For example, one can argue under the heading of 'Platonism' whether the objects of mathematical thought have any kind of actual 'existence' or whether it is just the concept of mathematical 'truth' which is absolute. I have not chosen to make an issue of such distinctions here. In my own mind, the absoluteness of mathematical truth and the Platonic existence of mathematical concepts are essentially the same thing. The 'existence' that must be attributed to the Mandelbrot set, for example, is a feature of its 'absolute' nature. Whether a point of the Argand plane does, or does not, belong to the Mandelbrot set is an absolute question, independent of which mathematician, or which computer, is examining it. It is the Mandelbrot set's 'mathematician-independence' that gives it its Platonic existence. Moreover, its finest details lie beyond what is accessible to us by use of computers. Those devices can yield only approximations to a structure that has a deeper and 'computer-independent' existence of its own. I do appreciate, however, that there can be many other viewpoints that are reasonable to hold on this question. We need not worry too much about these distinctions here.

There are also differences of viewpoint concerning the lengths to which one may be prepared to carry one's Platonism—if, indeed one claims to be a Platonist. Gödel, himself, was a very strong Platonist. The types of mathematical statement that I have been considering so far are rather 'mild' ones as such things go.[5] More controversial statements can arise, particularly in the theory of sets. When all the ramifications of set theory are considered, one comes across sets which are so wildly enormous and nebulously constructed that even a fairly determined Platonist such as myself may begin to have doubts that their existence, or otherwise, is indeed an 'absolute' matter.[6] There may come a stage at which the sets have such convoluted and conceptually dubious definitions that the question of the truth or falsity of mathematical statements concerning them may begin to take on a somewhat 'matter-of-opinion' quality rather than a 'God-given' one. Whether one is prepared to go the whole way with Platonism, along with Gödel, and demand that the truth or falsity of mathematical statements concerning such enormous sets is always an absolute or 'Platonic' matter, or whether one stops somewhere short of this and demands an absolute truth or falsity only when the sets are reasonably constructive and not so wildly enormous, is not a matter that will have a great relevance to our discussion here. The sets (finite or infinite) that will have significance for us are, by the standards of the ones that I have just been referring to, ridiculously tiny! Thus the distinctions between these various Platonistic views will not greatly concern us.

There are, however, other mathematical standpoints, such as that known as *intuitionism* (and another called *finitism*), which go to the other extreme, where one refuses to accept the completed existence of any infinite set whatever.* Intuitionism was initiated, in 1924, by the Dutch mathematician L. E. J. Brouwer as an alternative response—distinct from that of formalism—to the paradoxes (such as Russell's) that can arise when too free a use of infinite sets is made in mathematical reasoning. The roots of such a viewpoint can be traced back to Aristotle, who had been Plato's pupil but who had rejected Plato's views about the absolute existence of mathematical entities and about the acceptability of infinite sets. According to intuitionsim, sets (infinite or otherwise) are not thought of as having an 'existence' in themselves but are thought of merely in terms of the rules which might determine their membership.

A characteristic feature of Brouwer's intuitionism is the rejection of 'the law of the excluded middle'. This law asserts that the denial of the negation of a statement is equivalent to the assertion of that statement. (In symbols: $\sim(\sim P) \Longleftrightarrow P$, a relation we encountered above.) Perhaps Aristotle might

* Intuitionism was so called because it was supposed to mirror human thought.

have been unhappy with the denial of something so logically 'obvious' as this! In ordinary 'common-sense' terms, the law of the excluded middle may be regarded as a self-evident truth: if it is false that something is not true, then that thing is surely true! (This law is the basis of the mathematical procedure of '*reductio ad absurdum*', cf. p. 59.) But the intuitionists find themselves able to deny this law. This is basically because they are taking a different attitude to the concept of *existence*, demanding that a definite (mental) construction be presented before it is accepted that a mathematical object actually exists. Thus, to an intuitionist, 'existence' means 'constructive existence'. In a mathematical argument which proceeds by *reductio ad absurdum* one puts forward some hypothesis with the intention of showing that its consequences lead one into a contradiction, this contradiction providing the desired proof that the hypothesis in question is false. The hypothesis could take the form of a statement that a mathematical entity with certain required properties does not exist. When this leads to a contradiction, one infers, in *ordinary mathematics*, that the required entity does indeed exist. But such an argument, in itself, provides no means for actually *constructing* such an entity. To an intuitionist, this kind of existence is no existence at all; and it is in this kind of sense that they refuse to accept the law of the excluded middle and the procedure of *reductio ad absurdum*. Indeed, Brouwer was profoundly dissatisfied by such a non-constructive 'existence'.[7] Without an actual construction, he asserted, such a concept of existence is meaningless. In Brouwerian logic, one cannot deduce from the falsity of the non-existence of some object that the object actually exists!

In my own view, although there is something commendable about seeking constructiveness in mathematical existence, Brouwer's point of view of intuitionism is much too extreme. Brouwer first put forward his ideas in 1924, which was more than ten years before the work of Church and Turing. Now that the concept of constructiveness—in terms of Turing's idea of computability—can be studied within the *conventional* framework of mathematical philosophy, there is no need to go to the extremes to which Brouwer would wish to take us. We can discuss constructiveness as a separate issue from the question of mathematical existence. If we go along with intuitionism, we must deny ourselves the use of very powerful kinds of argument within mathematics, and the subject becomes somewhat stifled and impotent.

I do not wish to dwell on the various difficulties and seeming absurdities that the intuitionistic point of view leads one into; but it will perhaps be helpful if I refer to just a few of the problems. An example often referred to by Brouwer concerns the decimal expansion of π:

$$3.141592653589793\ldots$$

Does there exist a succession of ten consecutive sevens somewhere in this expansion, i.e.

$$\pi = 3.141592653589793\ .\ .\ .\ .\ .\ 7777777777\ .\ .\ .\ .\ .\ ,$$

or does there not? In ordinary mathematical terms, all that we can say, as of now, is that either there does or there does not—and we do not know which! This would seem to be a harmless enough statement. However, the intuitionists would actually deny that one can validly say 'either there exists a succession of ten consecutive sevens somewhere in the decimal expansion of π, or else there does not'—unless and until one has (in some constructive way acceptable to the intuitionists) either established that there is indeed such a succession, or else established that there is none! A direct calculation could suffice to show that a succession of ten consecutive sevens actually does exist somewhere in the decimal expansion of π, but some sort of mathematical theorem would be needed to establish that there is no such succession. No computer has yet proceeded far enough in the computation of π to determine that there is indeed such a succession. One's expectation on probabilistic grounds would be that such a succession does actually exist, but that even if a computer were to produce digits consistently at the rate of, say, one per second, it would be likely to take something of the order of between one hundred and one thousand years to find the sequence! It seems to me to be much more likely that, rather than by direct computation, the existence of such a sequence will someday be established mathematically (probably as a corollary of some much more powerful and interesting result)—although perhaps not in a way acceptable to the intuitionists!

This particular problem is of no real mathematical interest. It is given only as an example which is easy to state. In Brouwer's extreme form of intuitionism, he would assert that, at the present time, the assertion 'there exists a succession of ten consecutive sevens somewhere in the decimal expansion of π' is neither true nor false. If at some later date the appropriate result is established one way or the other, by computation or by (intuitionistic) mathematical proof, then the assertion would become 'true', or else 'false', as the case may be. A similar example would be 'Fermat's last theorem'. According to Brouwer's extreme intuitionism, this also is neither true nor false at the present time, but it might become one or the other at some later date. To me, such subjectiveness and time-dependence of mathematical truth is abhorrent. It is, indeed, a very subjective matter whether, or when, a mathematical result might be accepted as officially 'proved'. Mathematical truth should not rest on such society-dependent criteria. Also, to have a concept of mathematical truth which changes with time is, to say the least, most awkward and unsatisfactory for a mathematics which one hopes to be able to employ reliably for a description of the physical world. Not all intuitionists would take such a strong line as Brouwer's. Nevertheless the intuitionist point of view is a distinctly awkward one, even for those sympathetic to the aims of constructivism. Few present-day mathematicians would go along with intuitionsim wholeheartedly, if only because it is very limiting with regard to the type of mathematical reasoning one is allowed to use.

I have briefly described the three main streams of present-day mathematical philosophy: formalism, Platonism, and intuitionism. I have made no secret of the fact that my sympathies lie strongly with the Platonistic view that mathematical truth is absolute, external, and eternal, and not based on man-made criteria; and that mathematical objects have a timeless existence of their own, not dependent on human society nor on particular physical objects. I have tried to make my case for this view in this section, in the previous section, and at the end of Chapter 3. I hope that the reader is prepared to go most of the way with me on this. It will be important for a good deal of what we shall encounter later.

Gödel-type theorems from Turing's result

In my presentation of Gödel's theorem, I have omitted many details, and I have also left out what was historically perhaps the most important part of his argument; that referring to the 'undecidability' of the consistency of the axioms. My purpose here has *not* been to emphasize this 'axiom-consistency-provability problem', so important to Hilbert and his contemporaries, but to show that a specific Gödel proposition—neither provable nor disprovable using the axioms and rules of the formal system under consideration—is clearly *seen*, using our insights into the meanings of the operations in question, to be a *true* proposition!

I have mentioned that Turing developed his own later argument establishing the insolubility of the halting problem after studying the work of Gödel. The two arguments had a good deal in common and, indeed, key aspects of Gödel's result can be directly derived using Turing's procedure. Let us see how this works, and thereby obtain a somewhat different insight into what lies behind Gödel's theorem.

An essential property of a formal mathematical system is that it should be a computable matter to decide whether or not a given string of symbols constitutes a proof, within the system, of a given mathematical assertion. The whole point of formalizing the notion of mathematical proof, after all, is that no further judgements have to be made about what is valid reasoning and what is not. It must be possible to check in a completely mechanical and previously determined way whether or not a putative proof is indeed a proof;

that is, there must be an *algorithm* for checking proofs. On the other hand, we do not demand that it should necessarily be an algorithmic matter to *find* proofs (or disproofs) of suggested mathematical statements.

In fact, it turns out that there always *is* an algorithm for finding a proof within any formal system whenever some proof exists. For we must suppose that our system is formulated in terms of some symbolic language, this language being expressible in terms of some finite 'alphabet' of symbols. As before, let us order our strings of symbols *lexicographically*, which we recall means alphabetically for each fixed length of string, taking all strings of length one to be ordered first, those of length two next, then length three, and so on (p. 106). Thus we have all the correctly constructed proofs numerically ordered according to this lexicographical scheme. Having our list of proofs, we also have a list of all the *theorems* of the formal system. For the theorems are precisely the propositions that appear as last lines of correctly constructed proofs. The listing is perfectly computable: for we can consider the lexicographical list of *all* strings of symbols of the system, whether or not they make sense as proofs, and then test the first string with our proof-testing algorithm to see if it is a proof and discard it if it is not; then we test the second in the same way and discard it if it is not a proof, and then the third, then the fourth, and so on. By this means, if there is a proof, we shall eventually find it somewhere along the list.

Thus if Hilbert had been successful in finding his mathematical system—a system of axioms and rules of procedure strong enough to enable one to decide, by formal proof, the truth or falsity of any mathematical proposition correctly formulated within the system—then there *would* be a general algorithmic method of deciding the truth of any such proposition. Why is this? Because if, by the procedure outlined above, we eventually come across the proposition we are looking for as the final line in the proof, then we have *proved* this proposition. If, instead, we eventually come across the *negation* of our proposition as the final line, then we have *disproved* it. If Hilbert's scheme were complete, either one or the other of these eventualities would always occur (and, if consistent, they would never both occur together). Thus our mechanical procedure would always terminate at some stage and we should have a universal algorithm for deciding the truth or otherwise of all propositions of the system. This would contradict Turing's result, as presented in Chapter 2, that there is no general algorithm for deciding mathematical propositions. Consequently we have, in effect, proved Gödel's theorem that *no* scheme of Hilbert's intended type can be complete in the sense that we have been discussing.

In fact Gödel's theorem is more specific than this, since the type of formal system that Gödel was concerned with was required to be adequate only for propositions of arithmetic, not for propositions of mathematics in general. Can we arrange that all the necessary operations of Turing machines can be carried out using just arithmetic? To put this another way, can all the *computable* functions of natural numbers (i.e. recursive, or algorithmic functions—the results of Turing machine action) be expressed in terms of ordinary arithmetic? In fact it is almost true that we can, but not quite. We need one extra operation to be adjoined to the standard rules of arithmetic and logic (including $\exists$ and $\forall$). This operation simply selects

'the smallest natural number x such that $K(x)$ is true',

where $K(\ )$ is any given arithmetically calculable propositional function—for which it is assumed that there *is* such a number, i.e. that $\exists x[K(x)]$ is true. (If there were no such number, then our operation would 'run on forever'* trying to locate the required non-existent x.) In any case, the foregoing argument does establish, on the basis of Turing's result, that Hilbert's programme of reducing whole branches of mathematics to calculations within some formal system is indeed untenable.

As it stands, this procedure does not so readily show that we have a Gödel proposition (like $P_k(k)$) that is *true*, but not provable within the system. However, if we recall the argument given in Chapter 2 on 'how to outdo an algorithm' (cf. p. 64), we shall see that we can do something very similar. In that argument we were able to show that, given any algorithm for deciding whether a Turing machine action stops, we can produce a Turing machine action that *we* see does not stop, yet the algorithm cannot. (Recall that we insisted that the algorithm must correctly inform us when a Turing machine action *will* stop though it can sometimes fail to tell us that the Turing machine action will not stop—by running on forever itself.) Thus, as with the situation with Gödel's theorem above, we have a proposition that we can *see*, by the use of insight, must actually be *true* (non-stopping of the Turing machine action), but the given algorithmic action is not capable of telling us this.

Recursively enumerable sets

There is a way of describing the basic ingredients of Turing's and Gödel's results in a graphic way, in terms of the language of *set theory*. This enables us to get away from arbitrary descriptions in terms of specific symbolisms or formal systems, so that the essential issues are brought out. We shall consider only sets (finite or infinite) of the *natural numbers* 0, 1, 2, 3, 4, . . ., so we shall be examining collections of these, such as {4, 5, 8}, {0, 57, 100003}, {6}, {0}, {1, 2, 3, 4, . . ., 9999}, {1, 2, 3, 4, ...}, {0, 2, 4, 6, 8, ...}, or even the

* It is actually essential that such unfortunate possibilities be allowed to occur, so that we may have the potential to describe *any* algorithmic operation. Recall that to describe Turing machines generally, we must allow for Turing machines which never actually stop.

entire set $\mathbb{N} = \{0, 1, 2, 3, 4, \ldots\}$ or the empty set $\emptyset = \{\}$. We shall be concerned only with *computability* questions, namely; 'Which kinds of sets of natural numbers can be generated by algorithms and which cannot?'

In order to address such issues we may, if we like, think of each single natural number n as denoting a specific string of symbols in a particular formal system. This would be the 'nth' string of symbols, say Q_n, according to some lexicographical ordering of (the 'syntactically correctly' expressed) propositions in the system. Then each natural number represents a proposition. The set of *all* propositions of the formal system would be represented as the entire set $\mathbb{N}$ and, for example, the *theorems* of the formal system would be thought of as constituting some smaller set of natural numbers, say the set P. However, the details of any particular numbering system for propositions is not important. All that we should need, in order to set up a correspondence between natural numbers and propositions, would be a known algorithm for obtaining each proposition Q_n (written in the appropriate symbolic notation) from its corresponding natural number n, and another known algorithm for obtaining n from Q_n. Taking these two known algorithms as given, we are at liberty to *identify* the set of natural numbers $\mathbb{N}$ with the set of propositions of a specific formal system.

Let us choose a formal system which is consistent and broad enough to include all actions of all Turing machines—and, moreover, 'sensible' in the sense that its axioms and rules of procedure are things that may be taken to be 'self-evidently *true*'. Now, *some* of the propositions $Q_0, Q_1, Q_2, Q_3, \ldots$ of the formal system will actually have *proofs* within the system. These 'provable' propositions will have numbers which constitute some set in $\mathbb{N}$, in fact the set P of 'theorems' considered above. We have, in effect, already seen that there is an *algorithm* for generating, one after the other, all the propositions with proofs in some given formal system. (As outlined earlier, the 'nth proof' Π_n is obtained algorithmically from n. All we have to do is look at the last line of the nth proof to find the 'nth proposition provable within the system, i.e. the nth 'theorem'.) Thus we have an algorithm for generating the elements of P one after the other (perhaps with repetitions—that makes no difference).

A set, such as P, which can be generated in this way by some algorithm, is called *recursively enumerable*. Note that the set of propositions which are *dis*provable within the system—i.e. propositions whose negations are provable—is likewise recursively enumerable, for we can simply enumerate the provable propositions, taking their negations as we go along. There are many other subsets of $\mathbb{N}$ which are recursively enumerable, and we need make no reference to our formal system in order to define them. Simple examples of recursively enumerable sets are the set of even numbers

$$\{0, 2, 4, 6, 8, \ldots\},$$

the set of squares

$$\{0, 1, 4, 9, 16, \ldots.\},$$

and the set of primes

$$\{2, 3, 5, 7, 11, \ldots.\}.$$

Clearly we can generate each of these sets by means of an algorithm. In each of these three examples it will *also* be the case that the *complement* of the set—that is, the set of natural numbers *not* in the set—is recursively enumerable. The complementary sets in these three cases are, respectively:

$$\{1, 3, 5, 7, 9, \ldots\};$$
$$\{2, 3, 5, 6, 7, 8, 10, \ldots\};$$

and

$$\{0, 1, 4, 6, 8, 9, 10, 12, \ldots.\}.$$

It would be a simple matter to provide an algorithm for these complementary sets also. Indeed, we can algorithmically decide, for any given natural number n, whether or not it is an even number, or whether or not it is a square, or whether or not it is a prime. This provides us with an algorithm for generating *both* the set and the complementary set, for we can run through the natural numbers in turn and decide, in each case, whether it belongs to the original set or to the complementary set. A set which has the property that both it and its complementary set are recursively enumerable is called a *recursive* set. Clearly the complement of a recursive set is also a recursive set.

Now, are there any sets which are recursively enumerable but *not* recursive? Let us pause, for a moment, to note what that would entail. Since the elements of such a set can be generated by an algorithm, we shall have a means of deciding, for an element suspected of being in the set—and which, let us suppose for the moment, is actually in the set—that, indeed, it *is* in the set. All we need do is to allow our algorithm to run over all the elements of the set until it eventually finds the particular element that we are examining. But suppose our suspected element is actually *not* in the set. In that case our algorithm will do us no good at all, for it will just keep running forever, never coming to a decision. For that we need an algorithm for generating the *complementary* set. If *that* finds our suspect then we know for sure that the element is not in the set. With both algorithms we should be in good shape. We could simply alternate between the two algorithms and catch the suspect either way. That happy situation, however, is what happens with a *recursive* set. Here our set is supposed to be merely recursively enumerable but *not* recursive: our suggested algorithm for generating the complementary set does not exist! Thus we are presented with the curious situation that we can algorithmically decide, for an element *in* the set, that it *is* actually in the set;

but we cannot guarantee, by means of any algorithm, to decide this question for elements that happen *not* to be in the set!

Does such a curious situation ever arise? Are there, indeed, any recursively enumerable sets that are not recursive? Well, what about the set P? Is *that* a recursive set? We know that it is recursively enumerable, so what we have to decide is whether the complementary set is also recursively enumerable. In fact it is not! How can we tell that? Well, recall that the actions of Turing machines are supposed to be among the operations allowed within our formal system. We denote the nth Turing machine by T_n. Then the statment

'$T_n(n)$ stops'

is a proposition—let us write it $S(n)$—that we can express in our formal system, for each natural number n. The proposition $S(n)$ will be true for some values of n, and false for others. The set of *all* $S(n)$, as n runs through the natural numbers 0, 1, 2, 3, . . ., will be represented as some subsets S of $\mathbb{N}$. Now, recall Turing's fundamental result (Chapter 2, p. 59) that there is no algorithm that asserts '$T_n(n)$ does not stop' in precisely these cases in which $T_n(n)$ in fact does not stop. This shows that the set of *false* $S(n)$ is *not* recursively enumerable.

We observe that the part of S that lies in P consists precisely of those $S(n)$ which are *true*. Why is this? Certainly if any particular $S(n)$ is provable, then it must be true (because our formal system has been chosen to be 'sensible'!), so the part of S that lies in P must consist solely of *true* propositions $S(n)$. Moreover, no true proposition $S(n)$ can lie outside P, for if $T_n(n)$ stops then we can provide a proof within the system that it actually does so.*

Now, suppose that the complement of P were recursively enumerable. Then we should have some algorithm for generating the elements of this complementary set. We can run this algorithm and note down every proposition $S(n)$ that we come across. These are all the false $S(n)$, so our procedure would actually give us a recursive enumeration of the set of false $S(n)$. But we noted above that the false $S(n)$ are *not* recursively enumerable. This contradiction establishes that the complement of P cannot be recursively enumerated after all; so the *set P is not recursive*, which is what we have sought to establish.

These properties actually demonstrate that our formal system cannot be complete: i.e. there must be propositions which are neither provable nor disprovable within the system. For if there were no such 'undecidable' propositions, the complement of the set P would have to be the set of *disprovable* propositions (anything not provable would have to be disprovable). But we have seen that the disprovable propositions constitute a recursively enumerable set, so this would make P *recursive*. However, P is *not*

* The proof could consist, in effect, of a succession of steps which would mirror the action of the machine running until it stopped. The proof would be completed as soon as the machine stops.

recursive—a contradiction which establishes the required incompleteness. This is the main thrust of Gödel's theorem.

Now what about the subset T of $\mathbb{N}$ which represents the *true* propositions of our formal system? Is T recursive? Is T recursively enumerable? Is the complement of T recursively enumerable? In fact the answer to all these questions is 'No'. One way of seeing this is to note that the false propositions of the form

$$\text{`}T_n(n) \text{ stops'}$$

cannot be generated by an algorithm, as we have noted above. Therefore the false propositions as a *whole* cannot be generated by an algorithm, since any such algorithm would, in particular, enumerate all the above false '$T_n(n)$ stops' propositions. Similarly, the set of all *true* propositions cannot be generated by an algorithm (since any such algorithm could be trivially modified to yield all the false propositions, simply by making it take the *negation* of each proposition that it generates). Since the true propositions are thus not recursively enumerable (and nor are the false ones) they constitute a vastly more complicated and profound array than do the propositions provable within the system. This again illustrates aspects of Gödel's theorem: that the concept of mathematical *truth* is only partly accessible by the means of formal argument.

There are certain simple classes of true arithmetical proposition that do form recursively enumerable sets, however. For example, true propositions of the form

$$\exists w, x \ldots, z[f(w, x, \ldots, z) = 0],$$

where $f(\ )$ is some function constructed from the ordinary arithmetical operations of adding, subtracting, multiplying, and raising to a power, constitute a recursively enumerable set (which I shall label A), as is not too hard to see.[8] An instance of a proposition of this form—although we do not know if it is true—is the negation of 'Fermat's last theorem', for which we can take $f(\ )$ to be given by

$$f(w, x, y, z) = (x + 1)^{w+3} + (y + 1)^{w+3} + (z + 1)^{w+3}.$$

However, the set A turns out not to be recursive (a fact which is *not* so easy to see—though it is a consequence of Gödel's actual original argument). Thus we are not presented with any algorithmic means which could, even in principle, decide the truth or otherwise of 'Fermat's last theorem'!

In Fig. 4.1, I have tried schematically to represent a recursive set as a region with a nice simple boundary, so that one can imagine it being a direct matter to tell whether or not some given point belongs to the set or not. Each point in the picture is to be thought of as representing a natural number. The complementary set then also represented as a simple-looking region. In Fig.

4.2, I have tried to represent a recursively enumberable but *non*-recursive set as a set with a complicated boundary, where the set on one side of the boundary—the recursively enumerable side—is supposed to look simpler than that on the other. The figures are very schematic, and are not intended to be in any sense 'geometrically accurate'. In particular, there is no significance in the fact that these pictures have been represented as though on a flat two-dimensional plane!

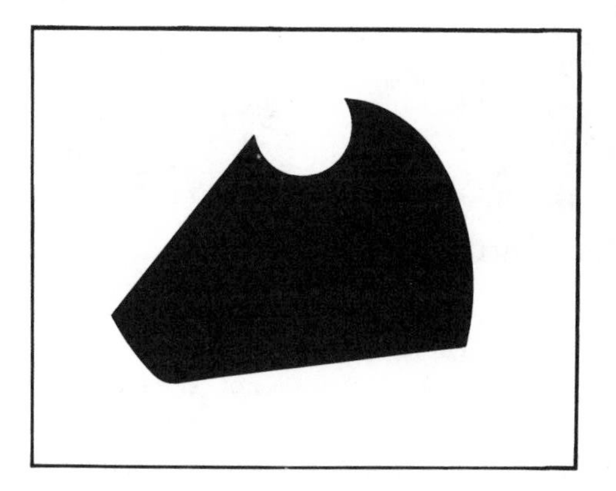

Fig. 4.1. Highly schematic representation of a recursive set.

Fig. 4.2. Highly schematic representation of a recursively enumerable set (black region) which is not recursive. The idea is that the white region is defined only as 'what's left' when the computably generated black region is removed; and it is not a computable matter to ascertain that a point is actually in the white region itself.

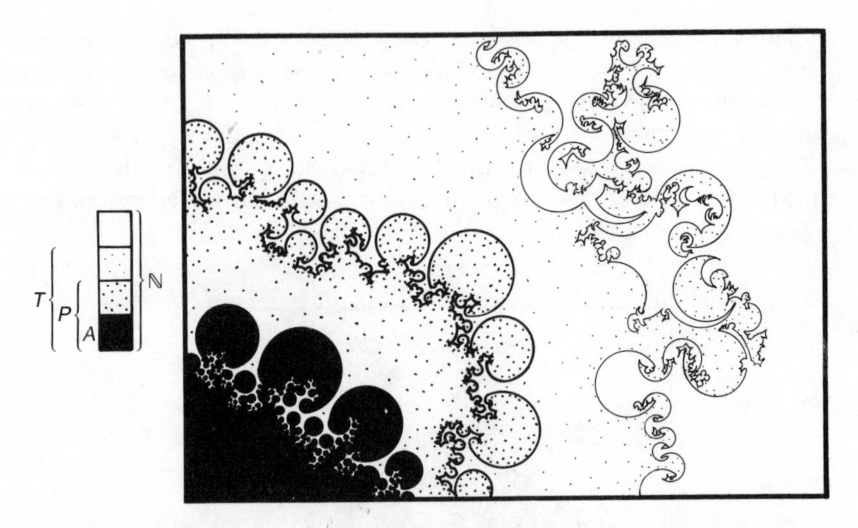

Fig. 4.3. Highly schematic representation of various sets of propositions. The set P of propositions that are provable within the system is recursively enumerable but not recursive; the set T of true propositions is not even recursively enumerable.

In Fig. 4.3 I have schematically indicated how the regions P, T, and A lie within the set $\mathbb{N}$.

Is the Mandelbrot set recursive?

Non-recursive sets must have the property that they are complicated in a very essential way. Their complication must, in some sense, defy all attempts at systematization, otherwise that very systematization would lead to some appropriate algorithmic procedure. For a non-recursive set, there is no general algorithmic way of deciding whether or not an element (or 'point') belongs to the set. Now we witnessed, at the beginning of Chapter 3, a certain extraordinarily complicated-looking set, namely the Mandelbrot set. Although the rules which provide its definition are surprisingly simple, the set itself exhibits an endless variety of highly elaborate structure. Could this be an example of a non-recursive set, truly exhibited before our mortal eyes?

The reader will not be slow to point out, however, that this paradigm of complication has been conjured up, for our eyes to see, by the magic of modern high-speed electronic computer technology. Are not electronic computers the very embodiments of algorithmic action? Indeed, that is certainly true, but we must bear in mind the way that the computer actually produces these pictures. To test whether a point of the Argand plane—a complex number c—belongs to the Mandelbrot set (coloured black) or to the complementary set (coloured white), the computer would start with 0, then apply the map

$$z \rightarrow z^2 + c$$

to $z = 0$ to obtain c, and then to $z = c$ to obtain $z^2 + c$ and then to $z = z^2 + c$ to obtain $z^4 + 2z^3 + z^2 + z$, and so on. If this sequence $0, z, z^2 + z, z^4 + 2z^3 + z^2 + z, \ldots$ remains bounded, then the point represented by c is coloured black; otherwise it is coloured white. How does the machine tell whether or not such a sequence remains bounded? In principle, the question involves knowing what happens after an *infinite* number of terms of the sequence! That, by itself, is not a computable matter. Fortunately, there are ways of telling after only a finite number of terms when the sequence has become unbounded. (In fact, as soon as it reaches the circle of radius $1 + \sqrt{2}$ centred at the origin, then one can be sure that the sequence is unbounded.)

Thus, in a certain sense, the *complement* of the Mandelbrot set (i.e. the *white* region) is recursively enumerable. If the complex number c is in the white region, then there is an algorithm for ascertaining that fact. What about the Mandelbrot set itself—the black region? Is there an algorithm for telling for sure that a point suspected to be in the black region is in fact in the black region? The answer to this question seems to be unknown, at present. I have consulted various colleagues and experts, and none seems to be aware of such an algorithm. Nor had they come across any demonstration that no such algorithm exists. At least, there appears to be no *known* algorithm for the black region. Perhaps the complement of the Mandelbrot set is, indeed, an example of a recursively enumerable set which is not recursive!

Before exploring this suggestion further, it will be necessary to address certain issues that I have glossed over. These issues will have some importance to us in our later discussions of computability in physics. I have actually been somewhat inexact in the preceding discussion. I have applied such terms as 'recursively enumerable' and 'recursive' to sets of points in the Argand plane, i.e. to sets of complex numbers. These terms should strictly only be used for the natural numbers or for other *countable* sets. We have seen in Chapter 3 (p. 84) that the real numbers are not countable, and so the complex numbers cannot be countable either—since real numbers can be regarded as particular kinds of complex numbers, namely complex numbers with vanishing imaginary parts (cf. p. 88). In fact there are precisely 'as many' complex numbers as real ones, namely '$\boldsymbol{C}$' of them. (To establish a one-to-one relation between the complex numbers and real numbers, one can, roughly speaking, take the decimal expansions of the real and imaginary parts of each complex number and interleave them according to the odd and even digits of the corresponding real number: e.g. the complex number $3.6781\ldots + i512.975\ldots$ would correspond to the real number $50132.6977851\ldots$)

One way to evade this problem would be to refer only to *computable* complex numbers, for we saw in Chapter 3 that the computable real numbers—and therefore also the computable complex numbers—are indeed countable. However, there is a severe difficulty with this: there is in fact no general algorithm for deciding whether two computable numbers, given in terms of their respective algorithms, are equal to each other or not! (We can algorithmically form their difference, but we cannot algorithmically decide whether or not this difference is zero. Imagine two algorithms generating the digits 0.99999 . . . and 1.00000 . . ., respectively, but we might never know if the 9s, or the 0s, are going to continue indefinitely, so that the two numbers are equal, or whether some other digit will eventually appear, and the numbers are unequal.) Thus, we might never know whether these numbers are equal. One implication of this is that even with such a simple set as the *unit disc* in the Argand plane (the set of points whose distance from the origin is not greater than one unit, i.e. the black region in Fig. 4.4) there would be

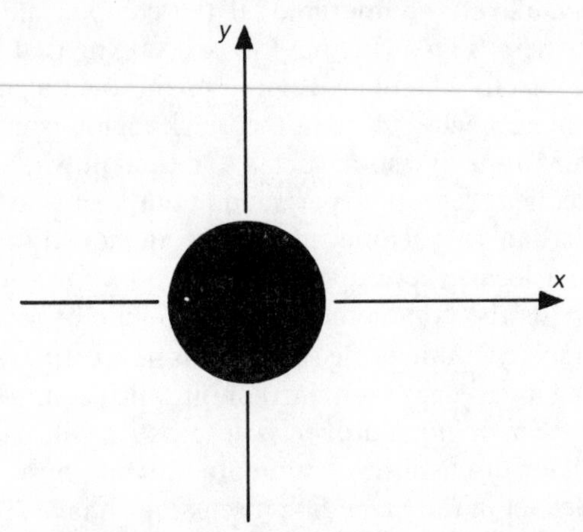

Fig. 4.4. The unit disc should surely count as 'recursive', but this requires an appropriate viewpoint.

no algorithm for deciding for sure whether a complex number actually lies on the disc. The problem does not arise with the points in the interior (or with points outside the disc) but with points which lie on the very edge of the disc—i.e. on the unit circle itself. The unit circle is considered to be part of the disc. Suppose that we are simply given an algorithm which generates the digits of the real and imaginary part of some complex number. If we suspect that this complex number actually lies on the unit circle, we cannot necessarily ascertain this fact. There is no algorithm for deciding whether the computable number

$$x^2 + y^2$$

is actually equal to 1 or not, this being the criterion for deciding whether or not the computable complex number $x + iy$ lies on the unit circle.

Clearly this is not what we want. The unit disc certainly *ought* to count as recursive! There are not many sets simpler than the unit disc! One way around this problem might be to *ignore* the boundary. For points actually in the interior or actually in the exterior, an algorithm for ascertaining these facts certainly exists. (Simply generate the digits of $x^2 + y^2$ one after the other, and eventually we find a digit other than 9 after the decimal point in 0.99999 . . . or other than 0 in 1.00000 . . .). In this sense the unit disc *is* recursive. But the point of view is rather an awkward one for mathematics since one often needs to phrase arguments in terms of what actually happens *at* boundaries. It is possible that such a point of view might be appropriate for physics, on the other hand. We shall need to reconsider this kind of issue again later.

There is another closely related point of view that one might adopt and this is not to refer to computable complex numbers at all. Instead of trying to enumerate the complex numbers inside or outside the set in question, we simply ask for an algorithm which decides, *given* the complex number, whether it lies in the set or whether it lies in the complement of the set. By 'given', I mean that, for each complex number that we are testing, successive digits of the real and imaginary parts are presented to us one after the other, for as long as we please—perhaps by some magical means. I do not require that there be any algorithm, known or unknown, for *presenting* these digits. A set of complex numbers would be considered to be 'recursively enumerable' if a single algorithm exists such that *whenever* it is presented with such a succession of digits in this way it would eventually say 'yes', after a finite number of steps, if and only if the complex number actually lies in that set. As with the first point of view suggested above, it turns out that this point of view 'ignores' boundaries. Thus the interior of the unit disc and the exterior of the unit disc would each count as recursively enumerable in this sense, whereas the boundary itself would not.

It is not altogether clear to me that either of these viewpoints is really what is needed.[9] When applied to the Mandelbrot set, the philosophy of 'ignoring the boundary' may miss a lot of the complication of the set. This set consists partly of 'blobs'—regions with interiors—and partly of 'tendrils'. The most extreme complication seems to lie in the tendrils, which can meander most wildly. However the tendrils do not lie in the interior of the set, and so they would be 'ignored' if we adopt either one of these two philosophies. Even so, it is still not clear whether the Mandelbrot set is 'recursive', where the blobs only are considered. The question seems to rest on a certain unproved conjecture concerning the Mandelbrot set: is it what is called 'locally connected'? I do not propose to explain the meaning or relevance of that term here. I merely wish to indicate that these are difficult issues, and they raise questions concerning the Mandelbrot set which are still not resolved, and some of which lie at the forefront of some current mathematical research.

There are also other points of view that one can adopt, in order to get around the problem that the complex numbers are not countable. Rather than consider *all* computable complex numbers, one may consider a suitable subset of such numbers having the property that it *is* a computable matter to decide whether or not two of them are equal. A simple such subset would be the '*rational*' complex numbers for which the real and imaginary parts of the numbers are both taken to be rational numbers. I do not think that this would give much in the way of tendrils on the Mandelbrot set, however, this point of view being very restrictive. Somewhat more satisfactory would be to consider the *algebraic* numbers—those complex numbers which are solutions of algebraic equations with integer coefficients. For example, all the solutions for z of

$$129z^7 - 33z^5 + 725z^4 + 16z^3 - 2z - 3 = 0$$

are algebraic numbers. Algebraic numbers are countable and computable, and it is actually a computable matter to decide whether or not two of them are equal. (It turns out that many of them lie on the boundary of the unit circle and on the tendrils of the Mandelbrot set.) We can, if desired, phrase the question as to whether or not the Mandelbrot set is recursive in terms of them.

It may be that algebraic numbers would be appropriate in the case of the two sets just considered, but they do not really resolve all our difficulties in general. For consider the set (the black region of Fig. 4.5) defined by the relation

$$y \geqslant e^x$$

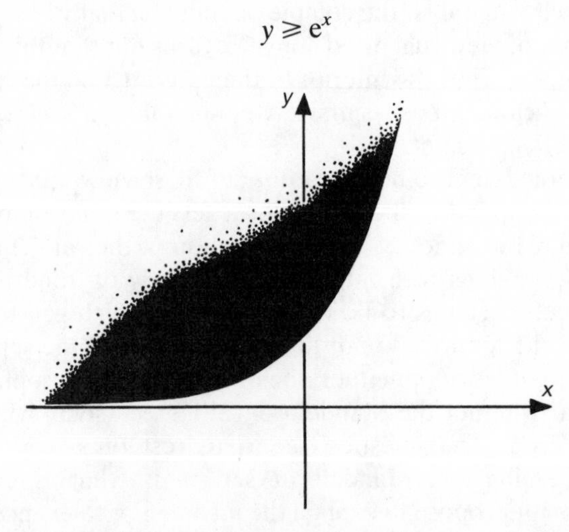

Fig. 4.5. The set defined by the exponential relation $y \geqslant e^x$ should also count as 'recursive'.

for $x + iy$ $(= z)$ in the Argand plane. The interior of the set and the interior of the complement of the set are both recursively enumerable according to any of the viewpoints expressed above, but (as follows from a famous theorem due to F. Lindemann, proved in 1882) the boundary, $y = e^x$, contains only *one* algebraic point, namely the point $z = i$. The algebraic numbers are no help to us in exploring the algorithmic nature of the boundary in this case! It would not be hard to find another subclass of computable numbers which would suffice in this particular case, but one is left with the strong feeling that the correct viewpoint has not yet been arrived at.

Some examples of non-recursive mathematics

There are very many areas of mathematics where problems arise which are not recursive. Thus, we may be presented with a class of problems to which the answer in each case is either 'yes' or 'no', but for which no general algorithm exists for deciding which of these two is actually the case. Some of these classes of problem are remarkably simple-looking.

First, consider the problem of finding integer solutions of systems of algebraic equations with integer coefficients. Such equations are known as *Diophantine* equations (after the Greek mathematician Diophantos, who lived in the third century BC and who studied equations of this type). Such a set of equations might be

$$z^3 - y - 1 = 0, \qquad yz^2 - 2x - 2 = 0, \qquad y^2 - 2xz + z + 1 = 0$$

and the problem is to decide whether or not they can be solved for *integer* values of x, y, and z. In fact, in this particular case they can, a solution being given by

$$x = 13, \qquad y = 7, \qquad z = 2.$$

However, there is no algorithm for deciding this question for an arbitrary set* of Diophantine equations: Diophantine arithmetic, despite the elementary nature of its ingredients, is part of non-algorithmic mathematics!

(A somewhat less elementary example is the *topological equivalence of manifolds*. I mention this only briefly, because it has some conceivable relevance to the issues discussed in Chapter 8. To understand what a 'manifold' is consider first a loop of string, which is a manifold of just *one* dimension, and then consider a closed surface, a manifold of *two* dimensions. Next try to imagine a kind of 'surface' which can have *three* or a higher number of dimensions. 'Topological equivalence' of two manifolds means that one of them can be deformed into the other by a continuous motion—

* This answers, negatively, Hilbert's tenth problem, mentioned on p. 34. (See, for example, Devlin 1988.) Here, the number of variables is not restricted. However, it is known that no more than nine are actually needed for this non-algorithmic property to hold.

without tearing or gluing. Thus a spherical surface and the surface of a cube are topologically equivalent, whereas they are both topologically *in*equivalent to the surface of a ring or a teacup—the latter two being actually topologically equivalent to each other. Now, for *two*-dimensional manifolds there is an algorithm for deciding whether or not two of them are topologically equivalent—amounting, in effect, to counting the number of 'handles' that each surface has. For three dimensions, the answer to the question is not known, at the time of writing, but for four or more dimensions, there is *no* algorithm for deciding equivalence. The four-dimensional case is conceivably of some relevance to physics, since according to Einstein's general relativity, space and time together constitute a 4-manifold; see Chapter 5, p. 207. It has been suggested by Geroch and Hartle 1987 that this non-algorithmic property might have relevance to 'quantum gravity'; cf. also Chapter 8.)

Let us consider a different type of problem, called the *word problem*[10]. Suppose that we have some alphabet of symbols, and we consider various strings of these symbols, referred to as *words*. The words need not in themselves have any meaning, but we shall be given a certain (finite) list of 'equalities' between them which we are allowed to use in order to derive further such 'equalities'. This is done by making substitutions of words from the initial list into other (normally longer) words which contain them as portions. Each such portion may be replaced by another portion which is deemed to be equal to it according to the list. The problem is then to decide, for some given pair of words, whether or not they are 'equal' according to these rules.

As an example, we might have, for our initial list:

$$
\begin{aligned}
\text{EAT} &= \text{AT}\\
\text{ATE} &= \text{A}\\
\text{LATER} &= \text{LOW}\\
\text{PAN} &= \text{PILLOW}\\
\text{CARP} &= \text{ME.}
\end{aligned}
$$

From these we can derive, for example,

$$\text{LAP} = \text{LEAP}$$

by use of successive substitutions from the second, the first, and again the second of the relations from the initial list:

$$\text{LAP} = \text{LATEP} = \text{LEATEP} = \text{LEAP.}$$

The problem now is, given some pair of words, can we get from one to the other simply using such substitutions? Can we, for example, get from CATERPILLAR to MAN, or, say, from CARPET to MEAT? The answer in the first case happens to be 'yes', while in the second it is 'no'. When the answer is 'yes', the normal way to show this would be simply to exhibit a string of equalities where each word is obtained from the preceding one by use of an allowed relation. Thus (indicating the letters about to be changed in bold type, and the letters which have just been changed in italics):

C**ATE**RPILLAR = C*A*RPILLAR = CARPIL**L*ATE*R** = CAR**PIL*LOW*** =
CAR*PAN* = *ME***A**N = **MEA*TE***N = M***AT***EN = M*A*N.

How can we tell that it is impossible to get from CARPET to MEAT by means of the allowed rules? For this, we need to think a little more, but it is not hard to see, in a variety of different ways. The simplest appears to be the following: in every 'equality' in our initial list, the number of As plus the number of Ws plus the number of Ms is the same on each side. Thus the total number of As, Ws, and Ms cannot change throughout any succession of allowed substitutions. However, for CARPET this number is 1 whereas for MEAT it is 2. Consequently, there is no way of getting from CARPET to MEAT by allowed substitutions.

Notice that when the two words are 'equal' we can show this simply by exhibiting an allowed formal string of symbols, using the rules that we had been given; whereas in the case where they are 'unequal', we had to resort to arguments *about* the rules that we had been given. There is a clear algorithm that we can use to establish 'equality' between words whenever the words *are* in fact 'equal'. All we need do is to make a lexicographical listing of all the possible sequences of words, and then strike from this list any such string for which there is a pair of consecutive words where the second does not follow from the first by an allowed rule. The remaining sequences will provide all the sought-for 'equalities' between words. However, there is no such obvious algorithm, in general, for deciding when two words are *not* 'equal', and we may have to resort to 'intelligence' in order to establish that fact. (Indeed, it took me some while before I noticed the above 'trick' for establishing that CARPET and MEAT are not 'equal'. With another example, quite a different kind of 'trick' might be needed. Intelligence, incidentally, is useful—although not necessary—also for establishing the *existence* of an 'equality'.)

In fact, for the *particular* list of five 'equalities' that constitute the initial list in the above case, it is not unduly difficult to provide an algorithm for ascertaining that two words are 'unequal' when they are indeed 'unequal'. However, in order to *find* the algorithm that works in this case we need to exercise a certain amount of intelligence! Indeed, it turns out that there is no single algorithm which can be used universally for *all* possible choices of initial list. In this sense there is no algorithmic solution to the word problem. The general word problem belongs to non-recursive mathematics!

There are even certain *particular* selections of initial list for which there is no algorithm for deciding when two words are unequal. One such is given by

AH = HA
OH = HO
AT = TA
OT = TO
TAI = IT
HOI = IH
THAT = ITHT

(This list is adapted from one given in 1955 by G. S. Tseitin and Dana Scott; see Gardner 1958, p. 144.) Thus this particular word problem by *itself* is an example of non-recursive mathematics, in the sense that using this particular initial list we cannot algorithmically decide whether or not two given words are 'equal'.

The general word problem arose from considerations of formalized mathematical logic ('formal systems' etc., as we considered earlier). The initial list plays the role of an axiom system and the substitution rule for words, the role of the formal rules of procedure. The proof of non-recursiveness for the word problem arises from such considerations.

As a final example of a problem in mathematics which is non-recursive, let us consider the question of covering the Euclidean plane with polygonal shapes, where we are given a finite number of different such shapes and we ask whether it is possible to cover the plane completely, without gaps or overlaps, using just these shapes and no others. Such an arrangement of shapes is called a *tiling* of the plane. We are familiar with the fact that such tilings are possible using just squares, or just equilateral triangles, or just regular hexagons (as illustrated in Fig. 10.2, Chapter 10), but not using just regular pentagons. Many other single shapes will tile the plane, such as each of the two *irregular* pentagons illustrated in Fig. 4.6. With a *pair* of shapes, the tilings can become more elaborate. Two simple examples are given in Fig. 4.7. All these examples, so far, have the property that they are *periodic*; which means that they are exactly repetitive in two independent directions. In mathematical terms, we say that there is a *period parallelogram*—a parallelogram which, when marked in some way and then repeated again and again in the two directions parallel to its sides, will reproduce the given tiling pattern. An example is shown in Fig. 4.8, where a periodic tiling with a thorn-shaped tile is depicted on the left, and related to a period parallelogram whose periodic tiling is indicated on the right.

Now there are many tilings of the plane which are *not* periodic. Figure 4.9 depicts three non-periodic 'spiral' tilings, with the same thorn-shaped tile as in Fig. 4.8. This particular tile shape is known as a 'versatile' (for obvious reasons!), and it was devised by B. Grünbaum and G. C. Shephard (1981, 1987), apparently based on an earlier shape due to H. Voderberg. Note that the versatile will tile *both* periodically and non-periodically. This property is

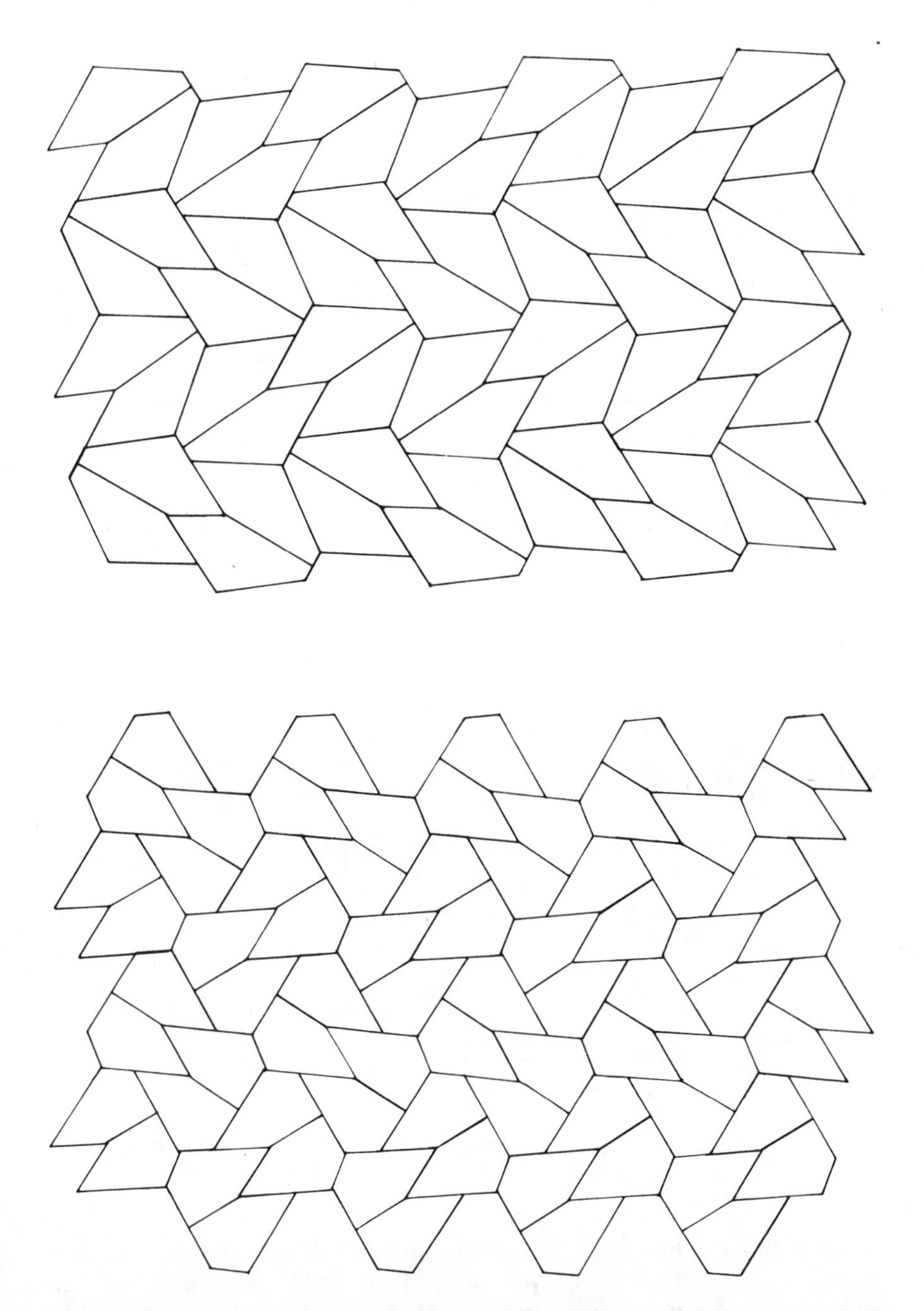

Fig. 4.6. Two examples of periodic tilings of the plane, each using a single shape (found by Majorie Rice in 1976).

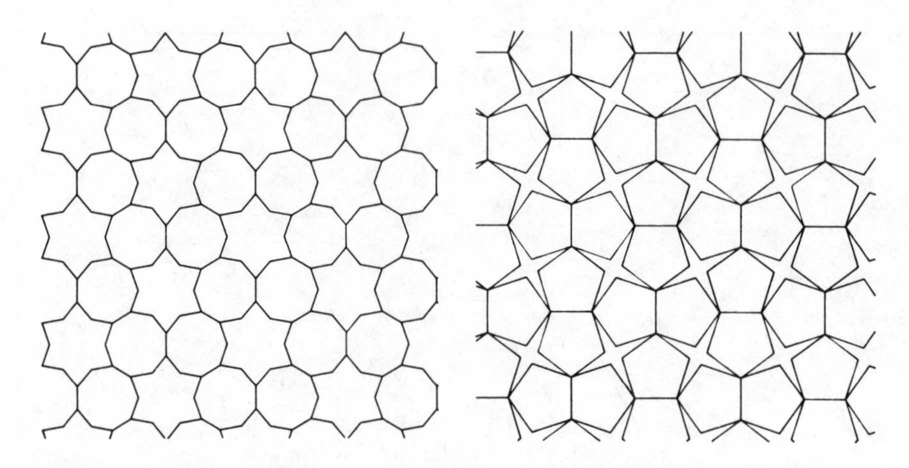

Fig. 4.7. Two examples of periodic tilings of the plane, each using two shapes.

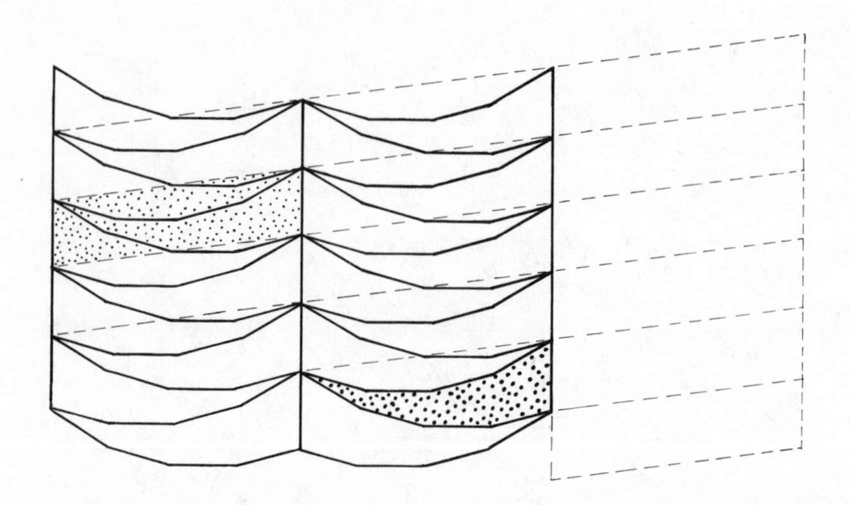

Fig. 4.8. A periodic tiling, illustrated in relation to its period parallelogram.

shared by many other single tile shapes and sets of tile shapes. Are there single tiles or sets of tiles which will tile the plane *only* non-periodically? The answer to this question is 'yes'. In Fig. 4.10, I have depicted a set of six tiles constructed by the American mathematician Raphael Robinson (1971) which will tile the entire plane, but only in a non-periodic way.

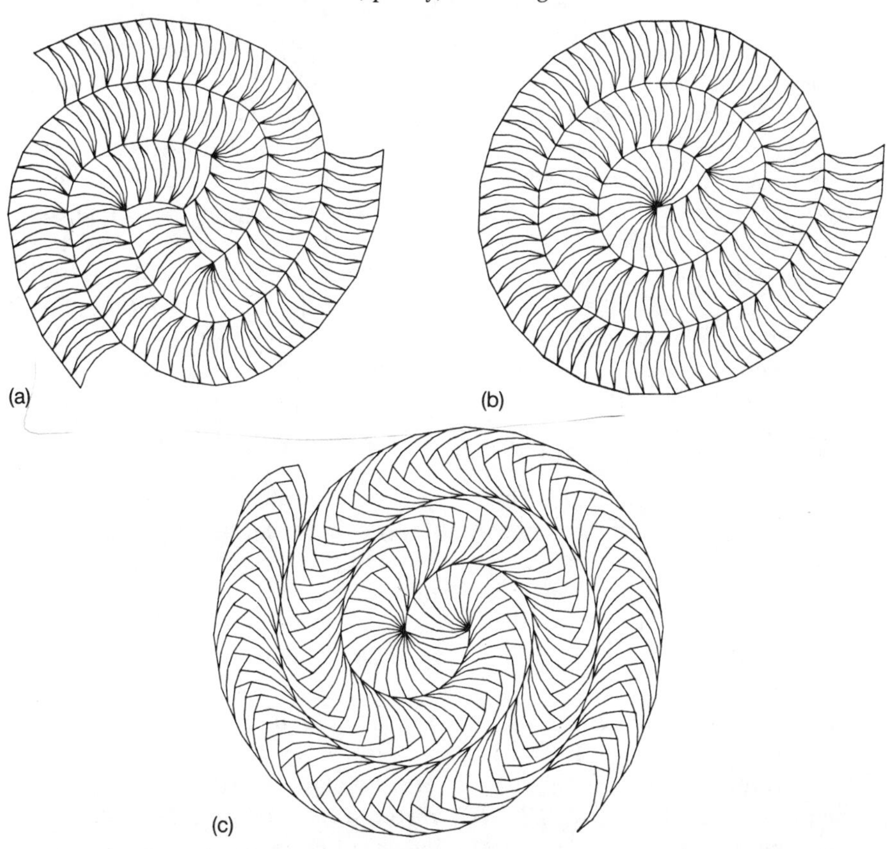

Fig. 4.9. Three non-periodic 'spiral' tilings, using the same 'versatile' shape that was used in Fig. 4.8.

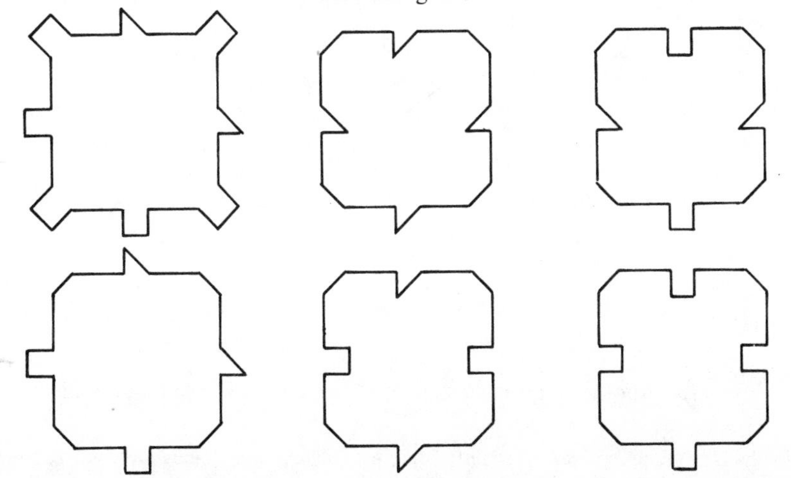

Fig. 4.10. Raphael Robinson's six tiles which tile the plane only non-periodically.

It is worthwhile to go into a little of the history of how this non-periodic set of tiles came about (cf. Grünbaum and Shephard 1987). In 1961 the Chinese-American logician Hao Wang addressed the question of whether or not there is a *decision procedure* for the tiling problem, that is to say, is there an *algorithm* for deciding whether or not a given finite set of different polygonal shapes will tile the entire plane!* He was able to show that there indeed would be such a decision procedure *if* it could be shown that every finite set of distinct tiles which will in some way tile the plane, will in fact also tile the plane periodically. I think that it was probably felt, at that time, that it would be unlikely that a set violating this condition—i.e. an 'aperiodic' set of tiles—could exist. However, in 1966, following some of the leads that Hao Wang had suggested, Robert Berger was able to show that there is in fact *no* decision procedure for the tiling problem: the tiling problem is also part of non-recursive mathematics![11]

Thus it follows from Hao Wang's earlier result that an aperiodic set of tiles must exist, and Berger was indeed able to exhibit the first aperiodic set of tiles. However, owing to the complication of this line of argument, his set involved an inordinately large number of different tiles—originally 20 426. By the use of some additional ingenuity, Berger was able to reduce his number to 104. Then in 1971, Raphael Robinson was able to get the number down to the six depicted in Fig. 4.10.

Another aperiodic set of six tiles is depicted in Fig. 4.11. This set I produced myself in about 1973 following a quite independent line of throught. (I shall return to this matter in Chapter 10 where an array tiled with these shapes is depicted in Fig. 10.3.) After Robinson's aperiodic set of six was brought to my attention, I began thinking about reducing the number; and by

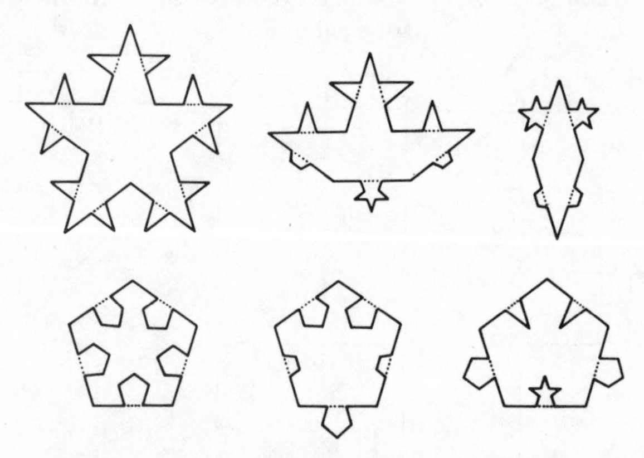

Fig. 4.11. Another set of six tiles which tile the plane only non-periodically.

* Actually, Hao Wang considered a slightly different problem—with square tiles, no rotation, and matching coloured edges, but the distinctions are not important to us here.

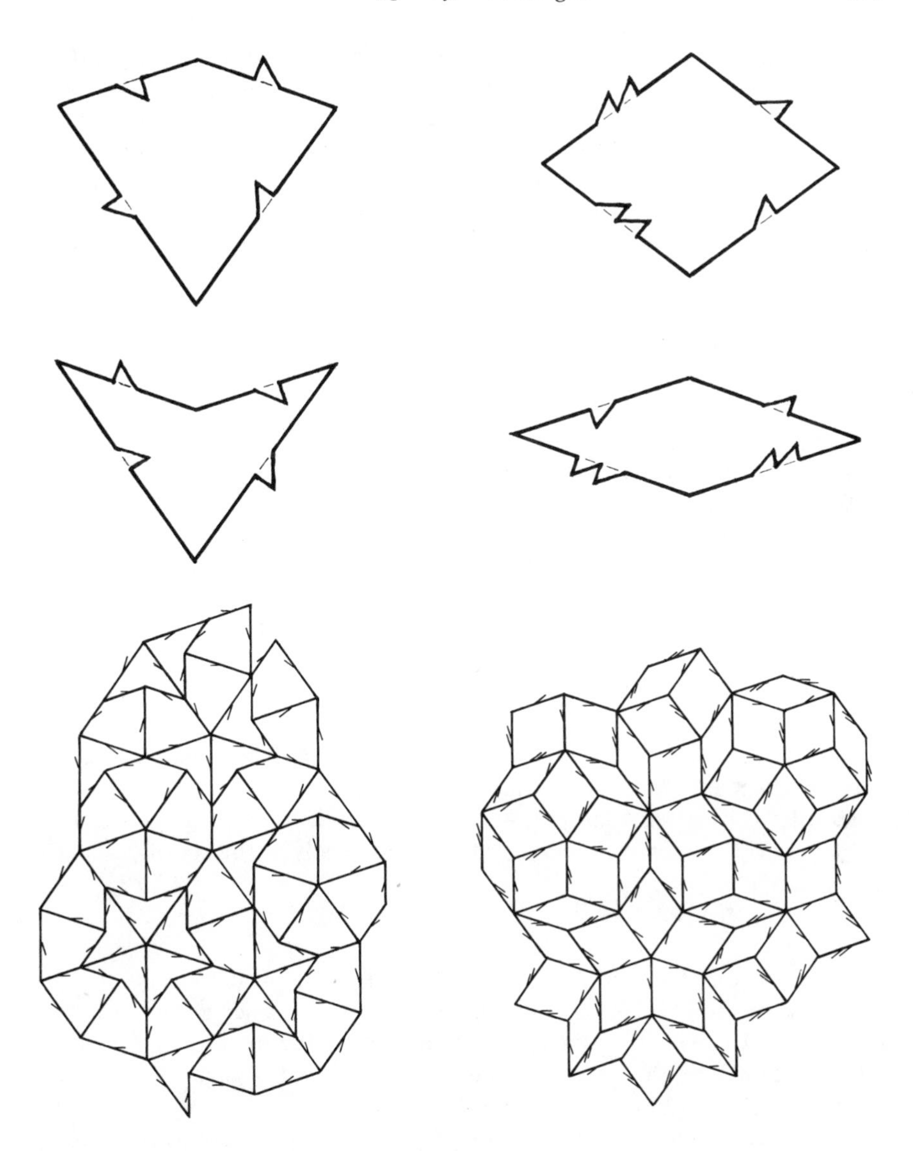

Fig. 4.12. Two pairs, each of which will tile only non-periodically ('Penrose tiles'); and regions of the plane tiled with each pair.

various operations of slicing and re-gluing, I was able to reduce it to two. Two alternative schemes are depticed in Fig. 4.12. The necessarily non-periodic patterns exhibited by the completed tilings have many remarkable properties, including a seemingly crystallographically impossible quasi-periodic structure with fivefold symmetry. I shall return to these matters later.

It is perhaps remarkable that such an apparently 'trivial' area of mathematics—namely covering the plane with congruent shapes—which seems almost like 'child's play' should in fact be part of non-recursive mathematics. In fact there are many difficult and unsolved problems in this area. It is not known, for example, if there is an aperiodic set consisting of a *single* tile.

The tiling problem, as treated by Wang, Berger, and Robinson, used tiles based on squares. I am here allowing polygons of general shape, and one needs some adequately computable way of displaying the individual tiles. One way of doing this would be to give their vertices as points in the Argand plane, and these points may perfectly adequately be given as algebraic numbers.

Is the Mandelbrot set like non-recursive mathematics?

Let us now return to our earlier discussion of the Mandelbrot set. I am going to assume, for the purposes of an illustration, that the Mandelbrot set is, in some appropriate sense, non-recursive. Since its complement is recursively enumerable, this would mean that the set itself would not be recursively enumerable. I think that it is likely that the form of the Mandelbrot set has some lessons to teach us as to the nature of non-recursive sets and non-recursive mathematics.

Let us return to Fig. 3.2, which we encountered early in Chapter 3. Notice that most of the set seems to be taken up with a large heart-shaped region, which I have labelled A, in Fig. 4.13. The shape is referred to as a *cardioid* and its interior region can be defined mathematically as the set of points c of the Argand plane which arise in the form

$$c = z - z^2,$$

where z is a complex number whose distance from the origin is less than 1/2. This set is certainly recursively enumerable in the sense suggested earlier: an algorithm exists which, when applied to a point of the interior of the region, will ascertain that the point is indeed in that interior region. The actual algorithm is easily obtained from the above formula.

Now consider the disc-like region just to the left of the main cardioid (region B in Fig. 4.13). Its interior is the set of points

$$c = z - 1$$

where z has distance from the origin less than 1/4. This region is indeed the interior of a disc—the set of points inside an exact circle. Again this region is recursively enumerable in the above sense. What about the other 'warts' on the cardioid? Consider the two next largest warts. These are the roughly circular blobs which appear approximately at the top and bottom of the

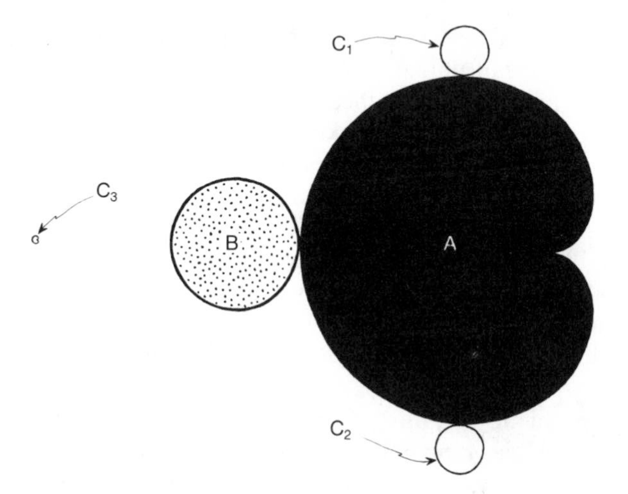

Fig. 4.13. The major parts of the interior of the Mandelbrot set can be defined by simple algorithmic equations.

cardioid in Fig. 3.2 and are marked C_1, C_2 in Fig. 4.13. They can be given in terms of the set

$$c^3 + 2c^2 + (1 - z)\,c + (1 - z)^2 = 0,$$

where now z ranges over the region which has distance 1/8 from the origin. In fact this equation provides us with not merely these two blobs (together) but also with the 'baby' cardioid-like shape which appears off to the left in Fig. 3.2—the main region of Fig. 3.1—and is the region marked C_3 in Fig. 4.13. Again, these regions (together or separately) constitute recursively enumerable sets (in the sense suggested earlier) by virtue of the existence of the above formula.

Despite the suggestion that I have been making that the Mandelbrot set may be non-recursive, we have been able to clean out the largest areas of the set already with some perfectly well-defined and not-too-complicated algorithms. It seems that this process should continue. All the most evident regions in the set—and certainly the overwhelming percentage of its area (if not all of it)—can be dealt with algorithmically. If, as I am supposing, the complete set is actually not recursive, then the regions that cannot be reached by our algorithms must be very delicate and hard to find. Moreover, when we have located such a region, the chances are that we can then see how to improve our algorithms so that those particular regions can also be reached. Yet then there would be *other* such regions (if my supposition of non-recursiveness is correct), hiding yet more deeply in the obscurities of subtlety and complication, that even our improved algorithm cannot reach. Again, by prodigious efforts of insight, ingenuity, and industry we might be able to locate such a region; but there would be yet others that would still escape, and so on.

I think that this is not unlike the way that mathematics often proceeds in areas where the problems are difficult, and presumably non-recursive. The most common problems that one is likely to come across in some specific area can often be handled by simple algorithmic procedures—procedures which may have been known for centuries. But some will escape the net, and more sophisticated procedures are needed to handle them. The ones that still escape would, of course, particularly intrigue the mathematicians and would goad them on to develop ever more powerful methods. These would need to be based upon deeper and deeper insights into the nature of the mathematics involved. Perhaps there is something of this in our understanding of the physical world.

In the word problems and tiling problems considered above, one can begin to catch a glimpse of this kind of thing (although these are not areas where the mathematical machinery has yet developed very far). We were able to use a very simple argument in one particular case to show that a certain word cannot be obtained from another by the allowed rules. It is not hard to imagine that much more sophisticated lines of reasoning can be brought into play to deal with more awkward cases. The likelihood would then be that these new lines of reasoning can be developed into an algorithmic procedure. We know that no one procedure can suffice for all instances of the word problem, but the examples which escape would need to be very carefully and subtly constructed. Indeed, as soon as we *know* how these examples are constructed—as soon as we know for sure that a particular case has eluded our algorithm—then we can improve our algorithm to include that case also. Only pairs of words that are not 'equal' can escape, so as soon as we know that they have escaped, we know they are not 'equal', and that fact can be tagged on to our algorithm. Our improved insight will lead to an improved algorithm!

Complexity theory

The arguments that I have given above, and in the preceding chapters, concerning the nature, existence, and limitations of algorithms have been very much at the 'in principle' level. I have not discussed at all the question of whether the algorithms arising are likely to be in any way practical. Even for problems where it is clear that algorithms exist and how such algorithms can be constructed, it may require much ingenuity and hard work to develop such algorithms into something usable. Sometimes a little insight and ingenuity will lead to considerable reductions in the complexity of an algorithm and sometimes to absolutely enormous improvements in its speed. These questions are often very detailed and technical, and a great deal of work has been done in many different contexts in recent years in the construction, understanding, and improvement of algorithms—a rapidly expanding and developing field of endeavour. It would not be pertinent for me to attempt to enter into a detailed discussion of such questions. However, there are various general things that are known, or conjectured, concerning certain *absolute* limitations on how much the speed of an algorithm can be increased. It turns out that even among mathematical problems that *are* algorithmic in nature, there are some classes of problem that are intrinsically vastly more difficult to solve algorithmically than others. The difficult ones can be solved only by very slow algorithms (or, perhaps, with algorithms which require an inordinately large amount of storage space, etc.). The theory which is concerned with questions of this kind is called *complexity theory*.

Complexity theory is concerned not so much with the difficulty of solving *single* problems algorithmically, but with infinite families of problems where there would be a general algorithm for finding answers to all the problems of one single family. The different problems in the family would have different 'sizes', where the size of a problem is measured by some natural number n. (I shall have more to say in a moment about how this number n actually characterizes the size of the problem.) The length of time—or more correctly, the number of elementary steps—that the algorithm would need for each particular problem of the class would be some natural number N which depends on n. To be a little more precise, let us say that among *all* of the problems of some particular size n, the greatest number of steps that the algorithm takes is N. Now, as n gets larger and larger, the number N is likely to get larger and larger too. In fact, N is likely to get large very much more rapidly than n. For example, N might be approximately proportional to n^2, or to n^3 or perhaps to 2^n (which, for large n, is very much greater than each of n, n^2, n^3, n^4, and n^5—greater, indeed, than n^r for every fixed number r), or N might even be approximately proportional to, say, 2^{2^n} (which is much greater still).

Of course, the number of 'steps' might depend on the type of computing machine on which the algorithm is to be run. If the computing machine is a Turing machine of the type described in Chapter 2, where there is just a single tape—which is rather inefficient—then the number N might increase more rapidly (i.e. the machine might run more slowly) than if two or more tapes are allowed. To avoid uncertainties of this kind, a broad categorization is made of the possible ways in which N gets large as a function of n, so that no matter what type of Turing machine is used, the measure of the rate of increase of N will always fall into the same category. One such category, referred to as **P** (which stands for 'polynomial time'), includes all rates which are, at most, fixed multiples* of one of n, n^2, n^3, n^4, n^5, ... That is to say, for any

* A 'polynomial' would really refer to a more general expression, like $7n^4 - 3n^3 + 6n + 15$, but

problem lying in the category **P** (where by a 'problem' I really mean a family of problems with a general algorithm for solving them), we have

$$N \leq K \times n^r,$$

the numbers K and r being *constants* (independent of r). This means that N is no larger than some multiple of n raised to some fixed power.

A simple type of problem that certainly belongs to **P** is that of multiplying two numbers together. To explain this, I should first describe how the number n characterizes the size of the particular pair of numbers that are to be multiplied. We can imagine that each number is written in the binary notation and that $n/2$ is simply the number of binary digits of each number, giving a *total* of n binary digits—i.e. n *bits*—in all. (If one of the numbers is longer than the other, we can simply start off the shorter one with a succession of zeros to make it up to the length of the longer one.) For example, if $n = 14$, we could be considering

$$1011010 \times 0011011$$

(which is 1011010×11011, but with added zeros on the shorter figure). The most direct way of carrying out this multiplication is just to write it out:

$$\begin{array}{r}
1011010 \\
\times\, 0011011 \\
\hline
1011010 \\
1011010 \\
0000000 \\
1011010 \\
1011010 \\
0000000 \\
0000000 \\
\hline
0100101111110
\end{array}$$

recalling that, in the binary system, $0 \times 0 = 0$, $0 \times 1 = 0$, $1 \times 0 = 0$, $1 \times 1 = 1$, $0 + 0 = 0$, $0 + 1 = 1$, $1 + 0 = 1$, $1 + 1 = 10$. The number of individual binary multiplications is $(n/2) \times (n/2) = n^2/4$, and there can be up to $(n^2/4) - (n/2)$ individual binary additions (including carrying). This makes $(n^2/2) - (n/2)$ individual arithmetical operations in all—and we should include a few extra for the logical steps involved in the carrying. The total number of steps is essentially $N = n^2/2$ (ignoring the lower order terms) which is certainly polynomial[12].

For a class of problems in general, we take the measure n of the 'size' of the problem to be the *total number of binary digits* (or *bits*) needed to specify the

this gives us no more generality. For any such expression, all the terms involving lower powers of n become unimportant when n becomes large (so in this particular example we can ignore all the terms except $7n^4$).

free data of the problem of that particular size. This means that, for given n, there will be up to 2^n different instances of the problem for that given size (because each digit can be one of two possibilities, 0 or 1, and there are n digits in all), and these have to be coped with uniformly by the algorithm, in not more than N steps.

There are many examples of (classes of) problems which are *not* in **P**. For example, in order to perform the operation of computing 2^{2^r} from the natural number r we would need about 2^n steps even just to *write down the answer*, let alone to perform the calculation, n being the number of binary digits in the binary representation of r. The operation of computing $2^{2^{2^r}}$ takes something like 2^{2^r} steps just to write down, etc.! These are much bigger than polynomials and so certainly not in **P**!

More interesting are problems where the answers can be written down, and even checked for correctness, in polynomial time. There is an important category of (algorithmically soluble classes of) problems characterized by this property. They are referred to as **NP** (classes of) problems. More precisely, if an individual problem of a class of problems in **NP** has a solution, then the algorithm will give that solution, and it must be possible to check in polynomial time that the proposed solution is indeed a solution. In the cases where the problem has no solution, the algorithm will say so, but one is not required to check—in polynomial time or otherwise—that there is indeed no solution.

NP problems arise in many contexts, both within mathematics itself and in the practical world. I shall just give one simple mathematical example: the problem of finding what is called a '*Hamiltonian circuit*' in a graph (a rather daunting name for an extremely simple idea). By a 'graph' is meant a finite collection of points, or 'vertices', a certain number of pairs of which are connected together by lines—called the 'edges' of the graph. (We are not interested in geometrical or 'distance' properties here, but only in which vertex is connected to which. Thus it does not really matter whether the vertices are all represented on one plane—assuming that we don't mind the edges crossing over one another—or in three-dimensional space.) A Hamiltonian circuit is simply a closed-up route (or loop) consisting solely of edges of the graph, and which passes exactly once through each vertex. An example of a graph, with a Hamiltonian circuit drawing on it, is depicted in Fig. 4.14. The Hamiltonian circuit problem is to decide, for any given graph, whether or not a Hamiltonian circuit exists, and to display one explicitly whenever one does exist.

There are various ways of presenting a graph in terms of binary digits. It does not matter a great deal which method is used. One procedure would be to number the vertices 1, 2, 3, 4, 5, . . ., and then to list the pairs in some appropriate fixed order:

$$(1,2), (1,3), (2,3), (1,4), (2,4), (3,4), (1,5), (2,5), (3,5), (4,5), (1,6), \ldots$$

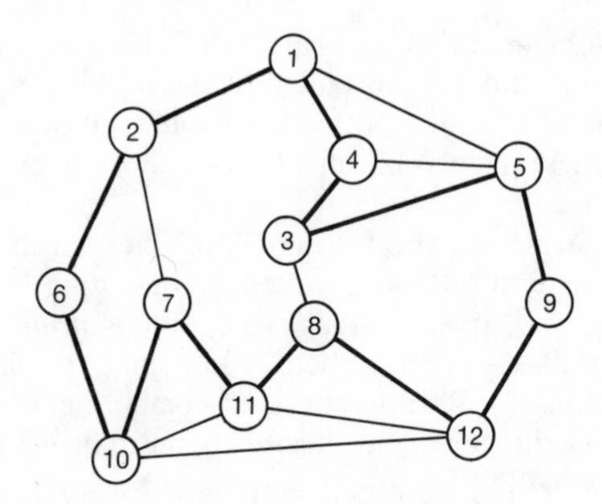

Fig. 4.14. A graph with a Hamiltonian circuit indicated (slightly darker lines). There is just one other Hamiltonian circuit, which the reader may care to locate.

Then we make an exactly matching list of '0's and '1's where we put a '1' whenever the pair corresponds to an edge of the graph and a '0' whenever it does not. Thus the binary sequence

$$10010110110\ldots$$

would designate that vertex 1 is joined to vertex 2, to vertex 4, and to vertex 5, . . . vertex 3 is joined to vertex 4 and to vertex 5, . . . , vertex 4 is joined to vertex 5, . . etc. (as in Fig. 4.14). The Hamiltonian circuit could be given, if desired, just as a sub-collection of these edges, which would be described as a binary sequence with many more zeros than before. The checking procedure is something that can be achieved much more rapidly than finding the Hamiltonian circuit in the first place. All that one needs to do is check that the proposed circuit is indeed a circuit, that its edges indeed belong to those of the original graph, and that each vertex of the graph is used exactly twice—once at the end of each of two edges. This checking procedure is something that can be easily achieved in polynomial time.

In fact this problem is not only **NP**, but what is known as **NP**-*complete*. This means that any other **NP** problem can be converted into it in polynomial time—so that if someone were clever enough to find an algorithm which solves the Hamiltonian circuit problem in *polynomial* time, i.e. to show that the Hamiltonian circuit problem is actually in **P**, then it would follow that *all* **NP** problems are actually in **P**! Such a thing would have momentous implications. In a general way, problems which are in **P** are regarded as being '*tractable*' (i.e. 'soluble in an acceptable length of time'), for reasonably large n, on a fast modern computer, while problems in **NP** which are not in **P** are regarded as being '*intractable*' (i.e. though soluble in principle, they are 'insoluble in practice') for reasonably large n—no matter what increases in operational computer speed, of any forseeable kind, are envisaged. (The actual time that would be taken, for large *n*, rapidly becomes longer than the age of the universe for an **NP** problem not in **P**, which is not much use for a practical problem!) Any clever algorithm for solving the Hamiltonian circuit problem in polynomial time could be converted into an algorithm for solving *any* other **NP** problem whatever, in polynomial time!

Another problem which is **NP**-complete is the 'travelling salesman problem', which is rather like the Hamiltonian circuit problem except that the various edges have numbers attached to them, and one seeks that Hamiltonian circuit for which the sum of the numbers (the 'distance' travelled by the salesman) is a *minimum*. Again a polynomial time solution of the travelling salesman problem would lead to polynomial time solutions to *all* other **NP** problems. (If such a solution were found, it would make headline news! For, in particular, there are the secret code systems, that have been introduced over the past several years, which depend on a problem of factorization of large integers, this being another **NP** problem. If that problem could be solved in polynomial time, then such codes could probably be cracked by the use of powerful modern computers, but if not, then the codes appear to be safe. See Gardner 1989.)

It is commonly believed by the experts that it is actually *impossible*, with any Turing machine-like device, to solve an **NP**-complete problem in polynomial time, and that, consequently, **P** and **NP** are *not* the same. Very likely this belief is correct, but as yet no-one has been able to prove it. This remains the most important unsolved problem of complexity theory.

Complexity and computability in physical things

Complexity theory is important for our considerations in this book because it raises another issue, somewhat separate from the question of whether or not things are algorithmic: namely, whether or not things that are known to be algorithmic are actually algorithmic in a *useful* way. In the later chapters, I shall have less to say concerning matters of complexity theory than of computability. For I am inclined to think (though, no doubt, on quite inadequate grounds) that unlike the basic question of computability itself, the issues of complexity theory are not quite the central ones in relation to mental phenomena. Moreover, I feel that the questions of practicality of algorithms are being only barely touched by complexity theory as it stands today.

However, I could well be wrong about the role of complexity. As I shall remark later (in Chapter 9, p. 402), the complexity theory for *actual physical objects* could perhaps be different in significant ways from that which we have just been discussing. For this possible difference to become manifest, it would be necessary to harness some of the magical properties of quantum mechanics—a mysterious yet powerfully accurate theory of the behaviour of atoms and molecules, and of many other phenomena, some of which are important on a much larger scale. We shall be coming to terms with this theory in Chapter 6. According to a recent set of ideas introduced by David Deutsch (1985), it is possible *in principle* to construct a 'quantum computer' for which there are (classes of) problems which are not in **P**, yet which could be solved by that device in polynomial time. It is not at all clear, as of now, how an actual physical device could be constructed which behaves (reliably) as a quantum computer—and, moreover, the particular class of problem so far considered is decidedly artificial—but the *theoretical* possibility that a quantum physical device may be able to improve on a Turing machine seems to be with us.

Can it be that a human brain, which I am taking for this discussion to be a 'physical device', albeit one of amazing subtlety and delicacy of design, as well as of complication, is itself taking advantage of the magic of quantum theory? Do we yet understand the ways in which quantum effects might be used beneficially in the solving of problems or the forming of judgements? Is it conceivable that we might have to go even 'beyond' present-day quantum theory to make use of such possible advantages? Is it really likely that actual physical devices might be able to improve upon the complexity theory for Turing machines? And what about the *computability* theory for actual physical devices?

To address such questions we must turn away from matters purely mathematical and ask, in the following chapters, how the physical world actually behaves!

1. In considering sets whose members may again be sets we must be careful to distinguish between the members of that set and the members of the *members* of that set. For example, suppose S is the set of *non-empty subsets* of a certain other set T, where the members of T are one apple and one orange. T has the property of 'twoness' not 'threeness', but S actually has the property 'threeness'; for the three members of S are: a set containing just one apple, a set containing just one orange, and a set containing one apple and one orange—three sets in all, these being the *three* members of S. Likewise, the set whose only member is the *empty set* possesses the property of 'oneness' not of 'zeroness'—it has *one* member, namely the empty set! The empty set *itself* has zero members, of course.
2. In fact the reasoning in Gödel's theorem can be presented in such a way that it does not depend upon a full external concept of 'truth' for propositions such as $P_k(k)$. However, it still depends upon an interpretation of the actual 'meaning' of *some* of the symbols: in particular that '$\sim\exists$' really *means* 'there is no (natural number) . . . such that . . .'.
3. In the following, small letters represent natural numbers and capital ones, finite sets of natural numbers. Let $m \to [n, k, r]$ stand for the statement 'If X is any m-element set of natural numbers whose k-element subsets are assigned to r

boxes, then there is a 'large' n-element subset Y of X such that, all k-element subsets of Y go into the same box.' Here 'large' means that Y has more elements than that natural number which is the smallest element of Y. Consider the proposition: 'For any choice of k, r, and n there exists an m_0 such that, for all m greater than m_0, the statement $m \rightarrow [n, k, r]$ always holds true'. This proposition has been shown by J. Paris and L. Harrington (1977) to be equivalent to a Gödel-type proposition for the standard (Peano) axioms for arithmetic, unprovable from those axioms, yet asserting something about those axioms which is 'obviously true' (namely, in this case, that propositions deducible from the axioms are themselves true).

4. The title was 'Systems of logic based on ordinals', and some readers will be familiar with the notation for Cantor's *ordinal numbers* that I have been employing in the subscripts. The hierarchy of logical systems that one obtains by the procedure that I have described above is characterized by *computable ordinal numbers*.

   There are some mathematical theorems that are quite natural and easy to state which, if one attempted to prove them using the standard (Peano) rules of arithmetic, would require using the above 'Gödelization' procedure to an outrageously huge degree (extending the procedure enormously beyond what I have outlined above). The mathematical proofs of these theorems, are not at all of the kind which depend on any vague or questionable reasoning that would seem to lie outside the procedures of normal mathematical argument. See Smorynski (1983).

5. The continuum hypothesis that was referred to in chapter 3, p. 85 (and which states that $\boldsymbol{C} = \aleph_0$) is the most 'extreme' mathematical statement that we have encountered here (although much more extreme statements than this are often considered). The continuum hypothesis is additionally interesting because Gödel himself, together with Paul J. Cohen, established that the continuum hypothesis is actually *independent* of the standard axioms and rules of procedure of set theory. Thus, one's attitude to the status of the continuum hypothesis distinguishes between the formalist and Platonist point of view. To a formalist, the continuum hypothesis is 'undecidable' since it cannot be established or refuted using the standard (Zermelo–Frankel) formal system, and it is 'meaningless' to call it either 'true' or 'false'. However, to a good Platonist, the continuum hypothesis is indeed either true or false, but to establish which is the case will require some new forms of reasoning—actually going beyond even employing Gödel-type propositions for the Zermelo–Frankel formal system. (Cohen (1966), himself suggested a reflection principle which makes the continuum hypothesis 'obviously false'!)

6. For a vivid and fairly non-technical account of these matters, see Rucker (1984).

7. Brouwer himself seems to have started on this line of thought partly because of nagging worries about a 'non-constructiveness' in his proof of one of his own theorems, 'the Brouwer fixed point theorem' of topology. The theorem asserts that if you take a disc—that is, a circle together with its interior—and move it in a continuous way to inside the region where it was originally located, then there is at least one point of the disc—called a fixed point—which ends up exactly where it started. One may have no idea exactly where this point is, or whether there might be many such points, it is merely the *existence* of some such point that the theorem asserts. (As mathematical existence theorems go, this is actually a fairly 'construc-

tive' one. Of a different order of non-constructiveness are existence theorems which depend on what is known as the 'Axiom of Choice' or 'Zorn's lemma' (cf. Cohen 1966, Rucker 1984). In Brouwer's case, the difficulty is similar to the following: if f is a real-valued continuous function of a real variable which takes both positive and negative values, find a place where f vanishes. The usual procedure involves repeatedly bisecting the interval where f changes sign, but it may not be 'constructive' in Brouwer's required sense, to decide whether the intermediate values of f are positive, negative, or zero.

8. We enumerate the sets $\{v, w, x, \ldots, z\}$, where v represents the function f according to some lexicographical scheme. We check (recursively) at each stage to see whether $f(w, x, \ldots, z) = 0$ and retain the proposition $\exists w, x, \ldots z[f(w, x, \ldots z) = 0]$ only if so.
9. There is a new theory of computability for real-valued functions of real numbers (as opposed to the conventional natural-number-valued functions of natural numbers), due to Blum, Shub, and Smale (1989), the details of which have come to my attention only very recently. This theory will also apply to complex-valued functions, and it could have a significant bearing on the issues raised in the text. Some of the results of this new work lend strong support to the conjecture that the Mandelbrot set is indeed non-recursive according to an appropriate definition.
10. This particular problem is more correctly called 'the word problem for semi-groups'. There are also other forms of the word problem, where the rules are slightly different. These will not concern us here.
11. Hanf (1974) and Myers (1974) have shown, moreover, that there is a single set of (a great number of tiles) which will tile the plane only in a *non-computable* way.
12. In fact, by the use of ingenuity, this number of steps can be reduced to something of the order of $n \log n \log \log n$ for large n—which is, of course, still in **P**. See Knuth (1981) for futher information about such matters.

5
The classical world

The status of physical theory

What need we know of the workings of Nature in order to appreciate how consciousness may be part of it? Does it really matter what are the laws that govern the constituent elements of bodies and brains? If our conscious perceptions are merely the enacting of algorithms, as many AI supporters would have us believe, then it would not be of much relevance what these laws actually are. Any device which *is* capable of acting out an algorithm would be as good as any other. Perhaps, on the other hand, there is more to our feelings of awareness than mere algorithms. Perhaps the detailed way in which we are constituted is indeed of relevance, as are the precise physical laws that actually govern the substance of which we are composed. Perhaps we shall need to understand whatever profound quality it is that underlies the very nature of matter, and decrees the way in which all matter must behave. Physics is not yet at such a point. There are many mysteries to be unravelled and many deep insights yet to be gained. Yet, most physicists and physiologists would judge that we *already* know enough about those physical laws that are relevant to the workings of such an ordinary-sized object as a human brain. While it is undoubtedly the case that the brain is exceptionally complicated as a physical system, and a vast amount about its detailed structure and relevant operation is not yet known, few would claim that it is in the *physical* principles underlying its behaviour that there is any significant lack of understanding.

I shall later argue an unconventional case that, on the contrary, we do *not* yet understand physics sufficiently well that the functioning of our brains can be adequately described in terms of it, even in principle. To make this case, it will be necessary for me first to provide some overview of the status of present physical theory. This chapter is concerned with what is called 'classical physics', which includes both Newton's mechanics and Einstein's relativity. 'Classical', here, means essentially the theories that held sway before the arrival, in about 1925 (through the inspired work of such physicists as Planck, Einstein, Bohr, Heisenberg, Schrödinger, de Broglie, Born, Jordan, Pauli, and Dirac), of *quantum theory*—a theory of uncertainty, indeterminism, and mystery, describing the behaviour of molecules, atoms, and subatomic particles. Classical theory is, on the other hand, *deterministic*, so the future is always completely fixed by the past. Even so, classical physics has much about it that is mysterious, despite the fact that the understanding that has been achieved over the centuries has led us to a picture of quite phenomenal accuracy. We shall also have to examine the quantum theory (in Chapter 6), for I believe that, contrary to what appears to be a majority view among physiologists, quantum phenomena *are* likely to be of importance in the operation of the brain—but that is a matter for chapters to follow.

What science has so far achieved has been dramatic. We have only to look about us to witness the extraordinary power that our understandings of Nature have helped us to obtain. The technology of the modern world has derived, in good measure, from a great wealth of empirical experience. However, it is physical *theory* that underlies our technology in a much more fundamental way, and it is physical theory that we shall be concerned with here. The theories that are now available to us have an accuracy which is quite remarkable. But it is not just their accuracy that has been their strength. It is also the fact that they have been found to be extraordinarily amenable to a precise and detailed mathematical treatment. It is these facts together that have yielded us a science of truly impressive power.

A good deal of this physical theory is not particularly recent. If one event can be singled out above all others, it is the publication, in 1687, of the *Principia* of Isaac Newton. This momentous work demonstrated how, from a few basic physical principles, one may comprehend, and often predict with striking accuracy, a great deal of how physical objects actually behave. (Much of the *Principia* was also concerned with remarkable developments in mathematical technique, though more practical methods were provided later by Euler and others.) Newton's own work, as he readily admitted, owed much to the achievements of earlier thinkers, the names of Galileo Galilei, René Descartes, and Johannes Kepler being pre-eminent among these. Yet there were important underlying concepts from more ancient thinkers still, such as the geometrical ideas of Plato, Eudoxos, Euclid, Archimedes, and Appolonios. I shall have more to say about these later.

Departures from the basic scheme of Newton's dynamics were to come later. First, there was the electromagnetic theory of James Clerk Maxwell, developed in the mid-nineteenth century. This encompassed not only the classical behaviour of electric and magnetic fields, but also that of light.[1] This remarkable theory will be the subject of our attentions later in this chapter. Maxwell's theory is of considerable importance to present-day technology, and there is no doubt that electromagnetic phenomena have relevance to the workings of our brains. What is less clear, however, is that there can be any significance for our thinking processes in the two great theories of relativity associated with the name of Albert Einstein. The *special* theory of relativity, which developed from a study of Maxwell's equations, was put forward by Henri Poincaré, Hendrick Antoon Lorentz, and Einstein (and later given an elegant geometrical description by Hermann Minkowski) to explain the puzzling behaviour of bodies when they move at speeds close to that of light. Einstein's famous equation '$E = mc^2$' was part of this theory. But the theory's impact on technology has been very slight so far (except where it impinges on nuclear physics), and its relevance to the workings of our brains would seem to be peripheral at best. On the other hand, special relativity tells us something deep about physical reality, in relation to the nature of *time*. We shall see in the next chapters that this leads to some profound puzzles concerning quantum theory which could have importance in relation to our perceived 'flow of time'. Moreover, we shall need to understand the special theory before we can properly appreciate Einstein's *general* theory of relativity—the theory which uses curved space–time to describe gravity. The impact of *this* theory on technology has been almost non-existent* so far, and it would seem fanciful in the extreme to suggest any relevance to the workings of our brains! But, remarkably, it is indeed the *general* theory that will have the greater relevance to our later deliberations, most particularly in Chapters 7 and 8, where we shall need to venture out to the farthest reaches of space and time in order to glean something of the changes that I claim are necessary before a properly coherent picture of quantum theory can come to light—but more of that later!

These are the broad areas of *classical* physics. What of quantum physics? Unlike relativity theory, quantum theory *is* beginning to have a really significant impact on technology. This is partly owing to the understandings that it has provided, in certain technologically important areas such as chemistry and metallurgy. Indeed, some would say that these areas have actually become subsumed into physics, by virtue of the detailed new insights that the quantum theory has given us. In addition, there are quite *new* phenomena that quantum theory has provided us with, the most familiar of which being, I suppose, the laser. Might not some essential aspects of quantum theory also be playing crucial roles in the physics that underlies our thought processes?

What about physical understandings of more recent origin? Some readers may have come across excitedly-expressed ideas, involving such names as 'quarks' (cf. p. 154), 'GUT' (Grand Unified Theories), the 'inflationary scenario' (see end-note 12 on p. 347), 'supersymmetry', '(super)string theory', etc. How do such new schemes compare with those that I have just been referring to? Shall we need to know about them also? I believe that in

* Almost, but not quite; the accuracy required for the behaviour of space probes actually requires that their orbits be calculated taking into account the effects of general relativity—and there are devices capable of locating one's position on the earth so accurately (to within a few feet, in fact) that the space–time curvature effects of general relativity are actually needed!

order to put things in a more appropriate perspective, I should formulate three broad categories of basic physical theory. I label these as follows:

1. SUPERB,
2. USEFUL,
3. TENTATIVE.

Into the SUPERB category must go all those that I have been discussing in the paragraphs preceding this one. To qualify as SUPERB, I do not deem it necessary that the theory should apply without refutation to the phenomena of the world, but I do require that the range and accuracy with which it applies should, in some appropriate sense, be *phenomenal*. The way that I am using the term 'superb', it is an extraordinary remarkable fact that there are any theories in this category at all! I am not aware of any basic theory in any other science which could properly enter this category. Perhaps the theory of natural selection, as proposed by Darwin and Wallace, comes closest, but it is still some good way off.

The most ancient of the SUPERB theories is the Euclidean geometry that we learn something of at school. The ancients may not have regarded it as a physical theory at all, but that is indeed what it was: a sublime and superbly accurate theory of physical space—and of the geometry of rigid bodies. Why do I refer to Euclidean geometry as a *physical* theory rather than a branch of mathematics? Ironically, one of the clearest reasons for taking that view is that we now know that Euclidean geometry is *not entirely accurate* as a description of the physical space that we actually inhabit! Einstein's general relativity now tells us that space(-time) is actually 'curved' (i.e. *not* exactly Euclidean) in the presence of a gravitational field. But that fact does not detract from Euclidean geometry's characterization as SUPERB. Over a metre's range, deviations from Euclidean flatness are tiny indeed, errors in treating the geometry as Euclidean amounting to less than the diameter of an atom of hydrogen!

It is reasonable to say that the theory of *statics* (which concerns bodies not in motion), as developed into a beautiful science by Archimedes, Pappos, and Stevin, would also have qualified as SUPERB. This theory is now subsumed by Newtonian mechanics. The profound ideas of *dynamics* (bodies in motion) introduced by Galileo in around 1600, and developed into a magnificent and comprehensive theory by Newton, must undoubtedly come into the SUPERB category. As applied to the motions of planets and moons, the observed accuracy of this theory is phenomenal—better than one part in ten million. The same Newtonian scheme applies here on earth—and out among the stars and galaxies—to some comparable accuracy. Maxwell's theory, likewise, is accurately valid over an extraordinary range, reaching inwards to the tiny scale of atoms and subatomic particles, and outwards, also, to that of galaxies, some million million million million million million times larger! (At the very small end of the scale, Maxwell's equations must be combined appropriately with the rules of quantum mechanics.) It surely also must qualify as SUPERB.

Einstein's special relativity (anticipated by Poincaré and elegantly reformulated by Minkowski) gives a wonderfully accurate description of phenomena in which the speeds of objects are allowed to come close to that of light—speeds at which Newton's descriptions at last begin to falter. Einstein's supremely beautiful and original theory of general relativity generalizes Newton's dynamical theory (of gravity) and improves upon its accuracy, inheriting all the remarkable precision of that theory concerning the motions of planets and moons. In addition, it explains various detailed observational facts which are incompatible with the older Newtonian scheme. One of these (the 'binary pulsar', cf. p.211) shows Einstein's theory to be accurate to about one part in 10^{14}. Both relativity theories—the second of which subsumes the first—must indeed be classified as SUPERB (for reasons of their mathematical elegance, almost as much as of their accuracy).

The range of phenomena which are explained according to the strangely beautiful and revolutionary theory of quantum mechanics, and the accuracy with which it agrees with experiment, clearly tells us that quantum theory, also, must certainly qualify as SUPERB. No observational discrepancies with that theory are known—yet its strength goes far beyond this, in the number of hitherto inexplicable phenomena that the theory now explains. The laws of chemistry, the stability of atoms, the sharpness of spectral lines (cf. p. 228) and their very specific observed patterns, the curious phenomenon of superconductivity (zero electrical resistance), and the behaviour of lasers are just a few amongst these.

I am setting high standards for the category SUPERB, but this is what we have become accustomed to in physics. Now, what about the more recent theories? In my opinion there is only one of them which can possible qualify as SUPERB and this is not a particularly recent one: a theory called *quantum electrodynamics* (or QED), which emerged from the work of Jordan, Heisenberg, and Pauli, was formulated by Dirac in 1926–1934, and made workable by Bethe, Feynman, Schwinger, and Tomonaga in 1947–1948. This theory arose as a combination of the principles of quantum mechanics with special relativity, incorporating Maxwell's equations and a fundamental equation governing the motion and spin of electrons, due to Dirac. The theory as a whole does not have the compelling elegance or consistency of the earlier SUPERB theories, but it qualifies by virtue of its truly phenomenal accuracy. A particularly noteworthy implication is the value of the magnetic moment of an electron. (Electrons behave like tiny magnets of spinning electric charge. The term 'magnetic moment' refers to the strength of this tiny magnet.) The value 1.001 159 652 46 (in appropriate units—with an allowance for error of about 20 in the last two digits) is what is calculated from QED for this magnetic moment, whereas the most recent experimental value is 1.001 159 652 193 (with a possible error of about 10 in the last two digits). As Feynman has pointed out, this kind of accuracy could determine the distance between New York and Los Angeles to within the width of a human hair! We shall not need to know about this theory here, but for completeness, I shall briefly mention some of its essential features towards the end of the next chapter.*

There are some current theories that I would place in the USEFUL category. Two of these we shall not need here, but they are worthy of a mention. The first is the Gell–Mann–Zweig *quark* model for the subatomic particles called *hadrons* (the protons, neutrons, mesons, etc. which constitute atomic nuclei—or, more correctly, the 'strongly interacting' particles) and the detailed (later) theory of their interactions, referred to as *quantum chromodynamics*, or QCD. The idea is that all hadrons are made up of constituents known as 'quarks' which interact with one another by a certain generalization of Maxwell's theory (called 'Yang–Mills theory'). Second, there is a theory (due to Glashow, Salam, Ward, and Weinberg—again using Yang–Mills theory) which combines electromagnetic forces with the 'weak' interactions that are responsible for radioactive decay. This theory incorporates a description of the so-called *leptons* (electrons, muons, neutrinos, W- and Z-particles—the 'weakly interacting' particles). There is some good experimental support for both of theories. However, they are, for various reasons, rather more untidy than one would wish (as is QED, but these are more so), and their observed accuracy and predictive power, at present, falls a very long way short of the 'phenomenal' standard required for their inclusion in the SUPERB category. These two theories together (the second including QED) are sometimes referred to as the *standard model*.

Finally, there is a theory of another type which I believe also belongs at least to the USEFUL category. This is what is called the theory of the *big bang* origin of the universe.† This theory will have an important role to play in the discussions of Chapters 7 and 8.

I do not think that anything else makes it into the USEFUL[2] category. There are many currently (or recently) popular ideas. The names of some of them are: 'Kaluza–Klein' theories, 'supersymmetry' (or 'supergravity'), and the still extremely fashionable 'string' (or 'superstring') theories, in addition to the 'GUT' theories (and certain ideas derived from them, such as the 'inflationary scenario', cf. note 12 on p. 347). All these are, in my opinion, firmly in the category TENTATIVE. (See Barrow 1988, Close 1983, Davies and Brown 1988, Squires 1985.) The important distinction between the

* See Feynman's (1985) book *QED* for a popular account of this theory.

† I refer here to what is known as the 'standard model' of the big bang. There are many variants of the big bang theory, the most currently popular providing what is known as the 'inflationary scenario'—firmly in the category TENTATIVE in my opinion!

categories USEFUL and TENTATIVE is the lack of any signficant experimental support for the theories in the latter category.[3] This is not to say that perhaps one of them might not be raised, dramatically, into the USEFUL or even SUPERB categories. Some of these theories indeed contain original ideas of some notable promise, but they remain ideas, as of now, without experimental support. The TENTATIVE category is a very broad-ranging one. The ideas involved in some of them could well contain the seeds of a new substantial advance in understanding, while some of the others strike me as being definitely misguided or contrived. (I was tempted to split off a fourth category from the respectable TENTATIVE one, and refer to it as, say, MISGUIDED—but then I thought better of it, since I do not want to lose half of my friends!)

One should not be surprised that the main SUPERB theories are ancient ones. Throughout history there must have been very many more theories that would have fallen in the TENTATIVE category, but most of these have been forgotten. Likewise, in the USEFUL category must have been many which have since faded; but there are also some which have been subsumed by theories which later came into their own as SUPERB. Let us consider a few examples. Before Copernicus, Kepler, and Newton produced a much better scheme, there was a wonderfully elaborate theory of planetary motion that the ancient Greeks had put forward, known as the *Ptolemaic system*. According to this scheme the motions of the planets are governed by a complicated composition of circular motions. It had been quite effective for making predictions, but became more and more over-complicated as greater accuracy was needed. The Ptolemaic system seems very artificial to us today. This was a good example of a USEFUL theory (for about ten centuries, in fact!) which subsequently *faded* altogether as a physical theory, though it played an organizational role of definite historical importance. For a good example of a USEFUL theory of the ultimately *successful* kind we may look, instead, to Kepler's brilliant conception of elliptical planetary motion. Another example was Mendeleev's periodic table of the chemical elements. In themselves they did not provide predictive schemes of the required 'phenomenal' character, but they later became 'correct' deductions within the SUPERB theories which grew out of them (Newtonian dynamics and quantum theory, respectively).

In the sections and chapters which follow, I shall not have much to tell about current theories which are merely USEFUL or TENTATIVE. There is enough to say about those which are SUPERB. It is indeed fortunate that we have such theories and can, in so remarkably complete a way, comprehend the world in which we live. Eventually we must try to decide whether even these theories are rich enough to govern the actions of our brains and minds. I shall broach this question in due course; but for now let us consider the SUPERB theories as we know them, and try to ponder upon their relevance to our purposes here.

Euclidean geometry

Euclidean geometry is simply that subject that we learn as 'geometry' at school. However, I expect that most people think of it as mathematics, rather than as a physical theory. It is also mathematics, of course, but Euclidean geometry is by no means the only conceivable mathematical geometry. The particular geometry that has been handed down to us by Euclid describes very accurately the physical space of the world in which we live, but this is *not* a logical necessity—it is just a (nearly exact) *observed* feature of the physcial world.

Indeed, there is another geometry called *Lobachevskian* (or *hyperbolic*) geometry* which is very like Euclidean geometry in most ways, but with certain intriguing differences. For example, recall that in Euclidean geometry the sum of the angles of any triangle is always 180°. In Lobachevskian geometry, this sum is always *less* than 180°, the difference being proportional to the area of the triangle (see Fig. 5.1).

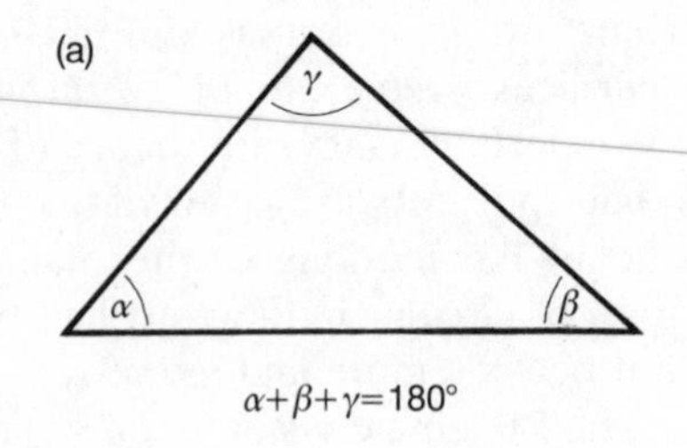

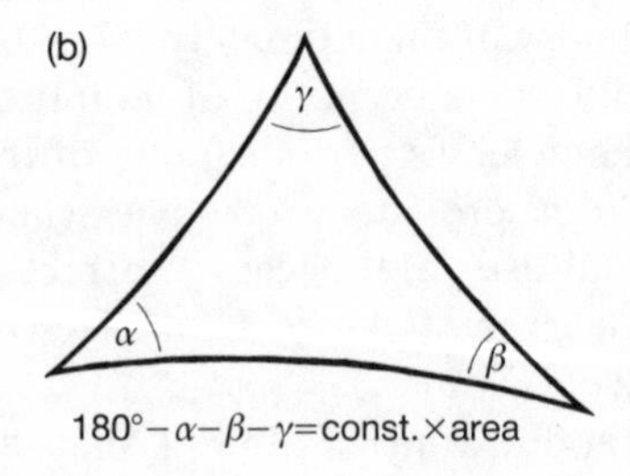

Fig. 5.1. (a) A triangle in Euclidean space, (b) a triangle in Lobachevskian space.

* Nicolai Ivanovich Lobachevsky (1792–1856), was one of several who independently discovered this kind of geometry, as an alternative to that of Euclid. Others were Carl Friedrich Gauss (1777–1855), Ferdinand Schweickard, and Janos Bolyai.

Fig. 5.2. Escher's depiction of Lobachevskian space. (Think of all the black fish as congruent and all the white fish as congruent.)

The remarkable Dutch artist Maurits C. Escher has produced some very fine and accurate representations of this geometry. One of his prints is reproduced in Fig. 5.2. Each black fish is to be thought of as being of the same size and shape as each other black fish, according to Lobachevskian geometry, and similarly for the white ones. The geometry cannot be represented completely accurately in the ordinary Euclidean plane, hence the apparent crowding just inside the circular boundary. Imagine yourself to be located inside the pattern but somewhere close to this boundary, then the Lobachevskian geometry is supposed to seem just the same to you as it would be in the middle, or anywhere else. What appears to be this 'boundary' of the pattern, according to this Euclidean representation, is really 'at infinity' in the Lobachevskian geometry. The actual boundary circle should not be thought of as part of the Lobachevsky space at all—and nor should any of the Euclidean region which lies *outside* this circle. (This ingenious representation of the Lobachevsky plane is due to Poincaré. It has the special virtue that very small shapes are not distorted in the representation—only the sizes are changed.) The 'straight lines' of the geometry (along certain of which Escher's fish are pointing) are circles meeting this boundary circle at right angles.

It could very well be the case that Lobachevskian geometry is actually true of our world on a cosmological scale (see Chapter 7, p. 324). However, the proportionality constant between the angle deficit for a triangle, and its area, would have to be *exceedingly* tiny in this case, and Euclidean geometry would be an excellent approximation to this geometry at any ordinary scale. In fact, as we shall be seeing later in this chapter, Einstein's general theory of relativity tells us that the geometry of our world *does* deviate from Euclidean geometry (though in an 'irregular' way that is more complicated than Lobachevskian geometry) at scales considerably less remote than cosmological ones, though the deviations are still exceedingly small at the ordinary scales of our direct experiences.

The fact that Euclidean geometry seems so accurately to reflect the structure of the 'space' of our world has fooled us (or our ancestors!) into thinking that this geometry is a logical necessity, or into thinking that we have an innate *a priori* intuitive grasp that Euclidean geometry *must* apply to the world in which we live. (Even the great philosopher Immanuel Kant claimed this.) The real break with Euclidean geometry only came with Einstein's general relativity, which was put forward many years later. Far from Euclidean geometry being a logical necessity, it is an *empirical observational fact* that this geometry applies so accurately—though not quite exactly—to the structure of our physical space! Euclidean geometry was indeed, all along, a (SUPERB) *physical* theory. This was in addition to its being an elegant and logical piece of pure mathematics.

In a sense, this was not so far from the philosophical viewpoint espoused by Plato (c. 360 BC; this was some fifty years before Euclid's *Elements*, his famous books on geometry). In Plato's view, the objects of pure geometry—straight lines, circles, triangles, planes, etc.—were only approximately realized in terms of the world of actual physical things. Those mathematically precise objects of pure geometry inhabited, instead, a different world—*Plato's ideal world* of mathematical concepts. Plato's world consists not of tangible objects, but of 'mathematical things'. This world is accessible to us not in the ordinary physical way but, instead, via the *intellect*. One's mind makes contact with Plato's world whenever it contemplates a mathematical truth, perceiving it by the exercise of mathematical reasoning and insight. This ideal world was regarded as distinct and more perfect than the material world of our external experiences, but just as real. (Recall our discussions of Chapters 3 and 4, pp. 97, 112 on the Platonic reality of mathematical concepts.) Thus, whereas the objects of pure Euclidean geometry can be studied by thought, and many properties of this ideal thereby derived, it would not be a necessity for the 'imperfect' physical world of external experience to adhere to this ideal exactly. By some miraculous insight Plato seems to have foreseen, on the basis of what must have been very sparse evidence indeed at that time that; on the one hand, mathematics must be studied and understood for its own sake, and one must not demand completely accurate applicability to the objects of physical experience; on the other hand, the workings of the actual external world can ultimately be understood only in terms of precise mathematics—which means in terms of Plato's ideal world 'accessible via the intellect'!

Plato founded an Academy in Athens aimed at the furtherance of such ideas. Among the elite who rose out of its membership was the exceedingly influential and famous philosopher Aristotle. But here we shall be concerned with another member of this Academy—somewhat less well-known than Aristotle, but in my view, a much finer scientist—one of the very great thinkers of antiquity: the mathematician and astronomer Eudoxos.

There is a profound and subtle ingredient to Euclidean geometry—indeed a most essential one—that we hardly think of as geometry at all nowadays! (Mathematicians would tend to call this ingredient 'analysis' rather than 'geometry'.) This was the introduction, in effect, of *real numbers*. Euclidean geometry refers to lengths and angles. To understand this geometry, we must appreciate what kind of 'numbers' are needed to describe such lengths and angles. The central new idea was put forward in the fourth century BC by Eudoxos (c. 408–355 BC.)* Greek geometry had been in a 'crisis' owing to the discovery by the Pythagoreans that numbers such as $\sqrt{2}$ (needed in order to express the length of the diagonal of a square in terms of its side) cannot be expressed as a fraction (cf. Chapter 3, p. 80). It had been important to the Greeks to be able to formulate their geometrical measures (ratios) in terms of (ratios of) integers, in order that geometrical magnitudes could be studied according to the laws of arithmetic. Basically Eudoxos's idea was to provide a method of describing ratios of lengths (i.e. real numbers!) in terms of *integers*. He was able to provide criteria, stated in terms of integer operations, for deciding when one length ratio exceeds another, or whether the two are actually to be regarded as exactly equal.

The idea was roughly as follows: If a, b, c, and d are four lengths, then a criterion for ascertaining that the ratio a/b is *greater* than the ratio c/d is that there exist integers M and N such that a added to itself N times exceeds b added to itself M times, whilst also d added to itself M times exceeds c added to itself N times.† A corresponding criterion can be used to ascertain that a/b

* Eudoxos was also the originator of that 2000-year-long USEFUL theory of planetary motion, later developed in more detail by Hipparchus and Ptolemy, known subsequently as the Ptolemaic system!

† In modern notation, this asserts the existence of a fraction, namely M/N, such that $a/b > M/N > c/d$. There will always be such a fraction lying between the two real numbers a/b and c/d whenever $a/b > c/d$, so the Eudoxan criterion is indeed satisfied.

is *less* than c/d. The sought-for criterion for the *equaltiy* $a/b = c/d$ is now simply that *neither* of these other two criteria can be satisfied!

A fully precise abstract mathematical theory of real numbers was not developed until the nineteenth century, by mathematicians such as Dedekind and Weierstrass. But their procedure actually followed very similar lines to that which Eudoxos had already discovered some twelve centuries earlier! There is no need to describe this modern development here. This modern theory was vaguely hinted at on p. 81 in Chapter 3, but for ease of presentation I preferred, in that chapter, to base the discussion of real numbers on the more familiar decimal expansions. (These expansions were, in effect, introduced by Stevinus in 1585.) It should be noted that decimal notation, though familiar to us, was actually unknown to the Greeks.

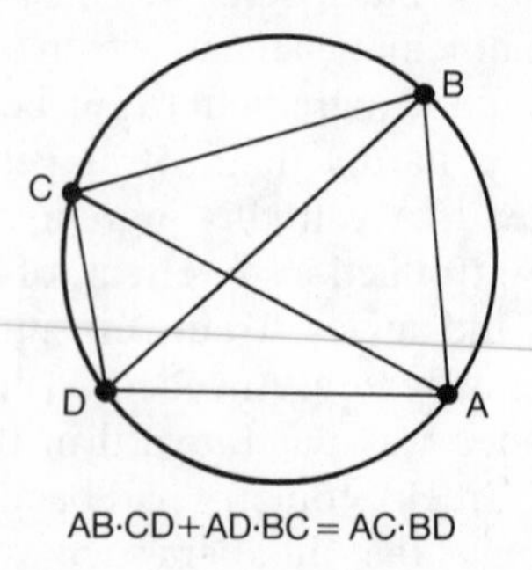

Fig. 5.3. Ptolemy's theorem.

There is an important difference, however, between Eudoxos's proposal and that of Dedekind and Weierstrass. The ancient Greeks thought of the real numbers as things *given*—in terms of (ratios of) geometrical magnitudes—that is, as properties of 'actual' space. It was necessary for the Greeks to be able to describe geometrical magnitudes in terms of arithmetic in order to be able to argue rigorously about them, and also about their sums and their products—essential ingredients of so many of the marvellous geometrical theorems of the ancients. (In Fig. 5.3 I have given, by way of illustration, the remarkable *theorem of Ptolemy*—though Ptolemy discovered it a good deal later than the time of Eudoxos—relating the distances between four points on a circle, which nicely illustrates how both sums and products are needed.) Eudoxos's criteria proved to be extraordinarily fruitful, and, in particular, they enabled the Greeks to compute areas and volumes in a rigorous way.

However, for the mathematicians of the nineteenth century—and, indeed, for the mathematicians of today—the role of geometry has changed. To the ancient Greeks, and to Eudoxos in particular, 'real' numbers were things to be *extracted* from the geometry of physical space. Now we prefer to think of the real numbers as logically more primitive than geometry. This allows us to construct all sorts of *different* types of geometry, each *starting* from the concept of number. (The key idea was that of coordinate geometry, introduced in the seventeenth century by Fermat and Descartes. Coordinates can be used to *define* other types of geometry.) Any such 'geometry' must be logically consistent, but need not have direct relevance to the physical space of our experiences. The particular physical geometry that we *do* seem to perceive is an *idealization* of experience (e.g. depending upon our extrapolations to indefinitely large or small sizes, cf. Chapter 3, p. 86), but experiments are now accurate enough that we must take it that our 'experienced' geometry *does* actually differ from the Euclidean ideal (cf. p. 211), and it is consistent with what Einstein's general theory of relativity tells us it should be. However, despite the changes in our view of the geometry of the physical world that have now come about, Eudoxos's twenty-three-century-old concept of real number has actually remained essentially unchanged and forms as much an essential ingredient of Einstein's theory as of Euclid's. Indeed it is an essential ingredient of all serious physical theories to this day!

The fifth book of Euclid's *Elements* was basically an exposition of the 'theory of proportion', described above, that Eudoxos introduced. This was deeply important to the work as a whole. Indeed, the entire *Elements*, first published in about 300 BC, must be rated as one of the most profoundly influential works of all time. It set the stage for almost all scientific and mathematical thinking thereafter. Its methods were deductive, starting from clearly stated axioms that were supposed to be 'self-evident' properties of space; and numerous consequences were derived, many of which were striking and important, and not at all self-evident. There is no doubt that Euclid's work was profoundly significant for the development of subsequent scientific thought.

The greatest mathematician of antiquity was undoubtedly Archimedes (287–212 BC). Using Eudoxos's theory of proportion in ingenious ways, he worked out the areas and volumes of many different kinds of shape, such as the sphere, or more complex ones involving parabolas or spirals. Today, we would use calculus to do this, but this was some nineteen centuries before calculus, as such, was finally introduced by Newton and Leibniz! (One could say that a good half—the 'integral' half—of the calculus was already known to Archimedes!) The degree of mathematical rigour that Archimedes achieved in his arguments was impeccable, even by modern standards. His writings deeply influenced many later mathematicians and scientists, most notably Galileo and Newton. Archimedes also introduced the (SUPERB?) physical theory of statics (i.e. the laws governing bodies in equilibrium, such as the law of the lever and the laws of floating bodies) and developed it as a deductive science, in a way similar to that in which Euclid had developed the science of geometrical space and the geometry of rigid bodies.

A contemporary of Archimedes whom I must also mention is Appolonios (*c*.262–200 BC), a very great geometer of profound insights and ingenuity, whose study of the theory of conic sections (i.e. ellipses, parabolas, and hyperbolas) had a very important influence on Kepler and Newton. These shapes turned out, quite remarkably, to be just what were needed for the descriptions of planetary orbits!

The dynamics of Galileo and Newton

The profound breakthrough that the seventeenth century brought to science was the understanding of *motion*. The ancient Greeks had a marvellous understanding of things static—rigid geometrical shapes, or bodies in *equilibrium* (i.e. with all forces balanced, so there is no motion)—but they had no good conception of the laws governing the way that bodies actually *move*. What they lacked was a good theory of *dynamics*, i.e. a theory of the beautiful way in which Nature actually controls the change in location of bodies from one moment to the next. Part (but by no means all) of the reason for this was an absence of any sufficiently accurate means of keeping time, i.e. of a reasonably good 'clock'. Such a clock is needed so that changes in position can be accurately timed, and so that the speeds and accelerations of bodies can be well ascertained. Thus, Galileo's observation in 1583 that a pendulum could be used as a reliable means of keeping time had a far-reaching importance for him (and for the development of science as a whole!) since the timing of motion could then be made precise. Some fifty-five years later, with the publication of Galileo's *Discorsi* in 1638, the new subject of dynamics was launched—and the transformation from ancient mysticism to modern science had begun!

Let me single out just *four* of the most important physical ideas that Galileo introduced. The first was that a force acting on a body determines *acceleration*, not velocity. What do the terms 'acceleration' and 'velocity' actually mean? The *velocity* of a particle—or point on some body—is the rate of change, with respect to time, of the position of that point. Velocity is normally taken to be a *vector* quantity, which is to say that its *direction* has to be taken account of as well as its magnitude (otherwise we use the term 'speed'; see Fig. 5.4). Acceleration (again a vector quantity) is the rate of change of this velocity with respect to time—so acceleration is actually *the rate of change of the rate of change* of position with respect to time! (This would have been difficult for the ancients to come to terms with, lacking both adequate 'clocks' and the relevant mathematical ideas concerning 'rates of change'.) Galileo ascertained that the force on a body (in his case, the force of gravity) controls the acceleration of that body but does *not* control its velocity directly—as the ancients, such as Aristotle, had believed.

In particular, if there is no force then the velocity is constant—hence,

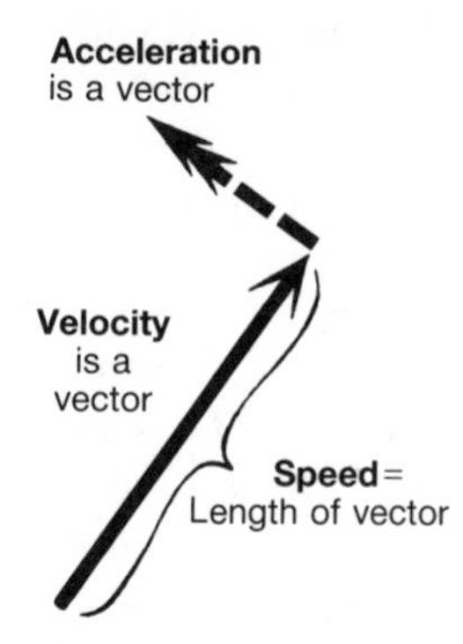

Fig. 5.4. Velocity, speed, and acceleration.

unchanging motion in a straight line would result from an *absence* of a force (which is Newton's first law). Bodies in free motion continue uniformly on their way, and need no force to keep them going. Indeed, one consequence of the dynamical laws that Galileo and Newton developed was that that uniform straight-line motion is physically completely indistinguishable from the state of rest (i.e. of absence of motion): there is no local way of telling uniform motion from rest! Galileo was particularly clear on this point (clearer even than was Newton) and gave a very graphic description in terms of a ship at sea (cf. Drake 1953, pp. 186–7):

> Shut yourself up with some friend in the main cabin below decks on some large ship, and have with you there some flies, butterflies, and other small flying animals. Have a large bowl of water with some fish in it; hang up a bottle that empties drop by drop into a wide vessel beneath it. With the ship standing still, observe carefully how the little animals fly with equal speed to all sides of the cabin. The fish swim indifferently in all directions; the drops fall into the vessel beneath; . . . When you have observed all these things carefully . . . have the ship proceed with any speed you like, so long as the motion is uniform and not fluctuating this way and that. You will discover not the least change in all the effects named, nor could you tell from any of them whether the ship was moving or standing still. . . . The droplets will fall as before into the vessel beneath without dropping toward the stern, although while the drops are in the air the ship runs many spans. The fish in their water will swim toward the front of their bowl with no more effort than toward the back, and will go with equal ease to bait placed anywhere around the edges of the bowl. Finally the butterflies and flies will continue their flights indifferently toward every side, nor will it ever happen that they are concentrated toward the stern, as if tired out from keeping up with the course of the ship, from which they will have been separated during long intervals by keeping themselves in the air.

This remarkable fact, called the *principle of Galilean relativity*, is actually crucial in order that the *Copernican* point of view can make dynamical sense.

Niccolai Copernicus (1473–1543, and the ancient Greek astronomer Aristarchus, *c*.310–230 BC—not to be confused with Aristotle!—eighteen centuries before him) had put forward the picture in which the sun remains at rest while the earth, as well as being in rotation about its own axis, moves in orbit about the sun. Why are we not aware of this motion, which would amount to some 100000 kilometres per hour? Before Galileo presented his dynamical theory, this indeed posed a genuine and deep puzzle for the Copernican point of view. If the earlier 'Aristotelian' view of dynamics had been correct, in which the actual *velocity* of a system in its motion through space would affect its dynamical behaviour, then the earth's motion would surely be something very directly evident to us. Galilean relativity makes it clear how the earth can be in motion, yet this motion is not something that we can directly perceive.*

Note that, with Galilean relativity, there is no local physical meaning to be attached to the concept of 'at rest'. This already has a remarkable implication with regard to the way that space and time should be viewed. The picture that we instinctively have about space and time is that 'space' constitutes a kind of arena in which physical events take place. A physical object may be at one point in space at one moment and at either the same or a different point in space at a later moment. We imagine that somehow the points in space persist from one moment to the next, so that it has meaning to say whether or not an object has actually changed its spatial location. But Galilean relativity tells us that there is no absolute meaning to the 'state of rest', so there is no meaning to be attached to 'the same point in space at two different times'. Which point of the Euclidean three-dimensional space of physical experience at one time is the 'same' point of our Euclidean three-dimensional space at another time? There is no way to say. It seems that we must have a completely *new* Euclidean space for each moment of time! The way to make sense of this is to consider a *four-dimensional space–time* picture of physical reality (see Fig. 5.5). The three-dimensional Euclidean spaces corresponding to different times are indeed regarded as separate from one another, but all these spaces are joined together to make the complete picture of a four-dimensional space–time. The histories of particles moving in uniform straight-line motion are described as straight lines (called world-lines) in the space–time. I shall return to the question of space–times, and the relativity of motion, later in the context of Einsteinian relativity. We shall find that the argument for four-dimensionality has a considerably greater force in that case.

The third of these great insights of Galileo was a beginning of an understanding of *conservation of energy*. Galileo was mainly concerned with

* Strictly speaking, this refers only to earth's motion in so far as it can be regarded as being approximately *uniform* and, in particular, without rotation. The rotational motion of the earth does indeed have (relatively small) dynamical effects that can be detected, the most noteworthy being the deflection of the winds in different ways in the northern and southern hemispheres. Galileo thought that this non-uniformity was responsible for the tides.

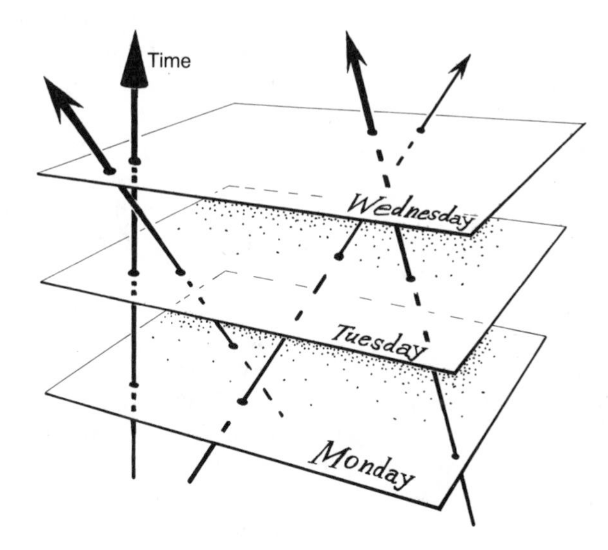

Fig. 5.5. Galilean space–time: particles in uniform motion are depicted as straight lines.

the motion of objects under gravity. He noticed that if a body is released from rest, then whether it simply drops freely, or swings on a pendulum of arbitrary length, or slides down a smooth inclined plane, its speed of motion always depends *only* upon the distance that it has reached below its point of release. Moreover this speed is always just sufficient to return it to the height from which it started. As we should now say, the energy stored in its height above the ground (gravitational potential energy) can be converted into the energy of its motion (kinetic energy, which depends upon the body's *speed*) and back again, but the energy as a whole is neither lost nor gained.

The law of energy conservation is a very important physical principle. It is not an independent physical requirement, but a *consequence* of the dynamical laws of Newton that we shall be coming to shortly. Increasingly comprehensive formulations of this law were made over the centuries by Descartes, Huygens, Leibniz, Euler and Kelvin. We shall come back to it later in this chapter and in Chapter 7. It turns out that when combined with Galileo's relativity principle, energy conservation yields further conservation laws of considerable importance: conservation of *mass* and of *momentum*. The momentum of a particle is the product of its mass with its velocity. Familiar examples of momentum conservation occur with rocket propulsion, where the increase of forward momentum of the rocket exactly balances the backward momentum of the (less massive, but compensatingly swifter) exhaust gases. The recoil of a gun is also a manifestation of momentum conservation. A further consequence of Newton's laws is conservation of *angular momentum* which describes the persistence of spin of a system. The earth's spin about its axis and a tennis ball's spin are both maintained through conservation of their angular momentum. Each constituent particle of any body contributes to that body's total angular momentum, where the magnitude of any particle's contribution is the product of its momentum with its perpendicular distance out from the centre. (As a consequence, the angular speed of a freely rotating object may be increased by making it more compact. This leads to a striking, but familiar, action often performed by skaters and trapeze artists. The act of drawing in the arms or the legs, as the case may be, causes the rotation rate spontaneously to increase, simply because of angular momentum conservation!) We shall find that mass, energy, momentum, and angular momentum are concepts that will have importance for us later.

Finally, I should remind the reader of Galileo's prophetic insight that, in the absence of atmospheric friction, all bodies fall at the same rate under gravity. (The reader may recall the famous story of Galileo's dropping various objects simultaneously from the Leaning Tower of Pisa.) Three centuries later, this very insight led Einstein to generalize the relativity principle to accelerating systems of reference, and it provided the cornerstone of his extraordinary general-relativistic theory of gravity, as we shall see, near the end of this chapter.

Upon the impressive foundations that Galileo had laid, Newton was able to erect a cathedral of superb grandeur. Newton gave three laws governing the behaviour of material objects. The first and second laws were essentially those given by Galileo: if no force acts on a body, it continues to move uniformly in a straight line; if a force does act on it, then its mass times its acceleration (i.e. the rate of change of its momentum) is equal to that force. One of Newton's own special insights was to realize the need for a third law: the force that body A exerts on body B is precisely equal and opposite to the force that body B exerts on body A ('to every action there is always opposed an equal reaction'). This provided the basic framework. The 'Newtonian universe' consists of particles moving around in a space which is subject to the laws of Euclid's geometry. The accelerations of these particles are determined by the forces which act upon them. The force on each particle is obtained by adding together (using the *vector addition law*; see Fig. 5.6) all the separate contributions to the force on that particle, arising from all the *other* particles. In order to make the system well defined, some definite rule is needed to tell us what the force on particle A should be that arises from some other particle B. Normally we require that this force acts in a direct line between A and B (see Fig. 5.7). If the force is a gravitational force, then it acts attractively between A and B and its strength is proportional to the product of the two masses and to the reciprocal of the square of the distance between them: the *inverse square law*. For other types of force, there might be a dependence on position different from this, and the force might depend on the particles according to some quality they possess different from their masses.

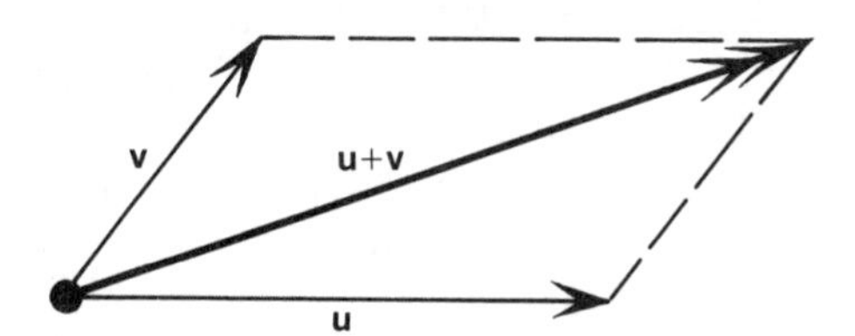

Fig. 5.6. The parallelogram law of vector addition.

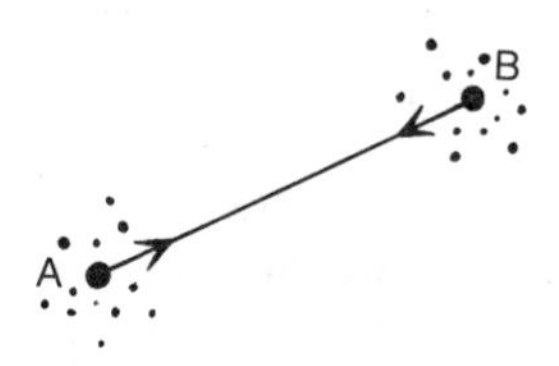

Fig. 5.7. The force between two particles is taken to be in a direct line between the two (and by Newton's third law, the force on A due to B is always equal and opposite to the force on B due to A).

The great Johannes Kepler (1571–1630), a contemporary of Galileo's, had noticed that the planets' orbits about the sun are *elliptical* rather than circular (with the sun being always at a focus, not the centre, of the ellipse) and he formulated two other laws governing the rates at which these ellipses are described. Newton was able to show that Kepler's three laws follow from his own general scheme of things (with an attractive inverse square law of force). Not only this, but he also obtained all sorts of detailed corrections to Kepler's elliptical orbits, as well as other effects, such as the precession of the equinoxes (a slow movement of the direction of the earth's rotation axis, which had been noticed by the Greeks over the centuries). In order to achieve all this, Newton had to develop many mathematical techniques—in addition to differential calculus. The phenomenal success of his efforts owed much to his supreme mathematical skills and to his equally superb physical insights.

The mechanistic world of Newtonian dynamics

With a specific law of force (such as the inverse square law of gravitation), the Newtonian scheme translates to a precise and determinate system of dynamical equations. If the positions, velocities, and masses of the various particles are specified at one time, then their positions and velocities (and their masses—these being taken to be *constant*) are mathematically determined for all later times. This form of *determinism*, as satisfied by the world of Newtonian mechanics, had (and still has) a profound influence on philosophical thought. Let us try to examine the nature of this Newtonian determinism a little more closely. What can it tell us about the question of 'free will'? Could a strictly Newtonian world contain minds? Can a Newtonian world even contain computing machines?

Let us try to be reasonably specific about this 'Newtonian' model of the world. We can suppose, for example, that the constituent particles of matter are all taken to be exact mathematical points, i.e. with no spatial extent whatever. As an alternative, we might take them all to be rigid spherical balls. In either case, we shall have to suppose that the laws of force are known to us, like the inverse square law of attraction of Newton's gravitational theory. We shall want to model other forces of nature also, such as *electric* and *magnetic* forces (first studied in detail by William Gilbert in 1600), or the strong *nuclear* forces which are now known to bind particles (protons and neutrons) together to form atomic nuclei. Electric forces are like gravitational ones in that they, also, satisfy the inverse square law, but for which similar particles *repel* each other (rather than attract, as in the gravitational case), and here it is not the masses of the particles that govern the strength of electric forces between them, but their *electric charges*. Magnetic forces are also 'inverse square' like electric ones,* but nuclear forces have a quite different dependence on distance, being extremely large at the very close separations that occur within the atomic nucleus, but negligible at greater distances.

Suppose that we adopt the rigid spherical ball picture, requiring that when two of the spheres collide they simply rebound perfectly *elastically*. That is to say, they separate again without any loss of energy (or of total momentum), as if they were perfect billiard balls. We also have to specify exactly how the *forces* are to act between one ball and another. For simplicity, we can assume that the force that each ball exerts on each other ball is along the line joining their centres, and its magnitude is a specified function of the length of this line. (For Newtonian *gravity* this assumption automatically holds true, by a remarkable theorem due to Newton; and for other laws of force it can be imposed as a consistent requirement.) Provided that the balls collide only in pairs, and no triple or higher-order collisions occur, then everything is well defined, and the outcome depends in a continuous way on the initial state (i.e. sufficiently tiny changes in the initial state lead to only small changes in the outcome). The behaviour of glancing collisions is continuous with the behaviour when the balls just miss one another. There is, however, a problem with what to do with triple or higher-order collisions. For example, if three balls A, B, and C come together at once, it makes a difference whether we consider A and B to come together just first, and C then to collide with B immediately afterwards, or whether we consider A and C to come together first, and B then to collide with A immediately afterwards (see Fig. 5.8). In

* The difference between the electric and the magnetic case is that individual 'magnetic charges' (i.e. north and south poles) do not seem to exist separately in nature, magnetic particles being what are called 'dipoles', i.e. tiny magnets (north and south together).

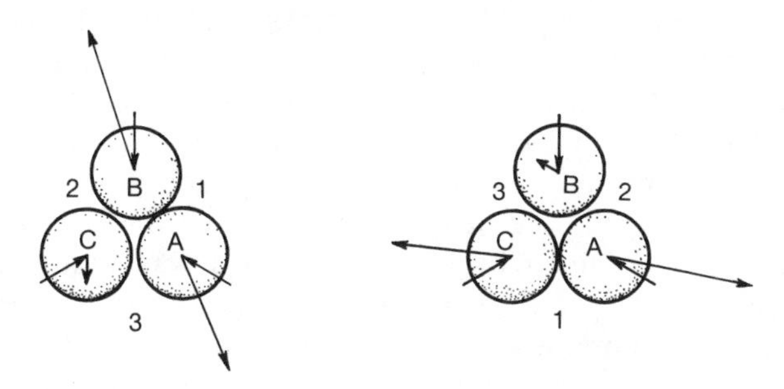

Fig. 5.8. A triple collision. The resulting behaviour depends critically upon which particles come together first, so the outcome depends discontinuously on the input.

our model, there is *indeterminism* whenever exact triple collisions occur! If we like, we can simply *rule out* exact triple or higher-order collisions as 'infinitely improbable'. This provides a reasonably consistent scheme, but the potential problem of triple collisions means that the resulting behaviour may *not* depend in a continuous way on the initial state.

This is a little unsatisfactory, and we may prefer a picture in terms of *point* particles. But in order to avoid certain theoretical difficulties raised by the point-particle model (infinite forces and infinite energies, when particles come towards coincidence), one must make other assumptions, such as that the forces between the particles always become very strongly repulsive at short distances. In this way we can ensure that no pair of particles ever actually do collide. (This also allows us to avoid the problem of deciding how point particles are supposed to *behave* when they collide!) However, for ease of visualization I shall prefer to phrase the discussion which follows entirely in terms of the rigid spheres. It seems that this kind of 'billiard-ball' picture is essentially the model of *reality* that a good many people have at the backs of their minds!

Now (ignoring the multiple collision problem), the Newtonian[4] billiard-ball picture of reality is indeed a *deterministic* model. The word 'deterministic' is to be taken in the sense that physical behaviour is mathematically completely determined for all times in the future (or past) by the positions and velocities of all the balls (assumed finite in number, say, to avoid certain problems), at any *one* time. It seems, then, that there is no room for a 'mind' to be influencing the behaviour of material things by the action of its 'free will' in this billiard-ball world. If we believe in 'free will', it would seem that we are forced to doubt that our *actual* world can be made up in this way.

The vexed question of 'free will' hovers at the background, throughout this book—though for most of what I shall have to say, it will remain only in the background. It will have one specific, but minor, role to play later in this chapter (in relation to the issue of faster-than-light signalling in relativity).

The question of free will is addressed directly in Chapter 10, and there the reader will doubtless be disappointed by what I have to contribute. I do indeed believe that there is a real issue here, rather than an imagined one, but it is profound and hard to formulate adequately. The issue of *determinism* in physical theory is important, but I believe that it is only part of the story. The world might, for example, be deterministic but *non-computable*. Thus, the future might be determined by the present in a way that is *in principle* non-calculable. In Chapter 10, I shall try to present arguments to show that the action of our conscious minds is indeed non-algorithmic (i.e. non-computable). Accordingly, the free will that we believe ourselves to be capable of would have to be intimately tied in with some non-computable ingredient in the laws that govern the world in which we actually live. It is an interesting question—whether or not one accepts this viewpoint with regard to free will—whether a given physical theory (such as Newton's) is indeed *computable*, not just whether it is deterministic. Computability is a different question from determinism—and the fact that it *is* a different question is something that I am trying to emphasize in this book.

Is life in the billiard-ball world computable?

Let me first illustrate, with an admittedly absurdly artificial example, that computability and determinism *are* different, by exhibiting a 'toy model universe' which is deterministic but not computable. Let the 'state' of this universe at any one 'time' be described as a pair of natural numbers (m, n). Let T_u be a fixed universal Turing machine, say the specific one defined in Chapter 2 (p. 56). To decide what the state of this universe is to be at the next 'instant of time', we must ask whether the action of T_u on m ultimately stops or does not stop (i.e. whether $T_u(m) \neq \square$ or $T_u(m) = \square$, in the notation of Chapter 2, p. 58). If it stops, then the state at this next 'instant' is to be $(m+1, n)$. If it does not stop, then it is to be $(n+1, m)$. We saw in Chapter 2 that there is no algorithm for the halting problem for Turing machines. It follows that there can be no algorithm for predicting the 'future' in this model universe, despite the fact that it is completely deterministic!

Of course, this is not a model to be taken seriously, but it shows that there *is* a question to be answered. We may ask of *any* deterministic physical theory whether or not it is computable. Indeed, is the Newtonian billiard-ball world computable?

The issue of physical computability depends partly on what kind of question we are proposing to ask of the system. I can think of a number of questions that could be asked for which my *guess* would be, for the Newtonian billiard-ball model, that it is *not* a computable (i.e. algorithmic) matter to ascertain the answer. One such question might be: does ball A ever collide with ball B? The idea would be that, as *initial data*, the positions and velocities of all the balls are given to us at some particular time ($t = 0$), and the problem is to work out from this data whether or not A and B will ever collide at any later time ($t>0$). To make the problem specific (although not particularly realistic), we may assume that all the balls have equal radius and equal mass and that there is, say, an inverse square law of force acting between each pair of balls. One reason for guessing that this particular question is not one that can be resolved algorithmically is that the model is somewhat like a 'billiard-ball model for a computation' that was introduced by Edward Fredkin and Tommaso Toffoli (1982). In their model (instead of having an inverse square law of force) the balls are constrained by various 'walls', but they bounce elastically off one another in a similar way to the Newtonian balls that I have just been describing (see Fig. 5.9). In the Fredkin–Toffoli model, all the basic logical operations of a computer can be performed by the balls. Any Turing machine computation can be imitated: the particular choice of Turing machine T_n defines the configuration of 'walls', etc. of the Fredkin–Toffoli machine; then an initial state of balls in motion codes the information of the input tape, and the output tape of the Turing machine is coded by the final state of the balls. Thus, in particular, one can pose the question: does such-and-such a Turing machine computation ever stop? 'Stopping' might be phrased in terms of ball A eventually colliding with

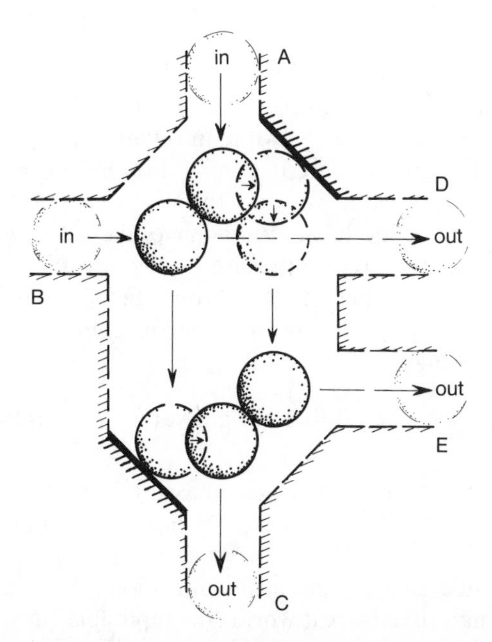

Fig. 5.9. A 'switch' (suggested by A. Ressler) in the Fredkin–Toffoli billiard-ball computer. If a ball enters at B then a ball subsequently leaves at D or at E depending upon whether another ball enters at A (where the entries at A and B are assumed to be simultaneous).

ball B. The fact that this question cannot be answered algorithmically (p. 63), at least *suggests* that the Newtonian question 'does ball A ever collide with ball B?', that I initially posed, cannot be answered algorithmically either.

In fact, the Newtonian problem is a much more awkward one than that put forward by Fredkin and Toffoli. They were able to specify the states of their model in terms of *discrete* parameters (i.e. in terms of 'on or off' statements like 'either the ball is in the channel or it is not'). But in the full Newtonian problem the initial positions and velocities of the balls have to be specified with infinite precision, in terms of coordinates which are *real numbers*, rather than in this discrete way. Thus, we are again faced with all the problems that we had to consider, when in Chapter 4 we addressed the question of whether the Mandelbrot set is recursive. What does 'computable' *mean* when continuously varying parameters are allowed for the input and output data?[5] The problem can, for the moment, be alleviated by supposing that all the initial position and velocity coordinates are given by *rational* numbers (although it cannot be expected that these coordinates will remain rational for later rational values of the time t). Recall that a rational number is a quotient of two integers, and is therefore specified in discrete finite terms. Using rational numbers, we can approximate, as closely as we like, whatever initial data sets we are choosing to examine. It is not altogether unreasonable to guess that, with rational initial data, there may be no algorithm for deciding whether or not A and B will eventually collide.

This, however, is not really what one would mean by an assertion such as: 'the Newtonian billiard-ball world is not computable'. The particular model that I have been comparing our Newtonian billiard ball world with, namely the Fredkin–Toffoli 'billiard-ball computer', does indeed proceed according to a computation. That, after all, was the essential point of Fredkin and Toffoli's idea—that their model should behave like a (universal) computer! The kind of issue that I am trying to raise is whether it is conceivable that a human brain can, by the harnessing of appropriate 'non-computable' physical laws, do 'better', in some sense, than a Turing machine. It is of no use trying to harness something like:

> 'If ball A never meets ball B then the answer to your problem is "No".'

One might have to wait forever in order to ascertain for sure that the balls in question never do meet! That, of course, is just the kind of way that Turing machines *do* behave.

In fact there would seem to be clear indications that, in an appropriate sense, the Newtonian billiard-ball world *is* computable (at least if we ignore the problem of multiple collisions). The way in which one would normally try to compute the behaviour of such a world would be to make some approximation. We might imagine that the centres of the balls are specified as lying on some grid of points, where the lattice points on the grid are just those whose coordinates are measured off, say, in hundredths of a unit. The time, also, is taken as 'discrete': all the allowed time-moments are to be multiples of some small unit (denoted by $\triangle t$, say). This gives rise to certain discrete possibilities for the 'velocities' (differences in position lattice-point values at two successive allowed time-moments, divided by $\triangle t$). The appropriate approximations to the accelerations are computed using the force law, and these are used to enable the 'velocities', and hence the new lattice-point positions to be computed to the required degree of approximation at the next allowed time-moment. The calculation proceeds for as many time-moments as will preserve the desired accuracy. It may well be that not many such time-moments can be computed before all accuracy is lost. The procedure is then to start again with a considerably finer space grid and somewhat finer division of allowed time-moments. This allows a greater accuracy to be achieved and the calculation can be carried farther into the future than previously, before accuracy is lost. With a finer space-grid still, and a finer division of time interval, the accuracy can be still further improved and the calculation projected still farther into the future. In this way, the Newtonian billiard-ball world can be computed as closely as desired (ignoring multiple collisions)—and, in this sense we can say that the Newtonian world is indeed computable.

There is a sense, however, in which this world is, 'non-computable' *in practice*. This arises from the fact that the accuracy with which the initial data can be *known* is always limited. In fact there is a very considerable 'instability' inherent in this kind of problem. A very tiny change in the initial data may rapidly give rise to an absolutely enormous change in the resulting behaviour. (Anyone who has tried to pocket a billiard ball, by hitting it with an intermediate ball that has to be hit first, will know what I mean!) This is particularly apparent when (successive) collisions are involved, but such instabilities in behaviour can also occur with Newtonian gravitational action at a distance (with more than two bodies). The term 'chaos', or 'chaotic behaviour' is often used for this type of instability. Chaotic behaviour is important, for example, with regard to the weather. Although the Newtonian equations governing the elements are well known, long-term weather prediction is notoriously unreliable!

This is not at all the kind of 'non-computability' that can be 'harnessed' in any way. It is just that since there is a limit on the accuracy with which the initial state can be known, the future state cannot reliably be computed from the initial one. In effect, a *random element* has been introduced into the future behaviour, but this is all. If the brain is indeed calling upon *useful* non-computable elements in physical laws, they must be of a completely different and much more positive character from this. Accordingly, I shall not refer to this kind of 'chaotic' behaviour as 'non-computability' at all, preferring to use the term 'unpredictability'. The presence of unpredictability is a very general phenomenon in the kind of deterministic laws that actually arise in (classical) physics, as we shall shortly see. Unpredictability is surely something that one would wish to *minimize* rather than 'harness' in the construction of a thinking machine!

In order to discuss questions of computability and unpredictability in a general way, it will be helpful to adopt a more comprehensive viewpoint than before, with regard to physical laws. This will enable us to consider not only the scheme of Newtonian mechanics, but also the later theories that have come to supersede it. We shall need to catch some glimpse of the remarkable *Hamiltonian* formulation of mechanics.

Hamiltonian mechanics

The successes of Newtonian mechanics resulted not only from its superb applicability to the physical world, but also from the richness of the mathematical theory that it gave rise to. It is remarkable that *all* the SUPERB theories of Nature have proved to be extraordinarily fertile as sources of mathematical ideas. There is a deep and beautiful mystery in this fact: that these superbly accurate theories are also extraordinarily fruitful simply as *mathematics*. No doubt this is telling us something profound about the connections between the real world of our physical experiences and the Platonic world of mathematics. (I shall attempt to address this issue later, in Chapter 10 p. 430.) Newtonian mechanics is perhaps supreme in this respect, since its birth yielded the calculus. Moreover, the specific Newtonian scheme has given rise to a remarkable body of mathematical ideas known as *classical mechanics*. The names of many of the great mathematicians of the eighteenth and nineteenth centuries are associated with its development: Euler, Lagrange, Laplace, Liouville, Poisson, Jacobi, Ostrogradski, Hamilton. What is called 'Hamiltonian theory'[6] summarizes much of this work, and a taste of this will be sufficient for our purposes here. The versatile and original Irish mathematician William Rowan Hamilton (1805–1865)—who was also responsible for the Hamiltonian circuits discussed on p. 143)—had developed this form of the theory in a way that emphasized an analogy with wave propagation. This hint of a relation between waves and particles—and the form of the Hamilton equations themselves—was highly important for the later development of *quantum mechanics*. I shall return to that aspect of things in the next chapter.

One novel ingredient of the Hamiltonian scheme lies in the 'variables' that one uses in the description of a physical system. Up until now, the *positions* of particles were taken as primary, the velocities being simply the rate of change of position with respect to time. Recall (p. 167) that in the specification of the initial state of a Newtonian system we needed the positions *and* the velocities of all the particles in order that the subsequent behaviour be determinate. With the Hamiltonian formulation we must select the *momenta* of the particles rather than the velocities. (We noted on p. 165 that the momentum of a particle is just its velocity multiplied by its mass.) This might seem a small change in itself, but the important thing is that the position and momentum of each particle are to be treated as though they are *independent* quantities, more or less on an equal footing with one another. Thus one 'pretends', at first, that the momenta of the various particles have nothing to do with the rates of change of their respective position variables, but are just a separate set of variables, so we can imagine that they 'could' have been quite independent of the position motions. In the Hamiltonian formulation, we now have *two* sets of equations. One of these tells us how the *momenta* of the various particles are changing with time, and the other tells us how the *positions* are changing with time. In each case, the rates of change are determined by the various positions and momenta *at* that time.

Roughly speaking, the first set of Hamilton's equations states Newton's crucial second law of motion (rate of change of momentum = force) while the second set of equations is telling us what the momenta actually *are*, in terms of the velocities (in effect, rate of change of position = momentum ÷ mass). Recall that the laws of motion of Galilei–Newton were described in terms of accelerations, i.e. rates of change of rates of change of position (i.e. 'second order' equations). Now, we need only to talk about rates of change of things ('first order' equations) rather than rates of change of rates of change of things. All these equations are derived from just one important quantity: the *Hamiltonian function H*, which is the expression for the *total energy* of the system in terms of all the position and momentum variables.

The Hamiltonian formulation provides a very elegant and symmetrical description of mechanics. Just to see what they look like, let us write down the equations here, even though many readers will not be familiar with the calculus notions required for a full understanding—which will not be needed here. All we really want to appreciate, as far as calculus is concerned, is that the 'dot' appearing on the left-hand side of each equation stands for *rate of change with respect to time* (of momentum, in the first case, and position, in the second):

$$\dot{p}_i = -\frac{\partial H}{\partial x_i}, \qquad \dot{x}_i = \frac{\partial H}{\partial p_i}.$$

Here the index i is being used simply to distinguish all the different momentum coordinates $p_1, p_2, p_3, p_4, \ldots$ and all the different position coordinates $x_1, x_2, x_3, x_4, \ldots$. For n unconstrained particles we shall have $3n$ momentum coordinates and $3n$ position coordinates (one, each, for the three independent directions in space). The symbol ∂ refers to 'partial differentiation' ('taking derivatives while holding all the other variables constant'), and

H is the Hamiltonian function, as described above. (If you don't know about 'differentiation', don't worry. Just think of the right-hand sides of these equations as some perfectly well-defined mathematical expressions, written in terms of the x_is and p_is.)

The coordinates $x_1, x_2, \ldots$ and $p_1, p_2, \ldots$ are actually allowed to be more general things than just ordinary Cartesian coordinates for particles (i.e. with the x_is being ordinary distances, measured off in three different directions at right angles). Some of the coordinates x_is could be *angles*, for example (in which case, the corresponding p_is would be *angular* momenta, cf. p. 166, rather than momenta), or some other completely general measure. Remarkably, the Hamiltonian equations still hold in exactly the same form. In fact, with suitable choices of H, Hamilton's equations still hold true for *any* system of classical equations whatever, not just for Newton's equations. In particular, this will be the case for the Maxwell(–Lorentz) theory that we shall be considering shortly. Hamilton's equations also hold true for special relativity. Even general relativity can, if due care is exercised, be subsumed into the Hamiltonian framework. Moreover, as we shall see later with Schrödinger's equation (p. 288), this Hamiltonian framework provides the taking-off point for the equations of quantum mechanics. Such unity of form in the structure of dynamical equations, despite all the revolutionary changes that have occurred in physical theories over the past century or so is truly remarkable!

Phase space

The form of the Hamiltonian equations allows us to 'visualize' the evolution of a classical system in a very powerful and general way. Try to imagine a 'space' of a large number of dimensions, one dimension for each of the coordinates $x_1, x_2, \ldots, p_1, p_2, \ldots$. (Mathematical spaces often have many more than three dimensions.) This space is called *phase space* (see Fig. 5.10). For n unconstrained particles, this will be a space of $6n$ dimensions (three position coordinates and three momentum coordinates for each particle). The reader may well worry that even for a *single* particle this is already twice as many dimensions as she or he would normally be used to visualizing! The secret is not to be put off by this. Whereas six dimensions are, indeed, more dimensions than can be readily(!) pictured, it would actually not be of much use to us if we were in fact able to picture it. For just a room full of air molecules, the number of phase-space dimensions might be something like

$$10\,000\,000\,000\,000\,000\,000\,000\,000\,000$$

There is not much hope in trying to obtain an accurate visualization of a space *that* big! Thus, the trick is not even to try—even in the case of the phase space for a single particle. Just think of some vague kind of three-dimensional (or

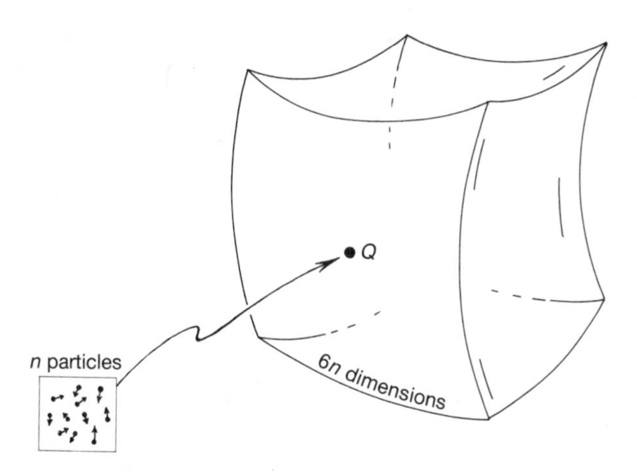

Fig. 5.10. Phase space. A single point Q of phase space represents the entire state of some physical system, including the instantaneous motions of all of its parts.

even just two-dimensional) region. Have another look at Fig. 5.10. That will do.

Now, how are we to visualize Hamilton's equations in terms of phase space? First, we should bear in mind what a *single point* Q of phase space actually represents. It corresponds to a particular set of values for all the position coordinates $x_1, x_2, \ldots$ and for all the momentum coordinates $p_1, p_2, \ldots$. That is to say, Q represents our *entire physical system*, with a particular state of motion specified for every single one of its constituent particles. Hamilton's equations tell us what the rates of change of all these coordinates are, when we know their present values; i.e. it governs how all the individual particles are to move. Translated into phase-space language, the equations are telling us how a single point Q in phase space must move, given the present location of Q in phase space. Thus, at each point of phase space, we have a little arrow—more correctly, a *vector*—which tells us the way that Q is moving, in order to describe the evolution of our entire system in time. The whole arrangement of arrows constitutes what is known as a *vector field* (Fig. 5.11). Hamilton's equations thus define a vector field on phase space.

Let us see how physical *determinism* is to be interpreted in terms of phase space. For initial data at time $t = 0$, we would have a particular set of values specified for all the position and momentum coordinates; that is to say, we have a particular choice of point Q in phase space. To find the evolution of the system in time, we simply follow the arrows. Thus, the entire evolution of our system with time—no matter how complicated that system might be—is described in phase space as just a single point moving along following the particular arrows that it encounters. We can think of the arrows as indicating the 'velocity' of our point Q in phase space. For a 'long' arrow, Q moves along swiftly, but if the arrow is 'short', Q's motion will be sluggish. To see what our physical system is doing at time t, we simply look to see where Q has moved to, by that time, by following arrows in this way. Clearly this is a deterministic procedure. The way that Q moves is completely determined by the Hamiltonian vector field.

What about computability? If we start from a computable point in phase space (i.e. from a point all of whose position and momentum coordinates are computable numbers, cf. Chapter 3, p. 82) and we wait for a computable time t, do we necessarily end up at a point which can be computably obtained from t and the values of the coordinates at the starting point? The answer would certainly depend upon the choice of Hamiltonian function H. In fact there would be *physical constants* appearing in H, such as Newton's gravitational constant or the speed of light—the exact values of these ones would depend on the choice of units, but others might be pure numbers—and it would be necessary to make sure that these constants are *computable numbers* if one is to have hope of obtaining an affirmative answer. If we assume that this *is* the case, then my *guess* would be that, for the usual Hamiltonians that are normally encountered in physics, the answer would indeed be in the affirmative. This *is* merely a guess, however, and the question is an interesting one that I hope will be examined further in the future.

On the other hand, it seems to me that, for reasons similar to those that I raised briefly in connection with the billiard-ball world, this is not quite the relevant issue. It would require *infinite precision* for the coordinates of a phase-space point—i.e. *all* the decimal places!—in order for it to make sense to say that the point is non-computable. (A number described by a *finite* decimal is always computable.) A finite portion of a decimal expansion of a number tells us nothing about the computability of the entire expansion of that number. But all physical measurements have a definite limitation on how accurately they can be performed, and can only give information about a finite number of decimal places. Does this nullify the whole concept of 'computable number' as applied to physical measurements?

Indeed, a device which could, in any *useful* way, take advantage of some (hypothetical) non-computable element in physical laws should presumably not have to rely on making measurements of unlimited precision. But it may be that I am taking too strict a line here. Suppose we have a physical device that, for known theoretical reasons, imitates some interesting non-algorithmic mathematical process. The exact behaviour of the device, if this behaviour could always be ascertained precisely, would then yield the correct answers to a succession of mathematically interesting yes/no questions for which there can be no algorithm (like those considered in Chapter 5). Any *given* algorithm would fail at some stage, and at *that* stage, the device would give us something new. The device might indeed involve examining some physical parameter to greater and greater accuracy, where more and more accuracy would be needed in order to go further and further down the list of questions. However, we *do* get something new from our device at a *finite* stage of accuracy, at least until we find an improved algorithm for the sequence of questions; then we should have to go to a greater accuracy in order to be able to achieve something that our *improved* algorithm cannot tell us.

Nevertheless, it would still seem that ever-increasing accuracy in a physical parameter is an awkward and unsatisfactory way of coding information. Much preferable would be to acquire our information in a *discrete* (or 'digital') form. Answers to questions further and further down a list could then be achieved by examining more and more of the discrete units, or perhaps by examining a *fixed* set of discrete units again and again, where the required unlimited information could be spread over longer and longer time intervals. (We could imagine these discrete units to be built up from parts, each capable of an 'on' or 'off' state, like the 0s and 1s of the Turing machine descriptions given in Chapter 2.) For this, we seem to require devices of some kind which can take up (distinguishably) discrete states and which, after evolving according to the dynamical laws, would again each take up one of a set of discrete states. If this were the case we could avoid having to examine each device to arbitrarily high accuracy.

Now, do Hamiltonian systems actually behave like this? A necessary ingredient would be some kind of stability of behaviour, so that it would be a clear-cut matter to ascertain which of these discrete states our device is actually in. Once it is in one of these states we shall want it to stay there (at

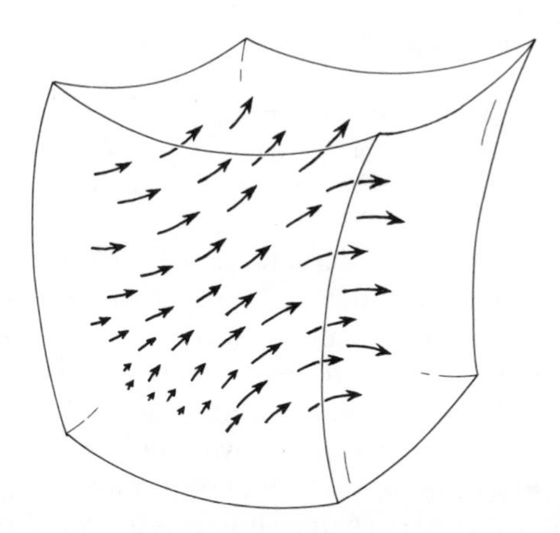

Fig. 5.11. A vector field on phase space, representing time-evolution according to Hamilton's equations.

least for a significant period of time) and not to drift from one of these states into another. Moreover, if the system arrives in these states a little inaccurately, we do not want these inaccuracies to build up; indeed, we really require that such inaccuracies should *die away* with time. Now our proposed device would have to be made up of particles (or other sub-units) which need to be described in terms of continuous parameters, and each distinguishable 'discrete' state would have to cover some *range* of these continuous parameters. (For example, one possible way of representing discrete alternatives would be to have a particle which can lie in one box or another. To specify that the particle indeed lies in one of the boxes we need to say that the position coordinates of the particle lie within some range.) What this means, in terms of phase space, is that each of our 'discrete' alternatives must correspond to a *region* in phase space, so that different phase-space points lying in the same region would correspond to the *same* one of these alternatives for our device (Fig. 5.12).

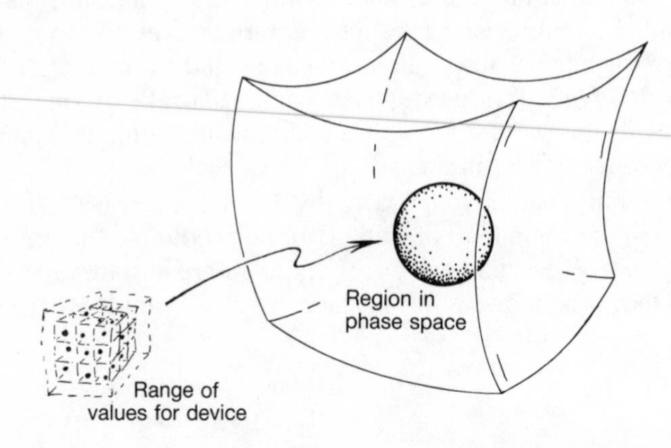

Fig. 5.12. A *region* in phase space corresponds to a range of possible values for the positions and momenta of all the particles. Such a region might represent a distinguishable state (i.e. an 'alternative') for some device.

Now suppose that the device starts off with its phase-space point in some region R_0 corresponding to a particular one of these alternatives. We think of R_0 as being dragged along the Hamiltonian vector field as time proceeds, until at time t the region becomes R_t. In picturing this, we are imagining, simultaneously, the time-evolution of our system for *all* possible starting states corresponding to this same alternative. (See Fig. 5.13.) The question of *stability* (in the sense we are interested in here) is whether, as t increases, the region R_t remains localized or whether it begins to spread out over the phase space. If such regions remain localized as time progresses, then we have a measure of stability for our system. Points of phase space which are close

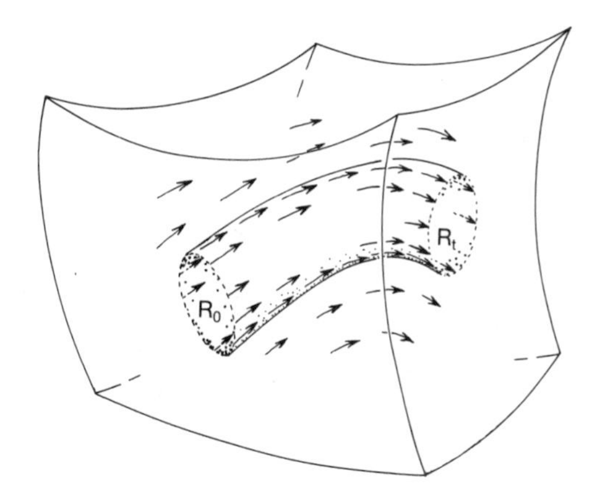

Fig. 5.13. As time evolves, a phase state region R_o is dragged along by the vector field to a new region R_t. This could represent the time-evolution of a particular alternative for our device.

together (so that they correspond to detailed physical states of the system which closely resemble one another) will remain close together in phase space, and inaccuracies in their specification will not become magnified with time. Any undue spreading would entail an effective unpredictability in the behaviour of the system.

What can be said about Hamiltonian systems generally? Do regions in phase space tend to spread with time or do they not? It might seem that for a problem of such generality, very little could be said. However, it turns out that there is a very beautiful theorem, due to the distinguished French mathematician Joseph Liouville (1809–1882), that tells us that the *volume* of any region of the phase space must remain constant under any Hamiltonian evolution. (Of course, since our phase space has some high dimension, this has to be a 'volume' in the appropriate high-dimensional sense.) Thus the volume of each R_t must be the *same* as the volume of the original R_0. At first sight this would seem to answer our stability question in the affirmative. For the *size*—in the sense of this phase-space volume—of our region *cannot grow*, so it would seem that our region cannot spread itself out over the phase space.

However, this is deceptive, and on reflection we see that the very reverse is likely to be the case! In Fig. 5.14 I have tried to indicate the sort of behaviour that one would expect, in general. We can imagine that the initial region R_0 is a small 'reasonably' shaped region, more roundish in shape than spindly—indicating that the states that belong to R_0 can be characterized in some way that does not require an unreasonable precision. However, as time unfolds, the region R_1 begins to distort and stretch—at first being perhaps somewhat amoeba-like, but then stretching out to great distances in the phase space and

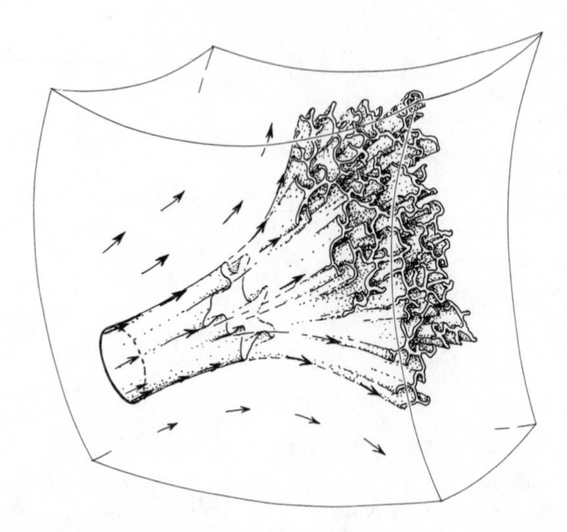

Fig. 5.14. Despite the fact that Liouville's theorem tells us that phase-space volume does not change with time-evolution, this volume will normally *effectively* spread outwards because of the extreme complication of this evolution.

winding about backwards and forwards in a very complicated way. The volume indeed remains the same, but this same small volume can get very thinly spread out over huge regions of the phase space. For a somewhat analogous situation, think of a small drop of ink placed in a large container of water. Whereas the actual volume of material in the ink remains unchanged, it eventually becomes thinly spread over the entire contents of the container. In the phase-space case, the region R_t is likely to behave in a similar way. It may not spread itself over the *whole* of phase space (which is the extreme situation referred to as 'ergodic'), but it is likely to spread over an enormously larger region than it started from. (For further discussion, see Davies 1974.)

The trouble is that preservation of volume does not at all imply preservation of *shape*: small regions will tend to get distorted, and this distortion gets magnified over large distances. The problem is a much more serious one in high dimensions than in low dimensions, since there are so many more 'directions' in which the region can locally spread. In fact, far from being a 'help', in keeping the region R_t under control, Liouville's theorem actually presents us with a fundamental problem! Without Liouville's theorem, one might envisage that this undoubted tendency for a region to spread out in phase space could, in suitable circumstances, be compensated by a reduction in overall volume. However, the theorem tells us that this is *impossible*, and we have to face up to this striking implication—a universal feature of all classical dynamical (Hamiltonian) systems of normal type![7]

We may ask, in view of this spreading throughout phase space, how is it possible at all to make predictions in classical mechanics? That is, indeed, a good question. What this spreading tells us is that, no matter how accurately we know the initial state of a system (within some reasonable limits), the uncertainties will tend to grow in time and our initial information may become almost useless. Classical mechanics is, in this kind of sense, essentially *unpredictable.* (Recall the concept of 'chaos' considered above.)

How is it, then, that Newtonian dynamics has been seen to be so successful? In the case of celestial mechanics (i.e. the motion of heavenly bodies under gravity), the reasons seem to be, *first*, that one is concerned with a comparatively small number of coherent bodies (the sun, planets, and moons) which are greatly segregated with regard to mass—so that to a first approximation one can ignore the perturbing effect of the less massive bodies and treat the larger ones as just a *few* bodies acting under each other's influence—and, *second*, that the dynamical laws that apply to the individual particles which constitute those bodies can be seen also to operate at the level of the bodies themselves—so that to a very good approximation, the sun, planets, and moons can themselves be actually treated as particles, and we do not have to worry about all the little detailed motions of the individual particles that actually compose these heavenly bodies![8] Again we get away with considering just a 'few' bodies, and the spread in phase space is not important.

Apart from celestial mechanics and the behaviour of projectiles (which is really just a special case of celestial mechanics), and the study of simple systems where small numbers of particles are involved, the main ways that Newtonian mechanics is used appear not to be in this detailed 'deterministically predictive' way at all. Rather, one uses the general Newtonian scheme to make models from which overall properties of behaviour can be inferred. Certain precise consequences of the laws, such as conservation of energy, momentum, and angular momentum do, indeed, have relevance at all scales. Moreover, there are statistical properties that can be combined with the dynamical laws governing the individual particles and which can be used to make overall predictions concerning behaviour. (See the discussion of thermodynamics in Chapter 7; the phase-space spreading effect that we have been discussing has some intimate relation to the second law of thermodynamics, and by exercising due care one can use these ideas in genuinely predictive ways.) Newton's own remarkable calculation of the speed of sound in air (subtly corrected over a century later by Laplace) was a good example of this. However, it is very rare indeed that the determinism inherent in Newtonian (or, more generally, Hamiltonian) dynamics is actually used.

This spreading effect in phase space has another remarkable implication. It tells us, in effect, that *classical mechanics cannot actually be true of our world*! I am somewhat overstating this implication, but perhaps not greatly so. Classical mechanics can account well for the behaviour of fluid bodies—

particularly gases, but also liquids to a good extent—where one is concerned only with overall 'averaged' properties of systems of particles, but it has problems in accounting for the structure of solids, where a more detailed organized structure is needed. There is a problem of how a solid body can hold its shape when it is composed of myriads of point-like particles whose organizational arrangement is continually being reduced, because of the spreading in phase space. As we now know, quantum theory is needed in order that the actual structure of solids can be properly understood. Somehow, quantum effects can prevent this phase-space spreading. This is an important issue to which we shall have to return later (see Chapters 8 and 9).

This issue is a matter of relevance, also, to the question of the construction of a 'computing machine'. The phase-space spreading is something that needs to be controlled. A region in phase space that corresponds to a 'discrete' state for a computing device (such as R_0, as described above) must not be allowed to spread unduly. Recall that even the Fredkin–Toffoli 'billiard-ball computer' required some extraneous *solid walls* in order that it could function. 'Solidity' for an object itself composed of many particles is something that really needs quantum mechanics in order to work. It seems that even a 'classical' computing machine must borrow from the effects of quantum physics if it is to function effectively!

Maxwell's electromagnetic theory

In the Newtonian picture of the world, one thinks of tiny particles acting upon one another by forces which operate at a distance—where the particles, if not entirely point-like, may be considered to rebound off one another occasionally by actual physical contact. As I have stated before (p. 168), the forces of electricity and magnetism (the existence of both having been known since antiquity, and studied in some detail by William Gilbert in 1600 and Benjamin Franklin in 1752), act in a way similar to gravitational forces in that they also fall off as the inverse square of the distance, though repulsively rather than attractively—i.e. with like repelling like—and electric charge (and magnetic pole stength), rather than the mass, measures the strength of the force. At this level there is no difficulty about incorporating electricity and magnetism into the Newtonian scheme. The behaviour of light, also, can be roughly accommodated (though with some definite difficulties), either by regarding light as being composed of individual particles ('photons', as we should now call them) or by regarding light as a wave motion in some medium, where in the latter case this medium ('ether') must itself be thought of as being composed of particles.

The fact that moving electric charges can give rise to magnetic forces caused some additional complication, but it did not disrupt the scheme as a whole. Numerous mathematicians and physicists (including Gauss) had proposed systems of equations for the effects of moving electric charges which had seemed to be satisfactory within the general Newtonian framework. The first scientist to have made a serious challenge to the 'Newtonian' picture seems to have been the great English experimentalist and theoretician Michael Faraday (1791–1867).

To understand the nature of this challenge, we must first come to terms with the concept of a physical *field*. Consider a magnetic field first. Most readers will have come across the behaviour of iron filings when placed on a piece of paper lying on a magnet. The filings line up in a striking way along what are called 'magnetic lines of force'. We imagine that the lines of force remain present even when the filings are not there. They constitute what we refer to as a *magnetic field*. At each point in space, this 'field' is oriented in a certain direction, namely the direction of the line of force at that point. Actually, we have a *vector* at each point, so the magnetic field provides us with an example of a vector field. (We may compare this with the Hamiltonian vector field that we considered in the previous section, but now this is a vector field in ordinary space rather than in phase space.) Similarly, an electrically charged body will be surrounded by a different kind of field, known as an *electric field*, and a *gravitational field* likewise surrounds any massive body. These, also, will be vector fields in space.

Such ideas were known a long time before Faraday, and they had become very much part of the armoury of theorists in Newtonian mechanics. But the prevailing viewpoint was not to regard such 'fields' as constituting, in themselves, actual physical substance. Rather, they would be thought of as providing the necessary 'bookkeeping' for the forces that would act, were a suitable particle to be placed at various different points. However, Faraday's profound experimental findings (with moving coils, magnets, and the like) led him to believe that electric and magnetic fields are *real* physical 'stuff' and, moreover, that varying electric and magnetic fields might sometimes be able to 'push' each other along through otherwise empty space to produce a kind of disembodied wave! He conjectured that light itself might consist of such waves. Such a view would have been at variance with the prevailing 'Newtonian wisdom', whereby such fields were not thought of as 'real' in any sense, but merely convenient mathematical auxiliaries to the 'true' Newtonian point-particle–action-at-a-distance picture of 'actual reality'.

Confronted with Faraday's experimental findings, together with earlier ones by the remarkable French physicist André Marie Ampère (1775–1836) and others, and inspired by Faraday's vision, the great Scottish physicist and mathematician James Clerk Maxwell (1831–1879) puzzled about the mathematical form of the equations for electric and magnetic fields that arose from those findings. With a remarkable stroke of insight he proposed a change in the equations—seemingly perhaps rather slight, but fundamental in its implications. This change was not at all suggested by (although it was consistent with) the known experimental facts. It was a result of Maxwell's own theoretical requirements, partly physical, partly mathematical, and partly aesthetic. One implication of Maxwell's equations was that electric and magnetic fields would indeed 'push' each other along through empty space. An oscillating magnetic field would give rise to an oscillating electric field (this was implied by Faraday's experimental findings), and this oscillating electric field would, in turn, give rise to an oscillating magnetic field (by Maxwell's theoretical inference), and this again would give rise to an electric field and so on. (See Figs 6.26, 6.27 on p. 271 for detailed pictures of such waves.) Maxwell was able to calculate the speed that this effect would propagate through space—and he found that it would be the speed of light! Moreover these so-called *electromagnetic* waves would exhibit the interference and puzzling polarization properties of light that had long been known (we shall come to these in Chapter 6, pp. 235, 270). In addition to accounting for the properties of visible light, for which the waves would have a particular range of wavelengths ($4–7 \times 10^{-7}$m), electromagnetic waves of other wavelengths were predicted to occur and to be produced by electric currents in wires. The existence of such waves was established experimentally by the remarkable German physicist Heinrich Hertz in 1888. Faraday's inspired hope had indeed found a firm basis in the marvellous equations of Maxwell!

Although it will not be necessary for us to appreciate the details of Maxwell's equations here, there will be no harm in our just taking a peek at them:

$$\frac{1}{c^2}\cdot\frac{\partial \boldsymbol{E}}{\partial t} = \text{curl}\,\boldsymbol{B} - 4\pi \boldsymbol{j}, \qquad \frac{\partial \boldsymbol{B}}{\partial t} = -\,\text{curl}\,\boldsymbol{E}$$

$$\text{div}\,\boldsymbol{E} = 4\pi\,\rho, \qquad \text{div}\,\boldsymbol{B} = 0$$

Here, $\boldsymbol{E}$, $\boldsymbol{B}$, and $\boldsymbol{j}$ are vector fields describing the electric field, the magnetic field, and the electric current, respectively; ρ describes the density of electric charge, and c is just a constant; the speed of light. Do not worry about the terms 'curl' and 'div', which simply refer to different kinds of spatial variation. (They are certain combinations of partial derivative operators, taken with respect to the space coordinates. Recall the 'partial derivative' operation, with symbol ∂, that we encountered in connection with Hamilton's equations.) The operators $\partial/\partial t$ appearing on the left-hand sides of the first two equations are, in effect, the same as the 'dot' that was used in Hamilton's equations, the difference being just technical. Thus, $\partial \boldsymbol{E}/\partial t$ means 'the rate of change of the electric field' and $\partial \boldsymbol{B}/\partial t$ means 'the rate of change of the magnetic field'. The first equation* tells how the electric field is changing with

* It was the presence of $\partial \boldsymbol{E}/\partial t$ in this equation which was Maxwell's master-stroke of theoretical

time, in terms of what the magnetic field and electric current are doing at the moment; while the second equation tells how the magnetic field is changing with time in terms of what the electric field is doing at the moment. The third equation is, roughly speaking, a coded form of the inverse square law telling how the electric field (at the moment) must be related to the distribution of charges; while the fourth equation says what would be the same thing about the magnetic field, except that in this case there are no 'magnetic charges' (separate 'north pole' or 'south pole' particles).

These equations are somewhat like Hamilton's in that they tell what must be the rate of change, with time, of the relevant quantities (here the electric and magnetic fields) in terms of what their values are at any given time. Thus Maxwell's equations are *deterministic*, just as are ordinary Hamiltonian theories. The only difference—and it *is* an important difference—is that Maxwell's equations are *field* equations rather than particle equations, which means that one needs an *infinite* number of parameters to describe the state of the system (the field vectors at every single point in space), rather than just the finite number that is needed for a particle theory (three coordinates of position and three of momenta for each particle). Thus the phase space for Maxwell theory is a space of an *infinite* number of dimensions! (As I have mentioned earlier, Maxwell's equations can actually be encompassed by the general Hamiltonian framework, but this framework must be extended slightly, because of this infinite-dimensionality.)[9]

The fundamentally *new* ingredient in our picture of physical reality as presented by Maxwell theory, over and above what had been the case previously, is that now *fields* must be taken seriously in their own right and cannot be regarded as mere mathematical appendages to what were the 'real' particles of Newtonian theory. Indeed, Maxwell showed that when the fields propagate as electromagnetic waves they actually carry definite amounts of *energy* with them. He was able to provide an explicit expression for this energy. The remarkable fact that energy can indeed be transported from place to place by these 'disembodied' electromagnetic waves was, in effect, experimentally confirmed by Hertz's detection of such waves. It is now something familiar to us—though still a very striking fact—that radio waves *can* actually carry energy!

Computability and the wave equation

Maxwell was able to deduce directly from his equations that, in regions of space where there are no charges or currents (i.e. where $\boldsymbol{j} = 0$, $\rho = 0$ in the

inference. All the remaining terms in all the equations were, in effect known from direct experimental evidence. The coefficient $1/c^2$ is very tiny, which is why that term had not been experimentally observed.

equations above), all the components of the electric and magnetic fields must satisfy an equation known as the *wave equation*.* The wave equation may be regarded as a 'simplified version' of the Maxwell equations since it is an equation for a *single* quantity, rather than for all the six components of the electric and magnetic fields. Its solutions exemplify wavelike behaviour without added complications, such as of the 'polarization' of Maxwell theory (direction of the electric field vector, see p. 270).

The wave equation has an additional interest for us here because it has been explicitly studied in relation to its *computability* properties. In fact Marian Boykan Pour-El and Ian Richards (1979, 1981, 1982) have been able to show that even though solutions of the wave equation behave *deterministically* in the ordinary sense—i.e. data provided at an initial time will determine the solution at all other times—there exist *computable* initial data, of a certain 'peculiar' kind, with the property that for a later computable time the determined value of the field is actually *not computable*. Thus the equations of a plausible physical field theory (albeit not quite the Maxwell theory that actually holds true in our world) can, in the sense of Pour-El and Richards, give rise to a non-computable evolution!

On the face of it this is a rather startling result—and it would seem to contradict what I had been conjecturing in the last section concerning the probable computability of 'reasonable' Hamiltonian systems. However, while the Pour-El–Richards result is certainly striking and mathematically relevant, it does not really contradict that conjecture in a way that makes good physical sense. The reason is that their 'peculiar' kind of initial data is not 'smoothly varying',[10] in a way that one would normally require for a physically sensible field. Indeed, Pour-El and Richards actually prove that non-computability *cannot arise* for the wave equation if we disallow this kind of field. In any case, even if fields of this kind were permitted, it would be hard to see how any physical 'device' (such as a human brain?) could make use of such 'non-computability'. It could have relevance only when measurements of arbitrarily high precision are allowed, which, as I described earlier, is not very realistic physically. Nevertheless, the Pour-El–Richards results represent an intriguing beginning to an important area of investigation in which little work has been done so far.

The Lorentz equation of motion; runaway particles

As they stand, Maxwell's equations are not really quite complete as a system of equations. They provide a wonderful description of the way that electric and magnetic fields propagate if we are *given* the distribution of electric

* The wave equation (or D'Alembert equation) can be written $\{(1/c^2)(\partial/\partial t)^2 - (\partial/\partial x)^2 - (\partial/\partial y)^2 - (\partial/\partial z)^2\}\varphi = 0$.

charges and currents. These charges are physically given to us as *charged particles*—mainly electrons and protons, as we now know—and the currents arise from the motions of such particles. If we know where these particles are and how they are moving, then Maxwell's equations tell us how the electromagnetic field is to behave. What Maxwell's equations do *not* tell us is how the particles themselves are to behave. A partial answer to this question had been known in Maxwell's day, but a satisfactory system of equations had not been settled upon until, in 1895 the remarkable Dutch physicist Hendrick Antoon Lorentz used ideas related to those of special relativity theory in order to derive what are now known as the *Lorentz equations of motion* for a charged particle (cf. Whittaker 1910, pp. 310, 395). These equations tell us how the velocity of a charged particle continuously changes, owing to the electric and magnetic fields at the point where the particle is located.[11] When the Lorentz equations are adjoined to those of Maxwell, one obtains rules for the time-evolution *both* of the charged particles and of the electromagnetic field.

However, all is not entirely well with this system of equations. They provide excellent results if the fields are very uniform down at the scale of size of the diameters of the particles themselves (taking this size as the electron's 'classical radius'—about 10^{-15}m), and the particles' motions are not too violent. However, there is a difficulty of *principle* here which can become important in other circumstances. What the Lorentz equations tell us to do is to examine the electromagnetic field at the precise *point* at which the charged particle is located (and, in effect, to provide us with a 'force' at that point). Where is that point to be taken to be if the particle has a finite size? Do we take the 'centre' of the particle, or else do we average the field (for the 'force') over all points at the surface? This could make a difference if the field is *not* uniform at the scale of the particle. More serious is another problem: what actually *is* the field at the particle's surface (or centre)? Remember that we are considering a *charged* particle. There will be an electromagnetic field *due to the particle itself* which must be added to the 'background field' in which the particle is sitting. The particle's own field gets enormously strong very close to its 'surface' and will easily swamp all the other fields in its neighbourhood. Moreover, the particle's field will point more-or-less directly outwards (or inwards) all around it, so the resulting *actual* field, to which the particle is supposed to be responding, will not be uniform at all, but will point in various different directions at different places on the particle's 'surface', let alone in its 'interior' (Fig. 5.15). Now we have to start worrying about whether the differing forces on the particle will tend to rotate it or distort it, and we must ask what elastic properties it has, etc. (and there are particularly problematic issues here in connection with *relativity* that I shall not worry the reader about). Clearly the problem is very much more complicated than it had seemed to be before.

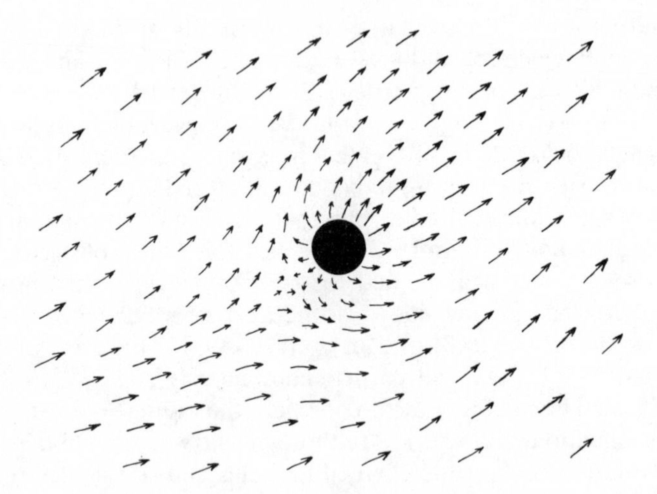

Fig. 5.15. How do we strictly apply the Lorentz equations of motion? The force on a charged particle cannot be obtained simply by examining the field 'at' the particle's location since the particle's own field dominates there.

Perhaps we are better off regarding the particle as a *point* particle. But this leads to other kinds of problems, for then the particle's own electric field becomes *infinite* in its immediate neighbourhood. If, according to the Lorentz equations, it is to respond to the electromagnetic field where it itself resides, then it would seem that it must respond to an infinite field! To make sense of the Lorentz force law, it is necessary to find a way of *subtracting the particle's own field* so as to leave a finite background field to which the particle can unambiguously respond. The problem of how to do this was solved in 1938 by Dirac (of whom we shall hear again later). However, Dirac's solution led to some alarming conclusions. He found that in order for the behaviour of the particles and fields to be determined by their initial data, it is necessary that not only must each particle's initial position and velocity be known but its initial *acceleration* must be known too (a somewhat anomalous situation in the context of standard dynamical theories). For most values of this initial acceleration the particle eventually behaves in a completely crazy way, spontaneously accelerating into the distance at a speed that very rapidly approaches the speed of light! These are Dirac's 'runaway solutions', and they do not correspond to anything that actually happens in Nature. One must find a way of ruling out the runaway solutions by choosing the initial accelerations in just the right way. This can always be done, but only if one applies 'foreknowledge'—that is to say, one must specify the initial accelerations in a way that anticipates which solutions eventually become runaway solutions, and avoids them. This is not at all the way that initial data are to be specified in a standard deterministic physical problem. With conventional determinism those data can be given arbitrarily, unconstrained by any requirement that the future is to behave itself. Here, not only is the future completely determined by data that can be specified at one time in the past, but the very specification of those data is very precisely constrained by the requirement that the future behaviour is, indeed, 'reasonable'!

This is as far as we get with fundamental classical equations. The reader will realize that the issue of determinism and computability in classical physical laws has become disturbingly muddied. Do we actually have a *teleological* element in physical laws, where the future somehow influences what is allowed to happen in the past? In fact physicists do not normally take these implications of *classical electrodynamics* (the theory of classical charged particles, and electric and magnetic fields) as serious descriptions of reality. Their usual answer to the above difficulties is to say that with charged individual particles one is properly in the realm of *quantum electrodynamics*, and one cannot expect to obtain sensible answers using a strictly classical procedure. This is undoubtedly true, but as we shall see later, the quantum theory *itself* has problems in this regard. In fact, Dirac had considered the classical problem of the dynamics of a charged particle precisely *because* he thought that it might yield insights for resolving the even greater fundamental difficulties of the (more physically appropriate) quantum problem. We must face the problems of *quantum* theory later!

The special relativity of Einstein and Poincaré

Recall the principle of Galilean relativity, which tells us that the physical laws of Galileo and Newton remain totally unchanged if we pass from a stationary to a moving frame of reference. This implies that we cannot ascertain, simply by examining the dynamical behaviour of objects in our neighbourhood, whether we are stationary or whether we are moving with uniform velocity in some direction. (Remember Galileo's ship at sea, p. 163.) But suppose we adjoin Maxwell's equations to these laws. Does Galilean relativity still remain true? Recall that Maxwell's electromagnetic waves propagate with a fixed speed c—the speed of light. Common sense would seem to tell us that if we were to travel very rapidly in some direction, then the speed of light in that direction ought to appear to us to be *reduced* to below c (because we are moving towards 'catching the light up' in that direction) and the apparent speed of light in the opposite direction ought to be correspondingly *increased* to above c (because we are moving away from the light)—which is different from the *fixed* value c of Maxwell's theory. Indeed, common sense would be right: the combined Newton and Maxwell equations would *not* satisfy Galilean relativity.

It was through worrying about such matters that Einstein was led, in

1905—as, in effect, was Poincaré before him (in 1898–1905)—to the special theory of relativity. Poincaré and Einstein independently found that Maxwell's equations *also* satisfy a relativity principle (cf. Pais 1982); i.e. the equations have a similar property of remaining unchanged if we pass from a stationary to a moving frame of reference, although the rules for this are *incompatible* with those for Galilean–Newtonian physics! To make the two compatible, it would be necessary to modify one set of equations or the other—or else abandon the relativity principle.

Einstein had no intention of abandoning the relativity principle. His superb physical instincts were insistent that such a principle must indeed hold for the physical laws of our world. Moreover, he was well aware that, for virtually all known phenomena, the physics of Galilei–Newton had been tested just for speeds very tiny compared with the speed of light, where this incompatibility would not be significant. Only *light itself* was known to involve velocities large enough that such discrepancies would be important. It would be the behaviour of light, therefore, that would inform us which relativity principle we must indeed adopt—and the equations that govern light are those of Maxwell. Thus the relativity principle for Maxwell's theory should be the one to preserve; and the laws of Galilei–Newton must, accordingly, be modified!

Lorentz, before Poincaré and Einstein, had also addressed and partially answered these questions. By 1895, Lorentz had adopted the view that the forces that bind matter together are electromagnetic in nature (as indeed they turned out to be) so that the behaviour of actual material bodies should satisfy laws derived from Maxwell's equations. One implication of this turned out to be that a body moving with a speed comparable with that of light would contract, slightly, in the direction of motion (the 'FitzGerald–Lorentz contraction'). Lorentz had used this to explain a puzzling experimental finding, that of Michelson and Morley in 1887, which seemed to indicate that electromagnetic phenomena could not be used to determine an 'absolute' rest frame. (Michelson and Morley showed that the apparent speed of light on the earth's surface is not influenced by the earth's motion about the sun—very much contrary to expectations.) Does matter always behave in such a way that its (uniform) motion cannot be locally detected? This was the *approximate* conclusion of Lorentz; moreover he was limited to a specific theory of matter, where no forces other than electromagnetic ones were regarded as significant. Poincaré, outstanding mathematician that he was, was able to show (in 1905) that there is an *exact* way for matter to behave, according to the relativity principle underlying Maxwell's equations, so that uniform motion cannot be locally detected at all. He also obtained much understanding of the physical implications of this principle (including the 'relativity of simultaneity' that we shall be considering shortly). He appears to have regarded it as just *one* possibility, and did not share Einstein's conviction that some relativity principle *must* hold.

The relativity principle satisfied by the Maxwell equations—that has become known as *special relativity*—is somewhat difficult to grasp, and it has many non-intuitive features that are, at first, hard to accept as actual properties of the world in which we live. In fact, special relativity cannot properly be made sense of without the *further* ingredient, introduced in 1908 by the highly original and insightful Russian/German geometer Hermann Minkowski (1864–1909). Minkowski had been one of Einstein's teachers at the Zürich Polytechnic. His fundamental new idea was that space and time had to be considered together as a single entity: a *four-dimensional space–time*. In 1908, Minkowski announced, in a famous lecture at the University of Göttingen:

> Henceforth space by itself, and time by itself, are doomed to fade away into mere shadows, and only a kind of union of the two will preserve an independent reality.

Let us try to understand the basics of special relativity in terms of the magnificent space–time of Minkowski.

One of the difficulties in coming to terms with the space–time concept is that it is *four-dimensional*, which makes it hard to visualize. However, having survived our encounter with phase space, we should have no trouble with a mere four dimensions! As before, we shall 'cheat' and picture a space of smaller dimension—but now the degree of cheating is incomparably less severe, and our picture will be correspondingly more accurate. Two dimensions (one space and one time) would suffice for many purposes, but I hope that the reader will allow me to be a little more adventurous and go up to three (two space and one time). This will provide a very good picture, and it should not be hard to accept that in principle the ideas will extend, without much change, to the full four-dimensional situation. The thing to bear in mind about a space–time diagram is that each point in the picture represents an *event*—that is, a point in space at a single moment, a point having only an *instantaneous* existence. The entire diagram represents the whole of history, past, present, and future. A particle, since it persists with time, is represented not by a point but by a line, called the *world-line* of the particle. This world-line—straight if the particle moves uniformly and curved if it accelerates (i.e. *non*-uniformly)—describes the entire history of the particle's existence.

In Fig. 5.16, I have depicted a space–time with two space and one time dimension. We imagine that there is a standard time-coordinate t, measured in the vertical direction, and two space-coordinates x/c and z/c, measured horizontally.* The cone at the centre is the (future) *light cone* of the space-time origin O. To appreciate its significance, imagine an explosion taking place at the event O. (Thus the explosion occurs at the spatial origin, at

* The reason for dividing the spatial coordinates by c—the speed of light—is in order that the world-lines of photons are angled at a convenient slope: 45° to the vertical, see later.

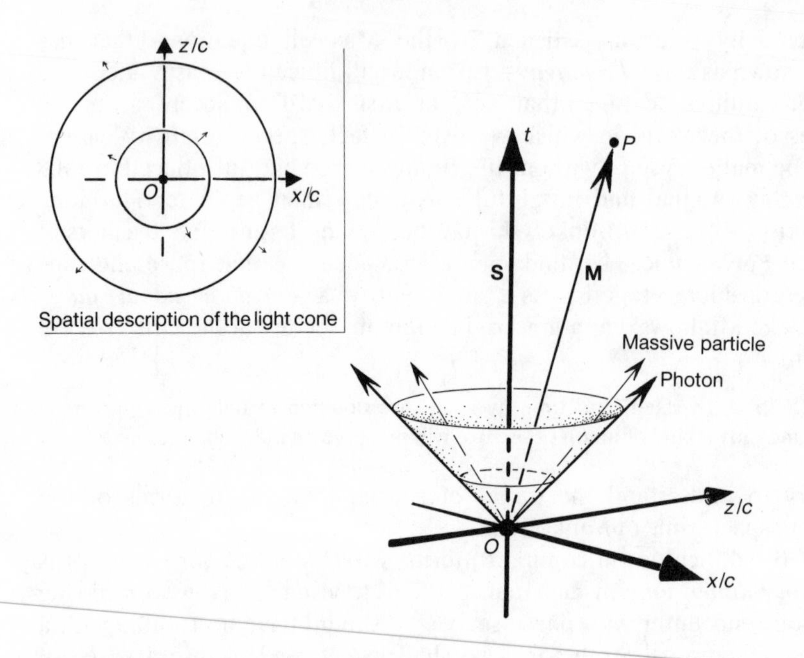

Fig. 5.16. A light cone in Minkowski's space–time (with just two spatial dimensions), describing the history of the light flash of an explosion that takes place at the event O, the space–time origin.

time $t = 0$.) The history of the light emanating from the explosion is this light cone. In two-dimensional spatial terms, the history of the light flash would be a circle moving outwards, with the fundamental light speed c. In full three-dimensional space this would be a *sphere* moving outwards at speed c—the spherical wave-front of the light—but here we are *suppressing* the spatial direction y, so we just get a circle, like the circular ripples emanating from the point of entry of a stone thrown in a pond. We can see this circle in the space–time picture if we take successive horizontal slices through the cone moving steadily upwards. These horizontal planes represent different spatial descriptions as the time-coordinate t increases. Now, one of the features of relativity theory is that it is impossible for a material particle to travel faster than light (more about this later). All the material particles coming from the explosion must lag behind the light. This means, in space–time terms, that the world-lines of all the particles emitted in the explosion must lie *inside* the light cone.

It is often convenient to describe light in terms of *particles*—called *photons*—rather than in terms of electromagnetic waves. For the moment we can think of a 'photon' as a small 'packet' of high-frequency oscillation of electromagnetic field. The term is physically more appropriate in the context of the *quantum* descriptions that we shall be considering in the next chapter, but 'classical' photons will also be helpful for us here. In free space, photons always travel in straight lines with the fundamental speed c. This means that in Minkowski's space–time picture the world-line of a photon is always depicted as a straight line tilted at 45° to the vertical. The photons produced at our explosion at O describe the light cone centred at O.

These properties must hold generally at all points of the space–time. There is nothing special about the origin; the point O is no different from any other point. Thus there must be a light cone at every point of the space–time, with a significance the same as that for the light cone at the origin. The history of any light flash—or the world-lines of photons, if we prefer to use a particle description of the light—is always along the light cone at each point, whereas the history of any material particle must always be inside the light cone at each point. This is illustrated in Fig. 5.17. The family of light cones at all the points may be regarded as part of the *Minkowskian geometry* of space–time.

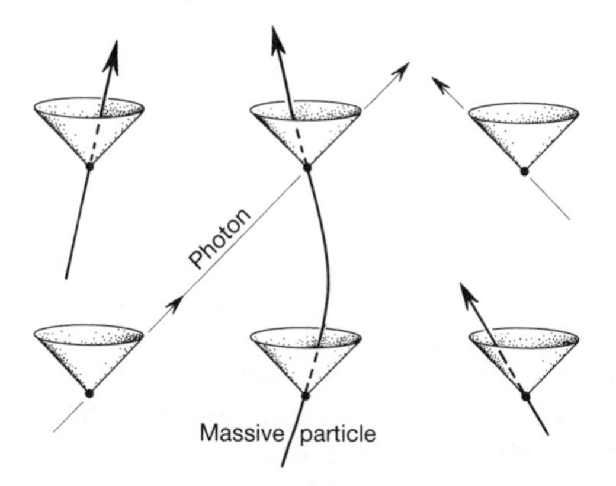

Fig. 5.17. A picture of Minkowskian geometry.

What is Minkowskian geometry? The light-cone structure is its most important aspect, but there is more to Minkowskian geometry than that. There is a concept of 'distance' which has some remarkable analogies with distance in Euclidean geometry. In three-dimensional Euclidean geometry, the distance r of a point from the origin, in terms of standard Cartesian coordinates, is given by

$$r^2 = x^2 + y^2 + z^2.$$

(See Fig. 5.18a. This is just the Pythagorean theorem—the two-dimensional case being perhaps more familiar.) In our three-dimensional Minkowskian geometry, the expression is formally very similar (Fig. 5.18b), the essential difference being that we now have two *minus signs*:

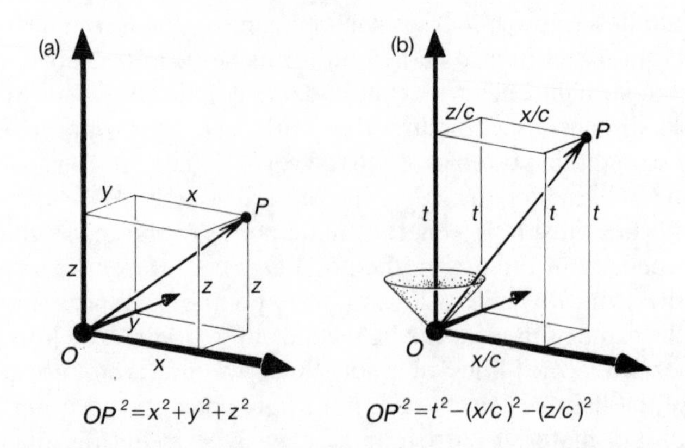

Fig. 5.18. A comparison between the 'distance' measures in (a) Euclidean geometry and (b) Minkowskian geometry (where 'distance' means 'time experienced').

$$s^2 = t^2 - (x/c)^2 - (z/c)^2.$$

More correctly we should have *four*-dimensional Minkowskian geometry, of course, and the 'distance' expression is

$$s^2 = t^2 - (x/c)^2 - (y/c)^2 - (z/c)^2.$$

What is the physical meaning of the 'distance' quantity s in this expression? Assume that the point in question—i.e. the point P, with coordinates $\{t, x/c, y/c, z/c\}$ (or $\{t, x/c, z/c\}$, in the three-dimensional case; see Fig. 5.16)—lies within the (future) light cone of O. Then the straight line segment OP can represent part of the history of some material particle—say a particular particle emitted by our explosion. The Minkowskian 'length' s of the segment OP has a direct physical interpretation. It is the *time interval* actually experienced by the particle between the events O and P! That is to say, if there were a very sturdy and accurate clock attached to the particle,[12] then the difference between the times it registers at O and P is precisely s. Contrary to expectations, the coordinate quantity t does *not* itself describe the time as measured by an accurate clock unless it is 'at rest' in our coordinate system (i.e. fixed coordinate values x/c, y/c, z/c), which means that the clock would have a world-line that is 'vertical' in the diagram. Thus, 't' means 'time' only for observers who are 'stationary' (i.e. with 'vertical' world-lines). The *correct* measure of time for a moving observer (uniformly moving away from the origin O), according to special relativity, is provided by the quantity s.

This is very remarkable—and quite at variance with the 'common-sense' Galileian–Newtonian measure of time, which would, indeed, be simply the coordinate value t. Note that the relativistic (Minkowskian) time-measure s is always somewhat *less* than t if there is any motion at all (since s^2 is less than t^2 whenever x/c, y/c, and z/c are not all zero, by the above formula). Motion (i.e. OP not along the t-axis) will tend to 'slow down' the clock, as compared with t—i.e. as viewed with respect to our coordinate frame. If the speed of this motion is small compared with c, then s and t will be almost the same, which explains why we are not directly aware of the fact that 'moving clocks run slow'. At the other extreme, when the speed is that of light itself, P then lies *on* the light cone; and we find $s = 0$. The light cone is precisely the set of points whose Minkowskian 'distance' (i.e. 'time') from O is actually zero. Thus, a photon would not 'experience' any passage of time at all! (We are not allowed the even *more* extreme case, where P is moved to just *outside* the cone, since that would lead to an imaginary s—the square root of a negative number—and would violate the rule that material particles or photons cannot travel faster than light.*)

This notion of Minkowskian 'distance' applies equally well to *any* pair of points in the space–time for which one lies within the light cone of the other—so that a particle could travel from one to the other. We simply consider O to be displaced to some other point of the space–time. Again, the Minkowski distance between the points measures the time interval experienced by a clock moving uniformly from one to the other. When the particle is allowed to be a photon, and the Minkowskian distance becomes zero, we get two points one of which is *on* the light cone of the other—and this fact serves to *define* the light cone of that point.

The basic structure of Minkowskian geometry, with this curious measure of 'length' for world lines—interpreted as the *time* measured (or 'experienced') by physical clocks—contains the very essence of special relativity. In particular, the reader may be acquainted with what is called 'the twin paradox' of relativity: one twin brother remains on the earth, while the other makes a journey to a nearby star, travelling there and back at great speed, approaching that of light. Upon his return, it is found that the twins have aged differently, the traveller finding himself still youthful, while his stay-at-home brother is an old man. This is easily described in terms of Minkowski's geometry—and one sees why, though a puzzling phenomenon, it is not actually a paradox. The world line AC, represents the twin who stays at home while the traveller has a world line composed of two segments AB and BC, these representing the outward and inward stages of the journey (see Fig. 5.19). The stay-at-home twin experiences a time measured by the Minkowski distance AC, while the traveller experiences a time given by the sum[13] of the two Minkowski distances AB and BC. These times are not equal, but we find:

$$AC > AB + BC,$$

* However, for events separated by negative values of s^2, the quantity $c\sqrt{(-s^2)}$ has a meaning, namely that of *ordinary* distance—for that observer to whom the events appear simultaneous (cf. later).

showing that indeed the time experienced by the stay-at-home is greater than that of the traveller.

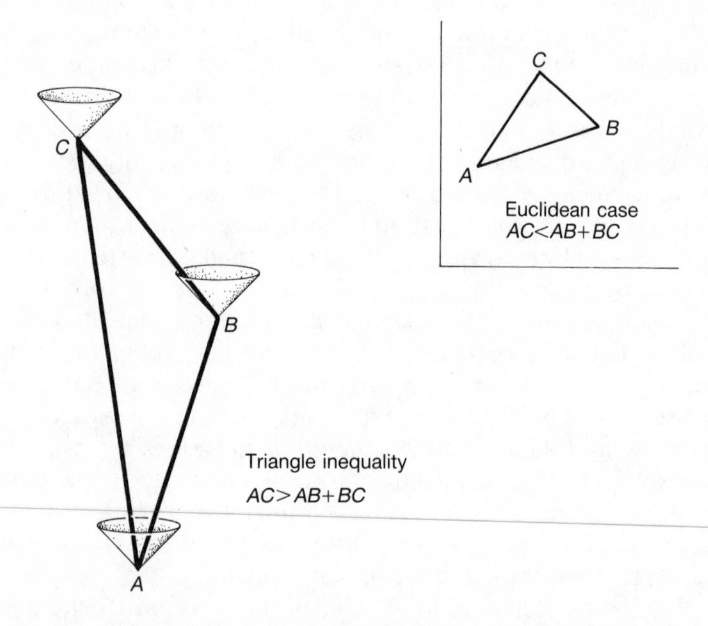

Fig. 5.19. The so-called 'twin paradox' of special relativity is understood in terms of a Minkowskian triangle inequality. (The Euclidean case is also given, for comparison.)

The above inequality looks rather similar to the well-known *triangle inequality* of ordinary Euclidean geometry namely (A, B, and C being now three points in *Euclidean* space):

$$AC < AB + BC,$$

which asserts that the sum of two sides of a triangle is always *greater* than the third. *This* we do not regard as a paradox! We are perfectly familiar with the idea that the Euclidean measure of distance along a path from one point to another (here from A to C) depends upon the actual path that we take. (In this case the two paths are AC and the longer angled route ABC.) This example is a particular instance of the fact that the shortest distance between two points (here A and C) is measured along the straight line joining them (the line AC). The reversal of the direction of the inequality sign in the *Minkowskian* case arises from the sign changes in the definition of 'distance', so the Minkowskian AC is '*longer*' than the jointed route ABC. Also, this Minkowskian 'triangle inequality' is a particular instance of a more general result: the *longest* (in the sense of greatest time experienced) among world-lines connecting two events is the straight (i.e. unaccelerated) one. If two twins start at the same event A and end at the same event C, where the first twin moves directly from A to C without accelerating, but the second twin does accelerate, then the first will always experience a longer time interval when they meet again.

It may seem outrageous to introduce such a strange concept of time-measure, at variance with our intuitive notions. However there is now an enormous amount of experimental evidence in favour of it. For example, there are many subatomic particles which decay (i.e. break up into other particles) in a definite time-scale. Sometimes such particles travel at speeds very close to that of light (e.g. in the cosmic rays that reach the earth from outer space, or in man-made particle accelerators), and their decay times become delayed in exactly the way that one would deduce from the above considerations. Even more impressive is the fact that clocks ('nuclear clocks') can now be made so accurate that these time-slowing effects are *directly* detectable by clocks transported by fast low-flying aeroplanes—agreeing with the Minkowskian 'distance' measure s, and *not* with t! (Strictly, taking the aeroplane's *height* into account, small additional gravitational effects of *general* relativity become involved, but these also agree with observation; see next section.) Moreover, there are many other effects, intimately connected with the whole framework of special relativity, which are constantly receiving detailed verification. One of these, Einstein's famous relation

$$E = mc^2,$$

which effectively equates energy and mass, will have some tantalizing implications for us at the end of this chapter!

I have not yet explained how the relativity principle is actually incorporated into this scheme of things. How is it that observers moving with different uniform velocities can be *equivalent* with respect to Minkowskian geometry? How can the time axis of Fig. 5.16 ('stationary observer') be completely equivalent to some other straight world-line, say OP extended ('moving observer')? Let us think about *Euclidean* geometry first. Clearly any two individual straight lines are quite equivalent to each other with regard to the geometry as a whole. One can imagine 'sliding' the whole Euclidean space 'rigidly over itself' until one straight line is taken to the position of the other one. Think of the *two*-dimensional case, a Euclidean *plane*. We can imagine moving a piece of paper rigidly over a plane surface, so that any straight line drawn on the paper comes into coincidence with a given straight line on the surface. This rigid motion preserves the structure of the geometry. A similar thing holds for Minkowskian geometry, although this is less obvious, and one has to be careful about what one means by 'rigid'. Now, in place of a piece of paper sliding, we must think of a peculiar kind of material—taking the two-dimensional case first, for simplicity—in which the 45° lines remain at 45° while the material can stretch in one 45°-direction and correspondingly squash in the other 45°-direction. This is illustrated in Fig. 5.20. In Fig. 5.21 I have tried to indicate what is involved in the three-dimensional case. This kind of 'rigid motion' of Minkowski space—called a *Poincaré motion* (or inhomogenious Lorentz motion)—may not look very 'rigid', but it preserves all the Minkowskian distances, and 'preserving all distances' is just what the word 'rigid' means in the Euclidean case. The principle of special relativity asserts that physics is unchanged under such Poincaré motions of space–time. In particular, the 'stationary' observer **S** whose world-line is the time-axis of our original Minkowski picture Fig. 5.16 has completely equivalent physics to the 'moving' observer **M** with world-line along OP.

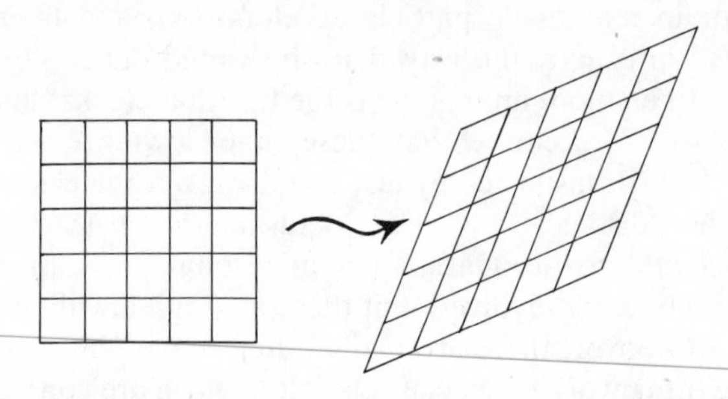

Fig. 5.20. A Poincaré motion in two space–time dimensions.

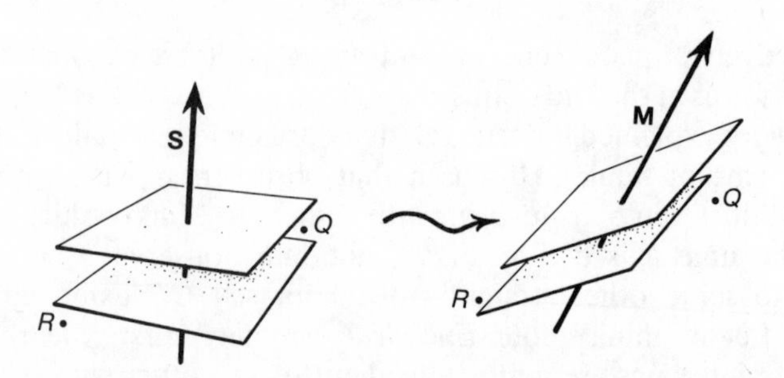

Fig. 5.21. A Poincaré motion in three space–time dimensions. The left diagram depicts simultaneous spaces for **S** and the right diagram, simultaneous spaces for **M**. Note that **S** thinks that R precedes Q, whereas **M** thinks that Q precedes R. (The motion is here thought of as *passive*, that is it only affects the different *descriptions* that the two observers **S** and **M** would make of one and the same space-time.)

Each coordinate plane t = constant represents 'space' at any one 'time' for the observer **S**, that is, it is a family of events which he would regard as *simultaneous* (i.e. taking place all at the 'same time'). Let us call these planes **S**'s *simultaneous spaces*. When we pass to another observer **M**, we must move our original family of simultaneous spaces to a new family by a Poincaré motion, so providing the simultaneous spaces for **M**.[14] Note that **M**'s simultaneous spaces look 'tipped up', in Fig. 5.21. This tipping up might seem to be in the wrong direction if we are thinking in terms of rigid motions in Euclidean geometry, but it is what we must expect in the Minkowskian case. While **S** thinks that all the events on any one t = constant plane occur simultaneously, **M** has a different view: to him, it is the events on each one of his 'tipped-up' simultaneous spaces that appear to be simultaneous! Minkowskian geometry does not, in itself, contain a unique concept of 'simultaneity'; but each uniformly moving observer carries with him his own idea of what 'simultaneous' means.

Consider the two events R and Q in Fig. 5.21. According to **S**, the event R takes place before the event Q, because R lies on an earlier simultaneous space than Q; but according to **M**, it is the other way around, Q lying on an earlier simultaneous space than R. Thus, to one observer, the event R takes place before Q, but for the other, it is Q that takes place before R! (This can only happen because R and Q are what are called *spatially separated*, which means that each lies outside the other's light cone, so that no material particle or photon can travel from one event to the other.) Even with quite slow relative velocities, significant differences in time-ordering will occur for events at great distances. Imagine two people walking slowly past each other in the street. The events on the Andromeda galaxy (the closest large galaxy to our own Milky Way, some 20000000000000000000 kilometres distant) judged by the two people to be simultaneous with the moment that they pass one another could amount to a difference of several days (Fig. 5.22). For one of the people, the space fleet launched with the intent to wipe out life on the planet Earth is already on its way; while for the other, the very decision about whether or not to launch that fleet has not yet even been made!

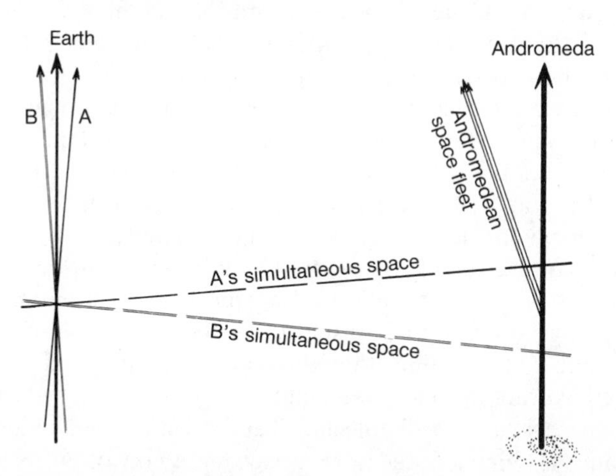

Fig. 5.22. Two people A and B walk slowly past each other, but they have differing views as to whether the Andromedean space fleet had been launched at the moment that they pass each other.

Einstein's general relativity

Recall Galileo's great insight that all bodies fall equally fast in a gravitational field. (It was an insight and not completely a direct observation for, owing to air resistance, feathers and rocks do *not* fall together! Galileo's insight was to realize that if air resistance could be reduced to zero, they *would* fall together.) It took three centuries before the deep significance of this insight was properly realized, and made the cornerstone of a great theory. That theory was Einstein's general relativity—an extraordinary description of gravity which, as we shall understand shortly, needed the concept of a *curved space–time* for its realization!

What has Galileo's insight to do with the idea of 'space–time curvature'? How could it be that such an idea, apparently so different from Newton's scheme whereby particles accelerate under ordinary gravitational forces, could reproduce, and even improve upon, all the superb precision of that theory? Moreover, can it really be true that Galileo's ancient insight contained something *not* already incorporated into Newton's theory?

Let me start with the last question, since it is the easiest to answer. What, according to Newton's theory, governs the acceleration of a body under gravity? First, there is the gravitational *force* on that body, which Newton's law of gravitational attraction tells us must be *proportional to the body's mass*. Second, there is the amount that the body accelerates *given* the force upon it, which by Newton's second law is *inversely proportional to the body's mass*. The thing that Galileo's insight depends upon is the fact that the 'mass' that occurs in Newton's gravitational force law is the *same* as the 'mass' in Newton's second law. ('Proportional to' would do, in place of 'same as'.) It is this that ensures that the body's acceleration under gravity is actually *independent* of its mass. There is nothing in Newton's general scheme that demands that these two concepts of mass be the same. This, Newton just *postulated*. Indeed, electric forces are similar to gravitational ones in that both are inverse square forces, but the force depends upon the *electric charge* which is totally different from the *mass* in Newton's second law. 'Galileo's insight' would not apply to electric forces: objects (charged objects, that is) 'dropping' in an electric field do *not* all 'fall' at the same speed!

For the moment, let us simply *accept* Galileo's insight—for motion under *gravity*—and ask for its implications. Imagine Galileo dropping two rocks from the Leaning Tower of Pisa. If there had been a video-camera on one of the rocks, pointing at the other, then the picture that it would provide would be of a rock hovering in space, seemingly *unaffected* by gravity (Fig. 5.23)! This holds precisely *because* all objects fall at the same speed under gravity.

Air resistance is being ignored here. Space flight now offers us a better test of these ideas, since there is effectively no air in space. Now, 'falling' in space means simply following the appropriate orbit under gravity. There is no need for this 'falling' to be straight downwards, towards the centre of the earth. There can be a horizontal component to the motion as well. If this horizontal component is large enough, then one can 'fall' right around the earth without getting any closer to the ground! Travelling in free orbit under gravity is just a sophisticated (and very expensive!) way of 'falling'. Now, just as with the video-camera picture above, an astronaut on a 'space walk' views his space vehicle as hovering before him, apparently unaffected by the gravitational force of the huge globe of the earth beneath him! (See Fig. 5.24.) Thus one can locally eliminate the effects of gravity by passing to the 'accelerating reference frame' of free fall.

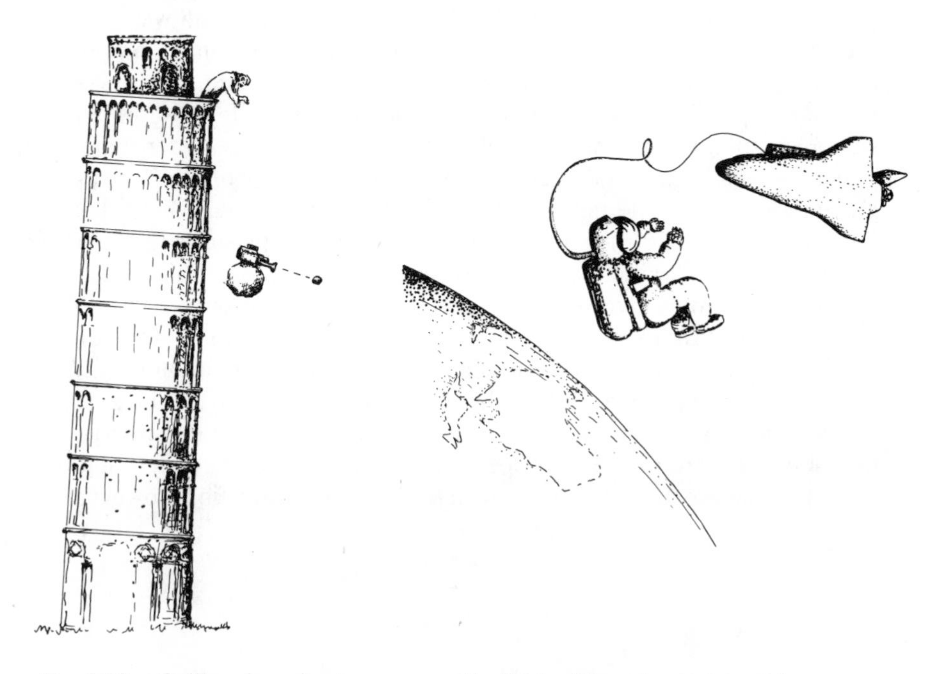

Fig. 5.23. Galileo dropping two rocks (and a video-camera) from the Leaning Tower of Pisa.

Fig. 5.24. The astronaut sees his space-vehicle hover before him, seemingly unaffected by gravity.

Gravity can be *cancelled* in this way by free fall because the effects of a gravitational field are just like those of an acceleration. Indeed, if you are inside a lift* which accelerates upwards, you simply experience an increase in the apparent gravitational field; and if downwards, then a decrease. If the cable suspending the lift were to break, then (ignoring air resistance and

* American readers: 'lift' means 'elevator'.

frictional effects) the resulting downward acceleration would completely cancel the effect of gravity, and the lift's occupants would appear to float freely—like the astronaut above—until it hits the ground! Even in a train or aeroplane, accelerations can be such that one's sensations as to the strength and direction of gravity may not coincide with where one's visual evidence suggests 'down' ought to be. This is because acceleration and gravitational effects *are* just like each other, so one's sensations are unable to distinguish one from the other. This fact—that the local effects of gravity are equivalent to those of an accelerating reference frame—is what Einstein referred to as *the principle of equivalence.*

The above considerations are 'local'. However, if one is allowed to make (not quite local) measurements of sufficient precision, one may, in principle, ascertain a *difference* between a 'true' gravitational field and a pure acceleration. In Fig. 5.25 I have shown, a little exaggerated, how an initially stationary spherical arrangement of particles, falling freely under the earth's gravity, would begin to be affected by the *non-uniformity* of the (Newtonian) gravitational field. The field is non-uniform for two reasons. First, because the centre of the earth lies some finite distance away, particles nearer to the earth's surface will accelerate downwards faster than those which are higher up (recall Newton's inverse square law). Second, for the same reason, there will be slight differences in the *direction* of this acceleration, for different horizontal displacements of the particles. Because of this non-uniformity, the spherical shape begins to distort slightly—into an 'ellipsoid'. It lengthens in the direction of the earth's centre (and also in the opposite direction), since the parts nearer to the centre experience a slightly greater acceleration than do the parts more distant; it narrows in the horizontal directions, owing to the

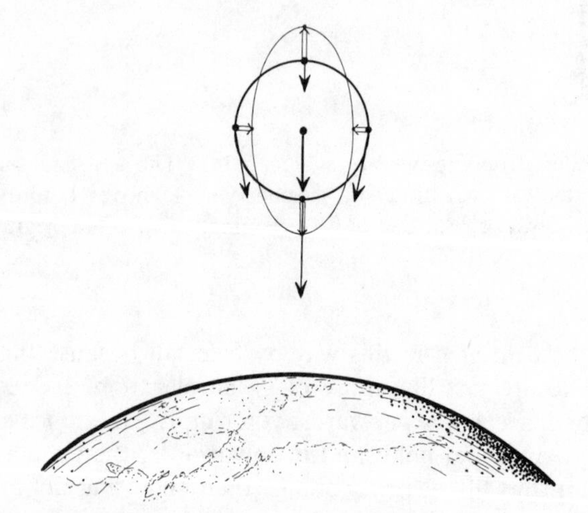

Fig. 5.25. The tidal effect. Double arrows show relative acceleration (**WEYL**).

fact that the accelerations there act slightly inwards, in the direction of the earth's centre.

This distortional effect is known as the *tidal* effect of gravity. If we replace the earth's centre by the moon, and the sphere of particles by the surface of the earth, then we have precisely the moon's influence in raising tides on the earth, with bulges being produced both in the direction towards the moon and away from it. The tidal effect is a general feature of gravitational fields which cannot be 'eliminated' by free fall. This effect measures non-uniformity in the Newtonian gravitational field. (The *magnitude* of the tidal distortion actually falls off as the inverse *cube* of the distance from the centre of attraction, rather than as the inverse square.)

Newton's inverse square law of gravitational force turns out to have a simple interpretation in terms of this tidal effect: the *volume* of the ellipsoid into which the sphere initially[15] distorts is *equal* to that of the original sphere—taking the sphere to surround a vacuum. This volume property is characteristic of the inverse square law; it holds for no other law of force. Next, suppose that the sphere surrounds not a vacuum but some matter of total mass M. There will now be an additional inward component of the acceleration due to the gravitational attraction of this matter. The volume of the ellipsoid, into which our sphere of particles initially distorts, *shrinks*—in fact by an amount that is *proportional to* M. An example of the volume-reducing effect would occur if we take our sphere to surround the earth at a constant height (Fig. 5.26). Then the ordinary downward (i.e. inward) acceleration due to the earth's gravity is what causes the volume of our sphere to reduce. This volume-reducing property codes the remaining part of Newton's gravitational force law, namely that the force is proportional to the mass of the *attracting body*.

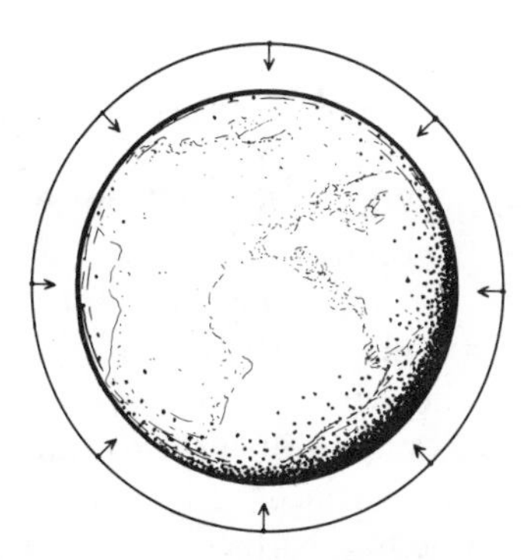

Fig. 5.26 When the sphere surrounds matter (here the earth), there is a net inwards acceleration (**RICCI**).

Let us try to obtain a space–time picture of this situation. In Fig. 5.27, I have indicated the world-lines of the particles of our spherical surface (drawn as a circle in Fig. 5.25), where I have made the description in a frame for which the point at the centre of the sphere appears to be at rest ('free fall'). The viewpoint of general relativity is to regard the freely falling motions as 'natural motions'—the analogues of 'uniform motion in a straight line' that one has in gravity-free physics. Thus, we *try* to think of free fall as described by 'straight' world-lines in the space-time! However, by the look of Fig. 5.27, it would be confusing to use the *word* 'straight' for this, and we shall, as a matter of terminology, call the world-lines of freely falling particles *geodesics* in space–time.

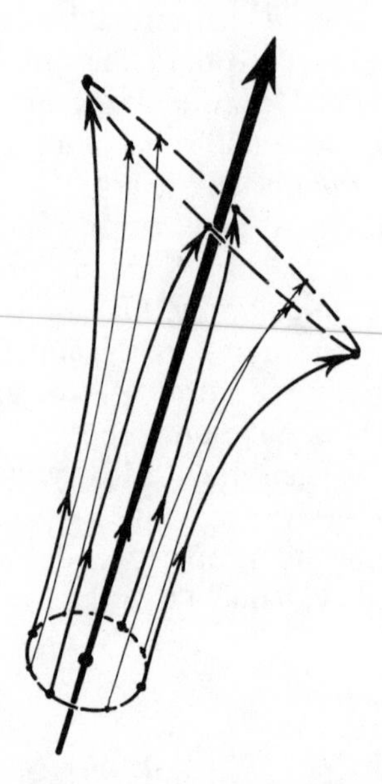

Fig. 5.27. Space–time curvature: the tidal effect depicted in space–time.

Is this a good terminology? What is normally meant by a 'geodesic'? Let us examine an analogy for a two-dimensional curved surface. The geodesics are the curves on that surface which (locally) are 'shortest routes'. Thus, if we think of a piece of string stretched over the surface (and not too long, or it might slip away) then the string will lie along a geodesic on the surface. In Fig. 5.28, I have indicated two examples of surfaces, the first with what is called 'positive curvature' (like the surface of a sphere) and the second with 'negative curvature' (a saddle-like surface). For the positive-curvature surface, two neighbouring geodesics which start off parallel to one another will, if we follow them along, begin to bend *towards* each other; for negative curvature, they begin to bend *away* from each other. If we imagine that the world-lines of freely falling particles are in some sense like geodesics on a surface, then we see that there is a close analogy between the gravitational tidal effect, discussed above, and the curvature effects of a surface—but now the positive and negative curvature effects are *both* present. Look at Figs 5.25, 5.27. We see that our space–time 'geodesics' begin to move *away* from each other in one direction (when they are lined up in the direction of the earth)—as with the *negative* curvature surface of Fig. 5.28—and they begin to move *towards* each other in other directions (when they are displaced horizontally, in relation to the earth)—as with the *positive* curvature surface of Fig. 5.28. Thus, our space–time indeed seems to possess a 'curvature', analogous to that of our two surfaces, but more complicated because of the higher dimensionality, and mixtures of both positive and negative curvatures are involved for different displacements.

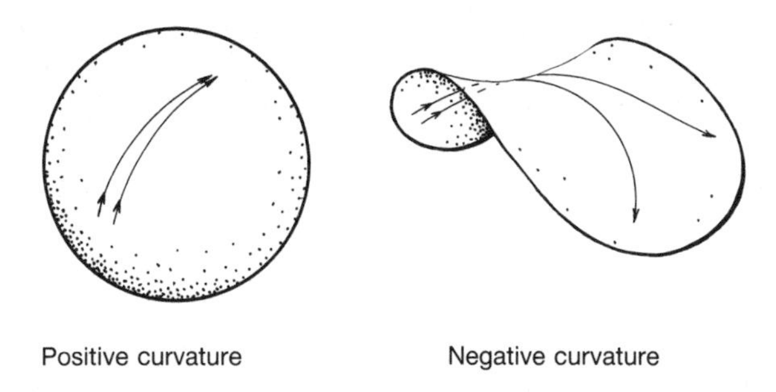

Fig. 5.28. Geodesics on a curved surface. With positive curvature the geodesics converge, whereas with negative curvature they diverge.

This shows how a concept of 'curvature' for space–time can be used to describe the action of gravitational fields. The possibility of using such a description follows ultimately from Galileo's insight (the principle of equivalence) and it allows us to eliminate gravitational 'force' by free fall. In fact, nothing that I have said so far requires us to go outside Newtonian theory. This new picture provides merely a *reformulation* of that theory.[16] However, new physics does come in when we try to combine this picture with what we have learnt from Minkowski's description of *special relativity*—the geometry of space–time that we now know applies in the *absence* of gravity. The resulting combination is Einstein's *general relativity*.

Recall what Minkowski has taught us. We have (in the absence of gravity) a space–time with a peculiar kind of 'distance' measure defined between points: if we have a world-line in the space–time, describing the history of some particle, then Minkowskian 'distance' measured along the world-line describes the *time* that is actually experienced by that particle. (In fact, in the previous section, we considered this 'distance' only along world-lines consisting of straight pieces, but the statement applies also to curved world-lines, where the 'distance' is measured along the curve.) Minkowski's geometry is taken to be exact if there is no gravitational field—i.e. no space–time curvature. But when gravity is present, we regard Minkowski's geometry as only approximate—in the same way as a flat surface gives only an approximate description to the geometry of a curved surface. If we imagine that we take a more and more powerful microscope to examine a curved surface—so that the geometry of the surface appears stretched out to greater and greater dimensions—then the surface appears to be flatter and flatter. We say that a curved surface is *locally* like a Euclidean plane.[17] In the same way, we can say that, in the presence of gravity, space–time is *locally* like Minkowski's geometry (which is *flat* space–time), but we allow some 'curviness' on a larger scale (see Fig. 5.29). In particular, any point of the space–time is the vertex of a *light cone*, just as it is in Minkowski space, but these light cones are not arranged in the completely uniform way that they are in Minkowski space. We shall see, in Chapter 7, some examples of space–time models for which this non-uniformity is manifest: cf. Figs 7.13, 7.14 on p. 333. Material particles have, for their world-lines, curves which are always directed *inside* the light cones, and photons have curves directed *along* the light cones. Also, along any such curve, there is a concept of Minkowski 'distance' which measures the time experienced by the particles, just as in Minkowski space. As with a curved surface, this distance measure defines a *geometry* for the surface which may differ from the geometry of the flat case.

Geodesics in the space–time can now be given an interpretation similar to that for those on two-dimensional surfaces considered above, where we must bear in mind the differences between the Minkowskian and Euclidean situations. Thus, rather than being curves of (locally) minimum length, our

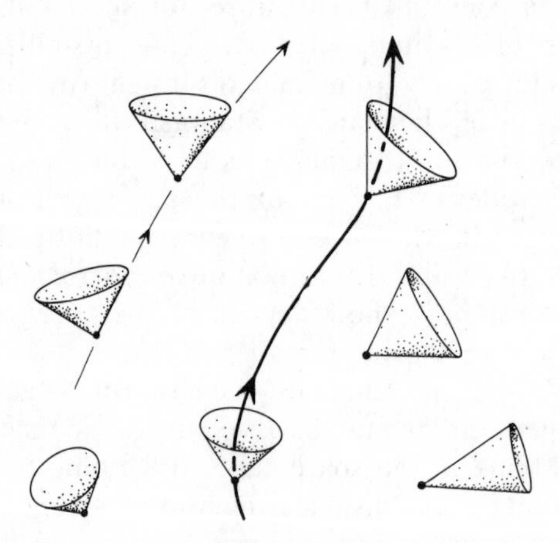

Fig. 5.29. A picture of curved space–time.

geodesic world-lines in space–time are curves which (locally) *maximize* the 'distance' (i.e. time) along the world-line. The world-lines of particles in free motion under gravity *are* actually geodesics according to this rule. Thus, in particular, heavenly bodies moving in a gravitational field are well described by such geodesics. Moreover, rays of light (world-lines of photons) in empty space are also geodesics, but this time geodesics of *zero* 'length'.[18] As an example, I have schematically indicated, in Fig. 5.30, the world-lines of the earth and sun, the earth's motion about the sun being a 'corkscrew'-like geodesic around the sun's world-line. I have also indicated a photon reaching the earth from a distant star. Its world-line appears slightly 'bent' owing to the fact that light, according to Einstein's theory, is actually *deflected* by the gravitational field of the sun.

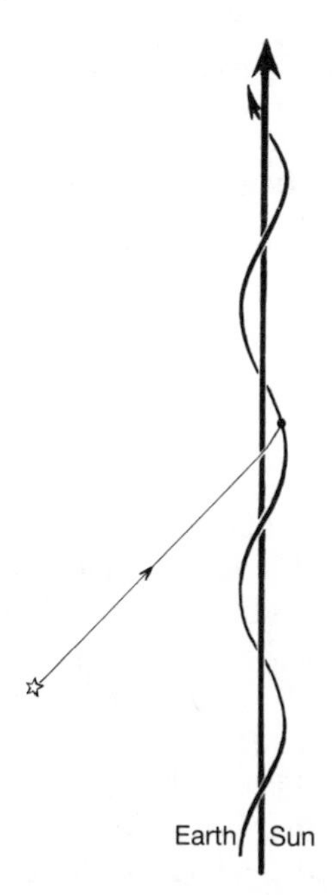

Fig. 5.30. World-lines of the earth and sun, and a light ray from a distant star being deflected by the sun.

We have yet to see how Newton's inverse square law is to be incorporated—and modified appropriately in accordance with Einstein's relativity. Let us return to our sphere of particles dropping in a gravitational field. Recall that if the sphere surrounds only vacuum, then according to Newton's theory, the volume of the sphere initially does not change; but if matter of total mass M is surrounded, then there is a volume reduction proportional to M. In Einstein's theory, the rules are just the same (for a small sphere), except that it is not exactly M that determines the volume change; there is an additional (normally very tiny) contribution from the *pressure* in the material surrounded.

The full mathematical expression for the curvature of four-dimensional space–time (which must describe the tidal effects for particles travelling in any possible direction at any given point) is given by something called the *Riemann curvature tensor*. This is a somewhat complicated object, needing twenty real numbers at each point in order to specify it. These twenty numbers are referred to as its *components*. The different components refer to the different curvatures in different directions in the space–time. The Riemann curvature tensor is usually written R_{ijkl}, but since I have no desire to explain, here, what all those little indices mean (nor, indeed, what a tensor actually *is*), I shall write it simply as:

RIEMANN.

There is a way in which this tensor can be split into two parts, called the *Weyl* tensor and the *Ricci* tensor (with ten *components each*). I shall write this splitting schematically as:

RIEMANN = WEYL + RICCI.

(The detailed expressions would not particularly help us here.) The Weyl tensor **WEYL** measures a *tidal distortion* of our sphere of freely falling particles (i.e. an initial change in shape, rather than in size), and the Ricci tensor **RICCI** measures its initial *change in volume*.[19] Recall that Newtonian gravitational theory demands that the *mass* surrounded by our falling sphere be proportional to this initial volume reduction. What this tells us, roughly speaking, is that the *mass* density of matter—or, equivalently the *energy* density (because of $E = mc^2$)—should be *equated* to the Ricci tensor.

In fact, this is basically what the field equations of general relativity—namely *Einstein's field equations*—actually assert.[20] However, there are certain technicalities about this which it is better for us not to get involved with. Suffice it to say that there is an object called the *energy–momentum* tensor, which organizes all the relevant information concerning the energy, pressure and momentum of matter and electromagnetic fields. I shall refer to this tensor as **ENERGY**. Then Einstein's equations become, very schematically:

RICCI = ENERGY.

(It is the presence of 'pressure' in the tensor **ENERGY**, together with some consistency requirements for the equations as a whole, that demand that the pressure also contributes to the volume-reducing effect described above.)

This equation seems to say nothing about the Weyl tensor. But it is an important quantity. The tidal effect that is experienced in empty space is entirely due to **WEYL**. In fact, the above Einstein's equations imply that there are *differential* equations connecting **WEYL** with **ENERGY**, rather like the Maxwell equations that we encountered earlier.[21] Indeed, a fruitful point of view is to regard **WEYL** as a kind of gravitational analogue of the electromagnetic field quantity (actually also a tensor—the Maxwell tensor) described by the pair $(\boldsymbol{E},\boldsymbol{B})$. Thus, in a certain sense, **WEYL** actually measures the *gravitational* field. The 'source' for **WEYL** is the tensor **ENERGY**, which is analogous to the fact that the source for the electromagnetic field $(\boldsymbol{E},\boldsymbol{B})$ is $(\rho,\boldsymbol{j})$, the collection of charges and currents of Maxwell's theory. This point of view will be helpful for us in Chapter 7.

It may seem remarkable, when we bear in mind such striking differences in formulation and underlying ideas, that it is difficult to find observational differences between Einstein's theory and the theory that had been put forward by Newton two and a half centuries earlier. But provided that the velocities under consideration are small compared with the speed of light c, and that the gravitational fields are not too strong (so that escape velocities are much smaller than c; cf. Chapter 7, p. 332) then Einstein's theory gives virtually identical results to that of Newton. However, Einstein's theory is more accurate in situations where the predictions of the two theories *do* differ. There are now several very impressive such experimental tests, and Einstein's newer theory is fully vindicated. Clocks run very slightly slow in a gravitational field, as Einstein maintained, this effect having now been directly measured in many different ways. Light and radio signals are indeed bent by the sun, and are slightly delayed by the encounter—again well-tested general relativity effects. Space-probes and planets in motion require small corrections to the Newtonian orbits, as demanded by Einstein's theory; these have also been experimentally verified. (In particular, the anomaly in the motion of the planet Mercury, known as the 'perihelion advance', which had been worrying astronomers since 1859, was explained by Einstein in 1915.) Perhaps most impressive of all is a set of observations on a system called the *binary pulsar*, consisting of a pair of tiny massive stars (presumably two 'neutron stars', cf. p. 331), which agrees very closely with Einstein's theory and indirectly verifies an effect that is totally absent in Newton's theory, the emission of *gravitational waves*. (A gravitational wave is the gravitational analogue of an electromagnetic wave, and it travels at light speed c.) No confirmed observations exist that contradict Einstein's general relativity. For all its initial strangeness, Einstein's theory is definitely with us!

Relativistic causality and determinism

Recall that, in relativity theory, material bodies cannot travel faster than light—in the sense that their world-lines must always lie within the light cones (cf. Fig. 5.29). (In general relativity, particularly, we need to state things in this local way. The light cones are not arranged uniformly, so it would not be very meaningful to say whether the velocity of a very *distant* particle exceeds the speed of light *here*.) The world-lines of photons lie *along* the light cones, but for no particle is it permitted that the world-line lie *outside* the cones. In fact, a more general statement must hold, namely that no *signal* is permitted to travel outside the light cone.

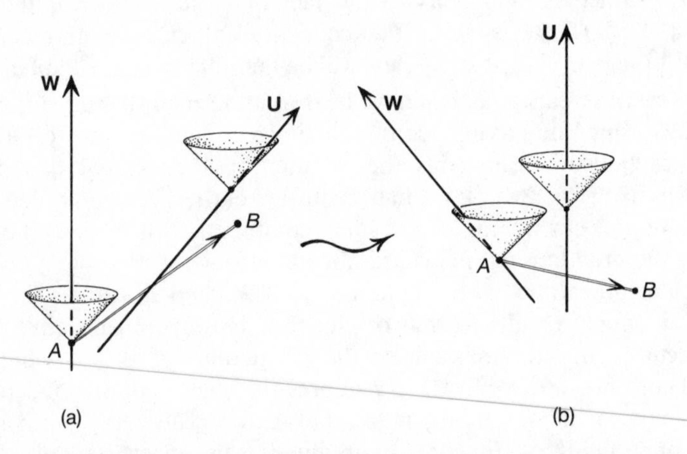

Fig. 5.31. A signal that is faster than light to the observer **W** will appear to travel backwards in time to the observer **U**. The right figure (b) is simply the left figure (a) redrawn from **U**'s point of view. (This redrawing can be thought of as a Poincaré motion. Compare Fig. 5.21.—but here the transformation from (a) to (b) is to be taken in an *active* rather than a passive sense.)

To appreciate why this should be so, consider our picture of Minkowski space (Fig. 5.31). Let us suppose that some device has been constructed which can send a signal at a speed a little greater than that of light. Using this device, the observer **W** sends a signal from an event A on his world-line to a distant event B, which lies just beneath A's light cone. In Fig. 5.31a this is drawn from **W**'s point of view, but in Fig. 5.31b, this is redrawn from the point of view of a second observer **U** who moves rapidly away from **W** (from a point between A and B, say), and for whom the event B appears to have occurred *earlier* than A! (This 'redrawing' is a Poincaré motion, as described above, p. 200.) From **W**'s viewpoint, the simultaneous spaces of **U** seem to be 'tipped up', which is why the event B can seem to **U** to be earlier than A. Thus, to **U**, the signal transmitted by **W** would seem to be travelling backwards in time!

This is not yet quite a contradiction. But by symmetry from **U**'s point of view (by the special relativity principle), a *third* observer **V**, moving away from **U** in the opposite direction from **W**, armed with the identical device to that of **W**, could also send a signal just faster than light, from *his* (i.e. **V**'s) viewpoint, back in the direction of **U**. This signal would also seem, to **U**, to be travelling backwards in time, now in the opposite spatial direction. Indeed **V** could transmit this second signal back to **W** the moment (B) that he receives the original one sent by **W**. This signal reaches **W** at an event C which is earlier, in **U**'s estimation, than the original emission event A (Fig. 5.32). But worse than this, the event C is actually earlier than the emission event A on **W**'s *own world-line*, so **W** actually *experiences* the event C to occur before he emits the signal at A! The message that the observer **V** sends back to **W** could, by prior arrangement with **W**, simply repeat the message he received at B. Thus, **W** receives, at an earlier time on his world-line, the very same message that he is to send out later! By separating the two observers to large enough distance, one can arrange that the amount by which the returning signal precedes the original signal is as great a time interval as we wish. Perhaps **W**'s original message is that he has broken his leg. He could receive the returning message *before* the accident has occurred and then (presumably), by the action of his free will, take action to avoid it!

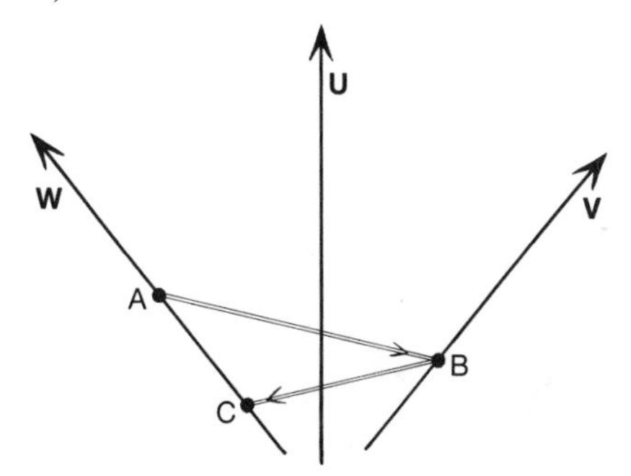

Fig. 5.32. If **V** is armed with a faster-than-light signalling device identical to that of **W**, but pointing in the opposite direction, it can be employed by **W** to send a message into his own past!

Thus, superluminary signalling, together with Einstein's relativity principle leads to a blatant contradiction with our normal feelings of 'free will'. Actually, the matter is even more serious than this. For we could imagine that perhaps 'observer **W**' is merely a mechanical device, programmed to send out the message 'YES' if it receives 'NO' and 'NO' if it receives 'YES'. Indeed, **V** may as well be a mechanical device too, but programmed to send back 'NO' if it receives 'NO' and 'YES' if it receives 'YES'. This leads to the same essential contradiction that we had before,[22] seemingly independent now of the question of whether or not the observer **W** actually has 'free will', and tells us that a faster-than-light signalling device is 'not on' as a physical possibility. This will have some puzzling implications for us later (Chapter 6, p. 286).

Let us accept, then, that *any* kind of signal—not merely signals carried by ordinary physical particles—must be constrained by the light cones. Actually the above argument uses *special* relativity, but in general relativity, the rules of the special theory do still hold locally. It is the local validity of special relativity that tells us that all signals are constrained by the light cones, so this should apply also in general relativity. We shall see how this affects the question of *determinism* in these theories. Recall that in the Newtonian (or Hamiltonian, etc.) scheme, 'determinism' means that *intial data* at *one particular time* completely fix the behaviour at all other times. If we take a space–time view in Newtonian theory, then the 'particular time' at which we specify data would be some three-dimensional 'slice' through the four-dimensional space–time (i.e. the whole of space at that one time). In relativity theory, there is no one global concept of 'time' which can be singled out for this. The usual procedure is to adopt a more flexible attitude. Anybody's 'time' will do. In special relativity, one can take some observer's simultaneous space on which to specify the initial data, in place of the 'slice' above. But in general relativity, the concept of a 'simultaneous space' is not very well defined. Instead one may use the more general notion of a *spacelike surface*.[23] Such a surface is depicted in Fig. 5.33; it is characterized by the fact that it lies completely outside the light cone at each of its points—so that *locally* it resembles a simultaneous space.

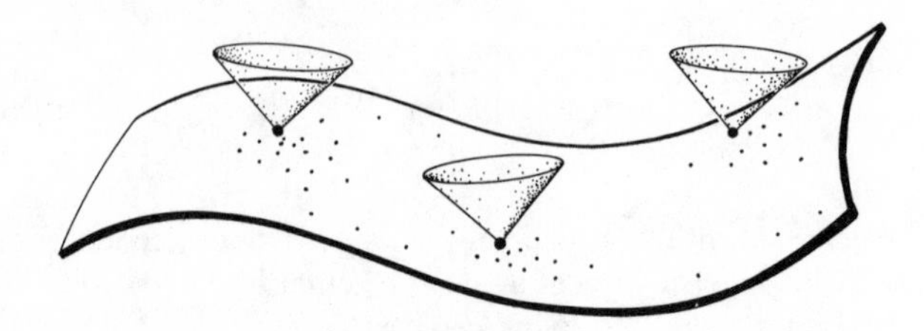

Fig. 5.33. A spacelike surface, for the specification of initial data in general relativity.

Determinism, in special relativity, can be formulated as the fact that initial data on any given simultaneous space **S** fixes the behaviour in the whole of the space–time. (This will be true, in particular, for Maxwell's theory—which is indeed a 'special-relativistic' theory.) There is a stronger statement that one can make, however. If we want to know what is going to happen at some event P lying somewhere to the future of S, then we only need the initial data in some bounded (finite) region of S, and not on the whole of S. This is because 'information' cannot travel faster than light, so any points of S which lie too far away for light signals to reach P can have no influence on P (see

Fig. 5.34).* This is actually much more satisfactory than the situation that arises in the Newtonian case, where one would, in principle, need to know what was going on over the entire *infinite* 'slice' in order to make any prediction at all about what is going to happen at any point a moment later. There is no restriction on the speed with which Newtonian information can propagate and, indeed, Newtonian forces are *instantaneous*.

'Determinism' in *general* relativity is a good deal more complicated a matter than in special relativity, and I shall make only a few remarks about it. In the first place, we must use a *spacelike surface S* for the specification of initial data (rather than just a simultaneous surface). Then it turns out that the Einstein equations do give a *locally* deterministic behaviour for the gravitational field, assuming (as is usual) that the matter fields contributing to the tensor **ENERGY** behave deterministically. However, there are considerable complications. The very geometry of the space–time—including its light-cone 'causality' structure—is now part of what is being actually determined. We do not know this light cone structure ahead of time, so we cannot tell which parts of *S* will be needed to determine the behaviour at some future event *P*. In some extreme situations it can be the case that even *all* of *S* may be insufficient, and global determinism is consequently lost! (Difficult questions are involved here and they relate to an important unsolved problem in general relativity theory called 'cosmic censorship'—which has to do with the formation of *black holes* (Tipler *et al.* 1980); cf. Chapter 7, p. 332, cf. also footnote on pp. 335, and p. 342.) It would seem to be highly unlikely that any possible such 'failure of determinism' that might occur with 'extreme'

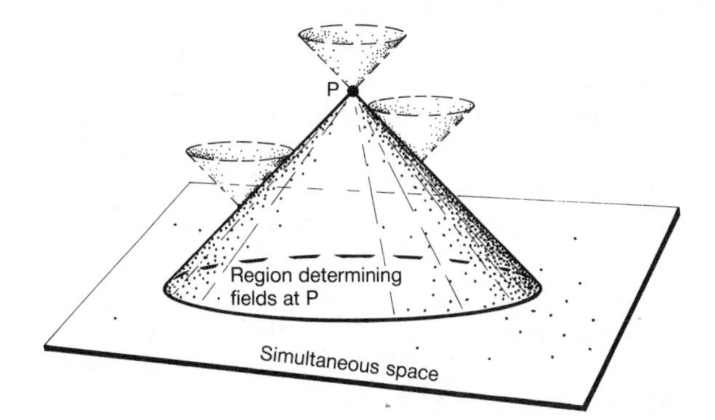

Fig. 5.34. In special relativity, what happens at P depends only on the data in a *finite* region of a simultaneous space. This is because effects cannot travel to P faster than light.

* It may be remarked that the wave equation (cf. footnote on p. 188) is, like Maxwell's equations, also a relativistic equation. Thus the Pour-El–Richards 'non-computability phenomenon' that we considered earlier is an effect which also refers only to initial data in bounded regions of *S*.

gravitational fields could have any direct bearing on matters at the human scale of things, but we see from this that the question of determinism in general relativity is not at all so clear-cut as one might wish it to be.

Computability in classical physics: where do we stand?

Throughout this chapter, I have tried to keep one eye on the issue of *computability*, as distinct from that of determinism, and I have tried to indicate that computability issues may be at least as important as those of determinism when it comes to the questions of 'free will' and mental phenomena. But determinism itself has turned out to be not quite so clear-cut, in classical theory, as we have been led to believe. We have seen that the classical Lorentz equation for the motion of a charged particle yields some disturbing problems. (Recall Dirac's 'runaway solutions'.) We have noted, also, that there are some difficulties for determinism in general relativity. When, in such theories, there is no determinism, there is certainly no computability. Yet in neither of the cases just cited would it seem that the lack of determinism has much direct philosophical relevance for ourselves. There is yet no 'home' for our free wills in such phenomena: in the first case, because the classical Lorentz equation for a point particle (as resolved by Dirac) is not thought to be physically appropriate at the level where these problems arise; and in the second, because the scales at which classical general relativity might lead to such problems (black holes, etc.) are totally different from the scales of our own brains.

Now, where do we stand with regard to *computability* in classical theory? It is reasonable to guess that, with general relativity, the situation is not significantly different from that of special relativity—over and above the differences in causality and determinism that I have just been presenting. Where the future behaviour of the physical system is determined from initial data, then this future behaviour would seem (by similar reasoning to that which I presented in the case of Newtonian theory) also to be *computably* determined by that data[24] (apart from the 'unhelpful' type of non-computability encountered by Pour-El and Richards for the wave equation, as considered above—and which does not occur for *smoothly* varying data). Indeed, it is hard to see that in *any* of the physical theories that I have been discussing so far there can be any significant 'non-computable' elements. It is certainly to be expected that 'chaotic' behaviour can occur in many of these theories, where very slight changes in initial data can give rise to enormous differences in resulting behaviour. (This seems to be the case in general relativity, cf. Misner 1969, Belinskii *et al.* 1970.) But, as I mentioned before, it is hard to see how *this* type of non-computability—i.e. 'unpredictability'—could be of any 'use' in a device which tries to 'harness' possible non-computable elements in physical laws. If the 'mind' can be in any way making use of non-computable elements, then it would appear that they must be elements lying outside classical physics. We shall need to re-examine this question later, after we have taken a look at quantum theory.

Mass, matter, and reality

Let us briefly take stock of that picture of the world that classical physics has presented us with. First, there is space–time, playing a primary role as the arena for all the varied activity of physics. Second, there are *physical objects*, indulging in this activity, but constrained by precise mathematical laws. The physical objects are of two kinds: *particles* and *fields*. Of the particles, little is said about their actual nature and distinguishing qualities, save that each has its own world-line and possesses an individual (rest-)mass and perhaps electric charge, etc. The fields, on the other hand, are given very specifically—the electromagnetic field being subject to the Maxwell equations and the gravitational field to the Einstein equations.

There is some ambivalence about how the particles are to be treated. If the particles have such tiny mass that their own influence on the fields can be neglected, then the particles are called *test particles*—and then their motion in *response* to the fields is unambiguous. The Lorentz force law describes the response of test particles to the electromagnetic field and the geodesic law, their response to the gravitational field (in an appropriate combination, when both fields are present). For this, the particles must be considered as *point* particles, i.e. having one-dimensional world-lines. However, when the effects of the particles on the fields (and hence on other particles) need to be considered—i.e. the particles act as *sources* for the fields—then the particles must be considered as objects spread out, to some extent, in space. Otherwise the fields in the immediate neighbourhood of each particle become infinite. These spread-out sources provide the charge-current distribution $(\rho, \boldsymbol{j})$ that is needed for Maxwell's equations and the tensor **ENERGY** that is needed for Einstein's equations. In addition to all this, the space–time—in which all these particles and fields reside—has a variable structure which itself directly describes gravity. The 'arena' joins in the very action taking place within itself!

This is what classical physics has taught us about the nature of physical reality. It is clear that a great deal has been learnt, and also that we should not be too complacent that the pictures that we have formed at any one time are not to be overturned by some later and deeper view. We shall see in the next chapter that even the revolutionary changes that relativity theory has wrought will pale almost to insignificance in comparison with those of the quantum theory. However, we are not quite finished with the classical theory, and with what it has to say to us about material reality. It has one further surprise for us!

What *is* 'matter'? It is the real substance of which actual physical objects—the 'things' of this world—are composed. It is what you, I and our houses are made of. How does one *quantify* this substance? Our elementary physics text-books provide us with Newton's clear answer. It is the *mass* of an object, or of a system of objects, which measures the quantity of matter that it contains. This indeed seems right—there is no other physical quantity that can seriously compete with mass as the true measure of total substance. Moreover it is *conserved*: the mass, and therefore the total matter content, of any system whatever must always remain the same.

Yet Einstein's famous formula from special relativity

$$E = mc^2$$

tells us that mass (m) and energy (E) are interchangeable into one another. For example, when a uranium atom decays, splitting up into smaller pieces, the total mass of each of these pieces, if they could be brought to rest, would be *less* than the original mass of the uranium atom; but if the *energy of motion—kinetic* energy, cf. p. 165*—of each piece is taken into account, and then converted to mass values by dividing by c^2 (by $E = mc^2$), then we find that the total is actually *unchanged*. Mass is indeed conserved but, being partly composed of energy, it now seems less clearly to be the measure of actual substance. Energy, after all, depends upon the speed with which that substance is travelling. The energy of motion in an express train is considerable, but if we happen to be sitting in that train, then according to our own viewpoint, the train possesses no motion at all. The energy of that motion (though not the *heat energy* of the random motions of individual particles) has been 'reduced to zero' by this suitable choice of viewpoint. For a striking example, where the effect of Einstein's mass–energy relation is at its most extreme, consider the decay of a certain type of subatomic particle, called a π^0-meson. It is certainly a *material* particle, having a well-defined (positive) mass. After some 10^{-16} of a second it disintegrates (like the uranium atom above, but very much more rapidly)—almost always into just *two photons* (Fig. 5.36). For an observer at rest with the π^0-meson, each photon carries away half the energy and, indeed, half of the π^0-meson's mass. Yet this photon 'mass' is of the nebulous kind: *pure energy*. For if we were to travel rapidly in the direction of one of the photons, then we could reduce its mass–energy to as small a value as we please—the intrinsic mass (or *rest*-mass, as we shall come to shortly) of a photon being actually *zero*. All this makes for a consistent picture of conserved mass, but it is not quite the one we had before. Mass can still, in a sense, measure 'quantity of matter', but there has

* In Newtonian theory, the kinetic energy is $\frac{1}{2}mv^2$ where m is the mass and v the speed; but in special relativity, the expression is somewhat more complicated.

been a distinct change of viewpoint: since mass is equivalent to energy, the mass of a system depends, like energy, upon the motion of the observer!

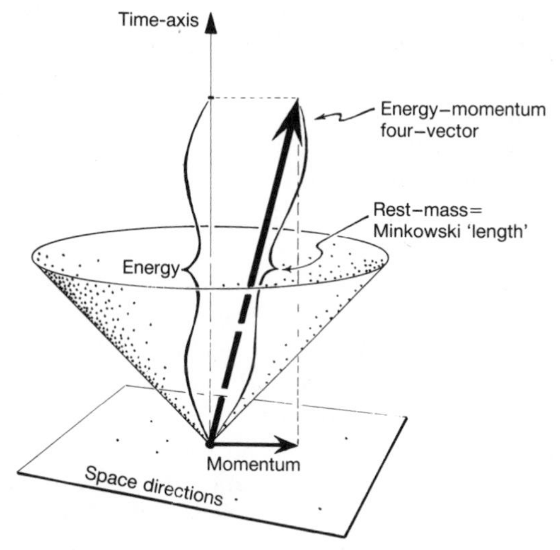

Fig. 5.35. The energy–momentum four–vector.

It is worth while to be somewhat more explicit about the viewpoint that we have been led to. The conserved quantity that takes over the role of mass is an entire object called the *energy–momentum four-vector*. This may be pictured as an arrow (vector) at the origin O in Minkowski space, pointing *inside* the future light cone at O (or, in the extreme case of a photon, *on* this cone); see Fig. 5.35. This arrow, which points in the same direction as the object's

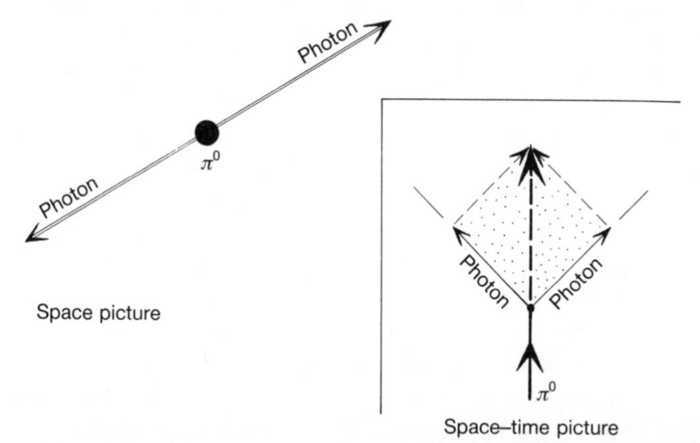

Fig. 5.36. A massive π°-meson decays into two massless photons. The space–time picture shows how the four-vector of energy–momentum is conserved: the π°-meson's four-vector is the sum of the two photon's four-vectors, added according to the parallelogram law (shown shaded).

world-line, contains all the information about its energy, mass, and momentum. Thus the 't-value' (or 'height') of the tip of this arrow, as measured in some observer's frame, describes the *mass* (or *energy* divided by c^2) of the object, according to that observer, while the spatial components provide the *momentum* (divided by c).

The Minkowskian 'length' of this arrow is an important quantity, known as the *rest-mass*. It describes the mass for an observer at rest with the object. One might try to take the view that *this* would be a good measure of 'quantity of matter'. However, it is not additive: if a system splits into two, then the original rest-mass is not the sum of the resulting two rest-masses. Recall the π^0-meson decay considered above. The π^0-meson has a positive rest-mass, while the rest-masses of each of the two resulting photons is zero. However, this additivity property does indeed hold for the whole arrow (four-vector), where we must now 'add' in the sense of the *vector* addition law depicted in Fig. 5.6. It is now this *entire arrow* which is our measure of 'quantity of matter'!

Think now of Maxwell's electromagnetic field. We have noted that it carries energy. By $E = mc^2$, it must also have mass. Thus, Maxwell's field is also matter! This must now certainly be accepted since Maxwell's field is intimately involved in the forces which bind particles together. There must be a substantial contribution[25] to any body's mass from the electromagnetic fields within it.

What about Einstein's gravitational field? In many ways it resembles Maxwell's field. Similarly to the way that, with Maxwell's theory, charged bodies in motion can emit *electromagnetic* waves, massive bodies in motion can (according to Einstein's theory) emit *gravitational* waves (cf. p. 211)—which, like electromagnetic waves, travel with the speed of light and carry energy. Yet this energy is not measured in the standard way, which would be by the tensor **ENERGY**, referred to before. In a (pure) gravitational wave, this tensor is actually *zero* everywhere! One might take the view, nevertheless, that somehow the *curvature* of space–time (now given entirely by the tensor **WEYL**) can represent the 'stuff' of gravitational waves. But gravitational energy turns out to be *non-local*, which is to say that one cannot determine what the measure of this energy is by merely examining the curvature of space–time in limited regions. The energy—and therefore the mass—of a gravitational field is a slippery eel indeed, and refuses to be pinned down in any clear location. Nevertheless, it must be taken seriously. It is certainly *there*, and has to be taken into account in order that the concept of mass can be conserved overall. There is a good (and positive) measure of mass (Bondi 1960, Sachs 1962) which applies to gravitational waves, but the non-locality is such that it turns out that this measure can sometimes be *non-zero* in *flat* regions of space–time—between two bursts of radiation (rather like the calm within the eye of a hurricane)—where the space–time is actually completely *free* of curvature (cf. Penrose and Rindler 1986, p. 427) (i.e. *both* **WEYL** and **RICCI** are zero)! In such cases, we seem to be driven to deduce that if this mass–energy is to be located at all, it must be in this *flat empty space*—a region completely free of matter or fields of any kind. In these curious circumstances, our 'quantity of matter' is now either *there*, in the emptiest of empty regions, or it is nowhere at all!

This seems to be pure paradox. Yet, it is a definite implication of what our best classical theories—and they are indeed superb theories—are telling us about the nature of the 'real' material of our world. Material reality according to classical theory, let alone in the quantum theory that we are about to explore, is a much more nebulous thing than one had thought. Its quantification—and even whether it is there or not—depends upon distinctly subtle issues and cannot be ascertained merely locally! If such non-locality seems puzzling, be prepared for much greater shocks to come.

1. It is a striking fact that *all* the established departures from the Newtonian picture have been, in some fundamental way, associated with the behaviour of *light*. First, there are the disembodied energy-carrying fields of Maxwell's electromagnetic theory. Second, there is, as we shall see, the crucial role that the speed of light plays in Einstein's special relativity. Third, the tiny deviations from Newtonian gravitational theory that Einstein's *general* relativity exhibits become significant only when speeds need to be compared with that of light. (Light deflection by the sun, Mercury's motion, escape velocities compared with that of light in black holes, etc.) Fourth, there is the wave-particle duality of the quantum theory, first observed in the behaviour of light. Finally there is quantum electrodynamics, which is the quantum field theory of light and charged particles. It is reasonable to speculate that Newton himself would have been ready to accept that deep problems for his picture of the world lay hidden in the mysterious behaviour of light cf. Newton (1730); also Penrose (1987*a*).
2. There is a magnificent body of well-established physical understanding—namely the *thermodynamics* of Carnot, Maxwell, Kelvin, Boltzmann, and others—that I have omitted to classify. This may be puzzling to some readers, but the omission has been deliberate. For reasons that may become clearer in Chapter 7, I should myself be rather reluctant to place thermodynamics, as it stands, in the actual category of a SUPERB theory. However, many physicists would probably regard it as *sacrilege* to place such a beautiful and fundamental body of ideas in a category so low as merely USEFUL! In my view, thermodynamics, as it is normally understood, being something that applies only to *averages*, and not to the individual constituents of a system—and being partly a deduction from other theories—is not quite a physical theory in the sense that I mean it here (the same applies to the underlying mathematical framework of *statistical mechanics*). I am using this fact as an excuse to evade the problem and leave it out of the classification altogether. As we shall see in Chapter 7, I am claiming an intimate

relation between thermodynamics and a topic that I have mentioned above as belonging to the USEFUL category, namely the standard model for the big bang. An appropriate union between these two sets of ideas (partly lacking at present) ought, I believe, to be regarded as a physical theory in the sense required—even belonging to the category SUPERB. This is something that we shall need to come back to later.

3. My colleagues have asked me where I would place 'twistor theory'—an elaborate collection of ideas and procedures with which I have myself been associated over a great number of years. Insofar as twistor theory *is* a different theory of the physical world, it cannot be other than in the category TENTATIVE; but to a large extent it is not one at all, being a mathematical transcription of earlier well-established physical theories.
4. Newton's name is being attached to this model—and indeed to 'Newtonian' mechanics as a whole—merely as a convenient *label*. Newton's own views as to the *actual* nature of the physical world seem to have been much less dogmatic and more subtle than this. (The person who most forcefully promoted this 'Newtonian' model appears to have been R. G. Boscovich 1711–1787.)
5. As suggested in Chapter 4 (note 9, p.148), the new Blum–Shub–Smale (1989) theory may turn out to provide a way of resolving some of these issues in a mathematically more acceptable way.
6. Hamilton's actual equations, though perhaps not quite his particular viewpoint, were already known to the great Italian/French mathematician Joseph C. Lagrange (1736–1813) about 24 years before Hamilton. An equally important earlier development was the formulation of mechanics in terms of the *Euler–Lagrange equations*, according to which Newton's laws can be seen to be derived from one over-riding principle; the *principle of stationary action* (P. L. M. de Maupertuis). In addition to their great theoretical significance, the Euler–Lagrange equations provide calculational procedures of remarkable power and practical value.
7. The situation is actually 'worse' in the sense that Liouville's phase-space volume is just one of a whole family of 'volumes' of differing dimensions (referred to as Poincaré invariants) which remain constant under Hamiltonian evolution. However, I have been a little unfair in the sweepingness of my assertions. One can imagine a system in which physical degrees of freedom (contributing to some of the phase-space volume) can be 'dumped' somewhere that we are not interested in (such as radiation escaping to infinity) so the phase-space volume of the part *of interest for us* might become reduced.
8. This second fact, in particular, is a matter of tremendous good fortune for science, for without it the dynamical behaviour of large bodies might have been incomprehensible and would have given little hint as to the precise laws which could be applying to the particles themselves. My guess is that one reason for Newton's strong insistence on his third law was that, without it, this transference of dynamical behaviour from microscopic to macroscopic bodies would simply not hold.

   Another 'miraculous' fact, that was vital for the development of science, is that the inverse square law is the only power law (falling off with distance) for which the general orbits about a central body are simple geometrical shapes. What would Kepler have done if the law or force had been an inverse distance law or an inverse cube?

9. In effect, we have an *infinite* number of x_is and p_is; but a further complication is that we cannot just use the field values as these coordinates, a certain 'potential' for the Maxwell field being needed in order that the Hamiltonian scheme can apply.
10. That is, not twice differentiable.
11. The Lorentz equations tell us what the *force* on a charged particle is, due to the electromagnetic field in which it resides; then, if we know its mass, Newton's second law tells us the particle's acceleration. However, charged particles often move at speeds close to that of light, and special relativity effects begin to be important, affecting what the mass of the particle must actually be taken to be (see next section). It was reasons of this kind that delayed the discovery of the correct force law on a charged particle until the birth of special relativity.
12. In fact, in a sense, any quantum-mechanical particle in Nature acts as such a clock entirely on its own. As we shall see in Chapter 6, there is an oscillation associated with any quantum particle, whose frequency is proportional to the particle's mass; see p. 230. Very accurate modern clocks (atomic clocks, nuclear clocks) ultimately depend upon this fact.
13. The reader might worry that since a 'corner' occurs in the traveller's world-line at B, the traveller, as depicted, undergoes an infinite acceleration at the event B. This is inessential. With a finite acceleration, the traveller's world-line simply has the corner at B smoothed off, this making very little difference to the total time that he experiences, which is still measured by the Minkowskian 'length' of the whole world-line.
14. These are the spaces of events that would be judged by **M** to be simultaneous according to *Einstein's definition of simultaneity*, which uses light signals sent out by **M** and reflected back to **M** from the events in question. See, for example, Rindler (1982).
15. This is the initial *second* derivative in time (or 'acceleration') of the shape. The rate of change (or 'velocity') of the shape is taken to be zero initially, since the sphere starts at rest.
16. The mathematical description of this reformulation of Newtonian theory was first carried out by the remarkable French mathematician Élie Cartan (1923)—which, of course, was *after* Einstein's general relativity.
17. Curved spaces which are locally Euclidean in this sense (also in higher dimensions) are called *Riemannian manifolds*—named after the great Bernhard Riemann (1826–1866), who first investigated such spaces, following some important earlier work by Gauss on the two-dimensional case. Here we need a significant modification of Riemann's idea, namely allowing the geometry to be locally *Minkowskian*, rather than Euclidean. Such spaces are frequently called *Lorentzian* manifolds (belonging to a class called *pseudo*-Riemannian or, less logically, *semi*-Riemannian manifolds).
18. The reader may be worrying how this *zero* value can represent the *maximum* value of 'length'! In fact it does, but in a vacuous sense: a zero-length geodesic is characterized by the fact that there are *no other* particle world-lines connecting any pair of its points (locally).
19. In fact this division into distortion effects and volume-change is not quite so clear-cut as I have presented it. Ricci tensor can itself give a certain amount of tidal distortion. (With light rays the division *is* completely clear-cut; cf. Penrose

and Rindler (1986), Chapter 7.) For a precise definition of the Weyl and Ricci tensors, see, for example Penrose and Rindler (1984) pp.240, 210. (The German-born Hermann Weyl was an oustanding mathematical figure of this century; the Italian Gregorio Ricci was a highly influential geometer who founded the theory of tensors in the last century.)

20. The correct form of the actual equations was found also by David Hilbert in November 1915, but the physical ideas of the theory are all solely due to Einstein.
21. For those who know about such matters, these differential equations are the full *Bianchi identities* with Einstein's equations substituted into them.
22. There are some (not very satisfying) ways around this argument, cf. Wheeler and Feynman (1945).
23. Technically, the term 'hypersurface' is more appropriate than 'surface' since it is three- rather than two-dimensional.
24. Rigorous *theorems* concerning these issues would be very helpful and interesting. These are, at present, lacking.
25. Incalculable, on present theory—which gives the (provisional) unhelpful answer: infinity!

6
Quantum magic and quantum mystery

Do philosophers need quantum theory?

In classical physics there is, in accordance with common sense, an objective world 'out there'. That world evolves in a clear and deterministic way, being governed by precisely formulated mathematical equations. This is as true for the theories of Maxwell and Einstein as it is for the original Newtonian scheme. Physical reality is taken to exist independently of ourselves; and exactly how the classical world 'is' is not affected by how we might choose to look at it. Moreover, our bodies and our brains are themselves to be part of that world. They, also, are viewed as evolving according to the same precise and deterministic classical equations. All our actions are to be fixed by these equations—no matter how we might feel that our conscious wills may be influencing how we behave.

Such a picture appears to lie at the background of most serious[1] philosophical arguments concerned with the nature of reality, of our conscious perceptions, and of our apparent free will. Some people might have an uncomfortable feeling that there should also be a role for *quantum theory*—that fundamental but disturbing scheme of things which, in the first quarter of this century, arose out of observations of subtle discrepancies between the actual behaviour of the world and the descriptions of classical physics. To many, the term 'quantum theory' evokes merely some vague concept of an 'uncertainty principle', which, at the level of particles, atoms or molecules, forbids precision in our descriptions and yields merely probabilistic behaviour. Actually, quantum descriptions *are* very precise, as we shall see, although radically different from the familiar classical ones. Moreover, we shall find, despite a common view to the contrary, that probabilities do *not* arise at the minute quantum level of particles, atoms, or molecules—those evolve *deterministically*—but, seemingly, via some mysterious larger-scale action connected with the emergence of a classical world that we can consciously perceive. We must try to understand this, and how quantum theory forces us to change our view of physical reality.

One tends to think of the discrepancies between quantum and classical theory as being very tiny, but in fact they also underlie many ordinary-scale physical phenomena. The very existence of solid bodies, the strengths and physical properties of materials, the nature of chemistry, the colours of substances, the phenomena of freezing and boiling, the reliability of inheritance—these, and many other familiar properties, require the quantum theory for their explanations. Perhaps, also, the phenomenon of consciousness is something that cannot be understood in entirely classical terms. Perhaps our minds are qualities rooted in some strange and wonderful feature of those physical laws which *actually* govern the world we inhabit, rather than being just features of some algorithm acted out by the so-called 'objects' of a *classical* physical structure. Perhaps, in some sense, this is 'why' we, as sentient beings, must live in a quantum world, rather than an entirely classical one, despite all the richness, and indeed mystery, that is already present in the classical universe. Might a quantum world be *required* so that thinking, perceiving creatures, such as ourselves, can be constructed from its substance? Such a question seems appropriate more for a God, intent on building an inhabited universe, than it is for us! But the question has relevance for us also. If a classical world is not something that consciousness could be part of, then our minds must be in some way dependent upon specific deviations from classical physics. This is a consideration I shall return to later in this book.

We must indeed come to terms with quantum theory—that most exact and mysterious of physical theories—if we are to delve deeply into some major questions of philosophy: how *does* our world behave, and what constitutes the 'minds' that are, indeed, 'us'? Yet, some day science may give us a *more* profound understanding of Nature than quantum theory can provide. It is my personal view that even quantum theory is a stop-gap, inadequate in certain essentials for providing a complete picture of the world in which we actually live. But that gives us no excuse; if we are to gain something of the philosophical insights we desire, we must comprehend the picture of the world according to existing quantum theory.

Unfortunately, different theorists tend to have very different (though observationally equivalent) viewpoints about the *actuality* of this picture. Many physicists, taking their lead from the central figure of Neils Bohr, would say that there is *no* objective picture at all. Nothing is actually 'out there', at the quantum level. Somehow, reality emerges only in relation to the results of 'measurements'. Quantum theory, according to this view, provides merely a calculational procedure, and does not attempt to describe the world as it actually 'is'. This attitude to the theory seems to me to be too defeatist, and I shall follow the more positive line which attributes *objective physical reality* to the quantum description: the *quantum state*.

There is a very precise equation, the *Schrödinger equation*, which provides a completely deterministic time-evolution for this state. But there is something very odd about the relation between the time-evolved quantum state and the actual behaviour of the physical world that is observed to take place. From time to time—whenever we consider that a 'measurement' has occurred—we must discard the quantum state that we have been laboriously evolving, and use it only to compute various probabilities that the state will 'jump' to one or another of a set of *new* possible states. In addition to the strangeness of this 'quantum jumping', there is the problem of deciding what it is about a physical arrangement that decrees that a 'measurement' has actually been made. The measuring apparatus itself is, after all, presumably built from quantum constituents, and so should also evolve according to the deterministic Schrödinger equation. Is the presence of a conscious being necessary for a 'measurement' *actually* to take place? I think that only a small minority of quantum physicists would affirm such a view. Presumably human observers are themselves also built from minute quantum constituents!

Later in this chapter we shall be examining some of the strange consequences of this 'jumping' of the quantum state—for example, how a 'measurement' in one place can seemingly cause a 'jump' to occur in a distant region! Earlier on, we shall encounter another strange phenomenon: sometimes two alternative routes that an object can take perfectly well, if either is traversed separately, will cancel one another out completely as soon as both are allowed together, so now *neither* can be traversed! We shall also be examining, in some detail, how quantum states are actually described. We shall see how greatly these descriptions differ from the corresponding classical ones. For example, particles can appear to be in two places at once! We shall also begin to get some feeling for how complicated are quantum descriptions when there are several particles to be considered together. It turns out that the particles do not each have individual descriptions, but must be considered as complicated superpositions of alternative arrangements of all of them together. We shall see how it is that different particles of the same type cannot have separate identities from one another. We shall examine, in detail, the strange (and fundamentally quantum-mechanical) property of *spin*. We shall consider the important issues raised by the paradoxical thought experiment of 'Schrödinger's cat', and of various different attitudes that theorists have expressed, partly as attempts to resolve this very basic puzzle.

Some of the material in this chapter may not be quite so readily comprehended as that of the preceding (or subsequent) chapters, and is sometimes a bit technical. In my descriptions I have tried not to cheat, and we shall have to work a little harder than otherwise. This is so that we can attain some genuine understanding of the quantum world. Where an argument remains unclear, I advise that you press on, and try to gain a flavour for the structure as a whole. But do not despair if a full understanding proves elusive. It is in the nature of the subject itself!

Problems with classical theory

How do we know that classical physics is not actually true of our world? The main reasons are experimental. Quantum theory was not wished upon us by theorists. It was (for the most part) with great reluctance that they found themselves driven to this strange and, in many ways, philosophically unsatisfying view of a world. Yet classical theory, despite its superb grandeur, has itself some profound difficulties. The root cause of these is that two kinds of physical object must coexist: *particles*, each described by a small *finite* number (six) of parameters (three positions and three momenta); and *fields*, requiring an *infinite* number of parameters. This dichotomy is not really physically consistent. For a system with both particles and fields to be in equilibrium (i.e. 'fully settled down'), all energy gets taken from the particles and put into the fields. This is a result of the phenomenon called 'equipartition of energy': at equilibrium, the energy is spread evenly among all the degrees of freedom of the system. Since the fields have infinitely many degrees of freedom, the poor particles get left with none at all!

In particular, classical atoms would not be stable, all motion of the particles getting transferred into the wave-modes of the fields. Recall the 'solar system' picture of the atom, as introduced by the great New Zealand/British experimental physicist Ernest Rutherford in 1911. In place of the planets would be the orbiting electrons, and instead of the sun, the central nucleus—on a tiny scale—held by electromagnetism rather than gravity. A fundamental and seemingly insurmountable problem is that as an orbiting electron circles the nucleus it should, in accordance with Maxwell's equations, emit electromagnetic waves of an intensity increasing rapidly to infinity, in a tiny fraction of a second, as it spirals inwards and plunges into the nucleus! However, nothing like this is observed. Indeed, what *is* observed is quite inexplicable on the basis of classical theory. Atoms can emit electromagnetic waves (light) but only in bursts of very specific discrete frequencies—the sharp *spectral lines* that are observed (Fig. 6.1). Moreover these frequencies satisfy 'crazy' rules[2] which have no basis from the point of view of classical theory.

Another manifestation of the instability of the co-existence of fields and particles, is the phenomenon known as 'black-body radiation'. Imagine an object at some definite temperature, electromagnetic radiation being in equilibrium with particles. In 1900, Rayleigh and Jeans had calculated that all the energy would be sucked up by the field—without limit! There is a physical absurdity involved in this (the 'ultraviolet catastrophe': energy keeps going into the field, to higher and higher frequencies, without stopping) and Nature herself behaves more prudently. At the *low* frequencies of field oscillation, the energy is as Rayleigh and Jeans had predicted, but at the *high* end, where they had predicted catastrophe, actual observation showed that the distribu-

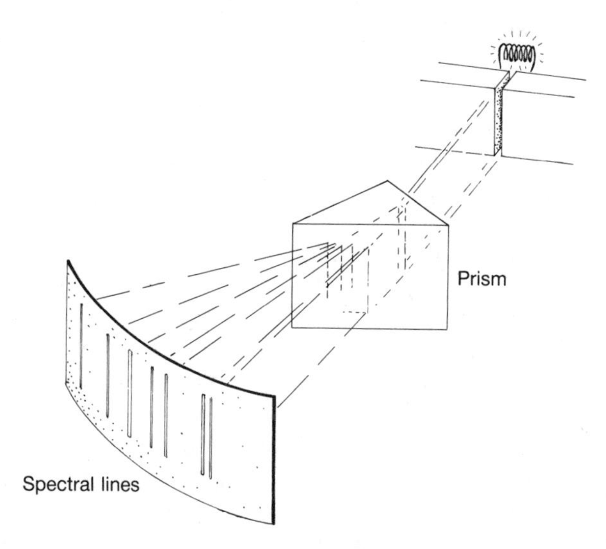

Fig. 6.1. Atoms in a heated material emit light which is often found to have only very specific frequencies. The different frequencies can be separated by use of a prism, and provide the spectral lines characteristic of the atoms.

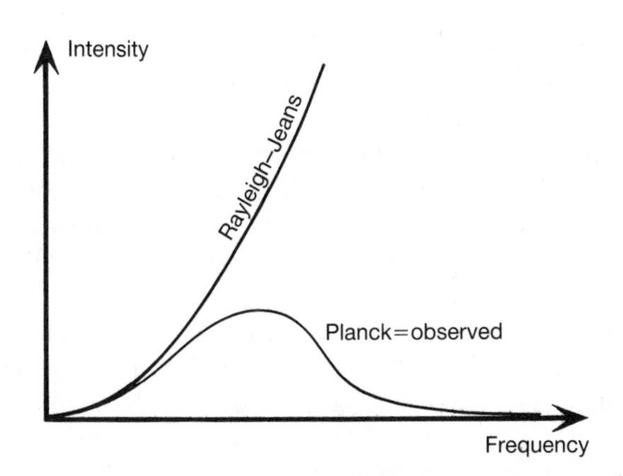

Fig. 6.2. The discrepancy between the classically calculated (Rayleigh–Jeans) and observed intensities of radiation for a hot body ('black body') led Planck to the beginnings of quantum theory.

tion of energy does *not* increase without limit, but instead falls off to zero as the frequencies increase. The greatest value of the energy occurs at a very specific frequency (i.e. colour) for a given temperature; see Fig. 6.2. (The red heat of a poker, or the yellow-white heat of the sun are, in effect, two familiar instances of this.)

The beginnings of quantum theory

How are these puzzles to be resolved? The original Newtonian scheme of particles certainly *needs* to be supplemented by Maxwell's field. Can one go to the other extreme and assume that *everything* is a field, particles being little finite-sized 'knots' of some kind of field? This has its difficulties also, for then particles could vary their shapes continuously, wriggling and oscillating in infinitely many different kinds of ways. But this is not what is seen. In the physical world, all particles of the same species appear to be *identical.* Any two electrons, for example, are just the same as each other. Even atoms or molecules can take on only discretely different arrangements.[3] If particles *are* fields, then some new ingredient is needed to enable *fields* to take on discrete characteristics.

In 1900, the brilliant but conservative and cautious German physicist, Max Planck, proposed a revolutionary idea for suppressing the high frequency 'black body' modes: that electromagnetic oscillations occur only in 'quanta', whole energy E bears a definite relation to the frequency ν, given by

$$E = h\nu,$$

h being a new fundamental constant of Nature, now known as *Planck's constant.* Astonishingly, with this outrageous ingredient, Planck was able to get theoretical agreement with the observed dependence of intensity with frequency—now called the *Planck radiation law.* (Planck's constant is very tiny by everyday standards, about 6.6×10^{-34} Joule seconds.) With this bold stroke Planck revealed the first glimmerings of the quantum theory to come, although it received scant attention until Einstein made another astonishing proposal: that the electromagnetic field can *exist* only in such discrete units! Recall that Maxwell and Hertz had shown that *light* consists of oscillations of electromagnetic field. Thus, according to Einstein—and as Newton had insisted over two centuries earlier—light itself must, after all, actually be *particles*! (In the early nineteenth century, the brilliant English theoretician and experimentalist Thomas Young had apparently established that light consists of waves.)

How is it that light can consist of particles and of field oscillations at the same time? These two conceptions seem irrevocably opposed. Yet some experimental facts clearly indicated that light is particles and others that it is waves. In 1923 the French aristocrat and insightful physicist Prince Louis de Broglie took this particle–wave confusion a stage further when, in his doctoral dissertation (for which approval was sought from Einstein!), he proposed that the very particles of *matter* should sometimes behave as waves! De Broglie's wave-frequency ν, for any particle of mass m, again satisfies the Planck relation. Combined with the Einstein $E = mc^2$, this tells us that ν is related to m by:

$$h\nu = E = mc^2.$$

Thus, according to de Broglie's proposal, the dichotomy between particles and fields that had been a feature of classical theory is *not* respected by Nature! Indeed, anything whatever which oscillates with some frequency ν can occur *only* in discrete units of mass $h\nu/c^2$. Somehow, Nature contrives to build a consistent world in which *particles and field-oscillations are the same thing*! Or, rather, her world consists of some more subtle ingredient, the words 'particle' and 'wave' conveying but partially appropriate pictures.

The Planck relation was put to another brilliant use (in 1913) by Neils Bohr, the Danish physicist and towering figure of twentieth-century scientific thought. Bohr's rules required that the *angular momentum* (see p. 166) of electrons in orbit about the nucleus can occur only in integer multiples of $h/2\pi$, for which Dirac later introduced the convenient symbol $\hbar$:

$$\hbar = h/2\pi.$$

Thus the only allowed values of angular momentum (about any one axis) are

$$0, \hbar, 2\hbar, 3\hbar, 4\hbar, \ldots .$$

With this *new* ingredient, the 'solar system' model for the atom now gave, to a considerable accuracy, many of the discrete stable energy levels and 'crazy' rules for spectral frequencies that Nature *actually* obeys!

Though strikingly successful, Bohr's brilliant proposal provided a somewhat provisional 'patchwork' scheme, referred to as the 'old quantum theory'. Quantum theory as we know it today arose out of two independent later schemes, which were initiated by a pair of remarkable physicists: a German, Werner Heisenberg, and an Austrian, Erwin Schrödinger. At first, their two schemes ('matrix mechanics' in 1925 and 'wave mechanics', in 1926, respectively) seemed quite different, but they were soon shown to be equivalent, and subsumed into a more comprehensive and general framework, primarily by the great British theoretical physicist Paul Adrien Maurice Dirac soon afterwards. In the succeeding sections we shall be catching a glimpse of this theory and of its extraordinary implications.

The two-slit experiment

Let us consider the 'archetypical' quantum-mechanical experiment, according to which a beam of electrons, or light, or some other species of 'particle–wave' is fired through a pair of narrow slits to a screen behind (Fig. 6.3). To be specific, let us take *light*, and refer to the light quanta as 'photons', in accordance with the usual terminology. The most evident manifestation of the light as *particles* (i.e. as photons) occurs at the screen. The light arrives there in discrete localized units of energy, this energy being invariably related to the light's frequency in accordance with Planck's formula: $E = h\nu$. Never is the energy of just 'half' a photon (or any other fraction) received. Light reception is an all-or-nothing phenomenon in photon units. Only whole numbers of photons are ever seen.

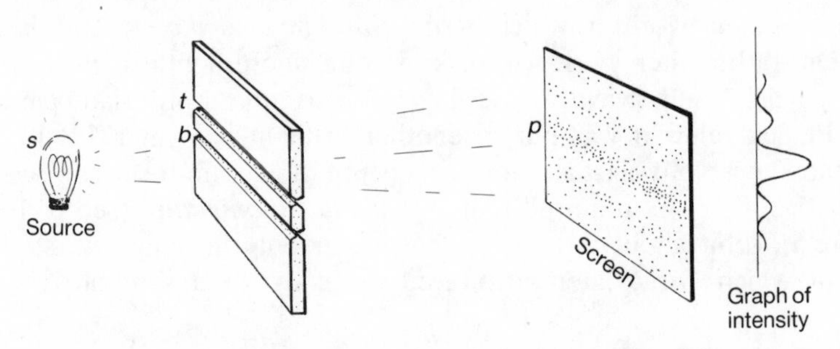

Fig. 6.3. The two-slit experiment, with monochromatic light.

However, a *wavelike* behaviour seems to arise as the photons pass through the slits. Suppose, first, that only one slit is open (the other being blocked off). After passing through, the light will spread out—by the phenomenon called *diffraction*, a feature of wave propagation. But one may still hold to a particle picture, and imagine that there is some influence exerted by the proximity of the edges of the slit which causes the photons to be deflected by some random amount, to one side or the other. When there is a reasonable intensity to the light passing through the slit, i.e. a large number of photons, then the illumination at the screen is seen to be very uniform. But if the light intensity is cut right down, then one can ascertain that the distribution of illumination is indeed made up of individual spots—in agreement with the particle picture—where the individual photons strike the screen. The smooth appearance of the illumination is a statistical effect, due to the very large number of photons involved (see Fig. 6.4). (For comparison, a sixty-Watt light bulb emits about 100 000 000 000 000 000 000 photons per second!) Photons indeed appear to be deflected in some random way as they pass through the slit—with different probabilities for different angles of deflection, yielding the observed distribution of illumination.

However, the key problem for the particle picture comes when we open the other slit! Suppose that the light is from a yellow sodium lamp, so that it is essentially of a pure unmixed colour—the technical term is *monochromatic*, i.e. of one definite wavelength or frequency, which in the particle picture, means that the photons all have the same energy. Here, the wavelength is about 5×10^{-7} m. Take the slits to be about 0.001 mm in width and about 0.15 mm apart, the screen being about one metre distant. For a reasonably strong light intensity we still get a regular-looking pattern of illumination, but

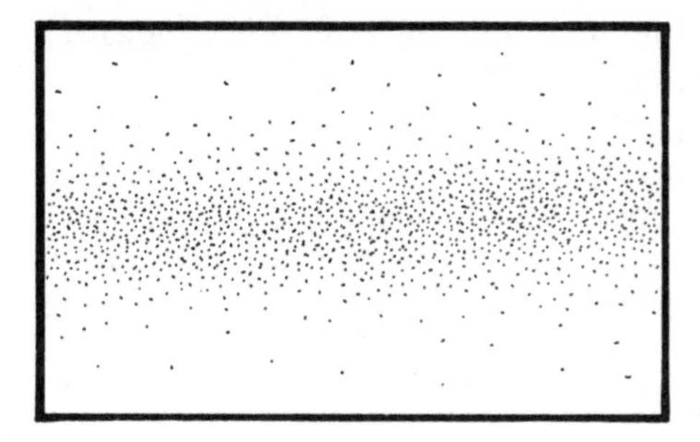

Fig. 6.4. Pattern of intensity at the screen when just one slit is open—a distribution of discrete tiny spots.

now there is a *waviness* to it, called an *interference pattern*, with bands on the screen about three millimetres in width near the centre of the screen (Fig. 6.5). We might have expected that by opening the second slit we would simply double the intensity of illumination on the screen. Indeed, this is true if we consider the *total* overall illumination. But now the detailed *pattern* of intensity is seen to be completely different from what it was with the single slit. At some points on the screen–where the pattern is at its brightest—the intensity of the illumination is *four* times what it was before, not just twice. At others—where the pattern is darkest—the intensity goes down to zero. The points of zero intensity perhaps present the greatest puzzle for the particle picture. These were points where a photon could quite happily arrive when just one of the slits was open. Now that we have opened the other, it suddenly turns out that the photon is somehow *prevented* from doing what it could do before. How can it be that by allowing the photon an *alternative* route, we have actually *stopped* it from traversing either route?

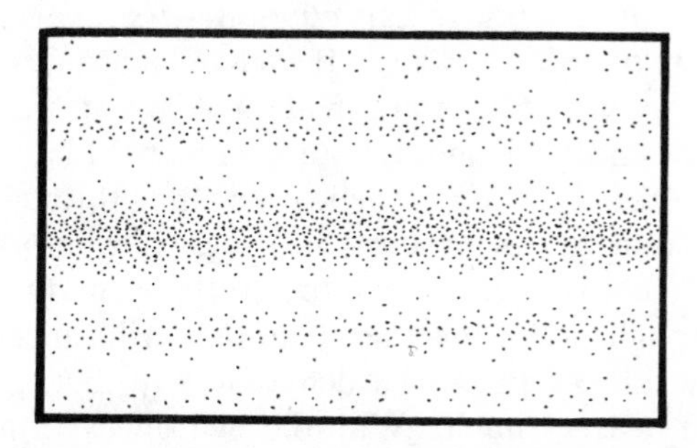

Fig. 6.5. Pattern of intensity when both slits are open—a wavy distribution of discrete spots.

On the scale of the photon, if we take its wavelength as a measure of its 'size', the second slit is some 300 'photon-sizes' away from the first (each slit being about a couple of wavelengths wide) (see Fig. 6.6), so how can the photon 'know', when it passes through one of the slits whether or not the other slit is actually open? In fact, in principle, there is no limit to the distance that the two slits can be from one another for this 'cancelling' or 'enhancing' phenomenon to occur.

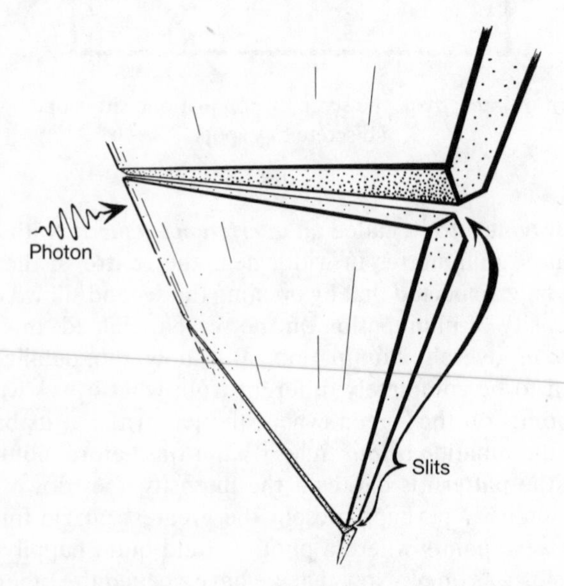

Fig. 6.6. The slits from the photon's point of view! How can it make any difference to it whether the second slit, some 300 'photon sizes' away, is open or closed?

As the light passes through the slit(s), it now seems to be behaving like a *wave* and *not* like a particle! Such cancellation—*destructive interference*—is familiar as a property of ordinary waves. If each of two routes can be travelled separately by a wave, and if *both* are made available to it, then it is quite possible for them to cancel one another out. In Fig. 6.7 I have illustrated how this comes about. When a portion of the wave coming through one of the slits meets a portion coming through the other, they will reinforce each other if they are 'in phase' (that is to say, when the crests of the two portions occur together and the troughs occur together), but they will cancel each other out if they are exactly 'out of phase' (which is to say that one is at a crest whenever the other is at a trough). With the two-slit experiment, the bright places on the screen occur whenever the distances to the two slits differ by an *integer* number of wavelengths, so that the crests and troughs do indeed occur together, and the dark places occur when the differences between the two distances come exactly midway between these values, so that crest meets with trough and trough with crest.

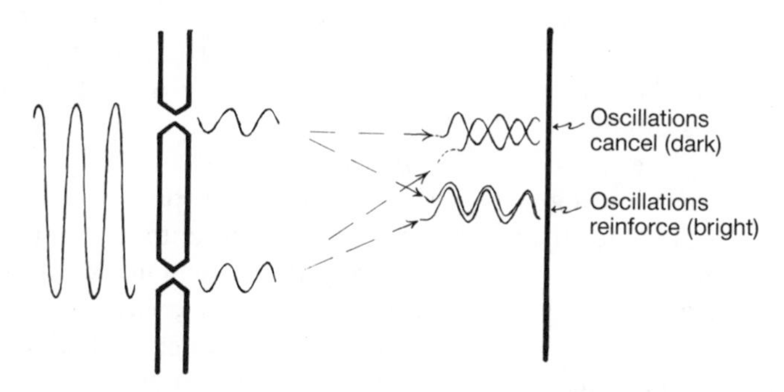

Fig. 6.7. With a purely wave picture we can understand the pattern of bright and dark bands at the screen (although not the discreteness) in terms of interference of waves.

There is nothing puzzling about an ordinary macroscopic classical wave travelling through two slits at once in this way. A wave, after all, is just a 'disturbance', either of some continuous medium (field), or of some substance composed of myriads of tiny point-like particles. A disturbance could pass partly through one slit and partly through the other. But here things are very different: each individual photon behaves like a wave entirely on its own! In some sense, each particle travels through *both slits at once* and it interferes with *itself*! For by cutting the overall intensity of the light down sufficiently, one can ensure that not more than one photon at a time is passing the vicinity of the slits. The phenomenon of destructive interference, whereby two alternative routes for the photon somehow contrive to cancel one another out as realized possibilities, is something that applies to *single* photons. If one of the two routes is alone open to the photon, then the photon can travel that route. If the other route is alone open, then the photon can travel that route instead. But if *both* are open to it, then the two possibilities miraculously cancel one another out, and the photon appears to be able to travel neither!

The reader should pause to consider the import of this extraordinary fact. It is really not that light sometimes behaves as particles and sometimes as waves. It is that *each individual particle* behaves in a wavelike way entirely on its own; and *different alternative possibilities open to a particle can sometimes cancel one another out*!

Does the photon actually split in two and travel partly through one slit and partly through the other? Most physicists would object to this way of phrasing things. They would insist that while the two routes open to the particle must both contribute to the final effect, these are just *alternative* routes, and the particle should not be thought of as splitting into two in order to get through the slits. As support for the view that the particle does not partly go through one slit and partly through the other, the modified situation may be considered in which a *particle detector* is placed at one of the slits or the other. Since when it is observed, a photon—or any other particle—always appears as a single whole and not as some fraction of a whole, it must be the case that our detector detects either a whole photon or no photon at all. However, when the detector is present at one of the slits, so that an observer can *tell* which slit the photon went through, the wavy interference pattern at the screen disappears. In order for the interference to take place, there must apparently be a 'lack of knowledge' as to which slit the particle 'actually' went through.

To get interference, the alternatives must *both* contribute, sometimes 'adding'—reinforcing one another by an amount which is double what one might expect—and sometimes 'subtracting'—so that alternatives can mysteriously 'cancel' one another out. In fact, according to the rules of quantum mechanics, what is happening is even more mysterious than that! Alternatives can indeed add (brightest points on the screen); and alternatives can indeed subtract (dark points); but they must actually also be able to combine together in other strange combinations, such as:

'alternative A' plus $i \times$ 'alternative B',

where 'i' is the 'square root of minus one' ($= \sqrt{-1}$) that we encountered in Chapter 3 (at points on the screen of intermediate intensity). In fact *any complex number* can play such a role in 'combining alternatives'!

The reader may recall my warning in Chapter 3 that complex numbers are 'absolutely fundamental to the structure of quantum mechanics'. These numbers are not just mathematical niceties. They forced themselves on the attentions of physicists through persuasive and unexpected experimental facts. To understand quantum mechanics, we must come to terms with complex-number weightings. Let us next consider what these entail.

Probability amplitudes

There is nothing special about using photons in the above descriptions. Electrons or any other kind of particle, or even whole atoms would do equally well. The rules of quantum mechanics appear even to insist that cricket balls* and elephants ought to behave in this odd way, where different alternative possibilities can somehow 'add up' in complex-number combinations! However, we never actually *see* cricket balls or elephants superposed in this strange way. Why do we not? This is a difficult and even a controversial issue, and I do not want to confront it just yet. For the moment, as a working rule, let us

* American readers unfamiliar with the term 'cricket ball' should read it 'quaint English version of a baseball'.

simply suppose that there are two different possible levels of physical description, which we call the *quantum level* and the *classical level.* We shall only use these strange complex-number combinations at the quantum level. Cricket balls and elephants are classical-level objects.

The quantum level is the level of molecules, atoms, subatomic particles, etc. It is normally thought of as being a level of very 'small-scale' phenomena, but this 'smallness' does not actually refer to physical size. We shall be seeing that quantum effects can occur over distances of many metres, or even light years. It would be a little closer to the mark to think of something as being 'at the quantum level' if it involves only very tiny differences of energy. (I shall try to be more precise later, especially in Chapter 8, p. 367). The classical level is the 'macroscopic' level of which we are more directly aware. It is a level at which our ordinary picture of 'things happening' holds true, and where we can use our ordinary ideas of probability. We shall be seeing that the complex numbers that we must use at the quantum level are closely related to classical probabilities. They are not really the same, but to come to terms with these complex numbers, it will be helpful first to remind ourselves how classical probabilities behave.

Consider a classical situation which is *uncertain*, so that we do not know which of two alternatives A or B takes place. Such a situation could be described as a 'weighted' combination of these alternatives:

$$p \times \text{'alternative A'} \quad \text{plus} \quad q \times \text{'alternative B'},$$

where p is the *probability* of A happening and q the probability of B happening. (Recall that a probability is a real number lying between 0 and 1. Probability 1 means 'certain to happen' and probability 0 means 'certain not to happen'. Probability 1/2 means 'equally likely to happen as not'.) If A and B are the *only* alternatives, then the sum of the two probabilities must be 1:

$$p + q = 1.$$

However, if there are further alternatives this sum could be less than 1. Then, the ratio $p{:}q$ gives the *ratio* of the probability of A happening to that of B happening. The actual probability of A and of B happening, out of just these two alternatives, would be $p/(p + q)$ and $q/(p + q)$ respectively. We could also use such an interpretation if $p + q$ is greater than 1. (This might be useful, for example, whenever we have an experiment that is performed many many times, p being the number of 'A' occurrences and q, the number of 'B' occurrences.) We shall say that p and q are *normalized* if $p + q = 1$, so that then they give the actual probabilities rather than just the ratios of probabilities.

In quantum physics we shall do something that *appears* to be very similar to this, except that p and q are now to be *complex* numbers—which I shall prefer to denote, instead, by w and z, respectively.

$$w \times \text{'alternative A'} \quad \text{plus} \quad z \times \text{'alternative B'}.$$

How are we to interpret w and z? They are certainly not ordinary probabilities (or ratios of probabilities) since each can independently be negative or complex, but they do behave in many ways like probabilities. They are referred to (when appropriately normalized—see later) as *probability amplitudes*, or simply *amplitudes*. Moreover, one often uses the kind of terminology that is suggested by probabilities such as: 'there is an amplitude w for A to happen and an amplitude z for B to happen'. They are *not* actually probabilities, but for the moment we shall try to pretend that they are—or rather, the quantum-level analogues of probabilities.

How do *ordinary* probabilities work? It will be helpful if we can think of a macroscopic object, say a ball, being struck through one of two holes to a screen behind—like the two-slit experiment described above (cf. Fig. 6.3), but where now a classical macroscopic ball replaces the photon of the previous discussion. There will be some probability $P(s,t)$ that the ball reaches the top hole t after it is struck at s, and some probability $P(s,b)$ that it reaches the bottom hole b. Moreover, if we select a particular point p on the screen, there will be some probability $P(t,p)$ that whenever the ball *does* pass through t it will reach the particular point p on the screen, and some probability $P(b,p)$ that whenever it passes through b it will reach p. If only the top hole t is open, then to get the probability that the ball actually reaches p via t after it is struck we multiply the probability that it gets from s to t by the probability that it gets from t to p:

$$P(s,t) \times P(t,p).$$

Similarly, if only the bottom hole is open, then the probability that the ball gets from s to p is:

$$P(s,b) \times P(b,p).$$

If *both* holes are open, then the probability that it gets from s to p *via* t is still the first expression $P(s,t) \times P(t,p)$, just as if only t were open, and the probability that it gets from s to p via b is still $P(s,b) \times P(b,p)$, so the *total* probability $P(s,p)$ that it reaches p from s is the *sum* of these two:

$$P(s,p) = P(s,t) \times P(t,p) + P(s,b) \times P(b,p).$$

At the *quantum* level the rules are just the same, except that now it is these strange complex *amplitudes* that must play the roles of the probabilities that we had before. Thus, in the two-slit experiment considered above, we have an amplitude $A(s,t)$ for a photon to reach the top slit t from the source s, and an amplitude $A(t,p)$ for it to reach the point p on the screen from the slit t, and we multiply these two to get the amplitude

$$A(s,t) \times A(t,p),$$

for it to reach the screen at p via t. As with probabilities, this is the correct amplitude, assuming the top slit is open, whether or not the bottom slit b is open. Similarly, assuming that b is open, there is an amplitude

$$A(s,b) \times A(b,p)$$

for the photon to reach p from s via b (whether or not t is open). If both slits are open, then we have a total amplitude

$$A(s,p) = A(s,t)\times A(t,p) + A(s,b)\times A(b,p)$$

for the photon to reach p from s.

This is all very well, but it is not of much use to us until we know how to interpret these amplitudes when a quantum effect is magnified until it reaches the classical level. We might, for example, have a photon detector, or *photo-cell* placed at p, which provides a way of magnifying an event at the quantum level—the arrival of a photon at p—into a classically discernible occurrence, say an audible 'click'. (If the screen acts as a photographic plate, so that the photon leaves a visible spot, then this would be just as good, but for clarity let's stick to using a photo-cell.) There must be an actual *probability* for the 'click' to happen, not just one of these mysterious 'amplitudes'! How are we to pass from amplitudes to probabilities when we go over from the quantum to the classical level? It turns out that there is a very beautiful but mysterious rule for this.

The rule is that we must take the *squared modulus* of the quantum complex amplitude to get the classical probability. What is a 'squared modulus'? Recall our description of complex numbers in the Argand plane (Chapter 3, p. 90). The *modulus* $|z|$ of a complex number z is simply the distance from the origin (i.e. from the point 0) of the point described by z. The squared modulus $|z|^2$ is simply the square of this number. Thus, if

$$z = x + \mathrm{i}y,$$

where x and y are real numbers, then (by the Pythagorean theorem, since the line from 0 to z is the hypotenuse of the right-angled triangle $0,x,z$) our required squared modulus is

$$|z|^2 = x^2 + y^2.$$

Note that for this to be an actual 'normalized' probability, the value of $|z|^2$ must lie somewhere between 0 and 1. This means that for a properly normalized amplitude, the point z in the Argand plane must then lie somewhere in the *unit circle* (see Fig. 6.8). Sometimes, however, we shall want to consider combinations

$$w \times \text{alternative A} \quad + \quad z \times \text{alternative B},$$

where w and z are merely *proportional* to the probability amplitudes, and

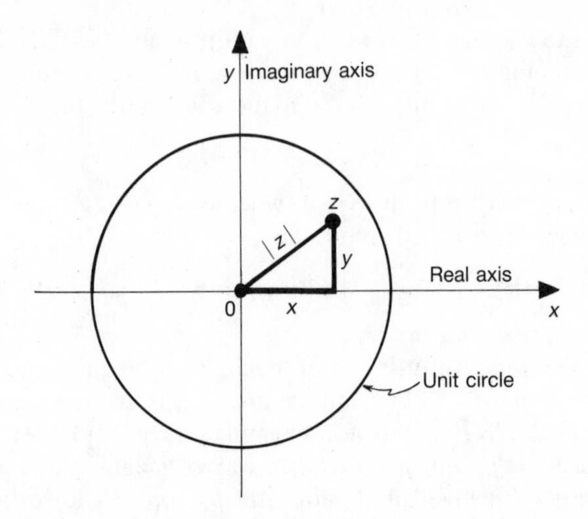

Fig. 6.8. A probability amplitude represented as a point z within the unit circle in the Argand plane. The squared distance $|z|^2$ from the centre can become an actual probability when effects become magnified to the classical level.

then they need not lie in this circle. The condition that they be *normalized* (and therefore provide actual probability amplitudes) is that their *squared moduli* should sum to unity:

$$|w|^2 + |z|^2 = 1.$$

If they are not so normalized, then the actual amplitudes for A and B, respectively, are $w/\surd(|w|^2 + |z|^2)$ and $z/\surd(|w|^2 + |z|^2)$, which *do* lie in the unit circle.

We now see that a probability amplitude is not really like a probability after all, but rather more like a 'complex square root' of a probability. How does this affect things when effects at the quantum level get magnified so as to reach the classical level? Recall that in manipulating probabilities and amplitudes we sometimes needed to multiply them and we sometimes needed to add them. The first point to note is that the operation of *multiplication* poses no problem in our passing from the quantum to the classical rules. This is because of the remarkable mathematical fact that the squared modulus of the product of two complex numbers is equal to the product of their individual squared moduli:

$$|zw|^2 = |z|^2|w|^2.$$

(This property follows at once from the geometrical description of the product of a pair of complex numbers, as given in Chapter 3; but in terms of real and imaginary parts $z = x + iy$, $w = u + iv$, it is a pretty little miracle. Try it!)

This fact has the implication that if there is only a single route open to a particle, e.g. when there is only a single slit (say t) open, in the two-slit experiment, then one can argue 'classically' and the probabilities come out the same whether or not an additional detection of the particle is made at the intermediate point (at t).* We can take the squared moduli at both stages or only at the end, e.g.

$$|A(s,t)|^2 \times |A(t,p)|^2 = |A(s,t) \times A(t,p)|^2,$$

and the answer comes out the same each way for the resulting probability.

However, if there is more than one route available (e.g. if both slits are open), then we need to form a *sum*, and it is here that the characteristic features of quantum mechanics begin to emerge. When we form the squared modulus of the sum, $w + z$, of two complex numbers w and z, we do *not* usually get the sum of their squared moduli separately; there is an additional 'correction term':

$$|w + z|^2 = |w|^2 + |z|^2 + 2|w|\,|z| \cos\theta.$$

Here, θ is the angle which the pair of points z and w subtend at the origin in the Argand plane (see Fig. 6.9). (Recall that the cosine of an angle is the ratio 'adjacent/hypotenuse' for a right-angled triangle. The keen reader who is unfamiliar with the above formula may care to derive it directly, using the geometry introduced in Chapter 3. In fact, this formula is none other than the familiar 'cosine rule', slightly disguised!) It is this correction term $2|w||z|\cos\theta$ that provides the *quantum interference* between quantum-mechanical alternatives. The value of $\cos\theta$ can range between -1 and 1. When $\theta = 0°$ we have $\cos\theta = 1$ and the two alternatives reinforce one another so that the total probability is greater than the sum of the individual probabilities. When

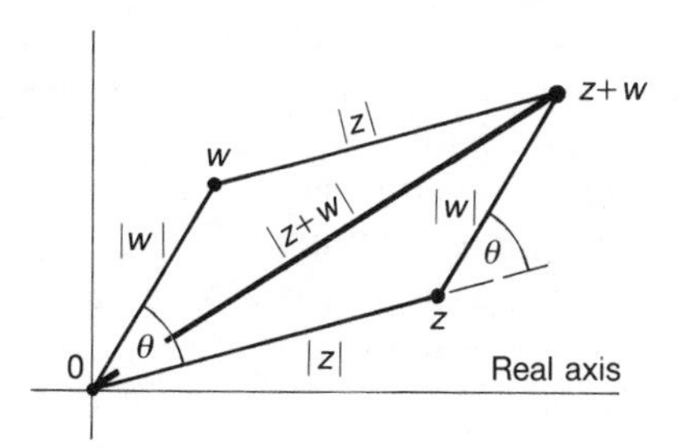

Fig. 6.9. Geometry relating to the correction term $2|w||z| \cos\theta$ for the squared modulus of the sum of two amplitudes.

* This detection must be made in such a way that it does not disturb the particle's passage through t. This could be achieved by placing detectors at *other* places around s and *inferring* the particle's passage through t when these other detectors do *not* click!

$\theta = 180°$ we have $\cos\theta = -1$ and the two alternatives tend to cancel one another out, giving a total probability less than the sum of the individual ones (destructive interference). When $\theta = 90°$ we have $\cos\theta = 0$ and we get an intermediate situation where the two probabilities do add. For large or complicated systems, the correction terms generally 'average out'—because the 'average' value of $\cos\theta$ is zero—and we are left with the ordinary rules of classical probability! But at the quantum level these terms give important interference effects.

Consider the two-slit experiment when both slits are open. The amplitude for the photon to reach p is a *sum*, $w + z$, where here

$$w = A(s,t) \times A(t,p) \quad \text{and} \quad z = A(s,b) \times A(b,p).$$

At the *brightest* points on the screen we have $w = z$ (so that $\cos\theta = 1$) whence

$$|w + z|^2 = |2w|^2 = 4|w|^2$$

which is four times the probability $|w|^2$ for when just the top slit is open—and therefore *four* times the intensity, when there is a large number of photons, in agreement with observation. At the *dark* points on the screen we have $w = -z$ (so that $\cos\theta = -1$) whence

$$|w + z|^2 = |w - w|^2 = 0,$$

i.e. it is *zero* (destructive interference!), again in agreement with observation. At the exactly intermediate points we have $w = \mathrm{i}z$ or $w = -\mathrm{i}z$ (so that $\cos\theta = 0$) whence

$$|w + z|^2 = |w \pm \mathrm{i}w|^2 = |w|^2 + |w|^2 = 2|w|^2$$

giving *twice* the intensity as for just one slit (which would be the case for classical particles). We shall see at the end of the next section how to calculate where the bright, dark, and intermediate places actually are.

One final point should be remarked upon. When both slits are open, the amplitude for the particle to reach p via t is indeed $w = A(s,t) \times A(t,p)$, but we cannot interpret its squared modulus $|w|^2$ as the probability that the particle 'actually' passed through the top slit to reach p. That would give us nonsensical answers, particularly if p is at a dark place on the screen. But if we choose to 'detect' the photon's presence at t, by magnifying the effect of its presence (or absence) *there* to the classical level, then we *can* use $|A(s,t)|^2$ for the probability for the photon being actually present at t. But such detection would then obliterate the wavy pattern. For interference to occur, we must ensure that the passage of the photon through the slits *remains at the quantum level*, so that the two alternative routes must *both* contribute and can sometimes cancel one another out. At the quantum level the individual alternative routes have only amplitudes, not probabilities.

The quantum state of a particle

What kind of picture of 'physical reality' does this provide us with at the quantum level, where different 'alternative possibilities' open to a system must always be able to coexist, added together with these strange complex-number weightings? Many physicists find themselves despairing of ever finding such a picture. They claim, instead, to be happy with the view that quantum theory provides merely a calculational procedure for computing probabilities and not an objective picture of the physical world. Some, indeed, assert that quantum theory proclaims no objective picture to be possible—at least none which is consistent with physcial facts. For my own part, I regard such pessimism as being quite unjustified. In any case it would be premature, on the basis of what we have discussed so far, to take such a line. Later on, we shall be addressing some of the more strikingly puzzling implications of quantum effects, and we shall perhaps begin to appreciate more fully the reasons for this despair. But for now, let us proceed more optimistically, and come to terms with the picture that quantum theory appears to be telling us we must face up to.

This picture is that presented by a *quantum state*. Let us try to think of a single quantum particle. Classically, a particle is determined by its position in space, and, in order to know what it is going to do next, we also need to know its velocity (or, equivalently, its momentum). Quantum mechanically, *every single position* that the particle might have is an 'alternative' available to it. We have seen that all alternatives must somehow be combined together, with complex-number weightings. This collection of complex weightings describes the quantum state of the particle. It is standard practice, in quantum theory, to use the Greek letter ψ (pronounced 'psi') for this collection of weightings, regarded as a complex function of position—called the *wavefunction* of the particle. For each position x, this wavefunction has a specific value, denoted $\psi(x)$, which is the amplitude for the particle to be at x. We can use the single letter ψ to label the quantum state as a whole. I am taking the view that the *physical reality* of the particle's location is, indeed, its quantum state ψ.

How do we picture the complex function ψ? This is a bit hard for all of three-dimensional space at once, so let us simplify a little and suppose that the particle is constrained to lie on a one-dimensional line—say along the x-axis of a standard (Cartesian) coordinate system. If ψ had been a real function, then we could have imagined a 'y-axis' perpendicular to the x-axis and plotted the *graph* of ψ (Fig. 6.10a). However, here we need a 'complex y-axis'—which would be an Argand plane—in order to describe the value of the *complex* function ψ. For this, in our imagination, we can use two other spatial dimensions: say the y-direction in space as the *real* axis of this Argand plane and the z-direction as the *imaginary* axis. For an accurate picture of the wavefunction, we can plot $\psi(x)$ as a point in this Argand plane (i.e. in the

(y,z)-plane through each position on the x-axis). As x varies, this point varies too, and its path describes a curve in space winding about in the neighbourhood of the x-axis (see Fig. 6.10b). Let us call this curve the ψ-*curve* of the particle. The probability of finding the particle at a particular point x, if a particle detector were to be placed at that point, is obtained by forming the squared modulus of the amplitude $\psi(x)$,

$$|\psi(x)|^2,$$

which is the square of the distance of the ψ-curve from the x-axis.*

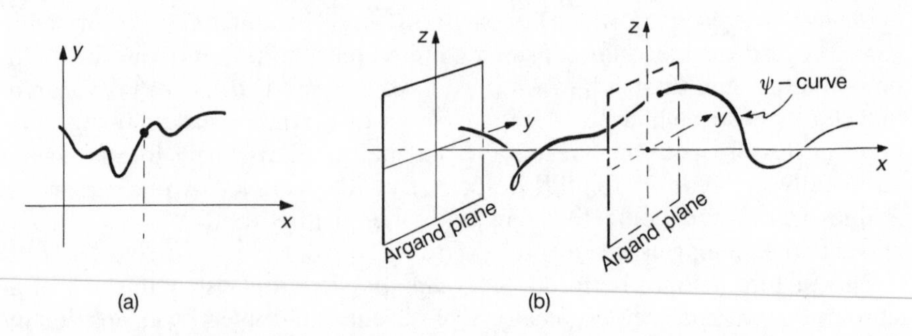

Fig. 6.10. (a) The graph of a real function of a real variable x.
(b) The graph of a complex function ψ of a real variable x.

To form a complete picture of this kind, for the wavefunction throughout all of three-dimensional physical space, *five* dimensions would be necessary: three dimensions for physical space plus another two for the Argand plane at each point in which to plot $\psi(x)$. However, our simplified picture is still helpful. If we choose to examine the behaviour of the wave function along any particular line in physical space we can do so simply by taking our x-axis along that line and provisionally using the other two spatial directions as providing the required Argand planes. This will prove useful in our understanding the two-slit experiment.

As I mentioned above, in classical physics one needs to know the velocity (or momentum) of a particle in order to determine what it is going to do next. Here, quantum mechanics provides us with a remarkable economy. The wavefunction ψ *already* contains the various amplitudes for the different possible momenta! (Some disgruntled readers may be thinking that it is 'about time' for a bit of economy, considering how much we have had to complicate the simple classical picture of a point particle. Though I have a

* A technical difficulty arises here because the actual probability of finding a particle at an *exact* point would be zero. Instead, we refer to $|\psi(x)|^2$ as defining a *probability density*, which means that the probability of finding the particle in some small fixed-sized interval about the point in question is what is defined. Thus $\psi(x)$ defines an *amplitude density*, rather than just an amplitude.

good deal of sympathy for such readers, I should warn them to take eagerly what morsels are thrown to them, for there is worse to come!) How is it that the velocity amplitudes are determined by ψ? Actually, it is better to think of momentum amplitudes. (Recall that momentum is velocity multiplied by the mass of the particle; cf. p. 165). What one does is to apply what is called *harmonic analysis* to the function ψ. It would be out of place for me to explain this in detail, here, but it is closely related to what one does with musical sounds. Any wave form can be split up as a sum of different 'harmonics' (hence the term 'harmonic analysis') which are the pure tones of different pitches (i.e. different pure frequencies). In the case of a wavefunction ψ, the 'pure tones' correspond to the different possible momentum values that the particle might have, and the size of each 'pure-tone' contribution to ψ provides the amplitude for that momentum value. The 'pure tones' themselves are referred to as *momentum states*.

What does a momentum state look like as a ψ-curve? It looks like a *corkscrew*, for which the official mathematical name is a *helix* (Fig. 6.11).* Those corkscrews that are tightly wound correspond to large momenta, and those that wind hardly at all give very small momenta. There is a limiting case for which it does not wind at all and the ψ-curve is a straight line: the case of zero momentum. The famous *Planck relation* is implicit in this. Tightly wound means short wavelength and high *frequency*, and hence high momentum and high *energy*; and loosely wound means low frequency and low energy, the energy E always being in proportion to the frequency ν ($E = h\nu$). If the Argand planes are oriented in the normal way (i.e. with the x,y,z-description given above, with the usual right-handed axes), then the momenta which point positively in the direction of the x-axis correspond to the right-handed corkscrews (which are the usual sorts of corkscrews).

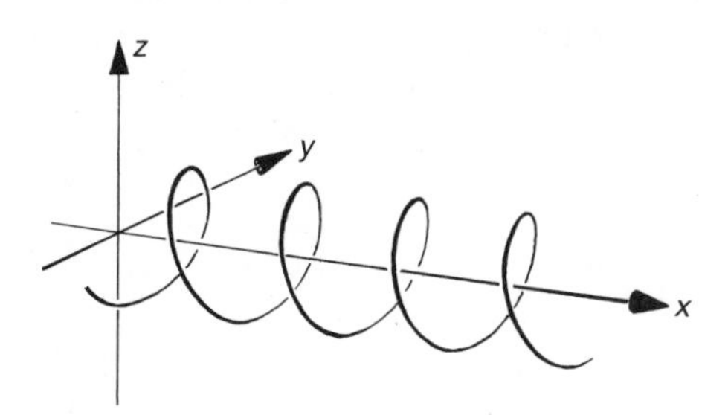

Fig. 6.11. A momentum state has a ψ-curve which is a corkscrew.

* In terms of a more standard analytic description, each of our corkscrews (i.e. momentum states) would be given by an expression $\psi = e^{ipx/\hbar} = \cos(ipx/\hbar) + i\sin(ipx/\hbar)$ (see Chapter 3, p. 89, where p is the momentum value in question.

Sometimes it is helpful to describe quantum states not in terms of ordinary wavefunctions, as was done above, but in terms of wavefunctions of *momentum*. This amounts to considering the decomposition of ψ in terms the various momentum states, and constructing a new function $\tilde{\psi}$, this time a function of momentum p rather than position x, whose value $\tilde{\psi}(p)$, for each p, gives the size of the contribution of the p-momentum state to ψ. (The space of ps is called *momentum space*.) The interpretation of $\tilde{\psi}$ is that, for each particular choice of p, the complex number $\tilde{\psi}(p)$ gives the *amplitude for the particle to have momentum p*.

There is a mathematical name for the relationship between the functions ψ and $\tilde{\psi}$. These functions are called *Fourier transforms* of one another—after the French engineer/mathematician Joseph Fourier (1768–1830). I shall make only a few comments about this relationship here. The first point is that there is a remarkable symmetry between ψ and $\tilde{\psi}$. To get back to ψ from $\tilde{\psi}$, we apply effectively the same procedure as we did to get from ψ to $\tilde{\psi}$. Now it is $\tilde{\psi}$ which is to be subjected to harmonic analysis. The 'pure tones' (i.e. the corkscrews in the momentum space picture) are now called *position states*. Each position x determines such a 'pure tone' in momentum space, and the size of the contribution of this 'pure tone' to $\tilde{\psi}$ gives the value of $\psi(x)$.

A position state itself corresponds, in the ordinary position-space picture, to a function ψ which is very sharply peaked at the value of x in question, all the amplitudes being zero except at that x-value itself. Such a function is called a (Dirac) *delta function*—though, technically, it is not quite a 'function' in the ordinary sense, the value at x being infinite. Likewise, the momentum states (corkscrews in the position-space picture) give delta functions in the momentum space picture (see Fig. 6.12.) Thus we see that the Fourier transform of a corkscrew is a delta function, and *vice versa*!

The position-space description is useful whenever one intends to perform a measurement of the particle's position, which amounts to doing something which magnifies the effects of the different possible particle positions to the classical level. (Roughly speaking, photo-cells and photographic plates effect

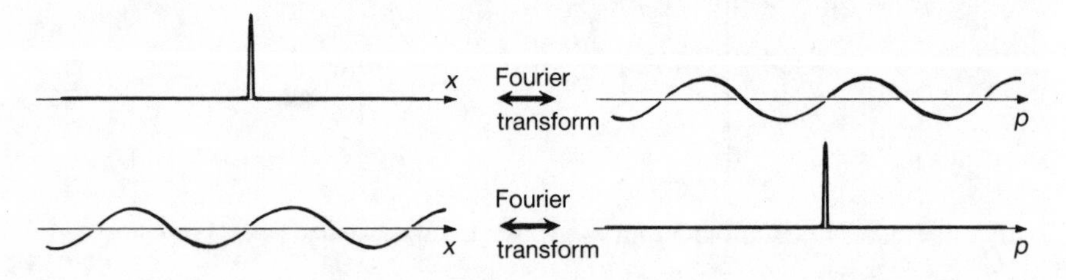

Fig. 6.12. Delta functions in position space go over to corkscrews in momentum space and *vice versa*.

position measurements for photons.) The momentum-space description is useful when one proposes to measure the particle's momentum, i.e. to magnify the effects of the different possible momenta to the classical level. (Recoil effects or diffraction from crystals can be used for momentum measurements.) In each case, the squared modulus of the corresponding wavefunction (ψ or $\tilde{\psi}$) gives the required probability for the result of the required measurement.

Let us end this section by returning once more to the two-slit experiment. We have learnt that, according to quantum mechanics, even a single particle must behave like a wave all by itself. This wave is described by the wavefunction ψ. The most 'wavelike' waves are the momentum states. In the two-slit experiment, we envisaged photons of a definite frequency; so the photon wavefunction is composed of momentum states in different directions, in which the distance between one corkscrew turn and the next is the same for all the corkscrews, this distance being the *wavelength*. (The wavelength is fixed by the frequency.)

Each photon wavefunction spreads out initially from the source s and (no detection being made at the slits) it passes through both slits on its way to the screen. Only a small part of this wavefunction emerges from the slits, however, and we think of each slit as acting as a new source from which the wavefunction separately spreads. These two portions of wavefunction interfere with one another so that, when they reach the screen, there are places where the two portions add up and places where they cancel out. To find out where the waves add and where they cancel, we take some point p on the screen and examine the straight lines to p from each of the two slits t and b. Along the line tp we have a corkscrew, and along the line bp we have another corkscrew. (We also have corkscrews along the lines st and sb, but if we assume that the source is the same distance from each of the slits, then *at* the slits the corkscrews will have turned by the same amounts.) Now the amounts by which the corkscrews will have turned by the time they reach the screen at p will depend on the lengths of the lines tp and bp. When these lengths differ by an integer number of wavelengths then, at p, the corkscrews will both be displaced in the *same* directions from their axes (i.e. $\theta = 0°$, where θ is as in the previous section), so the respective amplitudes will add up and we get a *bright* spot. When these lengths differ by an integer number of wavelengths plus half a wavelength then, at p, the corkscrews will both be displaced in the *opposite* directions from their axes ($\theta = 180°$), so the respective amplitudes will cancel and we get a *dark* spot. In all other cases, there will be some angle between the displacements of the corkscrews when they reach p, so the amplitudes add in some intermediate way, and we get a region of intermediate intensity (see Fig. 6.13).

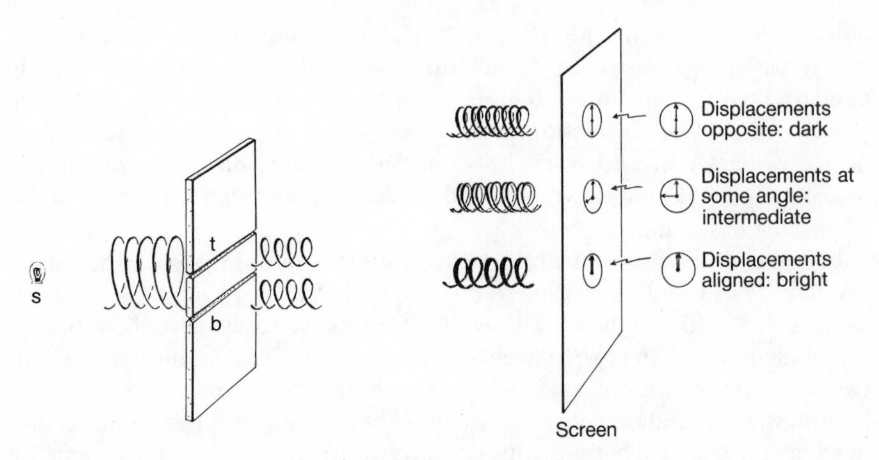

Fig. 6.13. The two-slit experiment analysed in terms of the corkscrew descriptions of the photon momentum states.

The uncertainty principle

Most readers will have heard of Heisenberg's *uncertainty principle*. According to this principle, it is not possible to measure (i.e. to magnify to the classical level) both the position and the momentum of a particle accurately at the same time. Worse than this, there is an *absolute limit* on the product of these accuracies, say Δx and Δp, respectively, which is given by the relation

$$\Delta x \, \Delta p \geq \hbar.$$

This formula tells us that the more accurately the position x is measured, the less accurately can the momentum p be determined, and *vice versa*. If the position were measured to *infinite* precision, then the momentum would become *completely* uncertain; on the other hand, if the momentum is measured exactly, then the particle's location becomes completely uncertain. To get some feeling for the size of the limit given by Heisenberg's relation, suppose that the position of an electron is measured to the accuracy of one part in a nanometre (10^{-9}m); then the momentum would become so uncertain that one could not expect that, one second later, the electron would be closer than 100 kilometres away!

In some descriptions, one is led to believe that this is merely some kind of inbuilt clumsiness inherent in the measurement process. Accordingly, in the case of the electron just considered, the attempt to localize it, according to this view, inevitably gives it a random 'kick' of such probable intensity that the electron is very likely to hurtle off at a great speed of the kind of magnitude indicated by Heisenberg's principle. In other descriptions one learns that the uncertainty is a property of the particle itself, and its motion has an inherent randomness about it which means that its behaviour is intrinsically unpredictable on the quantum level. In yet other accounts, one is informed that a quantum particle is something incomprehensible, to which the very concepts of classical position and classical momentum are inapplicable. I am not happy with any of these. The first is somewhat misleading; the second is certainly wrong; and the third, unduly pessimistic.

What does the wavefunction description actually tell us? First let us recall our description of a momentum state. This is the case where the momentum is specified exactly. The ψ-curve is a corkscrew, which remains the same distance from the axis all the way along. The amplitudes for the different position values therefore all have equal squared moduli. Thus, if a position measurement is performed, the probability of finding the particle at any one point is the same as finding it at any other. The position of the particle is indeed completely uncertain! What about a position state? Now, the ψ-curve is a delta function. The particle is precisely located—at the position of the delta function's spike—the amplitudes being zero for all other positions. The momentum amplitudes are best obtained by looking at the momentum space description, where it is now the $\tilde{\psi}$-curve which is a corkscrew, so that it is the different momentum amplitudes which now all have equal squared moduli. On performing a measurement of the particle's momentum, the result would now be completely uncertain!

It is of some interest to examine an intermediate case where the positions and momenta are both only partially constrained, albeit necessarily only to a degree consistent with Heisenberg's relation. The ψ-curve and corresponding $\tilde{\psi}$-curve (Fourier transforms of each other) for such a case are illustrated in Fig. 6.14. Notice that the distance of each curve from the axis is appreciable only in a quite small region. Far away, the curve hugs the axis very closely. This means that the squared moduli are of any appreciable size only in a very limited region, both in position space and in momentum space. In this way the particle can be rather localized in space, but there is a certain spread; likewise the momentum is fairly definite, so the particle moves off with a fairly definite speed and the spread of possible positions does not increase too greatly with

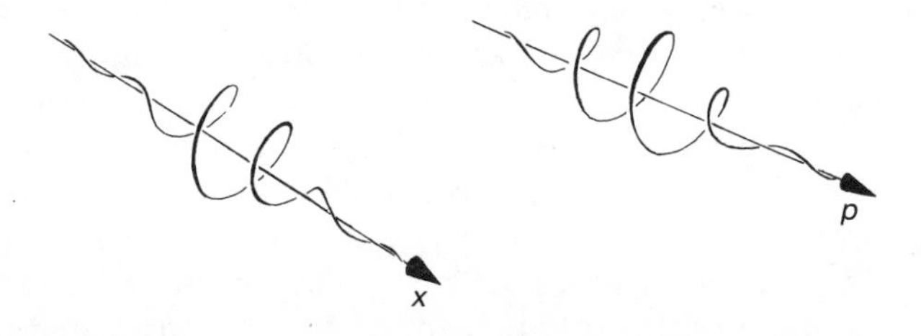

Fig. 6.14. Wave packets. These are localized both in position space and in momentum space.

time. Such a quantum state is referred to as a *wave packet*; it is often taken as quantum theory's best approximation to a classical particle. However, the spread in momentum (i.e. in velocity) values, implies that a wave packet *will* spread with time. The more localized in position that it is to start with, the more quickly it will spread.

The evolution procedures U and R

Implicit in this description of the time-development of a wave packet is *Schrödinger's equation*, which tells us how the wavefunction actually evolves in time. In effect, what Schrödinger's equation says is that if we decompose ψ into momentum states ('pure tones'), then each of these individual components will move off with a speed that is c^2 divided by the speed of a classical particle having the momentum in question. In fact Schrödinger's mathematical equation is written down more concisely than this. We shall be looking at its exact form later. It somewhat resembles Hamilton's or Maxwell's equations (having intimate relations with both) and, like those equations, gives a *completely deterministic* evolution of the wavefunction once the wavefunction is specified at any one time! (See p. 288.)

Regarding ψ as describing the 'reality' of the world, we have none of this indeterminism that is supposed to be a feature inherent in quantum theory—so long as ψ is governed by the deterministic Schrödinger evolution. Let us call this the evolution process **U**. However, whenever we 'make a measurement', magnifying quantum effects to the classical level, we change the rules. Now we do *not* use **U**, but instead adopt the completely different procedure, which I refer to as **R**, of forming the squared moduli of quantum amplitudes to obtain classical probabilities![4] It is the procedure **R**, and *only* **R**, that introduces uncertainties and probabilities into quantum theory.

The deterministic process **U** seems to be the part of quantum theory of main concern to working physicists; yet philosophers are more intrigued by the non-deterministic *state-vector reduction* **R** (or, as it is sometimes graphically described: *collapse of the wavefunction*). Whether we regard **R** as simply a change in the 'knowledge' available about a system, or whether we take it (as I do) to be something 'real', we are indeed provided with two completely *different* mathematical ways in which the state-vector of a physical system is described as changing with time. For **U** is totally deterministic, whereas **R** is a probabilistic law; **U** maintains quantum complex superposition, but **R** grossly violates it; **U** acts in a continuous way, but **R** is blatantly discontinuous. According to the standard procedures of quantum mechanics there is no implication that there be any way to 'deduce' **R** as a complicated instance of **U**. It is simply a *different* procedure from **U**, providing the other 'half' of the interpretation of the quantum formalism. All the non-determinism of the theory comes from **R** and not from **U**. *Both* **U** and **R** are needed for all the marvellous agreements that quantum theory has with observational facts.

Let us return to our wavefunction ψ. Suppose it is a momentum state. It will remain that momentum state happily for the rest of time so long as the particle does not interact with anything. (This is what Schrödinger's equation tells us.) At any time we choose to 'measure its momentum', we still get the same definite answer. There are no probabilities here. The predictability is as clear-cut as it is in the classical theory. However, suppose that at some stage we take it upon ourselves to measure (i.e. to magnify to the classical level) the particle's position. We find, now, that we are presented with an array of probability amplitudes, whose moduli we must square. At that point, probabilities abound, and there is complete uncertainty as to what result that measurement will produce. The uncertainty is in accord with Heisenberg's principle.

Let us suppose, on the other hand, that we start ψ off in a position state (or nearly in a position state). Now, Schrödinger's equation tells us that ψ will *not* remain in a position state, but will disperse rapidly. Nevertheless, the *way* in which it disperses is completely fixed by this equation. There is nothing indeterminate or probabilistic about its behaviour. In principle there would be experiments that we could perform to check up on this fact. (More of this later.) But if we unwisely choose to measure the momentum, then we find amplitudes for all different possible momentum values having equal squared moduli, and there is complete uncertainty as to the result of the experiment—again in accordance with Heisenberg's principle.

Likewise, if we start ψ off as a wave packet, its future evolution is completely fixed by Schrödinger's equation, and experiments could, in principle, be devised to keep track of this fact. But as soon as we choose to measure the particle in some *different* way from that—say to measure its position or momentum—then we find that uncertainties enter, again in accordance with Heisenberg's principle, with probabilities given by the squared moduli of amplitudes.

This is undoubtedly very strange and mysterious. But it is not an incomprehensible picture of the world. There is much about this picture that is governed by very clear and precise laws. There is, however, no clear rule, as yet, as to when the probabilistic rule **R** should be invoked, in place of the deterministic **U**. What constitutes 'making a measurement'? Why (and when) do squared moduli of amplitudes 'become probabilities'? Can the 'classical level' be understood quantum-mechanically? These are deep and puzzling questions which will be addressed later in this chapter.

Particles in two places at once?

In the above descriptions I have been adopting a rather more 'realistic' view of the wavefunction than is perhaps usual among quantum physicists. I have been taking the view that the 'objectively real' state of an individual particle is indeed described by its wavefunction ψ. It seems that many people find this a difficult position to adhere to in a serious way. One reason for this appears to be that it involves our regarding individual particles being spread out spatially, rather than always being concentrated at single points. For a momentum state, this spread is at its most extreme, since ψ is distributed equally all over the whole of space. Rather than thinking of the particle itself being spread out over space, people prefer to think of its position just being 'completely uncertain', so that all one can say about its position is that the particle is equally probable to be at any one place as it is to be at any other. However, we have seen that the wavefunction does not merely provide a probability distribution for different positions; it provides an *amplitude* distribution for different positions. If we know this amplitude distribution (i.e. the function ψ), then we know—from Schrödinger's equation—the precise way in which the state of the particle will evolve from moment to moment. We need this 'spread-out' view of the particle in order that its 'motion' (i.e. the evolution of ψ in time) be so determined; and if we *do* adopt this view, we see that the particle's motion *is* indeed precisely determined. The 'probability view' with regard to $\psi(x)$ would become appropriate if we performed a position measurement on the particle, and $\psi(x)$ would *then* be used only in its form as a squared modulus: $|\psi(x)|^2$.

It would seem that we must indeed come to terms with this picture of a particle which can be spread out over large regions of space, and which is likely to remain spread out until the next position measurement is carried out. Even when localized as a position state, a particle begins to spread at the next moment. A momentum state may seem hard to accept as a picture of the 'reality' of a particle's existence, but it is perhaps even harder to accept as 'real' the *two-peaked* state which occurs when the particle emerges from just having passed through a pair of slits (Fig. 6.15) In the vertical direction, the form of the wave function ψ would be sharply peaked at each of the slits, being the sum* of a wavefunction ψ_t, which is peaked at the top slit, and ψ_b peaked at the bottom slit:

$$\psi(x) = \psi_t(x) + \psi_b(x).$$

If we take ψ as representing the 'reality' of the state of the particle, then we must accept that the particle 'is' indeed in *two* places at once! On this view, the particle *has actually passed through both slits at once.*

Recall the standard objection to the view that the particle 'passes through both slits at once': if we perform a measurement *at* the slits in order to determine which slit it passed through, we always find that the *entire* particle

* The more usual quantum-mechanical description would be to divide this sum by a normalizing factor—here $\sqrt{2}$, to get $(\psi_t+\psi_b)/\sqrt{2}$—but there is no need to complicate the description in this way here.

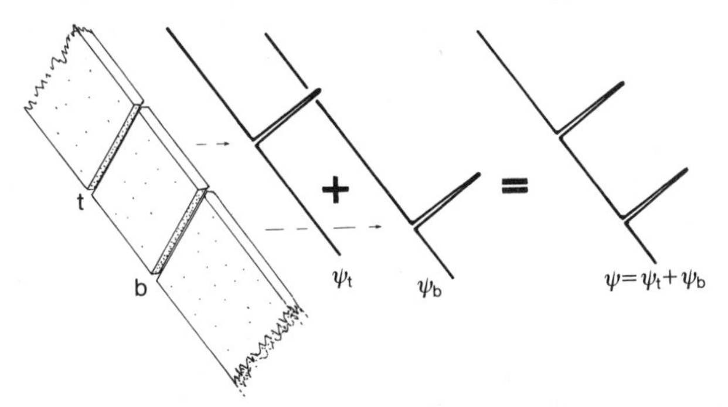

Fig. 6.15. As the photon wavefunction emerges from the pair of slits it is peaked at two places at once.

is at one or other of the slits. But this arises because we are performing a *position measurement* on the particle, so that ψ *now* merely provides a probability distribution $|\psi|^2$ for the particle's position in accordance with the squared-modulus procedure, and we indeed find it at just one place or the other. But there are also various types of measurement that one *could* perform at the slits, *other* than position measurements. For those, we should need to know the two-peaked wavefunction ψ, and not just $|\psi|^2$, for different positions x. Such a measurement might distinguish the two-peaked state

$$\psi = \psi_t + \psi_b,$$

given above, from other two-peaked states, such as

$$\psi_t - \psi_b$$

or

$$\psi_t + \mathrm{i}\psi_b.$$

(See Fig. 6.16, for the ψ-curves in each of these different cases.) Since there are indeed measurements which distinguish these various possibilities, they must all be *different* possible 'actual' ways in which the photon can exist!

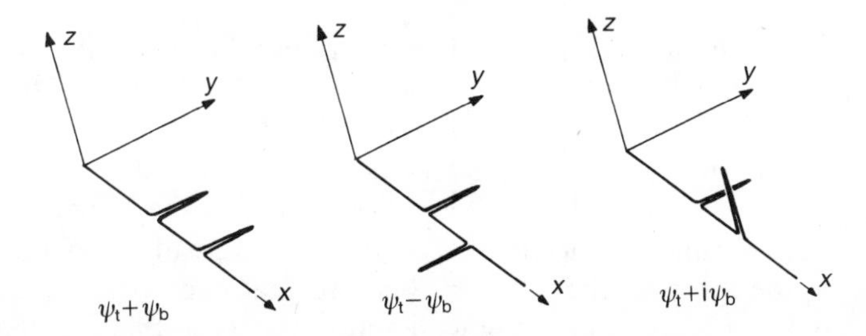

Fig. 6.16. Three different ways in which a photon wavefunction can be doubly peaked.

The slits do not have to be close together for a photon to pass through them 'both at once'. To see that a quantum particle can be in 'two places at once' no matter how distant the places are, consider an experimental set-up a little different from that of the two-slit experiment. As before, we have a lamp emitting monochromatic light, one photon at a time; but instead of letting the light pass through a pair of slits, let us reflect it off a half-silvered mirror, tilted at 45° to the beam. (A half-silvered mirror is a mirror which reflects exactly half of the light which impinges upon it, while the remaining half is transmitted directly through the mirror.) After its encounter with the mirror, the photon's wavefunction splits in two, with one part reflected off to the side and the other part continuing in the same direction in which the photon started. The wavefunction is again doubly peaked, as in the case of the photon emerging from the pair of slits, but now the two peaks are much more widely separated, one peak describing the photon reflected and the other peak, the photon transmitted. (See Fig. 6.17.) Moreover, as time progresses, the separation between the peaks gets larger and larger, and increases without any limit. Imagine that the two portions of the wavefunction escape into space and that we wait for a whole year. Then the two peaks of the photon's wavefunction will be over a light-year apart. Somehow, the photon has found itself to be in two places at once, more than a light-year distant from one another!

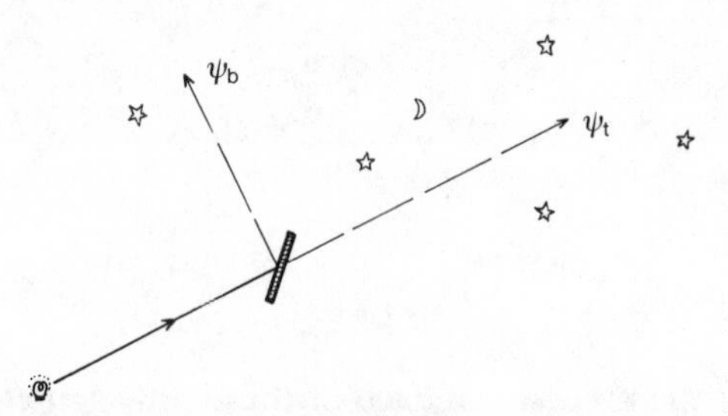

Fig. 6.17. The two peaks of a doubly peaked wavefunction could be light-years apart. This can be achieved by means of a half-silvered mirror.

Is there any reason to take such a picture seriously? Can we not regard the photon as simply having a 50 per cent probability that it is in one of the places and a 50 per cent probability that it is in the other? No, we cannot! No matter for how long it has travelled, there is always the possibility that the two parts of the photon's beam may be reflected back so that they encounter one another, to achieve interference effects that could not result from a probability weighting for the two alternatives. Suppose that each part of the beam encounters a fully silvered mirror, angled appropriately so that the beams are brought together, and at the meeting point is placed another half-silvered mirror, angled just as was the first one. Two photo-cells are placed in the direct lines of the two beams (see Fig. 6.18.) What do we find? If it were merely the case that there were a 50 per cent chance that the photon followed one route and a 50 per cent chance that it followed the other, then we should find a 50 per cent probability that one of the detectors registers the photon and a 50 per cent probability that the other one does. However, that is not what happens. If the two possible routes are exactly equal in length, then it turns out that there is a 100 per cent probability that the photon reaches the detector A, lying in the direction of the photon's initial motion and a 0 per cent probability that it reaches the other detector B—the photon is *certain* to strike the detector A! (We can see this by using the corkscrew description given above, as for the case of the two-slit experiment.)

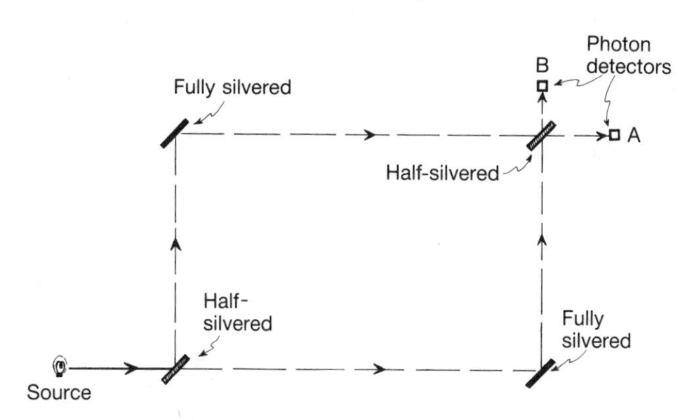

Fig. 6.18. The two peaks of a doubly peaked wavefunction cannot be thought of simply as probability weightings for the photon to be in one location or the other. The two routes taken by the photon can be made to interfere with one another.

Of course, such an experiment has never been carried out for path-lengths of the order of a light-year, but the stated result is not seriously doubted (by conventional quantum physicists!). Experiments of this very type have actually been carried out with path-lengths of many metres or so, and the results are indeed in complete agreement with the quantum-mechanical predictions (cf. Wheeler 1983). What does this tell us about the *reality* of the photon's state of existence between its first and last encounter with a half-reflecting mirror? It seems inescapable that the photon must, in some sense, have actually *travelled* both routes at once! For if an absorbing screen is placed in the way either of the two routes, then it becomes equally probable that A or B is reached; but when both routes are open (and of equal length) only A can be reached. Blocking off one of the routes actually *allows* B to be reached! With both routes open, the photon somehow 'knows' that is *not* permitted to reach B, so it must have actually felt out both routes.

Neils Bohr's view that no objective 'meaning' can be attached to the photon's existence between moments of measurement seems to me to be much too pessimistic a view to take concerning the reality of the state of the photon. Quantum mechanics gives us a *wavefunction* to describe the 'reality' of the photon's position, and between the half-silvered mirrors, the photon's wavefunction is just a doubly peaked state, the distance between the two peaks being sometimes very considerable.

We note, also, that just 'being in two specified places at once' is not a complete description of the photon's state: we need to be able to distinguish the state $\psi_t + \psi_b$ from the state $\psi_t - \psi_b$, say, (or from $\psi_t + i\psi_b$, say), where ψ_t and ψ_b now refer to photon positions in each of the two routes (now 'transmitted' and 'bounced', respectively!). It is this kind of distinction that determines whether the photon, when reaching the final half-silvered mirror, becomes certain to reach A or certain to reach B (or to reach A or B with an intermediate probability).

This puzzling feature of quantum reality—namely that we must take seriously that a particle may, in various (different!) ways 'be in two places at once'—arises from the fact that we must be allowed to add quantum states, using complex-number weightings, to get other quantum states. This kind of superposition of states is a general—and important—feature of quantum mechanics, referred to as *quantum linear superposition*. It is what allows us to compose momentum states out of position states, or position states out of momentum states. In these cases, the linear superposition applies to an *infinite* array of different states, i.e. to all the different position states, or to all the different momentum states. But, as we have seen, quantum linear superposition is quite puzzling enough when applied to just a *pair* of states. The rules are that *any* two states whatever, irrespective of how different from one another they might be, can coexist in the any complex linear superposition. Indeed, any physical object, itself made out of individual particles, ought to be able to exist in such superpositions of spatially widely separated states, and so 'be in two places at once'! The formalism of quantum mechanics makes no distinction, in this respect, between single particles and complicated systems of many particles. Why, then, do we not experience macroscopic bodies, say cricket balls, or even people, having two completely different locations at once? This is a profound question, and present-day quantum theory does not really provide us with a satisfying answer. For an object as substantial as a cricket ball, we must consider that the system is 'at the classical level'—or, as it is more usually phrased, an 'observation' or 'measurement' will have been made on the cricket ball—and then the complex probability amplitudes which weight our linear superpositions must now have their moduli squared and be treated as probabilities describing actual alternatives. However, this really begs the controversial question of *why* we are allowed to change our quantum rules from **U** to **R** in this way! I shall return to the matter later.

Hilbert space

Recall that in Chapter 5 the concept of *phase space* was introduced for the description of a classical system. A single point of phase space would be used to represent the (classical) state of an entire physical system. In the quantum theory, the appropriate analogous concept is that of a *Hilbert space.** A single point of Hilbert space now represents the *quantum* state of an entire system. We shall need to take a glimpse at the mathematical structure of a Hilbert space. I hope that the reader will not be daunted by this. There is nothing that is mathematically very complicated in what I have to say, though some of the ideas may well be unfamiliar.

The most fundamental property of a Hilbert space is that it is what is called a *vector space*—in fact, a *complex* vector space. This means that we are allowed to *add* together any two elements of the space and obtain another such element; and we are also allowed to perform these additions with complex-number weightings. We must be able to do this because these are the operations of *quantum linear superposition* that we have just been considering, namely the operations which give us $\psi_t + \psi_b$, $\psi_t - \psi_b$, $\psi_t + \mathrm{i}\psi_b$, etc., for the photon above. Essentially, all that we mean by the use of the phrase 'complex vector space', then, is that we are allowed to form weighted sums of this kind.[5]

It will be convenient to adopt a notation (essentially one due to Dirac) according to which the elements of the Hilbert space—referred to as *state vectors*—are denoted by some symbol in an angled bracket, such as $|\psi\rangle$, $|\chi\rangle$, $|\varphi\rangle$, $|1\rangle$, $|2\rangle$, $|3\rangle$, $|n\rangle$, $|\uparrow\rangle$, $|\downarrow\rangle$, $|\rightarrow\rangle$, $|\nearrow\rangle$, etc. Thus these symbols now denote quantum states. For the operation of addition of two state vectors we write

$$|\psi\rangle + |\chi\rangle$$

and with weightings with complex numbers w and z:

$$w|\psi\rangle + z|\chi\rangle$$

(where $w|\psi\rangle$ means $w \times |\psi\rangle$, etc.). Accordingly, we now write the above combinations $\psi_t + \psi_b$, $\psi_t - \psi_b$, $\psi_t + \mathrm{i}\psi_b$ as $|\psi_t\rangle + |\psi_b\rangle$, $|\psi_t\rangle - |\psi_b\rangle$, $|\psi_t\rangle + \mathrm{i}|\psi_b\rangle$, respectively. We can also simply multiply a *single* state $|\psi\rangle$ by a complex number w to obtain:

$$w|\psi\rangle.$$

(This is really a particular case of the above, given when $z = 0$.)

Recall that we allowed ourselves to consider complex-weighted combina-

* David Hilbert, whom we have encountered in earlier chapters, introduced this important concept—in the infinite-dimensional case—long before the discovery of quantum mechanics, and for a completely different mathematical purpose!

tions where w and z need not be the actual probability amplitudes but are merely *proportional* to these amplitudes. Accordingly, we adopt the rule that we can multiply an entire state-vector by a non-zero complex number and the physical state is unchanged. (This would change the actual values of w and z but the ratio w:z would remain unchanged.) Each of the vectors

$$|\psi\rangle, 2|\psi\rangle, -|\psi\rangle, i|\psi\rangle, \sqrt{2}|\psi\rangle, \pi|\psi\rangle, (1-3i)|\psi\rangle, \text{etc.}$$

represents the *same* physical state—as does any $z|\psi\rangle$, where $z \neq 0$. The only element of the Hilbert space that has *no* interpretation as a physical state is the *zero* vector **0** (or the *origin* of the Hilbert space).

In order to get some kind of geometrical picture of all this, let us first consider the more usual concept of a 'real' vector. One usually visualizes such a vector simply as an *arrow* drawn on a plane or in a three-dimensional space. Addition of two such arrows is achieved by means of the parallelogram law (see Fig. 6.19). The operation of multiplying a vector by a (real) number, in terms of the 'arrow' picture, is obtained by simply multiplying the length of the arrow by the number in question, keeping the direction of the arrow unchanged. If the number we multiply by is negative, then the direction of the arrrow is reversed; or if the number is zero, we get the zero vector **0**, which has no direction. (The vector **0** is represented by the 'null arrow' of zero length.) One example of a vector quantity is the force acting on a particle. Other examples are classical velocities, accelerations, and momenta. Also there are the momentum four-vectors that we considered at the end of the last chapter. Those were vectors in *four* dimensions rather than two or three. However, for a Hilbert space we need vectors in dimensions much higher still (often infinite, in fact, but that will not be an important one for us here).

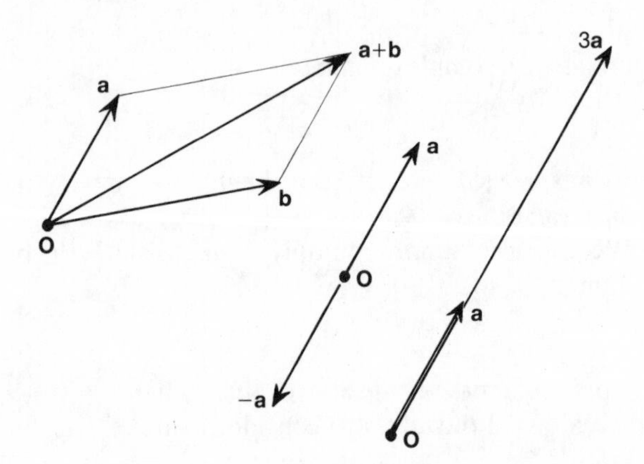

Fig. 6.19. Addition of Hilbert space vectors and multiplication by scalars can be visualized in the normal way, as for vectors in ordinary space.

Recall that arrows were also used to depict vectors in classical phase space—which could certainly be of very high dimension. The 'dimensions' in a phase space do not represent ordinary space-directions, and nor do the 'dimensions' of a Hilbert space. Instead, each Hilbert space dimension corresponds to one of the different independent physical states of a quantum system.

Because of the equivalence between $|\psi\rangle$ and $z|\psi\rangle$, a physical state actually corresponds to a whole *line through the origin* **0**, or *ray*, in the Hilbert space (described by all the multiples of some vector), not simply to a particular vector in that line. The ray consists of all possible multiples of a particular state-vector $|\psi\rangle$. (Bear in mind that these are *complex* multiples, so the line is actually a *complex* line, but it is better not to worry about that now!) (See Fig. 6.20). We shall shortly find an elegant picture of this space of rays for the case of a *two*-dimensional Hilbert space. At the other extreme is the case when the Hilbert space is infinite-dimensional. An infinite-dimensional Hilbert space arises even in the simple situation of the location of a single particle. There is then an entire dimension for each possible position that the particle might have! Every particle position defines a whole 'coordinate axis' in the Hilbert space, so with infinitely many different individual positions for the particle we have infinitely many different independent directions (or 'dimensions') in the Hilbert space. The momentum states will also be represented in this *same* Hilbert space. Momentum states are expressible as combinations of position states, so each momentum state corresponds to an axis going off 'diagonally', which is tilted with respect to the position-space axes. The set of all momentum states provides a new set of axes, and to pass from the position-space axes to the momentum-space axes involves a *rotation* in the Hilbert space.

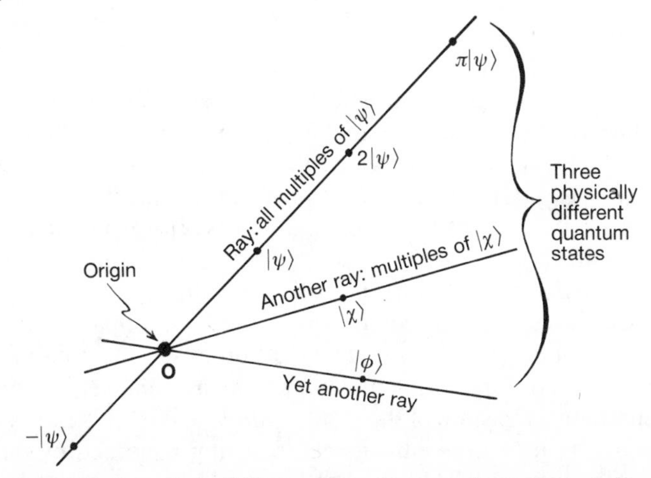

Fig. 6.20. Entire rays in Hilbert space represent physical quantum states.

One need not attempt to visualize this in any accurate way. That would be unreasonable! However, certain ideas taken from ordinary Euclidean geometry are very helpful for us. In particular, the axes that we have been considering (*either* all the position-space axes *or* all the momentum-space axes) are to be thought of as being all *orthogonal* to one another, that is, mutually at 'right angles'. 'Orthogonality' between rays is an important concept for quantum mechanics: orthogonal rays refer to states that are *independent* of one another. The different possible position states of a particle are all orthogonal to one another, as are all the different possible momentum states. But position states are not orthogonal to momentum states. The situation is illustrated, very schematically, in Fig. 6.21.

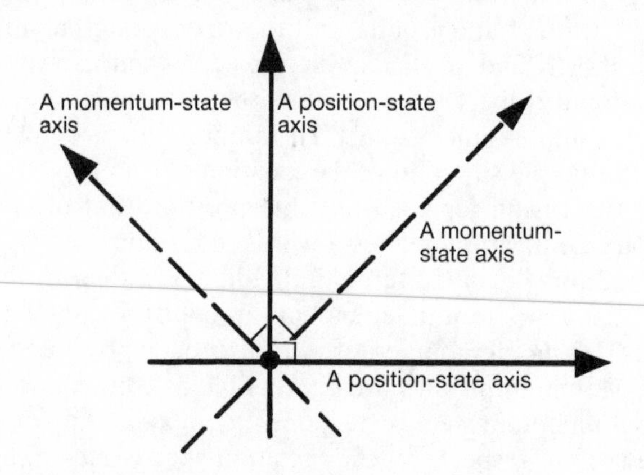

Fig. 6.21. Position states and momentum states provide different choices of orthogonal axes in the same Hilbert space.

Measurements

The general rule **R** for a *measurement* (or observation) requires that the different aspects of a quantum system that can be simultaneously magnified to the classical level—and between which the system must then choose—must always be *orthogonal*. For a *complete* measurement, the selected set of alternatives constitutes a set of orthogonal *basis* vectors, meaning that every vector in the Hilbert space can be (uniquely) linearly expressed in terms of them. For a *position* measurement—on a system consisting of a single particle—these basis vectors would define the position axes that we just considered. For *momentum*, it would be a different set, defining the momentum axes, and for a different kind of complete measurement, yet another set. After measurement, the state of the system *jumps* to one of the axes of the set determined by the measurement—its choice being governed by mere probability. There is no dynamical law to tell us which among the selected axes

Nature will choose. Her choice is random, the probability values being squared moduli of probability amplitudes.

Let us suppose that some complete measurement is made on a system whose state is $|\psi\rangle$, the basis for the selected measurement being

$$|0\rangle, |1\rangle, |2\rangle, |3\rangle, \ldots .$$

Since these form a complete set, any state-vector, and in particular $|\psi\rangle$, can be represented linearly* in terms of them:

$$|\psi\rangle = z_0|0\rangle + z_1|1\rangle + z_2|2\rangle + z_3|3\rangle + \ldots .$$

Geometrically, the components z_0, z_1, z_2, ... measure the *sizes of the orthogonal projections* of the vector $|\psi\rangle$ on the various axes $|0\rangle$, $|1\rangle$, $|2\rangle$, ... (see Fig. 6.22).

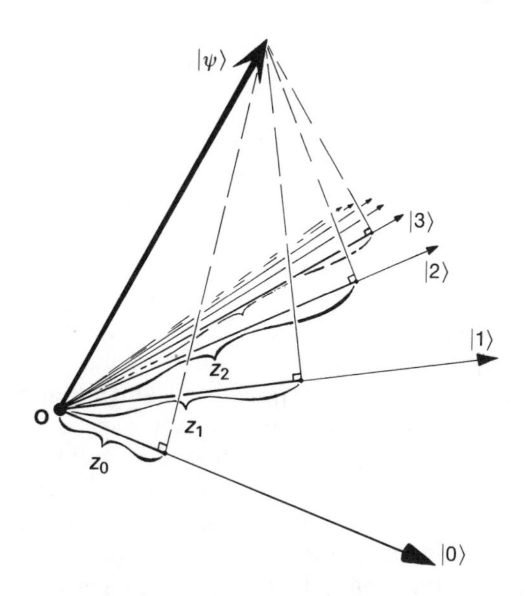

Fig. 6.22. The sizes of the orthogonal projections of the state $|\psi\rangle$ on the axes $|0\rangle$, $|1\rangle$, $|2\rangle$, ... provide the required amplitudes z_0, z_1, z_2, ...

We should like to be able to interpret the complex numbers z_0, z_1, z_2, ... as our required probability amplitudes, so that their squared moduli provide the various probabilities that, after the measurement, the system is found to be in the respective states $|0\rangle$, $|1\rangle$, $|2\rangle$, ... However, this will not quite do as it stands because we have not fixed the 'scales' of the various basis

* This has to be allowed to be taken in the sense that an *infinite* sum of vectors is permitted. The *full* definition of a Hilbert space (which is too technical for me to go into here) involves the rules pertaining to such infinite sums.

vectors $|0\rangle$, $|1\rangle$, $|2\rangle$, ... For this, we must specify that they are, in some sense, *unit vectors* (i.e. vectors of unit 'length'), and so, in mathematical terminology, they form what is called an *orthonormal* basis (mutually *ortho*gonal and *normal*ized to be unit vectors).[6] If $|\psi\rangle$ is also normalized to be a unit vector, then the required amplitudes will indeed now be the components $z_0, z_1, z_2, \ldots$. of $|\psi\rangle$, and the required respective probabilities will be $|z_0|^2, |z_1|^2, |z_2|^2, \ldots$. If $|\psi\rangle$ is not a unit vector, then these numbers will be *proportional* to the required amplitudes and probabilities, respectively. The actual amplitudes will be

$$\frac{z_0}{|\psi|}, \frac{z_1}{|\psi|}, \frac{z_2}{|\psi|}, \text{ etc.,}$$

and the actual probabilities

$$\frac{|z_0|^2}{|\psi|^2}, \frac{|z_1|^2}{|\psi|^2}, \frac{|z_2|^2}{|\psi|^2}, \text{ etc.,}$$

where $|\psi|$ is the 'length' of the state-vector $|\psi\rangle$. This 'length' is a positive real number defined for each state-vector (**0** has length zero), and $|\psi| = 1$ if $|\psi\rangle$ is a unit vector.

A complete measurement is a very idealized kind of measurement. The complete measurement of the position of a particle, for example, would require our being able to locate the particle with infinite precision, anywhere in the universe! A more elementary type of measurement is one for which we simply ask a *yes/no* question such as: 'does the particle lie to the left or to the right of some line?' or 'Does the particle's momentum lie within some range?', etc. Yes/no measurements are really the most fundamental type of measurement. (One can, for example, narrow down a particle's position or momentum as closely as we please, using only yes/no measurements.) Let us suppose that the result of a yes/no measurement turns out to be **YES**. Then the state-vector must find itself in the '**YES**' region of Hilbert space, which I shall call **Y**. If, on the other hand, the result of the measurement is **NO**, then the state-vector finds itself in the '**NO**' region **N** of Hilbert space. The regions **Y** and **N** are totally orthogonal to one another in the sense that any state-vector belonging to **Y** must be orthogonal to every state-vector belonging to **N** (and *vice-versa*). Moreover, any state-vector $|\psi\rangle$ can be expressed (in a unique way) as a sum of vectors, one from each of **Y** and **N**. In mathematical terminology, we say that **Y** and **N** are *orthogonal complements* of one another. Thus, $|\psi\rangle$ is expressed uniquely as

$$|\psi\rangle = |\psi_{\mathbf{Y}}\rangle + |\psi_{\mathbf{N}}\rangle,$$

where $|\psi_{\mathbf{Y}}\rangle$ belongs to **Y** and $|\psi_{\mathbf{N}}\rangle$ belongs to **N**. Here $|\psi_{\mathbf{Y}}\rangle$ is the *orthogonal projection* of the state $|\psi\rangle$ on **Y** and $|\psi_{\mathbf{N}}\rangle$, correspondingly, is the orthogonal projection of $|\psi\rangle$ on **N** (see Fig. 6.23).

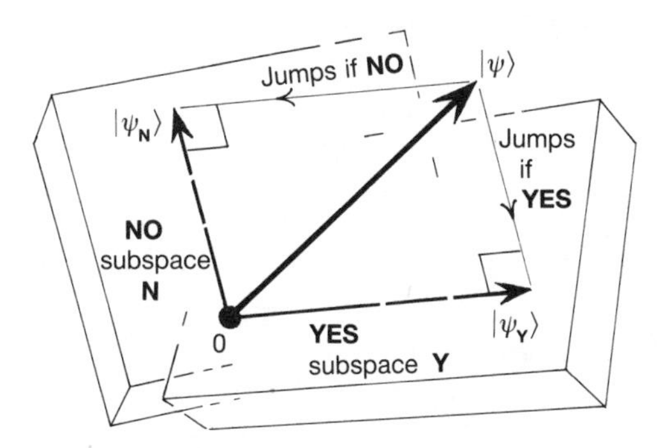

Fig. 6.23. State-vector reduction. A yes/no measurement can be described in terms of a pair of subspaces **Y** and **N** which are orthogonal complements of one another. Upon measurement, the state $|\psi\rangle$ jumps to its projection into one or other of these subspaces, with probability given by the factor by which the squared length of the state vector decreases in the projection.

Upon measurement, the state $|\psi\rangle$ *jumps*—and becomes (proportional to) either $|\psi_{\mathbf{Y}}\rangle$ or $|\psi_{\mathbf{N}}\rangle$. If the result is **YES**, then it jumps to $|\psi_{\mathbf{Y}}\rangle$, and if **NO**, it jumps to $|\psi_{\mathbf{N}}\rangle$. If $|\psi\rangle$ is normalized, the respective probabilities of these two occurrences are the *squared lengths*

$$|\psi_{\mathbf{Y}}|^2, \quad |\psi_{\mathbf{N}}|^2$$

of the projected states. If $|\psi\rangle$ is not normalized, we must divide each of these expressions by $|\psi|^2$. (The 'Pythagorean theorem', $|\psi|^2 = |\psi_{\mathbf{Y}}|^2 + |\psi_{\mathbf{N}}|^2$, asserts that these probabilities sum to unity, as they should!) Note that the probability that $|\psi\rangle$ jumps to $|\psi_{\mathbf{Y}}\rangle$ is given by the ratio by which its squared length is reduced upon this projection.

One final point should be made concerning such 'acts of measurement' that can be made on a quantum system. It is an implication of the tenets of the theory that for *any state whatever*—say the state $|\chi\rangle$—there is a yes/no measurement[7] that can *in principle* be performed for which the answer is **YES** if the measured state is (proportional to) $|\chi\rangle$, and **NO** if it is orthogonal to $|\chi\rangle$. Thus the region **Y**, above, could consist of all the multiples of any chosen state $|\chi\rangle$. This seems to have the strong implication that state vectors must be *objectively real*. Whatever the state of a physical system happens to be—and let us call that state $|\chi\rangle$—there is a measurement that can in principle be performed for which $|\chi\rangle$ is the *only* state (up to proportionality) for which the measurement yields the result **YES**, with *certainty*. For some states $|\chi\rangle$ this measurement might be extremely difficult to perform—perhaps 'impossible' in practice—but the fact that, according to the theory, such a measurement could *in principle* be made will have some startling consequences for us later in this chapter.

Spin and the Riemann sphere of states

The quantity referred to, in quantum mechanics, as '*spin*' is sometimes regarded as the most 'quantum-mechanical' of all physical quantities, so we shall be wise to pay some attention to it. What *is* spin? Essentially, it is a measure of rotation of a particle. The term 'spin' indeed suggests something like the spin of a cricket ball or baseball. Recall the concept of *angular momentum* which, like energy and momentum, is *conserved* (see Chapter 5, p. 166, and also p. 231). The angular momentum of a body persists in time so long as the body is not disturbed by frictional or other forces. This, indeed, is what quantum-mechanical spin is, but now it is the 'spinning' of a *single* particle itself that concerns us, not the orbiting motion of myriads of individual particles about their common centre of mass (which would be the case for a cricket ball). It is a remarkable physical fact that most of the particles found in nature actually do 'spin' in this sense, each according to its own very specific amount.[8] However, as we shall see, the spin of a single quantum-mechanical particle has some very peculiar properties that are not at all what we would expect from our experiences with spinning cricket balls and the like.

In the first place, the *amount* of the spin of a particle is always the *same*, for that particular type of particle. It is only the direction of the axis of that spin which can vary (in a very strange way, that we shall come to). This is in stark contrast with a cricket ball, which can spin by all manner of differing amounts, depending upon how it is bowled! For an electron, proton, or neutron, the amount of spin is always $\hbar/2$, that is just *one-half* of the smallest positive value that Bohr originally allowed for his quantized angular momenta of atoms. (Recall that these values were $0, \hbar, 2\hbar, 3\hbar, \ldots$.) Here we require one-half of the basic unit $\hbar$—and, in a sense, $\hbar/2$ is itself the more fundamental basic unit. This amount of angular momentum would not be allowed for an object composed solely of a number of orbiting particles, none of which was itself spinning; it can only arise because the spin is an *intrinsic* property of the particle itself (i.e. not arising from the orbital motion of its 'parts' about some centre).

A particle whose spin is an *odd*-number multiple of $\hbar/2$ (i.e. $\hbar/2$, $3\hbar/2$, or $5\hbar/2$ etc.) is called a *fermion*, and it exhibits a curious quirk of quantum-mechanical description: a complete rotation through 360° sends its state-vector not to itself but to *minus* itself! Many of the particles of Nature are indeed fermions, and we shall be hearing more about them and their odd ways—so vital to our very existence—later. The remaining particles, for which the spin is an *even* multiple of $\hbar/2$, i.e. a whole-number multiple of $\hbar$ (namely $0, \hbar, 2\hbar, 3\hbar, \ldots$), are called *bosons*. Under a 360° rotation, a boson state-vector goes back to *itself*, not to its negative.

Consider a *spin one-half* particle, i.e. with spin value $\hbar/2$. For definiteness I shall refer to the particle as an *electron*, but a proton or neutron would do just as well, or even an appropriate kind of atom. (A 'particle' is allowed to possess individual parts, so long as it can be treated quantum mechanically as a single whole, with a well defined total angular momentum.) We take the electron to be at rest and consider just its state of spin. The quantum-mechanical state space (Hilbert space) now turns out to be *two*-dimensional, so we can take a basis of just *two* states. These I label as $|\uparrow\rangle$ and $|\downarrow\rangle$, to indicate that for $|\uparrow\rangle$ the spin is right-handed about the *upward* vertical direction, while for $|\downarrow\rangle$, it is right-handed about the *downward* direction (Fig. 6.24). The states $|\uparrow\rangle$ and $|\downarrow\rangle$ are orthogonal to one another and we take them to be normalized ($|\uparrow|^2 = |\downarrow|^2 = 1$). Any possible state of spin of the electron is a linear superposition, say $w|\uparrow\rangle + z|\downarrow\rangle$, of just the *two* orthonormal states $|\uparrow\rangle$ and $|\downarrow\rangle$, that is, of *up* and *down*.

Fig. 6.24. A basis for the spin states of an electron consists of just two states. These can be taken to be *spin up* and *spin down*.

Now, there is nothing special about the directions 'up' and 'down'. We could equally well choose to describe the spin in (i.e. right-handed about) any other direction, say *right* $|\rightarrow\rangle$ as opposed to *left* $|\leftarrow\rangle$. Then (for suitable choice of complex scaling for $|\uparrow\rangle$ and $|\downarrow\rangle$), we find*

$$|\rightarrow\rangle = |\uparrow\rangle + |\downarrow\rangle \quad \text{and} \quad |\leftarrow\rangle = |\uparrow\rangle - |\downarrow\rangle.$$

This gives us a new view: any state of electron spin is a linear superposition of the two orthogonal states $|\rightarrow\rangle$ and $|\leftarrow\rangle$, that is, of *right* and *left*. We might choose, instead, some completely arbitrary direction, say that given by the state-vector $|\rightarrow\rangle$. This again is some complex linear combination of $|\uparrow\rangle$ and $|\downarrow\rangle$, say

$$|\nearrow\rangle = w|\uparrow\rangle + z|\downarrow\rangle,$$

and every state of spin will be a linear superposition of this state and the orthogonal state $|\swarrow\rangle$, which points in the opposite[9] direction to $|\nearrow\rangle$. (Note that the concept of 'orthogonal', in Hilbert space need not correspond to 'at right angles' in ordinary space. The orthogonal Hilbert space-vectors here

* As in various earlier places, I prefer not to clutter the descriptions with factors like $1/\sqrt{2}$ which would arise if we require $|\rightarrow\rangle$ and $|\leftarrow\rangle$ to be normalized.

correspond to diametrically opposite directions in space, rather than directions which are at right angles.)

What is the geometrical relation between the direction in space determined by $|\nearrow\rangle$ and the two complex numbers w and z? Since the physical state given by $|\nearrow\rangle$ is unaffected if we multiply $|\nearrow\rangle$ by a non-zero complex number, it will be only the *ratio* of z to w which will have significance. Write

$$q = z/w$$

for this ratio. Then q is just some complex number, except that the value '$q = \infty$' is also allowed, in order to cope with the situation $w = 0$, i.e. for when the spin direction is vertically downwards. Unless $q = \infty$, we can represent q as a point on the Argand plane, just as we did in Chapter 2. Let us imagine that this Argand plane is situated horizontally in space, with the direction of the real axis being off to the 'right' in the above description (i.e. in the direction of the spin-state $|\rightarrow\rangle$). Imagine a sphere of unit radius, whose centre is at the origin of this Argand plane, so that the points 1, i, -1, $-$i all lie on the equator of the sphere. We consider the point at the south pole, which we label ∞, and then project from this point so that the entire Argand plane is mapped to the sphere. Thus any point q on the Argand plane corresponds to a unique point q on the sphere, obtained by lining the two points up with the south pole (Fig. 6.25). This correspondence is called *stereographic* projection, and it has many beautiful geometrical properties (e.g. it preserves angles and maps circles to circles). The projection gives us a

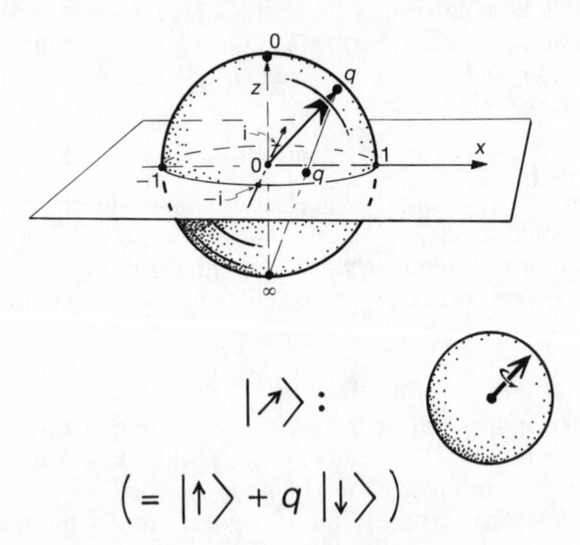

Fig. 6.25. The Riemann sphere, here represented as the space of physically distinct spin states of a spin-$\frac{1}{2}$ particle. The sphere is projected stereographically from its south pole (∞) to the Argand plane through its equator.

labelling of the points of the sphere by complex numbers together with ∞, i.e. by the set of possible complex ratios q. A sphere labelled in this particular way is called a *Riemann sphere*. The significance of the Riemann sphere, for the spin-states of an electron, is that the direction of spin given by $|\nearrow\rangle = w|\uparrow\rangle + z|\downarrow\rangle$ is provided by the actual direction from the centre to the point $q = z/w$, as marked on the Riemann sphere. We note that the north pole corresponds to the state $|\uparrow\rangle$, which is given by $z = 0$, i.e. by $q = 0$, and the south pole to $|\downarrow\rangle$, given by $w = 0$, i.e. by $q = \infty$. The rightmost point is labelled by $q = 1$, which provides the state $|\rightarrow\rangle = |\uparrow\rangle + |\downarrow\rangle$, and the leftmost point by $q = -1$, which provides $|\leftarrow\rangle = |\uparrow\rangle - |\downarrow\rangle$. The farthest point around the back of the sphere is labelled $q = \mathrm{i}$, corresponding to the state $|\uparrow\rangle + \mathrm{i}|\downarrow\rangle$ where the spin points directly away from us, and the nearest point, $q = -\mathrm{i}$, corresponds to $|\uparrow\rangle - \mathrm{i}|\downarrow\rangle$, where the spin is directly towards us. The general point, labelled by q, corresponds to $|\uparrow\rangle + q|\downarrow\rangle$.

How does all this tie in with measurements that one might perform on the spin of the electron?[10] Select some direction in space; let us call this direction α. If we measure the electron's spin in that direction, the answer **YES** says that the electron is (now) indeed spinning in (i.e. right-handed about) α, whereas **NO** says that it spins in the direction opposite to α.

Suppose the answer is **YES**; then we label the resulting state $|\alpha\rangle$. If we simply repeat the measurement, using precisely the same direction α as before, then we find that the answer must again be **YES**, with 100 per cent probability. But if we change the direction, for the second measurement, to a new direction β, then we find that there is some smaller probability for the **YES** answer, the state now jumping to $|\beta\rangle$, and there is some possibility that the answer to the second measurement might be **NO**, the state now jumping to the direction opposite to β. How do we calculate this probability? The answer is contained in the prescriptions given at the end of the previous section. The probability of **YES**, for the second measurement turns out to be

$$\tfrac{1}{2}(1 + \cos\theta),$$

where θ is the angle between the directions[11] α and β. The probability of **NO** for the second measurement is, accordingly,

$$\tfrac{1}{2}(1 - \cos\theta).$$

We can see from this that if the second measurement is made at right angles to the first then the probability is 50 per cent, either way ($\cos 90° = 0$): the result of the second measurement is completely random! If the angle between the two measurements is acute, then a **YES** answer is more likely than **NO**. If obtuse, then **NO** is more likely than **YES**. In the extreme case when β is opposite to α, the probabilities become 0 per cent for **YES** and 100 per cent for **NO**; i.e. the result of the second measurement is certain to be the reverse of the first. (See Feynman *et al.* 1965 for further information about spin.)

The Riemann sphere actually has a fundamental (but not always recognized) role to play in *any* two-state quantum system, describing the array of possible quantum states (up to proportionality). For a spin one-half particle, its geometrical role is particularly apparent, since the points of the sphere correspond to the possible spatial directions for the spin-axis. In many other situations, it is harder to see the role of the Riemann sphere. Consider a photon having just passed through a pair of slits, or having been reflected from a half-silvered mirror. The photon's state is some linear combination such as $|\psi_t\rangle + |\psi_b\rangle$, $|\psi_t\rangle - |\psi_b\rangle$, or $|\psi_t\rangle + \mathrm{i}|\psi_b\rangle$, of two states $|\psi_t\rangle$ and $|\psi_b\rangle$ describing two quite distinct locations. The Riemann sphere still describes the array of physically distinct possibilities, but now only *abstractly*. The state $|\psi_t\rangle$ is represented by the north pole ('top') and $|\psi_b\rangle$ by the south pole ('bottom'). Then $|\psi_t\rangle + |\psi_b\rangle$, $|\psi_t\rangle - |\psi_b\rangle$ and $|\psi_t\rangle + \mathrm{i}|\psi_b\rangle$ are respresented by various points around the equator, and in general, $w|\psi_t\rangle + z|\psi_b\rangle$ is represented by the point given by $q = z/w$. In many cases, such as this one, the 'Riemann sphere's worth' of possibilities is rather hidden, with no clear relation to spatial geometry.

Objectivity and measurability of quantum states

Despite the fact that we are normally only provided with probabilities for the outcome of an experiment, there seems to be something *objective* about a quantum-mechanical state. It is often asserted that the state-vector is merely a convenient description of 'our knowledge' concerning a physical system—or, perhaps, that the state-vector does not really describe a single system but merely provides probability information about an 'ensemble' of a large number of similarly prepared systems. Such sentiments strike me as unreasonably timid concerning what quantum mechanics has to tell us about the *actuality* of the physical world.

Some of this caution, or doubt, concerning the 'physical reality' of state-vectors appears to spring from the fact that what is physically measurable is strictly limited, according to theory. Let us consider an electron's state of spin, as described above. Suppose that the spin-state happens to be $|\alpha\rangle$, but we do not know this; that is, we do not know the *direction* α in which the electron is supposed to be spinning. Can we determine this direction by measurement? No, we cannot. The best that we can do is extract 'one bit' of information—that is, the answer to a single yes/no question. We may select some direction β in space and measure the electron's spin in that direction. We get either the answer **YES** or **NO**, but thereafter, we have lost the information about the original direction of spin. With a **YES** answer we know that the state is *now* proportional to $|\beta\rangle$, and with a **NO** answer we know that the state is *now* in the direction opposite to β. In neither case does this tell us the direction α of the state *before* measurement, but merely gives some probability information about α.

On the other hand, there would seem to be something completely *objective* about the direction α itself, in which the electron 'happened to be spinning' before the measurement was made.* For we *might* have chosen to measure the electron's spin in the direction α—and the electron has to be prepared to give the answer **YES** wth *certainty*, if we happened to have guessed right in this way! Somehow, the 'information' that the electron must actually give this answer is stored in the electron's state of spin.

It seems to me that we must make a distinction between what is 'objective' and what is 'measurable' in discussing the question of physical reality, according to quantum mechanics. The state-vector of a system is, indeed, *not measurable*, in the sense that one cannot ascertain, by experiments performed on the system, precisely (up to proportionality) what that state is; but the state-vector *does* seem to be (again up to proportionality) a completely *objective* property of the system, being completely characterized by the results that it must give to experiments that one *might* perform. In the case of a single spin one-half particle, such as an electron, this objectivity is not unreasonable because it merely asserts that there is *some* direction in which the electron's spin is precisely defined, even though we may not know what that direction is. (However, we shall be seeing later that this 'objective' picture is much stranger with more complicated systems—even for a system which consists merely of a *pair* of spin one-half particles.)

But need the electron's spin have any physically defined state *at all* before it is measured? In many cases it will *not* have, because it cannot be considered as a quantum system on its own; instead, the quantum state must generally be taken as describing an electron inextricably entangled with a large number of other particles. In particular circumstances, however, the electron (at least as regards its spin) *can* be considered on its own. In such circumstances, such as when its spin has actually previously been measured in some (perhaps unknown) direction and then the electron has remained undisturbed for a while, the electron *does* have a perfectly objectively defined direction of spin, according to standard quantum theory.

Copying a quantum state

The objectivity yet non-measurability of an electron's spin-state illustrates another important fact: *it is impossible to copy a quantum state whilst leaving*

* This objectivity is a feature of our taking the standard quantum-mechanical formalism seriously. In a *non*-standard viewpoint, the system might actually 'know', ahead of time, the result that it would give to *any* measurement. This could leave us with a *different*, apparently objective, picture of physical reality.

the original state intact! For suppose that we could make such a copy of an electron's spin-state $|\alpha\rangle$. If we could do it once, then we could do it again, and then again and again. The resulting system could have a huge angular momentum with a very well-defined direction. This direction, namely α, could then be ascertained by a macroscopic measurement. This would violate the fundamental *non*-measurability of the spin-state $|\alpha\rangle$.

However, it *is* possible to copy a quantum state if we are prepared to destroy the state of the original. For example, we could have an electron in an unknown spin-state $|\alpha\rangle$ and a neutron, say, in another spin-state $|\gamma\rangle$. It is quite legitimate to exchange these, so that the neutron's spin-state is now $|\alpha\rangle$ and the electron's is $|\gamma\rangle$. What we cannot do is *duplicate* $|\alpha\rangle$, (unless we already *know* what $|\alpha\rangle$ actually is)! (Cf. also Wooters and Zurek 1982.)

Recall the 'teleportation machine' discussed in Chapter 1 (p. 27). This depended upon it being possible, in principle, to assemble a complete copy of a person's body and brain on a distant planet. It is intriguing to speculate that a person's 'awareness' may depend upon some aspect of a quantum state. If so, quantum theory would forbid us making a copy of this 'awareness' without destroying the state of the original—and, in this way, the 'paradox' of teleportation might be resolved. The possible relevance of quantum effects to brain function will be considered in the final two chapters.

Photon spin

Let us next consider the 'spin' of a photon, and its relation to the Riemann sphere. Photons *do* possess spin, but because they always travel at light speed, one cannot consider the spin as being about a fixed point; instead, the spin axis is always the direction of motion. Photon spin is called *polarization*, which is the phenomenon on which the behaviour of 'polaroid' sunglasses depends. Take two pieces of polaroid, put one against the other and look through them. You will see that generally a certain amount of light comes through. Now rotate one of the pieces, holding the other fixed. The amount coming through will vary. In one orientation, where the transmitted light is at a maximum, the second polaroid subtracts virtually nothing from the amount coming through; while for an orientation chosen at right angles to that, the second polaroid cuts down the light virtually to zero.

What is happening is most easily understood in terms of the wave picture of light. Here we need Maxwell's description in terms of oscillating electric and magnetic fields. In Fig. 6.26, *plane-polarized* light is illustrated. The electric field oscillates backwards and forwards in one plane—called the *plane of polarization*—and the magnetic field oscillates in sympathy, but in a plane at right angles to that of the electric field. Each piece of polaroid lets through light whose polarization plane is aligned with the structure of the polaroid.

When the second polaroid has its structure pointing the same way as that of the first, then all the light getting through the first polaroid also gets through the second. But when the two polaroids have their structures at right angles, the second polaroid shuts off all the light getting through the first one. If the two polaroids are oriented at an angle φ to each other, then a fraction

$$\cos^2\varphi$$

is allowed through by the second polaroid.

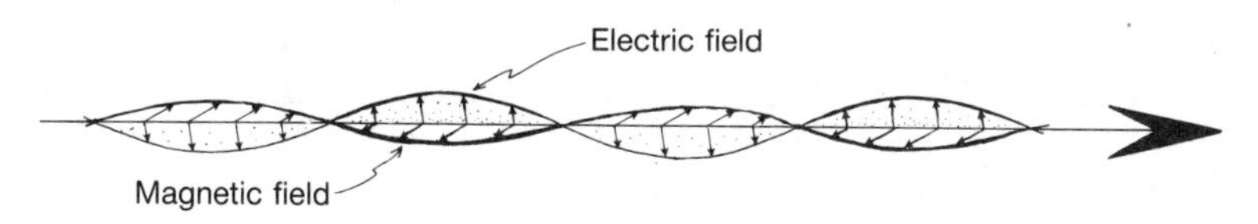

Fig. 6.26. A plane-polarized electromagnetic wave.

In the particle picture, we must think of *each individual photon* as possessing polarization. The first polaroid acts as a polarization measurer, giving the answer **YES** if the photon is indeed polarized in the appropriate direction, and the photon is then allowed through. If the photon is polarized in the orthogonal direction, then the answer is **NO** and the photon is absorbed. (Here 'orthogonal' in the Hilbert space sense *does* correspond to 'at right angles' in ordinary space!) Assuming that the photon gets through the first polaroid, then the second polaroid asks the corresponding question of it, but for some other direction. The angle between these two directions being φ, as above, we have $\cos^2\varphi$ now as the *probability* that the photon makes it through the second polaroid, given that it has passed through the first one.

Where does the Riemann sphere come in? To get the full complex-number array of polarization states, we must consider *circular* and *elliptical* polarization. For a classical wave, these are illustrated in Fig. 6.27. With circular polarization, the electric field *rotates*, rather than oscillates, and the magnetic

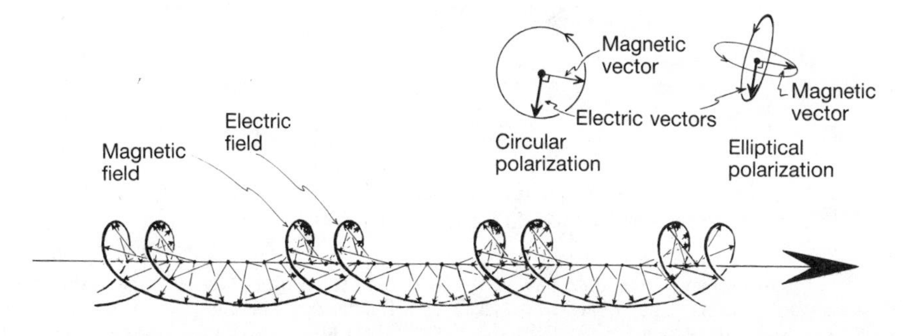

Fig. 6.27. A circularly polarized electromagnetic wave. (Elliptical polarization is intermediate between Figs 6.26 and 6.27.)

field, still at right angles to the electric field, rotates in sympathy. For elliptical polarization, there is a combination of rotational and oscillatory motion, and the vector describing the electric field traces out an *ellipse* in space. In the quantum description, each *individual* photon is allowed these alternative different ways of being polarized—the states of *photon spin*.

To see how the array of possibilities is again the Riemann sphere, imagine the photon to be travelling vertically upwards. The north pole now represents the state $|R\rangle$ of right-handed spin, which means that the electric vector rotates in an anticlockwise sense about the vertical (as viewed from above) as the photon passes by. The south pole represents the state $|L\rangle$ of *left*-handed spin. (We may think of photons as spinning like rifle bullets, either right-handed or left-handed.) The general spin state $|R\rangle + q|L\rangle$ is a complex linear combination of the two and corresponds to a point, labelled q, on the Riemann sphere. To find the connection between q and the ellipse of polarization we first take the *square root* of q, to get another complex number p:

$$p = \sqrt{q}.$$

Then we mark p, instead of q, on the Riemann sphere and consider the plane, through the centre of the sphere, which is perpendicular to the line joining the centre to the point marked p. This plane intersects the sphere in a circle, and we project this circle vertically downwards to get the ellipse of polarization (Fig. 6.28).* The Riemann sphere of q still describes the totality of photon polarization states, but the square root p of q provides its spatial realization.

To calculate probabilities, we can use the same formula $\frac{1}{2}(1 + \cos\theta)$ that we

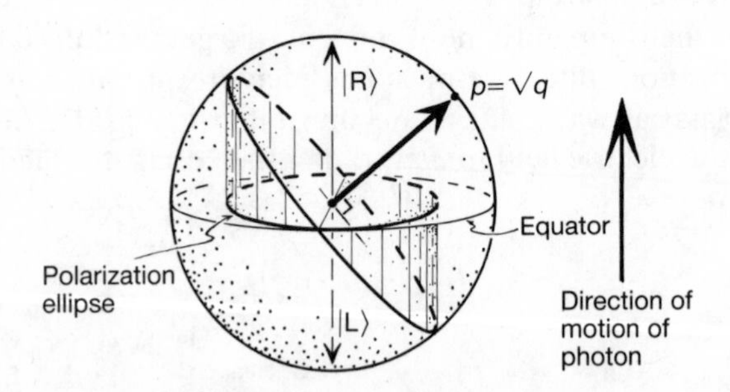

Fig. 6.28. The Riemann sphere (but now of $\sqrt{q}$) also describes polarization states of a photon. (The vector pointing to $\sqrt{q}$ is called a *Stokes vector*.)

* The complex number $-p$ would do just as well as p as a square root of q, and it gives the same ellipse of polarization. The square root has to do with the fact that the photon is a massless particle of *spin one*, i.e. *twice* the fundamental unit $\hbar/2$. For a *graviton*—the yet undetected massless quantum of gravity—the spin would be *two*, i.e. *four* times the fundamental unit, and we should need to take the *fourth* root of q in the above description.

used for the electron, so long as we apply this to q and not p. Consider *plane* polarization. We measure the photon's polarization first in one direction and then in another at angle φ to it, these two directions corresponding to two values of p on the equator of the sphere subtending φ at the centre. Because the ps are the square roots of the qs, the angle θ subtended at the centre by the q-points is *twice* that arising from the p-points: $\theta = 2\varphi$. Thus the probability of **YES** for the second measurement, given that it was **YES** for the first (i.e. that the photon gets through the second polaroid, given that it got through the first one) is $\frac{1}{2}(1 + \cos 2\varphi)$, which (by simple trigonometry) is the same as the $\cos^2\varphi$ asserted above.

Objects with large spin

For a quantum system where the number of basis states is greater than two, the space of physically distinguishable states is more complicated than the Riemann sphere. However in the case of spin, the Riemann sphere *itself* always has a direct geometric role to play. Consider a *massive* particle or atom, of spin $n \times \hbar/2$, taken at rest. The spin then defines an $(n + 1)$-state quantum system. (For a mass*less* spinning particle, i.e. one which travels at the speed of light, such as a photon, the spin is always a *two*-state system, as described above. But for a massive one the number of states goes up with the spin.) If we choose to measure this spin in some direction, then we find that there are $n + 1$ different possible outcomes, depending upon how much of the spin turns out to be oriented along that direction. In terms of the fundamental unit $\hbar/2$, the possible outcomes for the spin value in that direction are $n, n - 2, n - 4, \ldots, 2 - n$ or $-n$. Thus for $n = 2$, the values are 2, 0 or -2; for n = 3, the values are 3, 1, -1 or -3; etc. The *negative* values correspond to the spin pointing mainly in the *opposite* direction to that being measured. The case of spin one-half, i.e. $n = 1$, the value 1 corresponds to **YES** and the value -1 corresponds to **NO** in the descriptions given above.

Now it turns out, though I shall not attempt to explain the reasons (Majorana 1932, Penrose 1987*a*), that *every spin-state* (up to proportionality) for spin $n/2$ is uniquely characterized by an (unordered) *set of n points on the Riemann sphere*—i.e. by n (usually distinct) directions outwards from the centre (see Fig. 6.29). (These directions are characterized by measurements that one might perform on the system: if we measure the spin in one of them, then the result is certain not to be entirely in the opposite direction, i.e. it gives one of the values $n, n - 2, n - 4, \ldots, 2 - n$, but *not* $-n$.) In the particular case $n = 1$, like the electron above, we have *one* point on the Riemann sphere, and this is simply the point labelled q in the above descriptions. But for larger values of the spin, the picture is more elaborate,

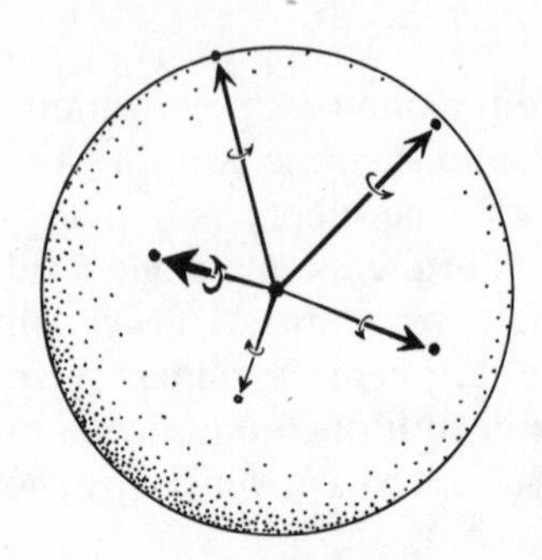

Fig. 6.29. A general state of higher spin, for a massive particle, can be described as a symmetrical combination of spin-$\frac{1}{2}$ states pointing in arbitrary directions.

and is as I have just described—although, for some reason, it is not very familiar to physicists.

There is something rather remarkable and puzzling about this description. One is frequently led to believe that, in some appropriate limiting sense, the quantum descriptions of atoms (or elementary particles or molecules) will necessarily go over to classical Newtonian ones when the system gets large and complicated. However, just as it stands, *this is simply not true*. For, as we have seen, the spin states of an object of large angular momentum will correspond to a large number of points peppered all over the Riemann sphere.* We may think of the spin of the object as being composed of a whole lot of spin one-halves pointing in all the different directions that these points determine. Only a very few of these combined states—namely where most of the points are concentrated together in one small region on the sphere (i.e. where most of the spin one-halves point in roughly the same direction)—will correspond to the actual states of angular momentum that one normally encounters with classical objects such as cricket balls. We might have expected that if we choose a spin state for which the total measure of spin is some very large number (in terms of $\hbar/2$), but otherwise 'at random', then something like a classical spin would begin to emerge. But that is not how things work at all. In general, quantum spin-states of large total spin are nothing whatever like classical ones!

How is it, then, that the correspondence with the angular momentum of classical physics is to be made? Whereas most large-spin quantum states actually do *not* resemble classical ones, they are linear combinations of (orthogonal) states *each* of which *does* resemble a classical one. Somehow a

* More correctly, the angular momentum is described by a complex linear combination of such arrangements with differing numbers of points, since there may be several different total spin values superposed, in the case of a complicated system. This only makes the total picture even *less* like that of a classical angular momentum!

'measurement' gets itself performed on the system, and the state 'jumps' (with some probability) to one or another of these classical-like states. The situation is similar with any other classically measurable property of a system, not just angular momentum. It is this aspect of quantum mechanics which must come into play whenever a system 'reaches the classical level'. I shall have more to say about this later, but before we can discuss such 'large' or 'complicated' quantum systems, we shall have to have some feeling for the odd way that quantum mechanics actually deals with systems involving more than one particle.

Many-particle systems

Quantum-mechanical descriptions of many-particle states are, unfortunately, rather complicated. In fact they can get *extremely* complicated. One must think in terms of superpositions of *all* the different possible locations of all the particles separately! This gives a vast space of possible states—much more than for a *field* in classical theory. We have seen that even the quantum state for a *single* particle, namely a wavefunction, has the sort of complication that an entire classical field has. This picture (requiring an *infinite* number of parameters to specify it) is already a good deal more complicated than the classical picture of a particle (which just needs a few numbers to specify its state—actually six, if it has no internal degrees of freedom, such as spin; cf. Chapter 5, p. 176). This may seem bad enough, and one might think that to describe a quantum state for two particles, *two* 'fields' would be needed, each describing the state of one of the particles. Not a bit of it! With two or more particles the description of the state is far more elaborate than this, as we shall see.

The quantum state of a *single* (spinless) particle is defined by a complex number (amplitude) for each possible position that the particle might occupy. The particle has an amplitude to be at point A and an amplitude to be at point B and an amplitude to be at point C, etc. Now think of *two* particles. The first particle might be at A and the second particle at B, say. There would have to be an amplitude for that possibility. Alternatively, the first particle might be at B and the second at A, and that also would need an amplitude; or the first might be at B and the second at C; or perhaps both particles might be at A. Every one of these alternatives would need an amplitude. Thus the wavefunction is not just a couple of functions of position (i.e. a couple of fields); it is one function of *two* positions!

To get some idea of how much more complicated it is to specify a function of two positions than two functions of position, let us imagine a situation in which there is only a finite set of possible positions available. Suppose there are just ten allowed positions, given by the (orthonormal) states

$$|0\rangle, |1\rangle, |2\rangle, |3\rangle, |4\rangle, |5\rangle, |6\rangle, |7\rangle, |8\rangle, |9\rangle.$$

The state $|\psi\rangle$ of a single particle would be some combination

$$|\psi\rangle = z_0|0\rangle + z_1|1\rangle + z_2|2\rangle + z_3|3\rangle + \ldots + z_9|9\rangle$$

where the various components $z_0, z_1, z_2, z_3, \ldots, z_9$ provide the respective amplitudes for the particle to be at each point in turn. Ten complex numbers specify the state of the particle. For a *two*-particle state we need an amplitude for each *pair of positions*. There are

$$10^2 = 100$$

different (ordered) pairs of positions, so we need *one hundred* complex numbers! If we merely had two one-particle states (i.e. 'two functions of position' rather than the above 'one function of two positions'), we should have needed only *twenty* complex numbers.

We can label these one-hundred numbers as

$$z_{00}, z_{01}, z_{02}, \ldots, z_{09}, z_{10}, z_{11}, z_{12}, \ldots, z_{20}, \ldots, z_{99}$$

and the corresponding (orthonormal) basis vectors as [12]

$$|0\rangle|0\rangle,\ |0\rangle|1\rangle,\ |0\rangle|2\rangle, \ldots, |0\rangle|9\rangle,\ |1\rangle|0\rangle, \ldots, |9\rangle|9\rangle.$$

Then the general two-particle state $|\psi\rangle$, would have the form

$$|\psi\rangle = z_{00}|0\rangle|0\rangle + z_{01}|0\rangle|1\rangle + \ldots + z_{99}|9\rangle|9\rangle.$$

This 'product' notation for states has the following meaning: if $|\alpha\rangle$ is a possible state for the first particle (not necessarily a position state), and if $|\beta\rangle$ is a possible state for the second particle, then the state which asserts that the first particle's state is $|\alpha\rangle$ *and* the second one is $|\beta\rangle$ would be written

$$|\alpha\rangle|\beta\rangle.$$

'Products' can also be taken between any other pair of quantum states, not necessarily single-particle states. Thus, we always interpret the product state $|\alpha\rangle|\beta\rangle$ (not necessarily states of single particles) as describing the conjunction:

'the first system has state $|\alpha\rangle$ *and* the second system has state $|\beta\rangle$'.

(A similar interpretation would hold for $|\alpha\rangle|\beta\rangle|\gamma\rangle$, etc.; see below.) However, the *general* two-particle state actually does not have this 'product' form. For example, it could be

$$|\alpha\rangle|\beta\rangle + |\rho\rangle|\sigma\rangle,$$

where $|\rho\rangle$ is another possible state for the first system and $|\sigma\rangle$, another possible state for the second. This state is a *linear superposition*; namely the first conjunction ($|\alpha\rangle$ *and* $|\beta\rangle$) *plus* the second conjuction ($|\rho\rangle$ *and* $|\sigma\rangle$), and it cannot be re-expressed as a simple product (i.e. as the conjunction of two states). The state $|\alpha\rangle|\beta\rangle - \mathrm{i}|\rho\rangle|\sigma\rangle$, as another example, would describe a different such linear superposition. Note that quantum mechanics requires a clear distinction to be maintained between the meanings of the words 'and' and 'plus'. There is a very unfortunate tendency in modern parlance—such as in insurance brochures—to misuse 'plus' in the sense of 'and'. Here we must be a good deal more careful!

The situation with three particles is very similar. To specify a general three-particle state, in the above case where only ten alternative positions are available, we now need *one-thousand* complex numbers! The complete basis for the three-particle states would be

$$|0\rangle|0\rangle|0\rangle, \quad |0\rangle|0\rangle|1\rangle, \quad |0\rangle|0\rangle|2\rangle, \ldots, \quad |9\rangle|9\rangle|9\rangle.$$

Particular three-particle states have the form

$$|\alpha\rangle|\beta\rangle|\gamma\rangle$$

(where $|\alpha\rangle$, $|\beta\rangle$, and $|\gamma\rangle$ need not be position states), but for the general three-particle state one would have to superpose many states of this simple 'product' kind. The corresponding pattern for four or more particles should be clear.

The discussion so far has been for the case of *distinguishable* particles, where we take the 'first particle', the 'second particle', and the 'third particle', etc. to be all of *different* kinds. It is a striking feature of quantum mechanics, however, that for *identical* particles the rules are different. In fact, the rules are such that, in a clear sense, particles of a specific type have to be *precisely* identical not just, say, extremely closely identical. This applies to all electrons, and it applies to all photons. But, as it turns out, all electrons are identical with one another in a *different* way from the way in which all photons are identical! The difference lies in the fact that electrons are fermions whereas photons are bosons. These two general kinds of particles have to be treated rather differently from one another.

Before I confuse the reader completely with such verbal inadequacies, let me explain how fermion and boson states are actually to be characterized. The rule is as follows. If $|\psi\rangle$ is a state involving some number of fermions of a particular kind, then if any two of the fermions are interchanged, $|\psi\rangle$ must undergo

$$|\psi\rangle \rightarrow -|\psi\rangle.$$

If $|\psi\rangle$ involves a number of bosons of a particular kind, then if any two of the bosons are interchanged, then $|\psi\rangle$ must undergo

$$|\psi\rangle \rightarrow |\psi\rangle.$$

One implication of this is that *no two fermions can be in the same state*. For if they were, interchanging them would not affect the total state at all, so we would have to have: $-|\psi\rangle = |\psi\rangle$, i.e. $|\psi\rangle = \mathbf{0}$, which is not allowed for a quantum state. This property is known as *Pauli's exclusion principle*,[13] and its implications for the structure of matter are fundamental. All the principal constituents of matter are indeed fermions: electrons, protons and neutrons. Without the exclusion principle, matter would collapse in on itself!

Let us examine our ten positions again, and suppose now that we have a state consisting of two identical fermions. The state $|0\rangle|0\rangle$ is excluded, by Pauli's principle (it goes to itself rather than to its negative under interchange of the first factor with the second). Moreover, $|0\rangle|1\rangle$ will not do as it stands, since it does not go to its negative under the interchange; but this is easily remedied by replacing it by

$$|0\rangle|1\rangle - |1\rangle|0\rangle.$$

(An overall factor $1/\sqrt{2}$ could be included, if desired, for normalization purposes.) This state correctly changes sign under interchange of the first particle with the second, but now we do not have $|0\rangle|1\rangle$ and $|1\rangle|0\rangle$ as independent states. In place of those *two* states we are now allowed only *one* state! In all, there are

$$\tfrac{1}{2}(10 \times 9) = 45$$

states of this kind, one for each unordered pair of distinct states from $|0\rangle, |1\rangle, \ldots, |9\rangle$. Thus, 45 complex numbers are needed to specify a two-fermion state in our system. For three fermions, we need three distinct positions, and the basis states look like

$$|0\rangle|1\rangle|2\rangle + |1\rangle|2\rangle|0\rangle + |2\rangle|0\rangle|1\rangle - |0\rangle|2\rangle|1\rangle - |2\rangle|1\rangle|0\rangle - |1\rangle|0\rangle|2\rangle,$$

there being $(10\times9\times8)/6 = 120$ such states in all; so 120 complex numbers are needed to specify a three-fermion state. The situation is similar for higher numbers of fermions.

For a pair of identical bosons, the independent basis states are of two kinds, namely states like

$$|0\rangle|1\rangle + |1\rangle|0\rangle$$

and states like

$$|0\rangle|0\rangle$$

(which are now allowed), giving $10 \times 11/2 = 55$ in all. Thus, 55 complex numbers are needed for our two-boson states. For three bosons there are basis states of three different kinds and $(10\times11\times12)/6 = 220$ complex numbers are needed; and so on.

Of course I have been considering a simplified situation here in order to convey the main ideas. A more realistic description would require an entire continuum of position states, but the essential ideas are the same. Another slight complication is the presence of *spin*. For a spin-one-half particle (necessarily a fermion) there would be two possible states for each position. Let us label these by '$\uparrow$' (spin 'up') and '$\downarrow$' (spin 'down'). Then for a single particle we have, in our simplified situation, twenty basic states rather than ten:

$$|0\uparrow\rangle,\ |0\downarrow\rangle,\ |1\uparrow\rangle,\ |1\downarrow\rangle,\ |2\uparrow\rangle,\ |2\downarrow\rangle, \ldots, |9\uparrow\rangle,\ |9\downarrow\rangle,$$

but apart from that, the discussion proceeds just as before (so for two such fermions we need $(20\times19)/2 = 190$ numbers; for three, we need $(20\times19\times18)/6 = 1140$; etc.).

In Chapter 1, I referred to the fact that, according to modern theory, if a particle of a person's body were exchanged with a similar particle in one of the bricks of his house then nothing would have happened at all. If that particle were a boson, then, as we have seen, the state $|\psi\rangle$ would indeed be totally unaffected. If that particle were a fermion, then the state $|\psi\rangle$ would be replaced by $-|\psi\rangle$, which is physically identical with $|\psi\rangle$. (We can remedy this sign change, if we feel the need, by simply taking the precaution of rotating one of the two particles completely through 360° when the interchange is made. Recall that fermions change sign under such a rotation whereas boson states are unaffected!) Modern theory (as of around 1926) does indeed tell us something profound about the question of individual identity of bits of physical material. One cannot refer, strictly correctly, to 'this particular electron' or 'that individual photon'. To assert that 'the first electron is here and the second is over there' is to assert that the state has the form $|0\rangle|1\rangle$ which, as we have seen, is not allowed for a fermion state! We can, however, allowably assert 'there is a pair of electrons, one here and one over there'. It is legitimate to refer to the conglomerate of all electrons or of all protons or of all photons (although even this ignores the *interactions* between different kinds of particle). Individual electrons provide an approximation to this total picture, as do individual protons or individual photons. For most purposes this approximation works well, but there are various circumstances for which it does not, superconductivity, superfluidity and the behaviour of a laser being noteworthy counter-examples.

The picture of the physical world that quantum mechanics has presented us with is not at all what we had got used to from classical physics. But hold on to your hat—there is more strangeness yet in the quantum world!

The 'paradox' of Einstein, Podolsky, and Rosen

As has been mentioned at the beginning of this chapter, some of Albert Einstein's ideas were quite fundamental to the development of quantum theory. Recall that it was he who first put forward the concept of the 'photon'—the quantum of electromagnetic field—as early as 1905, out of which the idea of wave–particle duality was developed. (The concept of a 'boson' also was partly his, as were many other ideas, central to the theory.) Yet, Einstein could never accept that the theory which later developed from these ideas could be anything but provisional as a description of the physical world. His aversion to the probabilistic aspect of the theory is well known, and is encapsulated in his reply to one of Max Born's letters in 1926 (quoted in Pais 1982, p. 443):

> Quantum mechanics is very impressive. But an inner voice tells me that it is not yet the real thing. The theory produces a good deal but hardly brings us closer to the secret of the Old One. I am at all events convinced that *He* does not play dice.

However, it appears that, even more than this physical indeterminism, the thing which most troubled Einstein was an apparent *lack of objectivity* in the way that quantum theory seemed to have to be described. In my exposition of quantum theory I have taken pains to stress that the description of the world, as provided by the theory, is really quite an objective one, though often very strange and counter-intuitive. On the other hand, Bohr seems to have regarded the quantum state of a system (between measurements) as having no actual physical reality, acting merely as a summary of 'one's knowledge' concerning that system. But might not different observers have different knowledge of a system, so the wavefunction would seem to be something essentially *subjective*—or 'all in the mind of the physicist'? Our marvellously precise physical picture of the world, as developed over many centuries, must not be allowed to evaporate away completely; so Bohr needed to regard the world at the *classical level* as indeed having an objective reality. Yet there would be no 'reality' to the *quantum*-level states that seem to underlie it all.

Such a picture was abhorrent to Einstein, who believed that there must indeed be an objective physical world, even at the minutest scale of quantum phenomena. In his numerous arguments with Bohr he attempted (but failed) to show that there were inherent contradictions in the quantum picture of things, and that there must be a yet deeper structure beneath quantum theory, probably more akin to the pictures that classical physics had presented us with. Perhaps underlying the probabilistic behaviour of quantum systems would be the statistical action of smaller ingredients or 'parts' to the system, about which one had no direct knowledge. Einstein's followers, particularly David Bohm, developed the viewpoint of 'hidden variables', according to which there would indeed be some definite reality, but the parameters which precisely define a system would not be directly accessible to us, quantum probabilities arising because these parameter values would be unknown prior to measurement.

Can such a hidden-variable theory be consistent with all the observational facts of quantum physcis? The answer seems to be yes, but only if the theory is, in an essential way, *non-local*, in the sense that the hidden parameters must be able to affect parts of the system in arbitrarily distant regions instantaneously! *That* would not have pleased Einstein, particularly owing to difficulties with special relativity that arise. I shall consider these later. The most successful hidden-variable theory is that known as the de Broglie–Bohm model (de Broglie 1956, Bohm 1952). I shall not discuss such models here, since my purpose in this chapter is only to give an overview of standard quantum theory, and not of various rival proposals. If one desires physical objectivity, but is prepared to dispense with determinism, then the standard theory itself will suffice. One simply regards the state-vector as providing 'reality'—usually evolving according to the smooth deterministic procedure **U**, but now and again oddly 'jumping' according to **R**, whenever an effect gets magnified to the classical level. However, the problem of non-locality and the apparent difficulties with relativity remain. Let us take a look at some of these.

Suppose that we have a physical system consisting of two sub-systems A and B. For example, take A and B to be two different particles. Suppose that two (orthogonal) alternatives for the state of A are $|\alpha\rangle$ and $|\rho\rangle$, whereas B's state might be $|\beta\rangle$ or $|\sigma\rangle$. As we have seen above, the general combined state would not simply be a product ('and') of a state of A with a state of B, but a superposition ('plus') of such products. (We say that A and B are then *correlated*.) Let us take the state of the system to be

$$|\alpha\rangle|\beta\rangle + |\rho\rangle|\sigma\rangle.$$

Now perform a yes/no measurement on A that distinguishes $|\alpha\rangle$ (**YES**) from $|\rho\rangle$ (**NO**). What happens to B? If the measurement yields **YES**, then the resulting state must be

$$|\alpha\rangle|\beta\rangle,$$

while if it yields **NO**, then it is

$$|\rho\rangle|\sigma\rangle.$$

Thus our measurement of A causes the state of B to jump: to $|\beta\rangle$, in the event of a **YES** answer, and to $|\sigma\rangle$, in the event of a **NO** answer! The particle B need not be localized anywhere near A; they could be light-years apart. Yet B jumps simultaneously with the measurement of A!

But hold on—the reader may well be saying. What's all this alleged 'jumping'? Why aren't things like the following? Imagine a box which is known to contain one white ball and one black ball. Suppose that the balls are taken out and removed to two opposite corners of the room, without either being looked at. Then if one ball is examined and found to be white (like '$|\alpha\rangle$'

above)—hey presto!—the other turns out to be black (like '$|\beta\rangle$')! If, on the other hand, the first is found to be black ('$|\rho\rangle$'), then, in a flash, the second ball's uncertain state jumps to 'white, with certainty' ('$|\sigma\rangle$'). No-one in his or her right mind, the reader will insist, would attribute the sudden change of the second ball's 'uncertain' state to being 'black with certainty', or to being 'white with certainty', to some mysterious non-local 'influence' travelling to it instantaneously from the first ball the moment that ball is examined.

But Nature is actually much more extraordinary than this. In the above, one could indeed imagine that the *system* already 'knew' that, say, B's state was $|\beta\rangle$ and that A's was $|\alpha\rangle$ (or else that B's was $|\sigma\rangle$ and A's was $|\rho\rangle$) before the measurement was performed on A; and it was just that the *experimenter* did not know. Upon finding A to be in state $|\alpha\rangle$, he simply *infers* that B is in $|\beta\rangle$. That would be a 'classical' viewpoint—such as in a local hidden-variable theory—and no *physical* 'jumping' actually takes place. (All in the experimenter's mind!) According to such a view, each part of the system 'knows', beforehand, the results of any experiment that might be performed on it. Probabilities arise only because of a lack of knowledge in the experimenter. Remarkably, it turns out that this viewpoint just *won't work* as an explanation for all the puzzling apparently non-local probabilities that arise in quantum theory!

To see this, we shall consider a situation like the above, but where the *choice of measurement* on the system A is not decided upon until A and B are well separated. The behaviour of B then seems to be instantaneously influenced by this very choice! This seemingly paradoxical 'EPR' type of 'thought experiment' is due to Albert Einstein, Boris Podolsky, and Nathan Rosen (1935). I shall give a variant, put forward by David Bohm (1951). The fact that no local 'realistic' (e.g. hidden variable, or 'classical-type') description can give the correct quantum probabilities follows from a remarkable theorem, by John S. Bell. (See Bell 1987, Rae 1986, Squires 1986.)

Suppose that two spin-one-half particles—which I shall call an *electron* and a *positron* (i.e. an *anti*-electron)—are produced by the decay of a single spin-zero particle at some central point, and that the two move directly outwards in opposite directions (Fig. 6.30). By conservation of angular momentum, the spins of the electron and positron must add up to zero, since

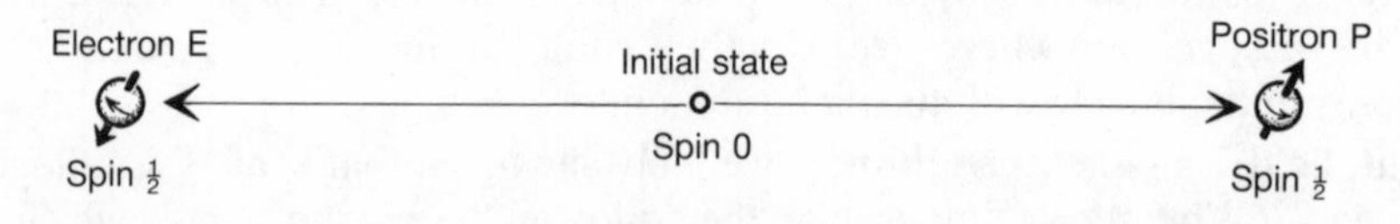

Fig. 6.30. A spin-zero particle decays into two spin-$\frac{1}{2}$ particles, an electron E and a positron P. Measurement of the spin of one of the spin-$\frac{1}{2}$ particles apparently *instantaneously* fixes the spin state of the other.

that was the angular momentum of the initial central particle. This has the implication that when we measure the spin of the electron in some direction, whatever direction we choose, the positron now spins in the *opposite* direction! The two particles could be miles or even light-years apart, yet that very *choice* of measurement on one particle seems *instantaneously* to have fixed the axis of spin of the other!

Let us see how the quantum formalism leads us to this conclusion. We represent the combined zero-angular-momentum state of the two particles by the state-vector $|\mathsf{Q}\rangle$, and we find a relation such as

$$|\mathsf{Q}\rangle = |\mathsf{E}\uparrow\rangle|\mathsf{P}\downarrow\rangle - |\mathsf{E}\downarrow\rangle|\mathsf{P}\uparrow\rangle,$$

where E refers to the electron and P to the positron. Here things have been described in terms of the up/down directions of spin. We find that the entire state is a linear superposition of the electron spinning up and the positron down, and of electron spinning down and the positron up. Thus if we measure the electron's spin in the up/down direction and find that it is indeed up, then we must jump to the state $|\mathsf{E}\uparrow\rangle\,|\mathsf{P}\downarrow\rangle$, so the positron's spin-state must be down. If, on the other hand, we find that the electron's spin is down, then the state jumps to $|\mathsf{E}\downarrow\rangle\,|\mathsf{P}\uparrow\rangle$, so the positron's spin is up.

Now suppose that we had chosen some other pair of opposite directions, say right and left, where

$$|\mathsf{E}\rightarrow\rangle = |\mathsf{E}\uparrow\rangle + |\mathsf{E}\downarrow\rangle, \qquad |\mathsf{P}\rightarrow\rangle = |\mathsf{P}\uparrow\rangle + |\mathsf{P}\downarrow\rangle$$

and

$$|\mathsf{E}\leftarrow\rangle = |\mathsf{E}\uparrow\rangle - |\mathsf{E}\downarrow\rangle, \qquad |\mathsf{P}\leftarrow\rangle = |\mathsf{P}\uparrow\rangle - |\mathsf{P}\downarrow\rangle;$$

then we find (check the algebra, if you like!):

$$\begin{aligned}
&|\mathsf{E}\rightarrow\rangle|\mathsf{P}\leftarrow\rangle - |\mathsf{E}\leftarrow\rangle|\mathsf{P}\rightarrow\rangle \\
&\quad = (|\mathsf{E}\uparrow\rangle + |\mathsf{E}\downarrow\rangle)\,(|\mathsf{P}\uparrow\rangle - |\mathsf{P}\downarrow\rangle) - (|\mathsf{E}\uparrow\rangle - |\mathsf{E}\downarrow\rangle)(|\mathsf{P}\uparrow\rangle + |\mathsf{P}\downarrow\rangle) \\
&\quad = |\mathsf{E}\uparrow\rangle|\mathsf{P}\uparrow\rangle + |\mathsf{E}\downarrow\rangle|\mathsf{P}\uparrow\rangle - |\mathsf{E}\uparrow\rangle|\mathsf{P}\downarrow\rangle - |\mathsf{E}\downarrow\rangle|\mathsf{P}\downarrow\rangle \\
&\quad\quad - |\mathsf{E}\uparrow\rangle|\mathsf{P}\uparrow\rangle + |\mathsf{E}\downarrow\rangle|\mathsf{P}\uparrow\rangle - |\mathsf{E}\uparrow\rangle|\mathsf{P}\downarrow\rangle + |\mathsf{E}\downarrow\rangle|\mathsf{P}\downarrow\rangle \\
&\quad = -2(|\mathsf{E}\uparrow\rangle|\mathsf{P}\downarrow\rangle - |\mathsf{E}\downarrow\rangle|\mathsf{P}\uparrow\rangle) \\
&\quad = -2|\mathsf{Q}\rangle.
\end{aligned}$$

which (apart from the unimportant factor -2) is the same state that we started from. Thus our original state can equally well be thought of as a linear superposition of the electron spinning right and the positron left, and of the electron spinning left and the positron right! This expression is useful if we choose to measure the spin of the electron in a right/left direction instead of up/down. If we find that it indeed spins right, then the state jumps to $|\mathsf{E}\rightarrow\rangle|\mathsf{P}\leftarrow\rangle$, so the positron spins left. If, on the other hand, we find that the electron spins left, then the state jumps to $|\mathsf{E}\leftarrow\rangle\,|\mathsf{P}\rightarrow\rangle$ so the positron spins right. Had we chosen to measure the electron's spin in any other direction, the story would be exactly corresponding: the state of spin of the positron would instantly jump to being either in that direction or the opposite one, depending upon the result of the measurement of the electron.

Why can we not model the spins of our electron and positron in a similar way to the example given above with a black and a white ball taken from a box? Let us be completely general. Instead of having a black and a white ball, we might have two pieces of machinery E and P being initially united, and then moving off in opposite directions. Suppose that each of E and P is able to evoke a response **YES** or **NO** to a spin measurement in any given direction. This response might be completely determined by the machinery for each choice of direction—or perhaps the machinery produces only probabilistic responses, the probability being determined by the machinery—but where we assume that, after separation, *each of* E *and* P *behaves completely independently of the other.*

We have spin measurers at each side, one which measures the spin of E and the other, the spin of P. Suppose that there are three settings for the direction of spin on each measurer, say A, B, C for the E-measurer and A′, B′, C′ for the P-measurer. The directions A′, B′ and C′ are to be parallel, respectively, to A, B, and C, and we take A, B, and C to be all in one plane and equally angled to one another, i.e. a 120° to each other. (See Fig. 6.31.) Now, imagine that the experiment is repeated many times with various different values of these settings at either side. Sometimes the E-measurer will register **YES** (i.e. the spin *is* in the measured direction: A, or B, or C) and sometimes it will register **NO** (spin in the opposite direction). Similarly the P-measurer will sometimes register **YES** and sometimes **NO**. We take note of two properties that the actual *quantum* probabilities must have:

(1) If the settings on the two sides are the *same* (i.e. A and A′, etc.) then the results produced by the two measurers always *disagree* (i.e. the E-

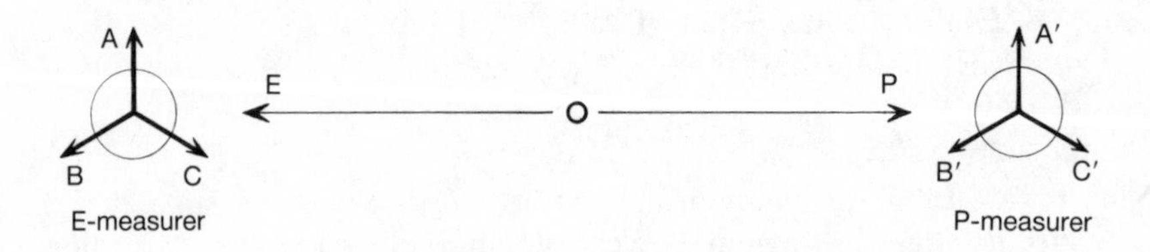

Fig. 6.31. David Mermin's simple version of the EPR paradox and Bell's theorem, showing that there is a contradiction between a local realistic view of nature and the results of quantum theory. The E-measurer and the P-measurer each independently have three settings for the directions in which they can measure the spins of their respective particles.

measurer registers **YES** whenever the P-measurer registers **NO**, and **NO** whenever P gives **YES**).

(2) If the dials for the settings are spun and set at *random*, completely independently of one another, then the two measurers are *equally likely to agree as to disagree.*

We can easily see that properties (1) and (2) follow directly from the quantum probability rules that have been given earlier. We can suppose that the E-measurer acts first. The P-measurer then finds a particle whose spin state is opposite to that measured by E-measurer, so property (1) follows immediately. To obtain property (2) we note that, for measured directions at 120° to one another, if the E-measurer gives **YES**, then the P-direction is at 60° to the spin state that it acts upon, and if **NO**, then it is at 120° to this spin state. Thus there is a probability $\frac{3}{4} = \frac{1}{2}(1 + \cos 60°)$ that the measurements agree and a probability $\frac{1}{4} = \frac{1}{2}(1 + \cos 120°)$ that they disagree. Thus the averaged probability for the three P-settings, if E gives **YES**, is $\frac{1}{3}(0 + \frac{3}{4} + \frac{3}{4}) = \frac{1}{2}$ for P giving **YES** and $\frac{1}{3}(1 + \frac{1}{4} + \frac{1}{4}) = \frac{1}{2}$ for P giving **NO**—i.e. equally likely to give agreement as disagreement—and similarly if E gives **NO**. This is indeed property (2).

It is a remarkable fact that (1) and (2) are *inconsistent* with any local realistic model (i.e. with any kind of machinery of the kind envisaged)! Suppose that we had such a model. The E-machine must be prepared for each of the possible measurements A, B, or C. Note that if it were prepared only to give a *probabilistic* answer, then the P-machine could not be *sure* to give disagreement with it, for A′, B′, and C′ respectively, in accordance with (1). Indeed, *both* machines must have their answers to each of the three possible measurements definitely prepared in advance. Suppose, for example, these answers are to be **YES**, **YES**, **YES**, respectively, for A, B, C; the right-hand particle must then be prepared to give **NO**, **NO**, **NO**, for the corresponding three right-hand settings. If, instead, the left-hand prepared answers are to be **YES**, **YES**, **NO**, then the right-hand answers must be **NO**, **NO**, **YES**. All other cases would be essentially similar to these. Now let us see whether this can be compatible with (2). The assignments **YES**, **YES**, **YES / NO**, **NO**, **NO** are not very promising, because this gives 9 cases of disagreement and 0 cases of agreement in all the possible pairings A/A′, A/B′, A/C′, B/A′, etc. What about the case, **YES**, **YES**, **NO / NO**, **NO**, **YES** and those like it? These give 5 cases of disagreement and 4 of agreement. (To check, just count them up: **Y/N**, **Y/N**, **Y/Y**, **Y/N**, **Y/N**, **Y/Y**, **N/N**, **N/N**, **N/Y**, five of which disagree and four agree.) That is a good deal closer to what is needed for (2), but it is *not* good enough, since we need as many agreements as disagreements! Every other pair of assignments consistent with (1) would again give 5 to 4 (except for **NO**, **NO**, **NO / YES**, **YES**, **YES** which is worse, again giving 9 to 0). There is *no* set of prepared answers which can produce the quantum-mechanical probabilities. *Local realistic models are ruled out*![14]

Experiments with photons: a problem for relativity?

We must ask whether actual experiments have borne out these astonishing quantum expectations. The precise experiment just described is a hypothetical one which has not actually been performed, but similar experiments *have* been performed using the polarizations of pairs of *photons*, rather than the spin of spin-one-half massive particles. Apart from this distinction, these experiments are, in their essentials, the same as the one described above—except that the angles concerned (since photons have spin one rather than one-half) would be just one-half of those for spin-one-half particles. The polarizations of the pairs of photons have been measured in various different combinations of directions, and the results are fully in agreement with the predictions of quantum theory, and inconsistent with any local realistic model!

The most accurate and convincing of the experimental results which have been obtained to date are those of Alain Aspect (1986) and his colleagues in Paris.[15] Aspect's experiments had another interesting feature. The 'decisions' as to which way to measure the polarizations of the photons, were made only after the photons were in full flight. Thus, if we think of some non-local 'influence' travelling from one photon detector to the photon at the opposite side, signalling the direction in which it intends to measure the polarization direction of the approaching photon, then we see that this 'influence' must travel faster than light! Any kind of realistic description of the quantum world which is consistent with the facts must apparently be *non-causal*, in the sense that effects must be able to travel faster than light!

But we saw in the last chapter that, so long as relativity holds true, the sending of signals faster than light leads to absurdities (and conflict with our feelings of 'free will', etc. cf. p. 212). This is certainly true, but the non-local 'influences' that arise in EPR-type experiments are not such that they can be used to send messages—as one can see, for the very reason that they would lead to such absurdities, if so. (A detailed demonstration that such 'influences' cannot be used to signal messages has been carried out by Ghirardi, Rimini, and Webber 1980.) It is of no use to be told that a photon is polarized 'either vertically or horizontally', (as opposed, say, to 'either at 60° or 150°') until one is informed *which* of the two alternatives it actually is. It is the first piece of 'information' (i.e. the *directions* of alternative polarization) which arrives faster than light ('instantaneously'), while the knowledge as to *which* of these two directions it must actually be polarized in arrives more slowly, via an ordinary signal communicating the *result* of the first polarization measurement.

Although EPR-type experiments do not, in the ordinary sense of sending messages, conflict with the *causality* of relativity, there is a definite conflict with the *spirit* of relativity in our picture of 'physical reality'. Let us see how the *realistic* view of the state-vector applies to the above EPR-type experiment (involving photons). As the two photons move outwards, the state-vector describes the situation as a photon *pair*, acting as a single unit. Neither photon individually has an objective state: the quantum state applies only to the two together. Neither photon individually has a direction of polarization: the polarization is a combined quality of the two photons together. When the polarization of one of these photons is measured, the state-vector *jumps* so that the unmeasured photon now *does* have a definite polarization. When *that* photon's polarization is subsequently measured, the probability values are correctly obtained by applying the usual quantum rules to its polarization state. This way of looking at the situation provides the correct answers; it is, indeed, the way that we ordinarily apply quantum mechanics. But it is an essentially non-relativistic view. For the two measurements of polarization are what are called *spacelike-separated*, which means that each lies outside the other's light cone, like the points R and Q in Fig. 5.21. The question of which of these measurements actually occurred *first* is not really physically meaningful, but depends on the 'observer's' state of motion (see Fig. 6.32). If the 'observer' moves rapidly enough to the right, then he considers the right-hand measurement to have occurred first; and if to the left, then it is the left-hand measurement! But if we regard the right-hand photon as having been measured first we get a completely different picture of physical reality from that obtained if we regard the left-hand photon as having been measured first! (It is a different measurement that causes the non-local 'jump'.) There is an essential conflict between our space–time picture of physical reality—even the correctly non-local quantum-mechanical one—and special relativity! This is a severe puzzle, which 'quantum realists' have not been able adequately to resolve (cf. Aharonov and Albert 1981). I shall need to return to the matter later.

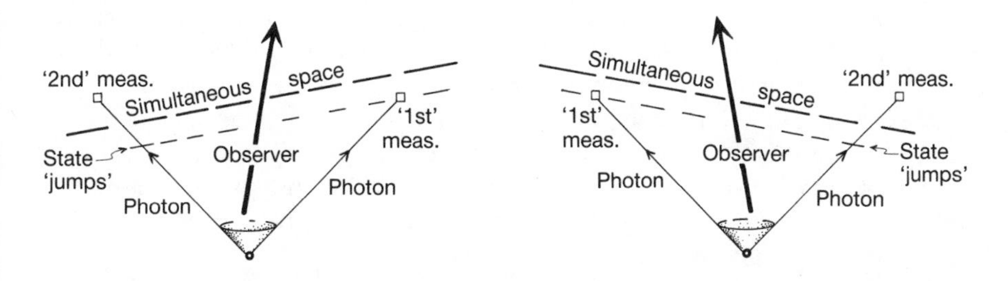

Fig. 6.32. Two different observers form mutually inconsistent pictures of 'reality' in an EPR experiment in which two photons are emitted in opposite directions from a spin-0 state. The observer moving to the right judges that the left-hand part of the state jumps *before* it is measured, the jump being caused by the measurement on the right. The observer moving to the left has the opposite opinion!

Schrödinger's equation; Dirac's equation

Earlier in this chapter, I referred to Schrödinger's equation, which is a perfectly well-defined deterministic equation, in many ways quite similar to the equations of classical physics. The rules say that so long as no 'measurements' (or 'observations') are made on a quantum system, Schrödinger's equation must hold true. The reader may care to witness its actual form:

$$i\hbar \frac{\partial}{\partial t} |\psi\rangle = H|\psi\rangle.$$

Recall that $\hbar$ is Dirac's version of Planck's constant ($h/2\pi$) (and $i = \sqrt{-1}$) and that the operator $\partial/\partial t$ (partial differentiation with respect to time) acting on $|\psi\rangle$ simply means the *rate of change* of $|\psi\rangle$ with respect to time. Schrödinger's equation states that '$H|\psi\rangle$' describes how $|\psi\rangle$ evolves.

But what is 'H'? It is the *Hamiltonian function* that we considered in the previous chapter, but with a fundamental difference! Recall that the classical Hamiltonian is the expression for the *total energy* in terms of the various position coordinates q_i and momentum coordinates p_i, for all the physical objects in the system. To obtain the *quantum* Hamiltonian we take the same expression, but substitute, for each occurrence of the momentum p_i, a multiple of the *differential operator* 'partial differentiation with respect to q_i'. Specifically, we replace p_i by $-i\hbar\partial/\partial q_i$. Our quantum Hamiltonian H then becomes some (frequently complicated) mathematical *operation* involving differentiations and multiplications, etc.—and not just a number! This looks like hocus-pocus! But it is not just mathematical conjuring; it is genuine *magic* which works! (There is a bit of 'art' in applying this process of generating a quantum Hamiltonian from a classical one, but it is remarkable, in view of its outlandish nature, how little the ambiguities inherent in the procedure seem to matter.)

An important thing to notice about Schrödinger's equation (whatever H is) is that it is *linear*, i.e. if $|\psi\rangle$ and $|\varphi\rangle$ both satisfy the equation, then so also does $|\psi\rangle + |\varphi\rangle$—or, indeed, any combination $w|\psi\rangle + z|\varphi\rangle$, where w and z are fixed complex numbers. Thus, complex linear superposition is maintained indefinitely by Schrödinger's equation. A (complex) linear superposition of two possible alternative states cannot get '*un*superposed' merely by the action of **U**! This is why the action of **R** is needed as a *separate* procedure from **U** in order that just *one* alternative finally survives.

Like the Hamiltonian formalism for classical physics, the Schrödinger equation is not so much a specific equation, but a framework for quantum-mechanical equations generally. Once one has obtained the appropriate quantum Hamiltonian, the time-evolution of the state according to Schrödinger's equation proceeds rather as though $|\psi\rangle$ were a classical field subject to some classical field equation such as Maxwell's. In fact, if $|\psi\rangle$ describes the state of a single *photon*, then it turns out that Schrödinger's equation actually *becomes* Maxwell's equations! The equation for a single photon is precisely the same as the equation* for an entire electromagnetic field. This fact is responsible for the Maxwell-field-wavelike behaviour and polarization of *single photons* that we caught glimpses of earlier. As another example, if $|\psi\rangle$ describes the state of a single *electron*, then Schrödinger's equation becomes Dirac's remarkable wave equation for the electron—discovered in 1928 after Dirac had supplied much additional originality and insight.

In fact, Dirac's equation for the electron must be rated, alongside the Maxwell and Einstein equations, as one of the Great Field Equations of physics. To convey an adequate impression of it here would require me to introduce mathematical ideas that would be too distracting. Suffice it to say that in the Dirac equation $|\psi\rangle$ has the curious 'fermionic' property $|\psi\rangle \rightarrow -|\psi\rangle$ under 360° rotation that we considered earlier (p. 264). The Dirac and Maxwell equations together constitute the basic ingredients of quantum electrodynamics, the most successful of the quantum field theories. Let us consider this briefly.

Quantum field theory

The subject known as 'quantum field theory' has arisen as a union of ideas from special relativity and quantum mechanics. It differs from standard (i.e. non-relativistic) quantum mechanics in that the number of particles, of any particular kind, need not be a constant. Each kind of particle has its *anti-particle* (sometimes, such as with photons, the same as the original particle). A massive particle and its anti-particle can annihilate to form energy, and such a pair can be created out of energy. Indeed, the number of particles need not even be definite; for linear superpositions of states with different numbers of particles are allowed. The supreme quantum field theory is 'quantum electrodynamics'—basically the theory of electrons and photons. This theory is remarkable for the accuracy of its predictions (e.g. the precise value of the magnetic moment of the electron, referred to in the last chapter, p. 153). However, it is a rather untidy theory—and a not altogether consistent one—because it initially gives nonsensical 'infinite' answers. These have to be removed by a process known as 'renormalization'. Not all quantum field theories are amenable to renormalization, and they are difficult to calculate with even when they are.

A popular approach to quantum field theory is via 'path integrals', which involve forming quantum linear superpositions not just of different particle

* However, there is an important difference in the type of *solution* for the equations that is allowed. Classical Maxwell fields are necessarily *real* whereas photon states are *complex*. There is also a so-called 'positive frequency' condition that the photon state must satisfy.

states (as with ordinary wavefunctions), but of entire space–time histories of physical behaviour (see Feynman 1985, for a popular account). However, this approach has additional infinities of its own, and one makes sense of it only via the introduction of various 'mathematical tricks'. Despite the undoubted power and impressive accuracy of quantum field theory (in those few cases where the theory can be fully carried through), one is left with a feeling that deeper understandings are needed before one can be confident of any 'picture of physical reality' that it may seem to lead to.[16]

I should make clear that the compatibility between quantum theory and special relativity provided by quantum field theory is only *partial*—referring only to **U**—and is of a rather mathematically formal nature. The difficulty of a consistent relativistic interpretation of the 'quantum jumps' occurring with **R**, that the EPR-type experiments leave us with, is not even touched by quantum field theory. Also, there is not yet a consistent or believable quantum field theory of gravity. I shall be suggesting, in Chapter 8, that these matters may not be altogether unrelated.

Schrödinger's cat

Let us finally return to an issue that has dogged us since the beginnings of our descriptions. Why do we not see quantum linear superpositions of classical-scale objects, such as cricket balls in two places at once? What is it that makes certain arrangements of atoms constitute a 'measuring device', so that the procedure **R** appears to take over from **U**? Surely any piece of measuring apparatus is itself part of the physical world, built up from those very quantum-mechanical constituents whose behaviour it may have been designed to explore. Why not treat the measuring apparatus *together* with the physical system being examined, as a *combined quantum system*. No mysterious 'external' measurement is now involved. The combined system ought simply to evolve according to **U**. But does it? The action of **U** on the combined system is completely deterministic, with no room for the **R**-type probabilistic uncertainties involved in the 'measurement' or 'observation' that the combined system is performing on itself! There is an apparent contradiction here, made especially graphic in a famous thought experiment introduced by Erwin Schrödinger (1935): *the paradox of Schrödinger's cat*.

Imagine a sealed container, so perfectly constructed that no physical influence can pass either inwards or outwards across its walls. Imagine that inside the container is a cat, and also a device that can be triggered by some quantum event. If that event takes place, then the device smashes a phial containing cyanide and the cat is killed. If the event does not take place, the cat lives on. In Schrödinger's original version, the quantum event was the decay of a radioactive atom. Let me modify this slightly and take our quantum event to be the triggering of a photo-cell by a photon, where the photon had been emitted by some light source in a predetermined state, and then reflected off a half-silvered mirror (see Fig. 6.33) The reflection at the mirror splits the photon wavefunction into two separate parts, one of which is reflected and the other transmitted through the mirror. The reflected part of the photon wavefunction is focused on the photo-cell, so if the photon *is* registered by the photo-cell it has been *reflected*. In that case, the cyanide is released and the cat killed. If the photo-cell does *not* register, the photon was *transmitted* through the half-silvered mirror to the wall behind, and the cat is saved.

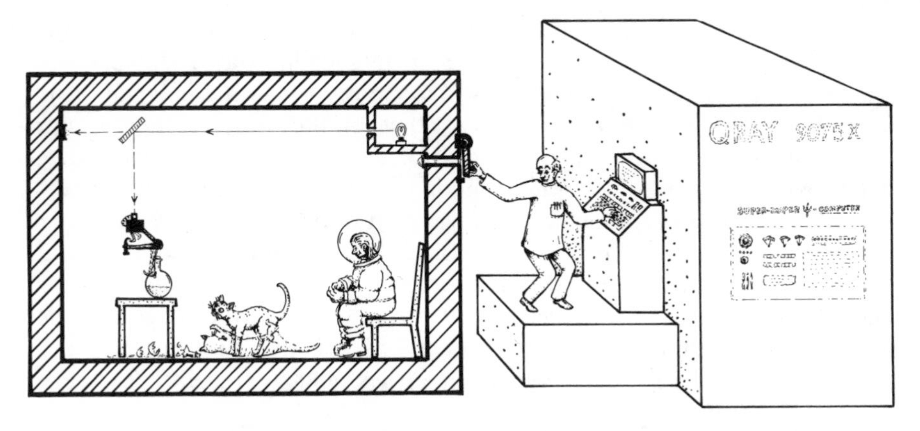

Fig. 6.33. Schrödinger's cat—with additions.

From the (somewhat hazardous) viewpoint of an observer *inside* the container, this would indeed be the description of the happenings there. (We had better provide this observer with suitable protective clothing!) *Either* the photon is taken as having been reflected, because the photo-cell is 'observed' to have registered and the cat killed, *or* the photon is taken as having been transmitted, because the photo-cell is 'observed' *not* to have registered and the cat is alive. Either one or the other *actually* takes place: **R** has been effected, and the probability of each alternative is 50 per cent (because it is a *half*-silvered mirror). Now, let us take the viewpoint of a physicist *outside* the container. We may take the *initial* state-vector of its contents to be 'known' to him before the container was sealed. (I do not mean that it could be known in practice, but there is nothing in quantum theory to say that it could not *in principle* be known to him.) According to the outside observer, no 'measurement' has actually taken place, so the entire evolution of the state-vector should have proceeded according to **U**. The photon is emitted from its source in its predetermined state—both observers are agreed about that—and its wavefunction is split into two beams, with an amplitude of, say, $1/\sqrt{2}$ for the photon to be in each (so that the squared modulus would indeed give a probability of 1/2). Since the entire contents are being treated as a single quantum system by the outside observer, linear superposition between alternatives must be maintained right up to the scale of the cat. There is an amplitude of $1/\sqrt{2}$ that the photo-cell registers and an amplitude of $1/\sqrt{2}$ that it does not. *Both* alternatives must be present in the state, equally weighted as part of a quantum linear superposition. According to the outside observer, the cat is in a linear superposition of being dead and being alive!

Do we really believe that this would be the case? Schrödinger himself made it clear that he did not. He argued, in effect, that the rule **U** of quantum mechanics should not apply to something so large or so complicated as a cat. Something must have gone wrong with the Schrödinger equation along the way. Of course Schrödinger had a right to argue this way about his own equation, but it is not a prerogative given to the rest of us! A great many (and probably most) physicists would maintain that, on the contrary, there is now so much experimental evidence in favour of **U**—and none at all against it—that we have no right whatever to abandon that type of evolution, even at the scale of a cat. If this is accepted, then we seem to be led to a very *subjective* view of physical reality. To the outside observer, the cat is indeed in a linear combination of being alive and dead, and only when the container is finally opened would the cat's state vector collapse into one or the other. On the other hand, to a (suitably protected) observer inside the container, the cat's state-vector would have collapsed much earlier, and the outside observer's linear combination

$$|\psi\rangle = \frac{1}{\sqrt{2}}\{|\text{dead}\rangle + |\text{alive}\rangle\}$$

has no relevance. It seems that the state-vector is 'all in the mind' after all!

But can we really take such a subjective view of the state-vector? Suppose that the outside observer did something much more sophisticated than merely 'looking' inside the container. Suppose that, from his knowledge of the initial state inside the container, he first uses some vast computing facility available to him to *compute*, using the Schrödinger equation, what the state must actually be inside the container, obtaining the ('correct'!) answer $|\psi\rangle$ (where $|\psi\rangle$ indeed involves the above linear superposition of a dead cat and a live cat). Suppose that he then performs that *particular* experiment on those contents which distinguishes the very state $|\psi\rangle$ from anything orthogonal to $|\psi\rangle$. (As has been described earlier, according to the rules of quantum mechanics, he can, *in principle*, perform such an experiment, even though it would be outrageously difficult in practice.) The probabilities for the two outcomes 'yes, it is in state $|\psi\rangle$' and 'no, it is orthogonal to $|\psi\rangle$' would have respective probabilities 100 per cent and 0 per cent. In particular, there is zero probability for the state $|\chi\rangle = |\text{dead}\rangle - |\text{alive}\rangle$, which *is* orthogonal to $|\psi\rangle$. The impossibility of $|\chi\rangle$ as a result of the experiment can only arise because *both* alternatives $|\text{dead}\rangle$ and $|\text{alive}\rangle$ *coexist*, and interfere with each other.

The same would be true if we adjusted the photon path-lengths (or amount of silvering) slightly so that, instead of the state $|\text{dead}\rangle + |\text{alive}\rangle$ we had some other combination, say $|\text{dead}\rangle - \mathrm{i}|\text{alive}\rangle$, etc. All these different combinations have distinct experimental consequences—in principle! So it is not even 'merely' a matter of some kind of coexistence between death and life that might be affecting our poor cat. All the different *complex* combinations are allowed, and they are, in principle, all distinguishable from one another! To the observer inside the container, however, all these combinations seem irrelevant. Either the cat *is* alive, or it *is* dead. How can we make sense of this kind of discrepancy? I shall briefly indicate a number of different points of view that have been expressed on this (and related) questions—though undoubtedly I shall not be fair to all of them!

Various attitudes in existing quantum theory

In the first place, there are obvious difficulties in performing an experiment like the one which distinguishes the state $|\psi\rangle$ from anything orthogonal to $|\psi\rangle$. There is no doubt that such an experiment is *in practice* impossible for the outside observer. In particular, he would need to know the precise state-vector of the entire contents (including the inside observer) before he could even begin to compute what $|\psi\rangle$, at the later time, actually would be! However, we require that this experiment be impossible *in principle*—not merely in practice—since otherwise we should have no right to remove one of the states '$|\text{alive}\rangle$' or '$|\text{dead}\rangle$' from physical reality. The trouble is that quantum theory, as it stands, makes no provision for drawing a clear line between measurements that are 'possible' and those that are 'impossible'. Perhaps there *should* be such a clear-cut distinction. But the theory as it stands does not allow for one. To introduce such a distinction would be to *change* quantum theory.

Second, there is the not uncommon point of view that the difficulties would disappear if we could adequately take the *environment* into account. It would, indeed, be a practical impossibility *actually* to isolate the contents completely from the outside world. As soon as the outside environment becomes involved with the state inside the container, the external observer cannot regard the contents as being given simply by a single state-vector. Even his *own* state gets correlated with it in a complicated way. Moreover, there will be an enormous number of different particles inextricably involved, the effects of the different possible linear combinations spreading out farther and farther into the universe over vast numbers of degrees of freedom. There is no *practical* way (say by observing suitable interference effects) of distinguishing these complex linear superpositions from mere probability-weighted alternatives. This need not even be a matter of the isolation of the contents from the outside. The cat itself involves a vast number of particles. Thus, the complex linear combination of a dead cat and a live one can be treated *as if* it were simply a probability mixture. However, I do not myself find this at all satisfactory. As with the previous view, we may ask at what stage is it officially deemed to be 'impossible' to obtain interference effects—so that the squared moduli of the amplitudes in the complex superposition can now be declared to provide a probability weighting of 'dead' and 'alive'? Even if the 'reality' of the world becomes,in some sense 'actually' a *real*-number probability weighting, how does this resolve itself into just one alternative or the other? I do not see how *reality* can ever transform itself from a complex (or real) linear *superposition* of two alternatives into *one or the other* of these alternatives, on the basis merely of the evolution **U**. We seem driven back to a subjective view of the world!

Sometimes people take the line that complicated systems should not really be described by 'states' but by a generalization referred to as *density matrices* (von Neumann 1955). These involve both classical probabilities and quantum amplitudes. In effect, many different quantum states are then taken together to represent reality. Density matrices are useful, but they do not in themselves resolve the deep problematic issues of quantum measurement.

One might try to take the line that the actual evolution is the deterministic **U**, but probabilities arise from the uncertainties involved in knowing what the quantum state of the combined system really *is*. This would be taking a very 'classical' view about the origin of the probabilities—that they all arise from uncertainties in the initial state. One might imagine that tiny differences in the initial state could give rise to enormous differences in the evolution, like the 'chaos' that can occur with classical systems (e.g. weather prediction; cf. Chapter 5, p. 173). However, such 'chaos' effects simply cannot occur with **U** by itself, since it is *linear*: unwanted linear superpositions simply persist forever under **U**! To resolve such a superposition into one alternative or the other, something *non*-linear would be needed, so **U** itself will not do.

For another viewpoint, we may take note of the fact that the only completely clear-cut discrepancy with observation, in the Schrödinger cat experiment, seems to arise because there are *conscious observers*, one (or two!) inside and one outside the container. Perhaps the laws of complex quantum linear superposition do *not* apply to consciousness! A rough mathematical model for such a viewpoint was put forward by Eugene P. Wigner (1961). He suggested that the linearity of Schrödinger's equation might fail for conscious (or merely 'living') entities, and be replaced by some non-linear procedure, according to which either one or the other alternative would be resolved out. It might seem to the reader that, since I am searching for some kind of role for quantum phenomena in our conscious thinking—as indeed I am—I should find this view to be a sympathetic possibility. However, I am not at all happy with it. It seems to lead to a very lopsided and disturbing view of the *reality* of the world. Those corners of the universe where consciousness resides may be rather few and far between. On this view, *only* in those corners would the complex quantum linear superpositions be resolved into actual alternatives. It may be that to *us*, such other corners would look the same as the rest of the universe, since whatever we, ourselves, actually *look* at (or otherwise observe) would, by our very acts of conscious observation, get 'resolved into alternatives', *whether or not* it had done so before. Be that as it may, this gross lopsidedness would provide a very disturbing picture of the *actuality* of the world, and I, for one, would accept it only with great reluctance.

There is a somewhat related viewpoint, called the participatory universe (suggested by John A. Wheeler 1983), which takes the role of consciousness to a (different) extreme. We note, for example, that the evolution of conscious life on this planet is due to appropriate mutations having taken place at various times. These, presumably, are quantum events, so they would exist only in linearly superposed form until they finally led to the evolution of a conscious being—whose very existence depends upon all the right mutations having 'actually' taken place! It is our own presence which, on this view, conjures our past into existence. The circularity and paradox involved in this picture has an appeal for some, but for myself I find it distinctly worrisome—and, indeed, barely credible.

Another viewpoint, also logical in its way, but providing a picture no less strange, is that of *many worlds*, first publicly put forward by Hugh Everett III (1957). According to the many-worlds interpretation, **R** never takes place at all. The entire evolution of the state-vector—which is regarded realistically—is always governed by the deterministic procedure **U**. This implies that poor Schrödinger's cat, together with the protected observer inside the container, must indeed exist in some complex linear combination, with the cat in some superposition of life and death. However the dead state is correlated with one state of the inside observer's consciousness, and the live one, with another (and presumably, partly, with the consciousness of the cat—and, eventually, with the outside observer's also, when the contents becomes revealed to him). The consciousness of each observer is regarded as 'splitting', so he now exists twice over, each of his instances having a different experience (i.e. one seeing a dead cat and the other a live one). Indeed, not just an observer, but the entire universe that he inhabits splits in two (or more) at each 'measurement' that he makes of the world. Such splitting occurs again and again—not merely because of 'measurements' made by observers, but because of the macroscopic magnification of quantum events generally—so that these universe 'branches' proliferate wildly. Indeed, every alternative possibility would coexist in some vast superposition. This is hardly the most economical of viewpoints, but my own objections to it do not spring from its lack of economy. In particular, I do not see why a conscious being need be aware of only 'one' of the alternatives in a linear superposition. What is it about consciousness that demands that one cannot be 'aware' of that tantalizing linear combination of a dead and a live cat? It seems to me that a theory of consciousness would be needed before the many-worlds view can be squared with what one actually observes. I do not see what relation there is between the 'true' (objective) state-vector of the universe and what we are supposed actually to 'observe'. Claims have been made that the 'illusion' of **R** can, in some sense, be effectively deduced in this picture, but I do not think that these claims hold up. At the very least, one needs further ingredients to make the scheme work. It seems to me that the many-worlds view introduces a multitude of problems of its own without really touching upon the *real* puzzles of quantum measurement. (Compare DeWitt and Graham 1973.)

Where does all this leave us?

These puzzles, in one guise or another, persist in *any* interpretation of quantum mechanics as the theory exists today. Let us briefly review what standard quantum theory has actually told us about how we should describe the world, especially in relation to these puzzling issues—and then ask: where do we go from here?

Recall, first of all, that the descriptions of quantum theory appear to apply sensibly (usefully?) only at the so-called *quantum level*—of molecules, atoms, or subatomic particles, but also at larger dimensions, so long as energy differences between alternative possibilities remain very small. At the quantum level, we must treat such 'alternatives' as things that can *coexist*, in a kind of complex-number-weighted superposition. The complex numbers that are used as weightings are called *probability amplitudes*. Each different totality of complex-weighted alternatives defines a different *quantum state*, and any quantum system must be described by such a quantum state. Often, as is most clearly the case with the example of *spin*, there is nothing to say which are to be 'actual' alternatives composing a quantum state and which are to be just 'combinations' of alternatives. In any case, so long as the system *remains* at the quantum level, the quantum state evolves in a completely *determinstic* way. This deterministic evolution is the process **U**, governed by the important *Schrödinger equation*.

When the effects of different quantum alternatives become magnified to the *classical level*, so that differences between alternatives are large enough that we might directly perceive them, then such complex-weighted superpositions seem no longer to persist. Instead, the squares of the moduli of the complex amplitudes must be formed (i.e. their squared distances from the origin in the complex plane taken), and these *real* numbers now play a new role as actual *probabilities* for the alternatives in question. Only *one* of the alternatives survives into the actuality of physical experience, according to the process **R** (called reduction of the state vector or collapse of the wavefunction; completely different from **U**). It is here, and only here, that the non-determinism of quantum theory makes its entry.

The quantum state may be strongly argued as providing an *objective* picture. But it can be a complicated and even somewhat paradoxical one. When several particles are involved, quantum states can (and normally 'do') get very complicated. Individual particles then do not have 'states' on their own, but exist only in complicated 'entanglements' with other particles, referred to as *correlations*. When a particle in one region is 'observed', in the sense that it triggers some effect that becomes magnified to the classical level, then **R** must be invoked—but this apparently *simultaneously* affects all the other particles with which that particular particle is correlated. Experiments of the Einstein–Podolsky–Rosen (EPR) type (such as that of Aspect, in which pairs of photons are emitted in opposite directions by a quantum source, and then separately have their polarizations measured many metres apart) give clear observational substance to this puzzling, but essential fact of quantum physics: it is *non-local* (so that the photons in the Aspect experiment cannot be treated as separate independent entities)! If **R** is considered to act in an objective way (and that would seem to be implied by the objectivity of the quantum state), then the spirit of special relativity is accordingly violated. *No objectively real space–time description* of the (reducing) state-vector seems to exist which is consistent with the requirements of relativity! However the *observational* effects of quantum theory do not violate relativity.

Quantum theory is silent about *when* and *why* **R** should actually (or appear to?) take place. Moreover, it does not, in itself, properly explain why the classical-level world 'looks' classical. 'Most' quantum states do not at all resemble classical ones!

Where does all this leave us? I believe that one must strongly consider the possibility that quantum mechanics is simply *wrong* when applied to macroscopic bodies—or, rather that the laws **U** and **R** supply excellent approximations, only, to some more complete, but as yet undiscovered, theory. It is the *combination* of these two laws together that has provided all the wonderful agreement with observation that present theory enjoys, not **U** alone. If the linearity of **U** were to extend into the macroscopic world, we should have to accept the physical realtiy of complex linear combinations of different positions (or of different spins, etc.) of cricket balls and the like. Common sense alone tells us that this is not the way that the world actually behaves!

Cricket balls are indeed well approximated by the descriptions of *classical* physics. They have reasonably well-defined locations, and are not seen to be in two places at once, as the linear laws of quantum mechanics would allow them to be. If the procedures **U** and **R** are to be replaced by a more comprehensive law, then, unlike Schrödinger's equation, this new law would have to be *non*-linear in character (because **R** itself acts non-linearly). Some people object to this, quite rightly pointing out that much of the profound mathematical elegance of standard quantum theory results from its linearity. However, I feel that it would be surprising if quantum theory were not to undergo some fundamental change in the future—to something for which this linearity would be only an approximation. There are certainly precedents for this kind of change. Newton's elegant and powerful theory of universal gravitation owed much to the fact that the forces of the theory add up in a *linear* way. Yet, with Einstein's general relativity, this linearity was seen to be only an (albeit excellent) approximation—and the elegance of Einstein's theory exceeds even that of Newton's!

I have made no bones of the fact that I believe that the resolution of the puzzles of quantum thoery must lie in our finding an improved theory. Though this is perhaps not the conventional view, it is not an altogether unconventional one. (Many of quantum theory's originators were also of such a mind. I have referred to Einstein's views. Schrödinger (1935), de Broglie (1956), and Dirac (1939) also regarded the theory as provisional.) But even if one believes that the theory is somehow to be modified, the constraints on *how* one might do this are enormous. Perhaps some kind of 'hidden variable' viewpoint will eventually turn out to be acceptable. But the non-locality that is exhibited by the EPR-type experiments severely challenges any 'realistic' description of the world that can comfortably occur within an ordinary space–time—a space–time of the particular type that has been given to us to accord with the principles of relativity—so I believe that a much more radical change is needed. Moreover, no discrepancy of any kind between quantum theory and experiment has ever been found—unless, of course, one regards the evident absence of linearly superposed cricket balls as contrary evidence. In my own view, the non-existence of linearly superposed cricket balls actually *is* contrary evidence! But this, in itself, is no great help. We know that at the sub-microscopic level of things the quantum laws do hold sway; but at the level of cricket balls, it is classical physics. Somewhere in between, I would maintain, we need to understand the new law, in order to see how the quantum world merges with the classical. I believe, also, that we shall need this new law if we are ever to understand minds! For all this we must, I believe, look for new clues.

In my descriptions of quantum theory in this chapter, I have been wholely conventional, though the emphasis has perhaps been more geometrical and 'realistic' than is usual. In the next chapter we shall try to search for some needed clues—clues that I believe must give us some hints about an improved quantum mechanics. Our journey will start close to home, but we shall be forced to travel far afield. It turns out that we shall need to explore very distant reaches of space, and to travel back, even to the very beginning of time!

1. I have taken for granted that any 'serious' philosophical viewpoint should contain at least a good measure of realism. It always surprises me when I learn of apparently serious-minded thinkers, often physicists concerned with the implications of quantum mechanics, who take the strongly subjective view that there is, in actuality, *no* real world 'out there' at all! The fact that I take a realistic line wherever possible is not meant to imply that I am unaware that such subjective views are often seriously maintained—only that I am unable to make sense of them. For a powerful and entertaining attack on such subjectivism, see Gardner (1983), Chapter 1.
2. In particular, J. J. Balmer had noted, in 1885, that the frequencies of the spectral lines of hydrogen had the form $R\,(n^{-2} - m^{-2})$ where n and m are positive integers (R being a constant).
3. Perhaps we should not dismiss this 'entirely field' picture too lightly. Einstein, who (as we shall see) was profoundly aware of the discreteness manifested by quantum particles, spent the last thirty years of his life trying to find a fully comprehensive theory of this general classical kind. But Einstein's attempts, like all others, were unsuccessful. Something else beside a classical field seems to be needed in order to explain the discrete nature of particles.
4. These two evolution procedures were described in a classic work by the remarkable Hungarian/American mathematician John von Neumann (1955). His 'process 1' is what I have termed **R**—'reduction of the state-vector'—and his process 2 is **U**—'unitary evolution' (which means, in effect that probability amplitudes are preserved by the evolution). In fact, there are other—though equivalent—descriptions of quantum-state evolution **U**, where one might not use the term 'Schrödinger's equation'. In the '*Heisenberg picture*', for example, the state is described so that it appears not to evolve at all, the dynamical evolution being taken up in a continual shifting of the meanings of the position/momentum coordinates. The various distinctions are not important for us here, the different descriptions of the process **U** being completely equivalent.
5. For completeness we should also specify all the required algebraic laws which, in the (Dirac) notation used in the text, are:

$$|\psi\rangle+|\chi\rangle = |\chi\rangle+|\psi\rangle, \qquad |\psi\rangle+(|\chi\rangle+|\varphi\rangle) = (|\psi\rangle+|\chi\rangle)+|\varphi\rangle,$$
$$(z+w)|\psi\rangle = z|\psi\rangle+w|\psi\rangle, \qquad z(|\psi\rangle+|\chi\rangle) = z|\psi\rangle+z|\chi\rangle,$$
$$z(w|\psi\rangle) = (zw)|\psi\rangle, \qquad 1|\psi\rangle = |\psi\rangle,$$
$$|\psi\rangle+\mathbf{0} = |\psi\rangle, \qquad 0|\psi\rangle = \mathbf{0}, \text{ and } z\mathbf{0} = \mathbf{0}.$$

6. There is an important operation, referred to as the *scalar product* (or inner product) of two vectors, which can be used to express the concepts of 'unit vector', 'orthogonality', and 'probability amplitude' very simply. (In ordinary vector

algebra, the scalar product is $ab\cos\theta$, where a and b are the lengths of the vectors and θ is the angle between their directions.) The scalar product between Hilbert space-vectors gives a *complex* number. For two state-vectors $|\psi\rangle$ and $|\chi\rangle$ we write this $\langle\psi|\chi\rangle$. There are algebraic rules $\langle\psi(|\chi\rangle + |\varphi\rangle) = \langle\psi|\chi\rangle + \langle\psi|\varphi\rangle$, $\langle\psi(q|\chi\rangle) = q\langle\psi|\chi\rangle$, and $\langle\psi|\chi\rangle = \overline{\langle\chi|\psi\rangle}$, where the bar denotes complex conjugation. (The complex conjugate of $z = x + iy$, is $\bar{z} = x - iy$, x and y being real; note that $|z|^2 = z\bar{z}$). Orthogonality between $|\psi\rangle$ and $|\chi\rangle$ is expressed as $\langle\psi|\chi\rangle = 0$. The squared length of $|\psi\rangle$ is $|\psi|^2 = \langle\psi|\psi\rangle$, so the condition for $|\psi\rangle$ to be normalized as a unit vector is $\langle\psi|\psi\rangle = 1$. If an 'act of measurement' causes a state $|\psi\rangle$ to jump either to $|\chi\rangle$ or to something orthogonal to $|\chi\rangle$, then the amplitude for it to jump to $|\chi\rangle$ is $\langle\chi|\psi\rangle$, assuming that $|\psi\rangle$ and $|\chi\rangle$ are both normalized. Without normalization, the probability of jumping from to $|\chi\rangle$ can be written $\langle\chi|\psi\rangle\langle\psi|\chi\rangle/\langle\chi|\chi\rangle\langle\psi|\psi\rangle$. (See Dirac 1947.)

7. For those familiar with the quantum-mechanical operator formalism, this measurement is defined (in the Dirac notation) by the bounded Hermitian operator $|\chi\rangle\langle\chi|$. The eigenvalue 1 (for normalized $|\chi\rangle$) means **YES** and the eigenvalue 0 means **NO**. (The vectors $\langle\chi|$, $\langle\psi|$, etc. belong to the *dual* of the original Hilbert space.) See von Neumann (1955), Dirac (1947.)
8. In my earlier descriptions of a quantum system consisting of a single particle, I have over-simplified, by ignoring spin and assuming that the state can be described in terms of its position alone. There *are* actually certain particles—called *scalar* particles, examples being those nuclear particles referred to as *pions* (π-mesons, cf. p. 218), or certain atoms—for which the spin value turns out to be zero. For these particles (but only for these) the above description in terms of position alone will actually suffice.
9. Take $|\swarrow\rangle = \bar{z}|\uparrow\rangle - \bar{w}|\downarrow\rangle$, where $\bar{z}$ and $\bar{w}$ are the complex conjugates of $\bar{z}$ and $\bar{w}$. (See note 6.)
10. There is a standard experimental device, known as a Stern–Gerlach apparatus, which can be used for measuring the spins of suitable atoms. Atoms are projected in a beam which passes through a highly inhomogeneous magnetic field, and the direction of the field's inhomogeneity provides the direction of the spin measurement. The beam splits into two (for a spin one-half atom, or into more that two beams for higher spin) one beam giving the atoms for which the answer to the spin measurement is **YES** and the other for which the answer is **NO**. Unfortunately, there are technical reasons, irrelevant for our purposes, why this apparatus cannot be used for the measurement of electron spin, and a more indirect procedure must be used. (See Mott and Massey 1965.) For this and other reasons, I prefer not to be specific about how our electron's spin is actually being measured.
11. The enterprising reader may care to check the geometry given in the text. It is easiest if we orient our Riemann sphere so that the α-direction is 'up' and the β-direction lies in the plane spanned by 'up' and 'right', i.e. given by $q = \tan(\theta/2)$ on the Riemann sphere, and then use the prescription $\langle\chi|\psi\rangle\langle\psi|\chi\rangle/\langle\chi|\chi\rangle\langle\psi|\psi\rangle$ for the probability of jumping from $|\psi\rangle$ to $|\chi\rangle$. See note 6.
12. Mathematically, we say that the space of two-particle states is the *tensor product* of the space of states of the first particle with that of the second particle. The state $|\chi\rangle|\varphi\rangle$ is then the tensor product of the state $|\chi\rangle$ with $|\varphi\rangle$.
13. Wolfgang Pauli, a brilliant Austrian physicist and prominent figure in the

development of quantum mechanics, put forward his exclusion principle as a hypothesis in 1925. The full quantum-mechanical treatment of what we now call 'fermions' was developed in 1926 by the highly influential and original Italian(–American) physicist Enrico Fermi and by the great Paul Dirac, whom we have encountered several times before. The statistical behaviour of fermions accords with 'Fermi–Dirac statistics', which term distinguishes it from 'Boltzmann statistics'—the classical statistics of distinguishable particles. The 'Bose–Einstein statistics' of bosons was developed for the treatment of photons by the remarkable Indian physicist S. N. Bose and Albert Einstein in 1924.

14. This is such a remarkable and important result that it is worth giving another version of it. Suppose that there are just *two* settings for the E-measurer, up [↑] and right [→], and two for the P-measurer, 45° up to the right [↗] and 45° down to the right [↘]. Take the *actual* settings to be [→] and [↗], for the E- and P-measurers, respectively. Then the probability that the E- and P-measurements give agreement is $\frac{1}{2}(1 + \cos 135°) = 0.146\ldots$, which is a little under 15 per cent. A long succession of experiments, with these settings gives, say,

E: **Y N N Y N Y Y Y N Y Y N N Y N N N N Y Y N . . .**
P: **N Y Y N N N Y N Y N N Y Y N Y Y N Y N N Y . . .**
(✓ under the 5th, 7th and 16th letters, marking agreement)

will give just under 15 per cent agreement. Now, let us suppose that the P-measurements are not influenced by the E-setting—so that *if* the E-setting had been [↑] rather than [→], then the run of P-results would have been exactly the same—and since the angle between [↑] and [↗], is the same as between [→] and [↗], there would again be just under 15 per cent agreement between the P-measurements and the new E-measurements, say E′. On the other hand, if the E-setting had been [→], as before, but the P-setting were [↘] rather than [↗], then the run of E-results would have been as before but the new P-results, say P′, would have been something with just under 15 per cent agreement with the original E-results. It follows that there could be no more than 45 per cent (= 15 per cent + 15 per cent + 15 per cent) agreement between the P′-measurement [↘] and the E′-measurement [↑] had *these* been the actual settings. But the angle between [↘] and [↑] is 135° not 45°, so the probability of agreement *ought* to be just over 85 per cent, not 45 per cent. This is a contradiction, showing that the assumption that the choice of measurement made on E cannot influence the results for P (and *vice-versa*) must be false! I am indebted to David Mermin for this example. The version given in the main text is taken from his article Mermin (1985).

15. Earlier results were due to Freedman and Clauser (1972) based on ideas suggested by Clauser, Horne, Shimony, and Holt (1969). There is still a point of argument, in these experiments, owing to the fact that the photon detectors that are used fall a good deal short of 100 per cent efficiency, so that only a comparatively small fraction of the emitted photons are actually detected. However, the agreement with the quantum theory is so perfect, with these comparatively inefficient detectors, that it is hard to see how making the detectors better is suddenly going to produce a *worse* agreement with that theory!

16. Quantum field theory appears to offer some scope for non-computability (cf. Komar 1964).

7
Cosmology and the arrow of time

The flow of time

Central to our feelings of awareness is the sensation of the progression of time. We *seem* to be moving ever forward, from a definite past into an uncertain future. The past is over, we feel, and there is nothing to be done with it. It is unchangeable, and in a certain sense, it is 'out there' still. Our present knowledge of it may come from our records, our memory traces, and from our deductions from them, but we do not tend to doubt the *actuality* of the past. The past was one thing and can (now) *be* only one thing. What has happened has happened, and there is now nothing whatever that we, nor anyone else can do about it! The future, on the other hand, seems yet undetermined. It could turn out to be one thing or it could turn out to be another. Perhaps this 'choice' is fixed completely by physical laws, or perhaps partly by our own decisions (or by God); but this 'choice' *seems* still there to be made. There appear to be merely *potentialities* for whatever the 'reality' of the future may actually resolve itself to be. As we consciously perceive time to pass, the most immediate part of that vast and seemingly undetermined future continuously becomes realized as actuality, and thus makes its entry into the fixed past. Sometimes we may have the feeling that *we* even have been personally 'responsible' for somewhat influencing that choice of particular potential future which in fact becomes realized, and made permanent in the actuality of the past. More often, we feel ourselves to be helpless spectators—perhaps thankfully relieved of responsibility—as, inexorably, the scope of the determined past edges its way into an uncertain future.

Yet physics, as we know it, tells a different story. All the successful equations of physics are symmetrical in time. They can be used equally well in one direction in time as in the other. The future and the past seem physically to be on a completely equal footing. Newton's laws, Hamilton's equations, Maxwell's equations, Einstein's general relativity, Dirac's equation, the Schrödinger equation—all remain effectively unaltered if we reverse the direction of time. (Replace the coordinate t which represents time, by $-t$.) The whole of classical mechanics, together with the '**U**' part of quantum mechanics, is entirely reversible in time. There is a question as to whether the

'**R**' part of quantum mechanics is actually time-reversible or not. This question will be central to the arguments I shall be presenting in the next chapter. For the moment, let us sidestep the issue by referring to what may be regarded as a 'conventional wisdom' of the subject—namely that, despite initial appearances, the operation of **R** must indeed be taken to be time-symmetrical also (cf. Aharonov, Bergmann and Lebowitz 1964). If we accept this, it seems that we shall need to look elsewhere if we are to find where our physical laws assert that the distinction between past and future must lie.

Before addressing that issue, we should consider another puzzling discrepancy between our perceptions of time and what modern physical theory tells us to believe. According to relativity, there is not really such a thing as the 'now' at all. The closest that we get to such a concept is an observer's 'simultaneous space' in space–time, as depicted in Figure 5.21, p. 200, but that depends on the *motion* of the observer! The 'now' according to one observer would not agree with that for another.[1] Concerning two space–time events A and B, one observer **U** might consider that B belongs to the fixed past and A to the uncertain future, while for a second observer **V**, it could be A that belongs to the fixed past and B to the uncertain future! (See Fig. 7.1). We cannot meaningfully assert that either one of the events A and B remains uncertain, so long as the other is definite.

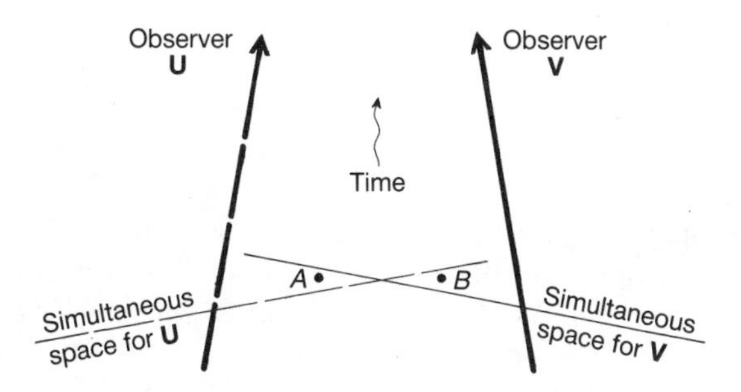

Fig. 7.1. Can time actually 'flow'? To observer **U**, B can be in the 'fixed' past while A lies yet in the 'uncertain' future. Observer **V** holds the contrary view!

Recall the discussion on p. 201 and Fig. 5.22. Two people pass each other on the street; and according to one of the two people, an Andromodean space fleet has already set off on its journey, while to the other, the decision as to whether or not the journey will actually take place has not yet been made. How can there still be some uncertainty as to the outcome of that decision? If to *either* person the decision has already been made, then surely there *cannot* be any uncertainty. The launching of the space fleet is an inevitability. In fact neither of the people can yet *know* of the launching of the space fleet. They can know only later, when telescopic observations from earth reveal that the fleet is indeed on its way. Then they can hark back to that chance encounter, and come to the conclusion that at *that* time, according to one of them, the decision lay in the uncertain future, while to the other, it lay in the certain past. Was there *then* any uncertainty about that future? Or was the future of *both* people already 'fixed'?

It begins to seem that if anything is definite at all, then the entire space–time must indeed be definite! There can be no 'uncertain' future. The *whole* of space–time must be fixed, without any scope for uncertainty. Indeed, this seems to have been Einstein's own conclusion (cf. Pais 1982, p. 444). Moreover, there is no flow of time at all. We have just 'space–time'—and no scope at all for a future whose domain is being inexorably encroached upon by a determined past! (The reader may be wondering what is the role of the 'uncertainties' of quantum mechanics in all this. I shall return to the questions raised by quantum mechanics later in the next chapter. For the moment, it will be better to think in terms of a purely classical picture.)

It seems to me that there are severe discrepancies between what we consciously feel, concerning the flow of time, and what our (marvellously accurate) theories assert about the reality of the physical world. These discrepancies must surely be telling us something deep about the physics that presumably must actually underlie our conscious perceptions—assuming (as I believe) that what underlies these perceptions can indeed be understood in relation to some appropriate kind of physics. At least it seems to be clearly the case that whatever physics is operating, it must have an essentially time-asymmetrical ingredient, i.e. it must make a distinction between the past and the future.

If the equations of physics seem to make no distinction between future and past—and if even the very idea of the 'present' fits so uncomfortably with relativity—then where in heaven do we look, to find physical laws more in accordance with what we seem to perceive of the world? In fact things are not so discrepant as I may have been seeming to imply. Our physical understanding actually contains important ingredients *other* than just equations of time-evolution—and some of these do indeed involve time-asymmetries. The most important of these is what is known as the *second law of thermodynamics*. Let us try to gain some understanding of what this law means.

The inexorable increase of entropy

Imagine a glass of water poised on the edge of a table. If nudged, it is likely to fall to the ground—no doubt to shatter into many pieces, with water splashed over a considerable area, perhaps to become absorbed into a carpet, or to fall between the cracks in the floorboards. In this, our glass of water has been merely following faithfully the equations of physics. Newton's descriptions will do. The atoms in the glass and in the water are each individually following Newton's laws (Fig. 7.2). Now let us run this picture in the reverse direction in time. By the time-reversibility of these laws, the water could equally well flow out from the carpet and from the cracks in the floorboards, to enter a glass which is busily reconstructing itself from numerous shattered pieces, the assembled whole then jumping from the floor exactly to table height, to come to rest poised on the edge of the table. All that is also in accordance with Newton's laws, just as was the falling and shattering glass!

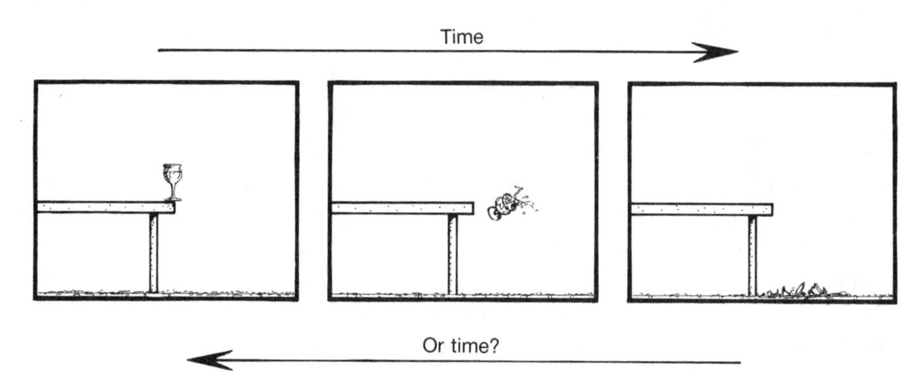

Fig. 7.2. The laws of mechanics are time-reversible; yet the time-ordering of such a scene from the right frame to the left is something that is never experienced, whereas that from the left frame to right would be commonplace.

The reader may perhaps be asking where the energy comes from, which raises the glass from floor to table. *That* is no problem. There cannot be a problem with energy, because in the situation in which the glass *falls* from the table, the energy that it gains from falling must *go* somewhere. In fact the energy of the falling glass goes into *heat*. The atoms in the glass fragments, in the water, in the carpet, and in the floorboards will be moving around in a random way just a little bit faster than they were, at the moment after the glass strikes the ground, i.e. the glass fragments, water, carpet, and floorboards will be just a little *warmer* than they would have been before (ignoring possible heat loss from evaporation—but that, also, is reversible in principle). By *energy conservation*, this heat energy is just equal to the energy lost by the glass of water in falling from the table. Thus that little bit of heat energy would be just enough to raise the glass back again to the table! It is important to realize that heat energy must be included when we consider energy conservation. The law of energy conservation, when the energy of heat is taken into account, is called the *first law of thermodynamics*. The first law of thermodynamics, being a deduction from Newtonian mechanics, is time-symmetrical. The first law does *not* constrain the glass and water in any way which rules out its assembling itself, filling with water and jumping back miraculously on to the table.

The reason that we do not see such things happen is that the 'heat' motion of the atoms in the glass fragments, water, floorboards, and carpet will be all higgledy-piggledy, so most of the atoms will be moving in all the wrong directions. An absurdly precise coordination in their motions would be needed in order to reassemble the glass, with all the splashes of water collected back in it, and hurl it delicately up on to the table. It is an effective certainty that such coordinated motion will *not* to be present! Such coordination could occur only by the most amazing fluke—of a kind that would be referred to as 'magic' if ever it did occur!

Yet in the other direction in time, such coordinated motion is commonplace. Somehow, we do not regard it as a fluke if the particles are moving in coordinated fashion, provided that they do it *after* some large-scale change in the physical state has taken place (here the shattering and spilling of the glass of water), rather than *before* such a change. The particle motions must indeed be highly coordinated after such an event; for these motions are of such a nature that if we were to reverse, in a precise way, the motion of every single atom, the resulting behaviour would be exactly what would be needed to reassemble, fill, and lift the glass to its precise starting configuration.

Highly coordinated motion is acceptable and familiar if it is regarded as being an *effect* of a large-scale change and not the *cause* of it. However, the words 'cause' and 'effect' somewhat beg the question of time-asymmetry. In our normal parlance we are used to applying these terms in the sense that the cause must precede the effect. But if we are trying to understand the physical difference between past and future, we have to be very careful not to inject our everyday feelings about past and future into the discussion unwittingly. I must warn the reader that it is extremely difficult to avoid doing this, but it is imperative that we try. We must try to use words in such a way that they do not prejudice the issue of the physical distinction between past and future. Accordingly, if the circumstances happened to deem it appropriate, we would have to allow ourselves to take the causes of things to lie in the future and the effects to lie in the past! The deterministic equations of classical physics (or the operation of **U** in quantum physics, for that matter) have no preference for evolving in the future direction. They can be used equally well to evolve into the past. The future determines the past in just the same way that the past determines the future. We can specify some state of a system in some arbitrary way in the future and then use this state to compute what it would have had to be like in the past. If we are allowed to view the past as 'cause' and the future as 'effect' when we evolve the equations for the system in the normal future direction in time, then when we apply the equally valid procedure of evolving the equations in the past direction in time we must apparently regard the future as 'cause' and the past as 'effect'.

However, there is something else involved in our use of the terms 'cause' and 'effect' which is not really a matter of which of the events referred to happen to lie in the past and which in the future. Let us imagine a hypothetical universe in which the same time-symmetric classical equations apply as those of our own universe, but for which behaviour of the familiar kind (e.g. shattering and spilling of water glasses) coexist with occurrences like the time-reverses of these. Suppose that, along with our more familiar experiences, sometimes water glasses *do* assemble themselves out of broken pieces, mysteriously fill themselves out of splashes of water, and then leap up on to tables; suppose also that, on occasion, scrambled eggs magically unscramble and de-cook themselves, finally to leap back into broken eggshells which perfectly assemble and seal themselves about their newly-acquired contents; that lumps of sugar can form themselves out of the dissolved sugar in sweetened coffee and then spontaneously jump from the cup into someone's hand. If we lived in a world where such occurrences were commonplace, surely we would ascribe the 'causes' of such events not to fantastically improbable chance coincidences concerning correlated behaviour of the individual atoms, but to some 'teleological effect' whereby the self-assembling objects sometimes strive to achieve some desired macroscopic configuration. 'Look!', we would say, 'It's happening again. That mess is going to assemble itself into another glass of water!' No doubt we would take the view that the atoms were aiming themselves so accurately *because* that was the way to produce the glass of water on the table. The glass on the table would be the 'cause' and the apparently random collection of atoms on the floor, the 'effect'—despite the fact that the 'effect' now occurs earlier in time than the 'cause'. Likewise, the minutely organized motion of the atoms in the scrambled egg is not the 'cause' of the leap up into the assembling eggshell, but the 'effect' of this future occurrence; and the sugar lump does not assemble itself and leap from the cup 'because' the atoms move with such extraordinary precision, but owing to the fact that someone—albeit in the future—will later hold that sugar lump in his hand!

Of course, in our world we do not see such things happen—or, rather, what we do not see is the *coexistence* of such things with those of our normal kind. If *all* that we saw were happenings of the perverse kind just described, then we should have no problem. We could just interchange the terms 'past' and 'future', 'before' and 'after', etc., in all our descriptions. Time could be deemed to be progressing in the reverse direction from that originally specified, and that world could be described as being just like our own. However, I am here envisaging a different possibility—just as consistent with the time-symmetrical equations of physics—where shattering and self-assembling water glasses can *coexist*. In such a world, we cannot retrieve our familiar descriptions merely by a reversal of our conventions about the direction of progression of time. Of course our world happens not to be like that, but why is it not? To begin to understand this fact, I have been asking you to try to imagine such a world and to wonder how we would describe the happenings taking place within it. I am asking that you accept that, in such a world, we should surely describe the gross macroscopic configurations—such as complete glasses of water, unbroken eggs, or a lump of sugar held in the hand—as providing 'causes' and the detailed, and perhaps finely correlated, motions of individual atoms as 'effects', whether or not the 'causes' lie to the future or to the past of the 'effects'.

Why is it that, in the world in which we happen to live, it is the causes which actually *do* precede the effects; or to put things in another way, why do precisely coordinated particle motions occur only *after* some large-scale change in the physical state and not *before* it. In order to get a better physical description of such things, I shall need to introduce the concept of *entropy*. In rough terms, the entropy of a system is a measure of its manifest *disorder*. (I shall be a little more precise later.) Thus, the smashed glass and spilled water on the floor is in a higher entropy state than the assembled and filled glass on the table; the scrambled egg has higher entropy than the fresh unbroken egg; the sweetened coffee has a higher entropy than the undissolved sugar lump sitting in unsweetened coffee. The low entropy state seems 'specially ordered', in some manifest way, and the high entropy state, less 'specially ordered'.

It is important to realize, when we refer to the 'specialness' of a low entropy state, that we are indeed referring to *manifest* specialness. For, in a more subtle sense, the higher entropy state, in these situations, *is* just as 'specially ordered' as the lower entropy state, owing to the very precise coordination of the motions of the individual particles. For example, the seemingly random motions of the water molecules which have leaked between the floorboards after the glass has smashed are indeed very special: the motions are so precise that if they were all exactly *reversed* then the original low entropy state, where the glass sits assembled and full of water on the table, would be recovered. (This must be the case since the reversal of all these motions would simply correspond to reversing the direction of time—according to which the glass would indeed assemble itself and jump back on to the table.) But such coordinated motion of all the water molecules is *not* the kind of 'specialness' that we refer to as low entropy. Entropy refers to *manifest* disorder. The order which is present in precise coordination of particle motions is not manifest order, so it does not count towards lowering the entropy of a system. Thus the order in the molecules in the spilled water does not count in this way, and the entropy is high. However, the *manifest* order in the *assembled* water glass gives a low entropy value. This refers to the fact that comparatively few different possible arrangements of particle motions are compatible with the manifest configuration of an assembled and filled water glass; whereas there are very many more motions which are compatible with the manifest configuration of the slightly heated water flowing between the cracks in the floorboards.

The *second law of thermodynamics* asserts that *the entropy of an isolated system increases with time (or remains constant, for a reversible system)*. It is well that we do not count coordinated particle motions as low entropy; for if we did, the 'entropy' of a system, according to that definition, would always remain a constant. The entropy concept must refer only to disorder which is indeed manifest. For a system in isolation from the rest of the universe, its total entropy increases, so if it starts off in some state with some kind of manifest organization, this organization will, in due course, become eroded, and these manifest special features will become converted into 'useless' coordinated particle motions. It might seem, perhaps, that the second law is like a council of despair, for it asserts that there is a relentless and universal physical principle, telling us that organization is necessarily continually breaking down. We shall see later that this pessimistic conclusion is not entirely appropriate!

What is entropy?

But what precisely *is* the entropy of a physical system? We have seen that it is some sort of measure of manifest disorder, but it would appear, by my use of such imprecise terms as 'manifest' and 'disorder', that the entropy concept could not really be a very clear-cut scientific quantity. There is also another aspect of the second law which seems to indicate an element of imprecision in the concept of entropy: it is only with so-called *irreversible* systems that the entropy actually increases, rather than just remaining constant. What does 'irreversible' mean? If we take into account the detailed motions of all the particles, then *all* systems are reversible! *In practice*, we should say that the glass falling from the table and smashing, or the scrambling of the egg, or the dissolving of the sugar in the coffee are all irreversible; whereas the bouncing of a small number of particles off one another would be considered reversible, as would various carefully controlled situations in which energy is not lost into heat. Basically, the term 'irreversible' just refers to the fact that it has not been possible to keep track of, nor to control, all the relevant details of the individual particle motions in the system. These uncontrolled motions are referred to as 'heat'. Thus, irreversibility seems to be merely a 'practical' matter. We cannot *in practice* unscramble an egg, though it is a perfectly allowable procedure according to the laws of mechanics. Does our concept of entropy depend upon what is practical and what is not?

Recall, from Chapter 5, that the physical concept of *energy*, as well as momentum and angular momentum, *can* be given precise mathematical definitions in terms of particle positions, velocities, masses, and forces. But how can we be expected to do as well for the concept of 'manifest disorder' that is needed for making the concept of entropy mathematically precise? Surely, what is 'manifest' to one observer may not be so to another. Would it not depend upon the precision with which each observer might be able to measure the system under examination? With better measuring instruments one observer might be able to get much more detailed information about the microscopic constituents of a system than would another observer. More of the 'hidden order' in the system might be manifest to one observer than to the other—and, accordingly, he would ascertain the entropy as being lower than would the other. It seems, also, that the various observers' aesthetic judgements might well get involved in what they deem to be 'order', rather than 'disorder'. We could imagine some artist taking the view that the collection of shattered glass fragments was far more beautifully ordered than was the hideously ugly glass that once stood on the edge of the table! Would entropy have actually been *reduced* in the judgement of such an artistically sensitive observer?

In view of these problems of subjectivity, it is remarkable that the concept of entropy is useful at all in precise scientific descriptions—which it certainly is! The reason for this utility is that the changes from order to disorder in a system, in terms of detailed particle positions and velocities, are utterly enormous, and (in almost all circumstances) will completely swamp any reasonable differences of viewpoint as to what is or is not 'manifest order' on the macroscopic scale. In particular, the artist's or scientist's judgement as to whether it is the assembled or shattered glass which is the more orderly arrangement is of almost no consequence whatever, with regard to its entropy measure. By far the main contribution to the entropy comes from the random particle motions which constitute the tiny raising of temperature, and dispersion of the water, as the glass and water hit the ground.

In order to be more precise about the entropy concept, let us return to the idea of *phase space* which was introduced in Chapter 5. Recall that the phase space of a system is a space, normally of enormous dimensions, each of whose points represents an entire physical state in all its minute detail. A *single* phase-space point provides all the position and momentum coordinates of all the individual particles which constitute the physical system in question. What we need, for the concept of entropy, is a way of grouping together all the states which look identical, from the point of view of their *manifest* (i.e. macroscopic) properties. Thus, we need to divide our phase space into a number of compartments (cf. Fig. 7.3), where the different points belonging to any particular compartment represent physical systems which, though different in the minute detail of their particle configurations and motions, are nevertheless deemed to be identical with regard to macroscopically observable features. From the point of view of what is manifest, all the points of a single compartment are to be considered as representing the *same* physical system. Such a division of the phase space into compartments is referred to as a *coarse-graining* of the phase space.

Now some of these compartments will turn out to be enormously more

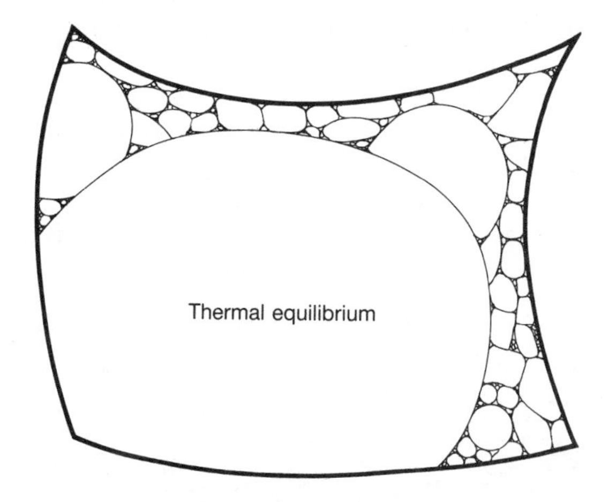

Fig. 7.3. A coarse-graining of phase space into regions corresponding to states that are macroscopically indistinguishable from one another. The *entropy* is proportional to the logarithm of the phase-space volume.

huge than others. For example, consider the phase space of a gas in a box. Most of the phase space will correspond to states in which the gas is very uniformly distributed in the box, with the particles moving around in a characteristic way that provides a uniform termperature and pressure. This characteristic type of motion is, in a sense, the most 'random' that is possible, and it is referred to as a *Maxwellian distribution*—after that same James Clerk Maxwell whom we have encountered before. When the gas is in such a random state it is said to be in *thermal equilibrium.* There is an absolutely vast volume of points of phase space corresponding to thermal equilibrium; the points of this volume describe all the different detailed arrangements of positions and velocities of individual particles which are consistent with thermal equilibrium. This vast volume is one of our compartments in phase space—easily the largest of all, and it occupies almost the entire phase space! Let us consider another possible state of the gas, say where all the gas is tucked up in one corner of the box. Again there will be many different individual detailed states, all of which describe the gas being tucked up in the same way in the corner of the box. All these states are macroscopically indistinguishable from one another, and the points of phase space that represent them constitute another compartment of the phase space. However, the volume of this compartment turns out to be far tinier than that of the states representing thermal equilibrium—by a factor of about $10^{10^{25}}$, if we take the box to be a metre cube, containing air at ordinary atmospheric pressure and temperature, when in equilibrium, and if we take the region in the corner to be a centimetre cube!

To begin to appreciate this kind of discrepancy between phase-space volumes, imagine a simplified situation in which a number of balls are to be distributed amongst several cells. Suppose that each cell is either empty or contains a single ball. The balls are to represent gas molecules and the cells, the different positions in the box that the molecules might occupy. Let us single out a small subset of the cells as *special*; these are to represent the gas molecule positions corresponding to the region in the corner of the box. Suppose, for definiteness, that exactly one-tenth of the cells are special—say there are n special cells and $9n$ non-special ones (see Fig. 7.4.) We wish to

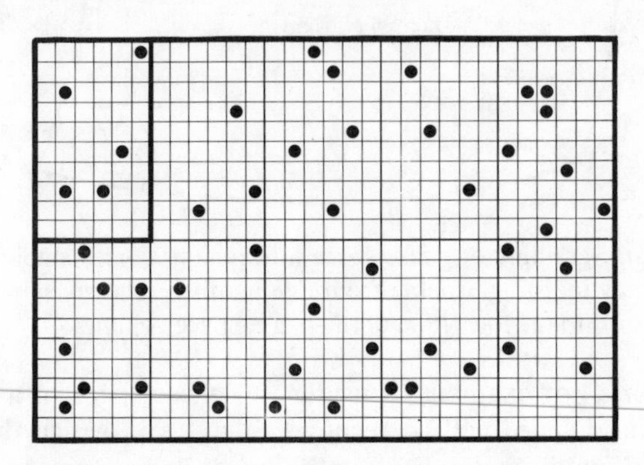

Fig. 7.4. A model for a gas in a box: a number of tiny balls are distributed among a much larger number of cells. One-tenth of the cells are labelled as *special*. These are the ones marked off in the upper left-hand corner.

distribute m balls among the cells at random and find the chance that all of them lie in the special cells. If there is just one ball and ten cells (so we have one special cell) this chance is clearly one-tenth. The same holds if there is one ball and any number $10n$ of cell (so n special cells). Thus, for a 'gas' with just *one* atom, the special compartment, corresponding to the gas 'tucked into the corner', would have a volume of just one *tenth* of the entire volume of the 'phase space'. But if we increase the number of balls, the chance that they *all* find their way into the special cells will decrease very considerably. For *two* balls, say with twenty cells* (two of which are special) ($m = 2$, $n = 2$), the chance is 1/190, or with one-hundred cells (with ten special) ($m = 2$, $n = 10$) it is 1/110; with a very large number of cells it becomes 1/100. Thus, the volume of the special compartment for a *two*-atom 'gas' is just one *hundredth* of the entire volume of the 'phase space'. For *three* balls and thirty cells ($m = 3$, $n = 3$), it is 1/4060; and with a very large number of cells it becomes 1/1000—so for a *three*-atom 'gas', the volume of the special compartment is now one *thousandth* of the volume of the 'phase space'. For four balls with a

* For general n, m the chance is ${}^{10n}C_m \div {}^{n}C_m = (10n)!(n-m)!/n!(10n-m)!$

very large number of cells, the chance becomes 1/10000. For five balls and a very large number of cells, the chance becomes 1/100000, and so on. For m balls and a very large number of cells, the chance becomes $1/10^m$, so for an m-atom 'gas', the volume of the special region is $1/10^m$ of that of the 'phase space'. (This is still true if 'momentum' is included.)

We can apply this to the situation considered above, of an actual gas in a box, but now, instead of being only one-tenth of the total, the special region occupies only one millionth (i.e. 1/1000000) of this total (i.e. one cubic centimetre in one cubic metre). This means that instead of the chance being $1/10^m$, it is now $1/(1000000)^m$, i.e. $1/10^{3m}$. For ordinary air, there would be about 10^{25} molecules in our box as a whole, so we take $m = 10^{25}$. Thus, the special compartment of phase space, representing the situation where all the gas lies tucked up in the corner, has a volume of only

$$1/10^{60\,000\,000\,000\,000\,000\,000\,000\,000}$$

of that of the entire phase space!

The *entropy* of a state is a measure of the *volume* V of the compartment containing the phase-space point which represents the state. In view of the enormous discrepancies between these volumes, as noted above, it is perhaps as well that the entropy is taken not to be proportional to that volume, but to the *logarithm* of the volume:

$$\text{entropy} = k \log V.$$

Taking a logarithm helps to make these numbers look more reasonable. The logarithm* of 10000000, for example, is only about 16. The quantity k is a constant, called *Boltzmann's constant.* Its value is about 10^{-23} Joules per degree Kelvin. The essential reason for taking a logarithm here is to make the entropy an *additive* quantity, for independent systems. Thus, for two completely independent physical systems, the total entropy of the two systems combined will be the *sum* of the entropies of each system separately. (This is a consequence of the basic algebraic property of the logarithm function: $\log AB = \log A + \log B$. If the two systems belong to compartment volumes A and B, in their respective phase spaces, then the phase space volume for the two together will be their product AB, because each possibility for one system has to be separately counted with each possibility for the other; hence the entropy of the combined system will indeed be the sum of the two individual entropies.)

The enormous discrepancies between the sizes of the compartments in phase space will look more reasonable in terms of entropy. The entropy of

* The logarithm used here is a *natural* logarithm, i.e. taken to the base $e = 2.7182818285\ldots$ rather than 10, but this distinction is quite unimportant. The natural logarithm, $x = \log n$, of a number n is the power to which we must raise e in order to obtain n, i.e. the solution of $e^x = n$ (see footnote on p. 88).

our cubic metre-sized box of gas, as described above, turns out to be only about $1400 \mathrm{J K}^{-1}$ ($= 14k\ 10^{-25}$) larger than the entropy of the gas concentrated in the cubic centimetre-sized 'special' region (since $\log_e (10^{6\times 10^{25}})$ is about 14×10^{25}).

In order to give the *actual* entropy values for these compartments we should have to worry a little about the question of the units that are chosen (metres, Joules, kilograms, degrees Kelvin, etc.). That would be out of place here, and, in fact, for the utterly stupendous entropy values that I shall be giving shortly it makes essentially no difference at all what units are in fact chosen. However, for definiteness (for the experts), let me say that I shall be taking *natural* units, as are provided by the rules of quantum mechanics, and for which Boltzmann's constant turns out to be *unity*:

$$k = 1.$$

The second law in action

Suppose, now, that we start a system off in some very special situation, such as with the gas all in one corner of the box. The next moment, the gas will spread, and it will rapidly occupy larger and larger volumes. After a while it will settle into thermal equilibrium. What is our picture of this in terms of phase space? At each stage, the complete detailed state of positions and motions of all the particles of the gas would be described by a single point in phase space. As the gas evolves, this point wanders about in the phase space, its precise wanderings describing the entire history of all the particles of the gas. The point starts off in a very tiny region—the region which represents the collection of possible initial states for which the gas is all in one particular corner of the box. As the gas begins to spread, our moving point will enter a considerably larger phase-space volume, corresponding to the states where the gas is spread out a little through the box in this way. The phase-space point keeps entering larger and larger volumes as the gas spreads further, where each new volume totally dwarfs all the ones that the point has been in before—by absolutely stupendous factors! (see Fig. 7.5). In each case, once the point has entered the larger volume, there is (in effect) no chance that it can find any of the previous smaller ones. Finally it loses itself in the hugest volume of all in the phase space—that corresponding to thermal equilibrium. This volume practically occupies the entire phase space. One can be virtually assured that our phase-space point, in its effectively random wanderings, will not find any of the smaller volumes in any plausible time. Once the state of thermal equilibrium has been reached then, to all intents and purposes, the state stays there for good. Thus we see that the entropy of the system, which simply provides a logarithmic measure of the volume of the appropriate

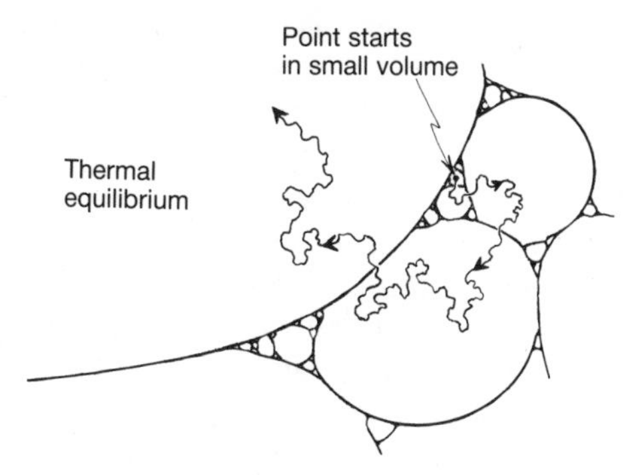

Fig. 7.5. The second law of thermodynamics in action: as time evolves, the phase-space point enters compartments of larger and larger volume. Consequently the entropy continually increases.

compartment in phase space, will have this inexorable tendency to increase* as time progresses.

We would seem now to have an *explanation* for the second law! For we may suppose that our phase-space point does not move about in any particularly contrived way, and if it starts off in a tiny phase-space volume, corresponding to a *small* entropy, then, as time progresses, it will indeed be overwhelmingly likely to move into successively larger and larger phase-space volumes, corresponding to gradually increasing entropy values.

However, there is something a little odd about what we seem to have deduced by this argument. We seem to have deduced a *time-asymmetric* conclusion. The entropy *increases* in the *positive* direction in time, and therefore it must *decrease* in the *reversed* time-direction. Where has this time-asymmetry come from? We have certainly not introduced any time-asymmetric physical laws. The time-asymmetry comes merely from the fact that the system has been *started off* in a very special (i.e. low-entropy) state; and having so started the system, we have watched it evolve in the *future* direction, and found that the entropy increases. This entropy increase is indeed in accordance with the behaviour of systems in our actual universe. But we could equally well have applied this very same argument in the reverse direction in time. We could again specify that the system is in a low-entropy state at some given time, but now ask what is the most likely sequence of states which *preceded* this.

* Of course it is not true that our phase-space point will *never* find one of the smaller compartments again. If we wait for long enough, these comparatively tiny volumes will eventually be re-entered. (This would be referred to as a *Poincaré recurrence*.) However, the timescales would be ridiculously long in most circumstances, e.g. about $10^{10^{26}}$ years in the case of the gas all reaching the centimetre cube in the corner of the box. This is far, far longer than the time that the universe has been in existence! I shall ignore this possibility in the discussion to follow, as being not really relevant to the problem at issue.

Let us try the argument in this reverse way. As before, take the low-entropy state to be the gas all in one corner of the box. Our phase-space point is now in the same very tiny region that we started it off in before. But now let us try to trace its *backwards* history. If we imagine the phase-space point jiggling around in a fairly random way as before, then we expect, as we trace the motion backwards in time, that it would soon reach the same considerably larger phase-space volume as before, corresponding to the gas being spread a little through the box but not in thermal equilbrium, and then the larger and larger volumes, each new volume totally dwarfing the previous ones, and back further in time we would find it in the hugest volume of all, representing thermal equilibrium. *Now* we seem to have deduced that given that the gas was, at one time, all tucked up in the corner of the box, then the most likely way that it could have got there was that it started in thermal equilibrium, then began to concentrate itself over at one end of the box, more and more, and finally collected itself in the small specified volume in the corner. All the time, the entropy would have to be *decreasing*: it would start from the high equilibrium value, and then gradually decrease until it reached the very low value corresponding to the gas tucked in the small corner of the box!

Of course, this is nothing like what actually happens in our universe! Entropy does not decrease in this way; it *increases*. If the gas were known to be all tucked in one corner of the box at some particular time, then a much more likely situation to *precede* this might have been the gas being held firmly in the corner by a partition, which was rapidly removed. Or perhaps the gas had been held there in a frozen or liquid state and was rapidly heated to become gaseous. For any of these alternative possibilities, the entropy was even *lower* in the previous states. The second law did indeed hold sway, and the entropy was increasing all the time—i.e. in the *reverse* time-direction it was actually *decreasing*. *Now* we see that our argument has given us completely the wrong answer! It has told us that the most likely way to get the gas into the corner of the box would be to start from thermal equilibrium and then, with entropy steadily reducing, the gas would collect in the corner; whereas in fact, in our actual world, this is an exceedingly *un*likely way for it to happen. In our world, the gas would start from an even *less* likely (i.e. lower-entropy) state, and the entropy would steadily *in*crease to the value it subsequently has, for the gas tucked in the corner.

Our argument seemed to be fine when applied in the future direction, although not in the past direction. For the *future* direction, we correctly anticipate that whenever the gas starts off in the corner the most likely thing to happen in the future is that thermal equilibrium *will* be reached and *not* that a partition will suddenly appear, or that the gas will suddenly freeze or become liquid. Such bizarre alternatives would represent just the kind of entropy-lowering behaviour in the future direction that our phase-space argument seems correctly to rule out. But in the *past* direction, such 'bizarre' alternatives are indeed what would be likely to happen—and they do not seem to us to be at all bizarre. Our phase-space argument gave us completely the wrong answer when we tried to apply it in the reverse direction of time!

Clearly this casts doubt on the original argument. We have *not* deduced the second law. What that argument actually showed was that for a given low-entropy state (say for a gas tucked in a corner of a box), then, *in the absence of any other factors constraining the system*, the entropy would be expected to increase in *both* directions in time away from the given state (see Fig. 7.6). The argument has not worked in the past direction in time precisely because there *were* such factors. There was indeed something constraining the system in the past. Something *forced* the entropy to be low in the past. The tendency towards high entropy in the future is no surprise. The high-entropy states are, in a sense, the 'natural' states, which do not need further explanation. But the low-entropy states in the past are a puzzle. What constrained the entropy of our world to be so low in the past? The common presence of states in which the entropy is absurdly low is an amazing fact of the actual universe that we inhabit—though such states are so commonplace and familiar to us that we do not normally tend to regard them as amazing. We ourselves are configurations of ridiculously tiny entropy! The above argument shows that we should not be surprised if, *given* a low-entropy state, the entropy turns out to be higher at a later time. What *should* surprise us is that entropy gets more and more ridiculously tiny the farther and farther that we examine it in the past!

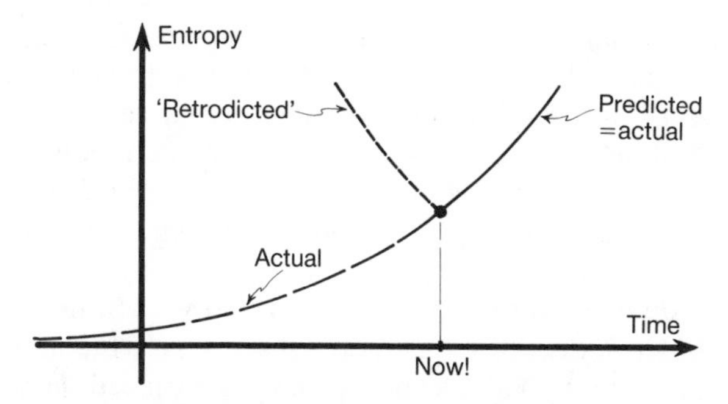

Fig. 7.6. If we use the argument depicted in Fig. 7.5 in the reverse direction in time we 'retrodict' that the entropy should also increase into the *past*, from its value now. This is in gross contradiction with observation.

The origin of low entropy in the universe

We shall try to understand where this 'amazing' low entropy comes from in the actual world that we inhabit. Let us start with ourselves. If we can understand where our own low entropy came from, then we should be able to see where the low entropy in the gas held by the partition came from—or in the water glass on the table, or in the egg held above the frying pan, or the lump of sugar held over the coffee cup. In each case a person or collection of people (or perhaps a chicken!) was directly or indirectly responsible. It was, to a large extent, some small part of the low entropy in ourselves which was actually made use of in setting up these other low-entropy states. Additional factors might have been involved. Perhaps a vacuum pump was used to suck the gas to the corner of the box behind the partition. If the pump was not operated manually, then it may have been that some 'fossil fuel' (e.g. oil) was burnt in order to provide the necessary low-entropy energy for its operation. Perhaps the pump was electrically operated, and relied, to some extent, on the low-entropy energy stored in the uranium fuel of a nuclear power station. I shall return to these other low-entropy sources later, but let us first just consider the low entropy in ourselves.

Where indeed *does* our own low entropy come from? The organization in our bodies comes from the food that we eat and the oxygen that we breathe. Often one hears it stated that we obtain *energy* from our intake of food and oxygen, but there is a clear sense in which that is not really correct. It is true that the food we consume does combine with this oxygen that we take into our bodies, and that this provides us with energy. But, for the most part, this energy leaves our bodies again, mainly in the form of heat. Since energy is conserved, and since the actual energy content of our bodies remains more-or-less constant throughout our adult lives, there is no need simply to *add* to the energy content of our bodies. We do not *need* more energy within ourselves than we already have. In fact we do add to our energy content when we put on weight—but that is usually not considered desirable! Also, as we grow up from childhood we increase our energy content considerably as we build up our bodies; that is not what I am concerned about here. The question is how we keep ourselves *alive* throughout our normal (mainly adult) lives. For that, we do *not* need to add to our energy content.

However, we do need to replace the energy that we continually lose in the form of heat. Indeed, the more 'energetic' that we are, the more energy we actually lose in this form. All this energy must be replaced. Heat is the most *disordered* form of energy that there is, i.e. it is the highest-entropy form of energy. We take in energy in a *low*-entropy form (food and oxygen) and we discard it in a *high*-entropy form (heat, carbon dioxide, excreta). We do not need to gain energy from our environment, since energy is *conserved*. But we are continually fighting against the second law of thermodynamics. Entropy is *not* conserved; it is *increasing* all the time. To keep ourselves alive, we need to keep lowering the entropy that is within ourselves. We do this by feeding on the low-entropy combination of food and atmospheric oxygen, combining them within our bodies, and discarding the energy, that we would otherwise have gained, in a high-entropy form. In this way, we can keep the entropy in our bodies from rising, and we can maintain (and even increase) our internal organization. (See Schrödinger 1967.)

Where does this supply of low entropy come from? If the food that we are eating happens to be meat (or mushrooms!), then *it*, like us, would have relied on a further external low-entropy source to provide and maintain its low-entropy structure. That merely pushes the problem of the origin of the external low entropy to somewhere else. So let us suppose that we (or the animal or mushroom) is consuming a *plant*. We must all be supremely grateful to the green plants—either directly or indirectly—for their cleverness: taking atmospheric carbon dioxide, separating the oxygen from the carbon, and using that carbon to build up their own substance. This procedure, *photosynthesis*, effects a large reduction in the entropy. We ourselves make use of this low-entropy separation by, in effect, simply recombining the oxygen and carbon within our own bodies. How is it that the green plants are able to achieve this entropy-reducing magic? They do it by making use of *sunlight*. The light from the sun brings energy to the earth in a comparatively *low*-entropy form, namely in the photons of visible light. The earth, including its inhabitants, does not *retain* this energy, but (after some while) re-radiates it all back into space. However, the re-radiated energy is in a *high*-entropy form, namely what is called 'radiant heat'—which means infra-red photons. Contrary to a common impression, the earth (together with inhabitants) does *not* gain energy from the sun! What the earth does is to take energy in a low-entropy form, and then spew it *all* back again into space, but in a high-entropy form (Fig. 7.7). What the sun has done for us is to supply us with a huge source of low entropy. We (via the plants' cleverness), make use of this, ultimately extracting some tiny part of this low entropy and converting it into the remarkable and intricately organized structures that are ourselves.

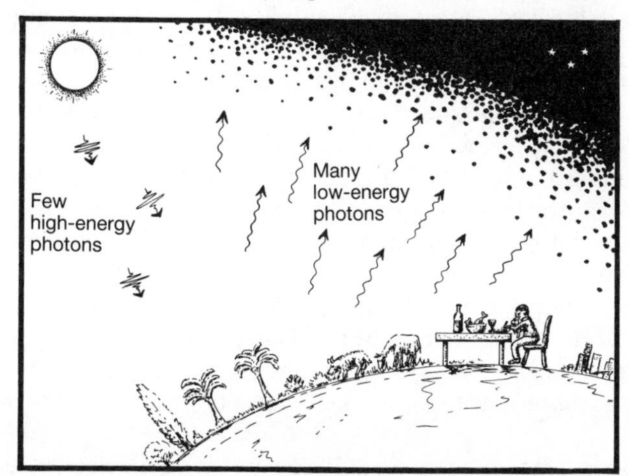

Fig. 7.7 How we make use of the fact that the sun is a hot spot within the darkness of space.

Let us see, from the overall point of view with regard to the sun and the earth, what has happened to the energy and entropy. The sun emits energy in the form of visible-light photons. Some of these are absorbed by the earth, their energy being re-radiated in the form of infra-red photons. Now, the crucial difference between the visible-light and the infra-red photons is that the former have a higher frequency and therefore have individually a higher energy than do the latter. (Recall Planck's formula $E = h\nu$, given on p. 230. This tells us that the higher the frequency of a photon, the higher will be its energy.) Since the visible-light photons each have higher energy than do each of the infra-red ones, there must be *fewer* visible-light photons reaching the earth than there are infra-red ones leaving the earth, so that the *energy* coming in to the earth balances that leaving it. The energy that the earth spews back into space is spread over many more degrees of freedom than is the energy that it receives from the sun. Since there are so many more degrees of freedom involved when the energy is sent back out again, the volume of phase space is much larger, and the *entropy* has gone up enormously. The green plants, by taking in energy in a low-entropy form (comparatively *few* visible-light photons) and re-radiating it in a high-entropy form (comparatively *many* infra-red photons) have been able to feed on this low entropy and provide us with this oxygen–carbon separation that we need.

All this is made possible by the fact that the sun is a *hot-spot* in the sky! The sky is in a state of temperature imbalance: one small region of it, namely that occupied by the sun, is at a very much higher temperature than the rest. This fact provides us with the required powerful low-entropy source. The earth gets energy from that hot-spot in a low entropy form (few photons), and it re-radiates to the cold regions in a high-entropy form (many photons).

Why is the sun such a hot-spot? How has it been able to achieve such a temperature imbalance, and thereby provide a state of low entropy? The answer is that it has formed by gravitational contraction from a previously uniform distribution of gas (mainly hydrogen). As it contracted, in the early stages of its formation, the sun heated up. It would have continued to contract and to heat up even further except that, when its temperature and pressure reached a certain point, it found another source of energy, besides that of gravitational contraction, namely *thermonuclear reactions*: the fusion of hydrogen nuclei into helium nuclei to provide energy. Without thermonuclear reactions, the sun would have got very much *hotter* and tinier than it is now, until finally it would have died out. Thermonuclear reactions have actually kept the sun from getting *too* hot, by stopping it from contracting further, and have stabilized the sun at a temperature that is suitable for ourselves, enabling it to continue shining for far longer than it could otherwise have done.

It is important to realize that although thermonuclear reactions are undoubtedly highly significant in determining the nature and the amount of the radiated energy from the sun, it is *gravitation* that is the crucial consideration. (In fact the potentiality for thermonuclear reactions *does* give a highly significant contribution to the lowness of the sun's entropy, but the questions raised by the entropy of fusion are delicate, and a full discussion of them would serve merely to complicate the argument without affecting the ultimate conclusion.)[2] Without gravity, the sun would not even exist! The sun would still shine without thermonuclear reactions—though not in a way suitable for us—but there could be *no* shining sun at all, without the gravity that is needed in order to hold its material together and to provide the temperatures and pressures that are needed. Without gravity, all we should have would be a cold, diffuse gas in place of the sun and there would be *no* hot-spot in the sky!

I have not yet discussed the source of the low entropy in the 'fossil fuels' in the earth; but the considerations are basically the same. According to the conventional theory, all the oil (and natural gas) in the earth comes from prehistoric plant life. Again it is the plants which are found to have been responsible for this source of low entropy. The prehistoric plants got their low entropy from the sun—so it is the gravitational action in forming the sun out of a diffuse gas that we must again turn to. There is an interesting alternative 'maverick' theory of the origin of the oil in the earth, due to Thomas Gold, which disputes this conventional view, suggesting that there is much more oil in the earth than can have arisen from prehistoric plants. Gold believes that the oil was trapped inside the earth when the earth formed, and it has been continually oozing outwards into subterranian pockets ever since.[3] According to Gold's theory, the oil would still have been synthesized by sunlight, however, though out in space, before the earth was formed. Again it is the gravitationally formed sun that would be responsible.

What about the low-entropy nuclear energy in the uranium-235 isotope that is used in nuclear power stations? This did not come originally from the sun (though it may well have passed through the sun at some stage) but from some other star, which exploded many thousands of millions of years ago in a supernova explosion! Actually, the material was collected from *many* such exploding stars. The material from these stars was spewed into space by the explosion, and some of it eventually collected together (through the agency of the sun) to provide the heavy elements in the earth, including all its uranium-235. Each nucleus, with its low-entropy store of energy, came from the violent nuclear processes which took place in some supernova explosion. The explosion occurred as the aftermath of the gravitational collapse[4] of a star that had been too massive to be able to hold itself apart by thermal pressure forces. As the result of that collapse and subsequent explosion, a small core remained—probably in the form of what is known as a *neutron star* (more about this later!). The star would have originally contracted gravitationally from a diffuse cloud of gas, and much of this original material, including our uranium-235, would have been thrown back into space. However, there was a huge gain in entropy due to gravitational contraction, because of the neutron-star core that remained. Again it was *gravity* that was ultimately responsible—this time causing the (finally violent) condensation of diffuse gas into a neutron star.

We seem to have come to the conclusion that all the remarkable lowness of entropy that we find about us—and which provides this most puzzling aspect of the second law—must be attributed to the fact that vast amounts of entropy can be gained through the gravitational contraction of diffuse gas into stars. Where has all this diffuse gas come from? It is the fact that this gas starts off as *diffuse* that provides us with an enormous store of low entropy. We are still living off this store of low entropy, and will continue to do so for a long while to come. It is the potential that this gas has for gravitationally clumping which has given us the second law. Moreover, it is not just the second law that this gravitational clumping has produced, but something very much more precise and detailed than the simple statement: 'the entropy of the world started off very low'. The entropy might have been given to us as 'low' in many *other* different ways, i.e. there might have been great 'manifest order' in the early universe, but quite different from that 'order' we have actually been presented with. (Imagine that the early universe had been a regular dodecahedron—as might have appealed to Plato—or some other improbable geometrical shape. This would indeed be 'manifest order', but not of the kind that we expect to find in the *actual* early universe!) We must understand where all this diffuse gas has come from—and for that we shall need to turn to our cosmological theories.

Cosmology and the big bang

As far as we can tell from using our most powerful telescopes—both optical and radio—the universe, on a very large scale, appears to be rather uniform; but, more remarkably, it is *expanding*. The farther away that we look, the more rapidly the distant galaxies (and even more distant quasars) appear to be receding from us. It is as though the universe itself was created in one gigantic explosion—an event referred to as the *big bang*, which occurred some ten thousand million years ago.* Impressive further support for this uniformity, and for the actual existence of the big bang, comes from what is known as the *black-body background radiation*. This is thermal radiation—photons moving around randomly, without discernible source—corresponding

* At the present time there is a dispute still raging as to the value of this figure, which ranges between about 6×10^9 and 1.5×10^{10} years. These figures are considerably larger than the figure of 10^9 years that originally seemed appropriate after Edwin Hubble's initial observations in around 1930 showed that the universe is expanding.

to a temperature of about 2.7° absolute (2.7 K), i.e. −270.3° Celsius, or 454.5° below zero Farenheit. This may seem like a *very* cold temperature—as indeed it is!—but it appears to be the leftover of the flash of the big bang itself! Because the universe has expanded by such a huge factor since the time of the big bang, this initial fireball has dispersed by an absolutely enormous factor. The temperatures in the big bang far exceeded any temperatures that can occur at the present time, but owing to this expansion, that temperature has cooled to the tiny value that the black-body background has now. The presence of this background was *predicted* by the Russian–American physicist and astronomer George Gamow in 1948 on the basis of the now-standard big-bang picture. It was first observed (accidentally) by Penzias and Wilson in 1965.

I should address a question that often puzzles people. If the distant galaxies in the universe are all receding from us, does that not mean that we ourselves are occupying some very special central location? No it does not! The same recession of distant galaxies would be seen *wherever* we might be located in the universe. The expansion is uniform on a large scale, and no particular location is preferred over any other. This is often pictured in terms of a balloon being blown up (Fig. 7.8). Suppose that there are spots on the balloon to represent the different galaxies, and take the two-dimensional surface of the balloon itself to represent the entire three-dimensional spatial universe. It is clear that from *each* point on the balloon, *all* the other points are receding from it. No point on the balloon is to be preferred, in this respect, over any other point. Likewise, as seen from the vantage point of each galaxy in the universe, all other galaxies appear to be receding from it, equally in all directions.

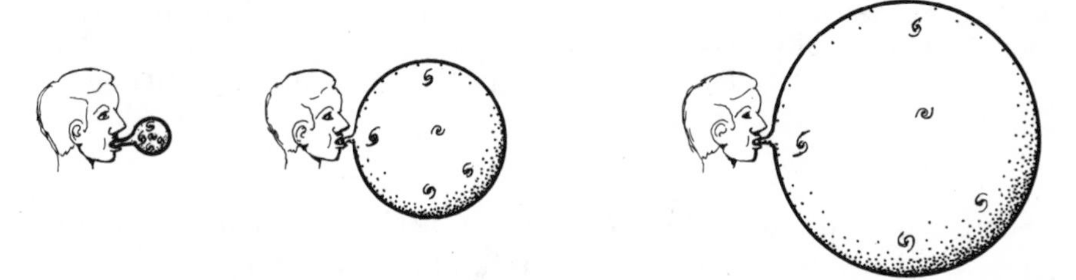

Fig. 7.8. The expansion of the universe can be likened to the surface of a balloon being blown up. The galaxies all recede from one another.

This expanding balloon provides quite a good picture of one of the three standard so-called *Friedmann–Robertson–Walker* (FRW) models of the universe—namely the spatially closed *positively curved* FRW-model. In the other two FRW-models (zero or negative curvature), the universe expands in the same sort of way, but instead of having a spatially finite universe, as the surface of the balloon indicates, we have an *infinite* universe with an infinite number of galaxies.

In the easier to comprehend of these two infinite models, the spatial geometry is *Euclidean*, i.e. it has *zero* curvature. Think of an ordinary flat plane as representing the entire spatial universe, where there are points marked on the plane to represent galaxies. As the universe evolves with time, these galaxies recede from one another in a uniform way. Let us think of this in *space–time* terms. Accordingly, we have a different Euclidean plane for each 'moment of time', and all these planes are imagined as being stacked one above the other, so that we have a picture of the entire space–time all at once (Fig. 7.9). The galaxies are now represented as *curves*—the *world-lines* of the galaxies' histories—and these curves move away from each other into the future direction. Again no particular galaxy world-line is preferred.

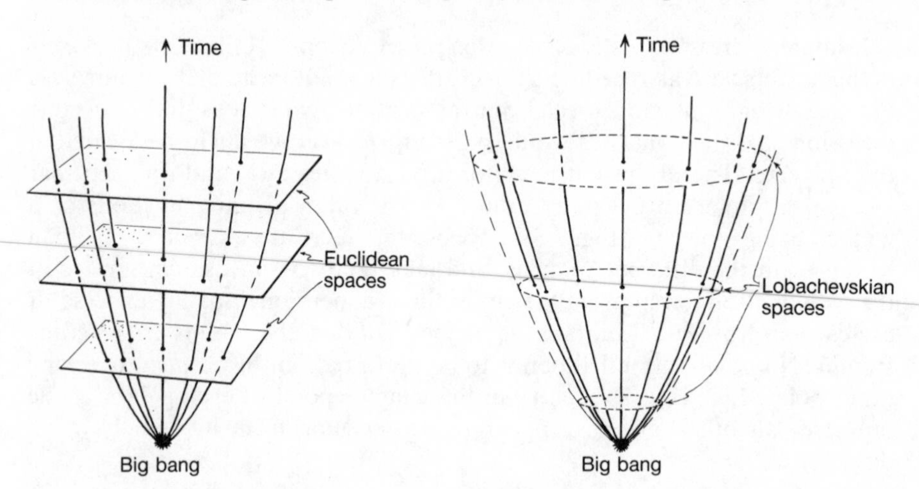

Fig. 7.9. Space–time picture of an expanding universe with Euclidean spatial sections (two space dimensions depicted).

Fig. 7.10. Space–time picture of an expanding universe with Lobachevskian spatial sections (two space dimensions depicted).

For the one remaining FRW-model, the *negative*-curvature model, the spatial geometry is the *non*-Euclidean *Lobachevskian* geometry that was described in Chapter 5 and illustrated by the Escher print depicted in Figure 5.2 (p. 157). For the space–time description, we need one of these Lobachevsky spaces for each 'instant of time', and we stack these all on top of one another to give a picture of the entire space–time (Fig. 7.10).[5] Again the galaxy world-lines are curves moving away from each other in the future direction, and no galaxy is preferred.

Of course, in all these descriptions we have suppressed one of the three spatial dimensions (as we did in Chapter 5, cf. p. 194) to give a more visualizable three-dimensional space–time than would be necessary for the complete four-dimensional space–time picture. Even so, it is hard to visualize the positive-curvature space–time without discarding yet another spatial dimension! Let us do so, and represent the positively curved closed spatial universe by a *circle* (one-dimensional), rather than the sphere (two-dimensional) which had been the balloon's surface. As the universe expands, this circle grows in size, and we can represent the space–time by stacking these circles (one circle for each 'moment of time') above one another to obtain a kind of curved cone (Fig. 7.11(a)). Now, it follows from the equations of Einstein's general relativity that this positively closed universe cannot continue to expand forever. After it reaches a stage of maximum expansion, it collapses back in on itself, finally to reach zero size again in a kind of big bang in reverse (Fig. 7.11(b)). This time-reversed big bang is sometimes referred to as the *big crunch*. The negatively curved and zero-curved (infinite) universe FRW-models do not recollapse in this way. Instead of reaching a big crunch, they continue to expand forever.

At least this is true for *standard* general relativity in which the so-called *cosmological constant* is taken to be zero. With suitable non-zero values of this cosmological constant, it is possible to have spatially infinite universe models which recollapse to a big crunch, or finite positively curved models

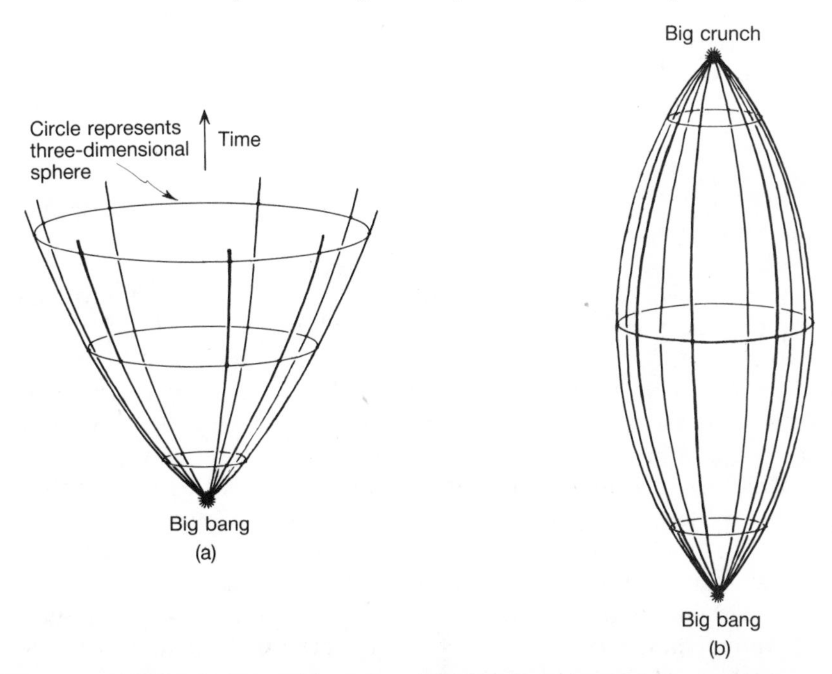

Fig. 7.11. (a) Space–time picture of an expanding universe with spherical spatial sections (only one space dimension depicted). (b) Eventually this universe recollapses to a final big crunch.

which expand indefinitely. The presence of a non-zero cosmological constant would complicate the discussion slightly, but not in any significant way for our purposes. For simplicity, I shall take the cosmological constant to be zero.* At the time of writing, the cosmological constant is known observationally to be very small, and the data are consistent with its being zero. (For further information on cosmological models, see Rindler 1977.)

Unfortunately the data are not yet good enough to point clearly to one or other of the proposed cosmological models (nor to determine whether or not the presence of a tiny cosmological constant might have a significant overall effect). On the face of it, it would appear that the data indicate that the universe is spatially negatively curved (with Lobachevsky geometry on the large scale) and that it will continue to expand indefinitely. This is largely based on observations of the amount of actual matter that seems to be present in visible form. However, there may be huge amounts of invisible matter spread throughout space, in which case the universe could be positively curved, and it could finally recollapse to a big crunch—though only on a time-scale far larger than the 10^{10} years, or so, for which the universe has been in existence. For this recollapse to be possible there would have to be some thirty times as much matter permeating space in this invisible form—the postulated so-called 'dark matter'—than can be directly discerned through telescopes. There is some good indirect evidence that a substantial amount of dark matter is indeed present, but whether there is *enough* of it 'to close the universe' (or make it spatially flat)—and recollapse it—is very much an open question.

The primordial fireball

Let us return to our quest for the origin of the second law of thermodynamics. We had tracked this down to the presence of diffuse gas from which the stars have condensed. What is this gas? Where did it come from? Mainly it is hydrogen, but there is also about 23 per cent helium (by mass) and small quantities of other materials. According to standard theory, this gas was just spewed out as a result of the explosion which created the universe: the big bang. However, it is important that we do not think of this as an ordinary explosion of the familiar kind, where material is ejected from some central point into a pre-existing space. Here, the space itself is *created* by the explosion, and there is, or was, no central point! Perhaps the situation is easiest to visualize in the positively curved case. Consider Fig. 7.11 again, or the blown-up balloon of Fig. 7.8. There is no 'pre-existing empty space' into which the material produced by the explosion pours. The space itself, i.e. the

* Einstein introduced the cosmological constant in 1917, but had retracted it again by 1931, referring to its earlier introduction as his 'greatest mistake'!

'balloon's surface', is brought into being by the explosion. We must appreciate that it is only for visualization purposes that our pictures, in the positvely curved case, have used an 'ambient space'—the Euclidean space in which the balloon sits, or the three-dimensional space in which the space-time of Fig. 7.11 is depicted—these ambient spaces are not to be taken as having physical reality. The space inside or outside the balloon is there only to help us visualize the balloon's surface. It is the balloon's surface *alone* which is to represent the physical space of the universe. We now see that there is no central point from which the material from the big bang emanates. The 'point' that appears to be at the centre of the balloon is not part of the universe, but is merely an aid to our visualization of the model. The material blasted out in the big bang is simply spread out uniformly over the *entire* spatial universe!

The situation is the same (though perhaps a little harder to visualize) for the other two standard models. The material was never concentrated at any one point in space. It uniformly filled the *whole* of space—right from the very start!

This picture underlies the theory of the *hot big bang* referred to as the *standard model.* In this theory, the universe, moments after its creation, was in an extremely hot thermal state—the *primordial fireball.* Detailed calculations have been performed concerning the nature and proportions of the initial constituents of this fireball, and how these constituents changed as the fireball (which was the entire universe) expanded and cooled. It may seem remarkable that calculations can be reliably performed for describing a state of the universe which was so different from that of our present era. However, the physics on which these calculations are based is not controversial, so long as we do not ask for what happened *before* about the first 10^{-4} of a second after creation! From that moment, one ten-thousandth of a second after creation, until about three minutes later, the behaviour has been worked out in great detail (cf. Weinberg 1977)—and, remarkably, our well-established physical theories, derived from experimental knowledge of a universe now in a very different state, are quite adequate for this.[6] The final implications of these calculations are that, spreading out in a uniform way throughout the entire universe, would be many photons (i.e. light), electrons and protons (the two constituents of hydrogen), some α-particles (the nuclei of helium), still smaller numbers of deuterons (the nuclei of deuterium, a heavy isotope of hydrogen), and traces of other kinds of nuclei—with perhaps also large numbers of 'invisible' particles such as neutrinos, that would barely make their presence known. The *material* constituents (mainly protons and electrons) would combine together to produce the gas from which stars are formed (largely hydrogen) at about 10^8 years after the big bang.

Stars would not be formed at once, however. After some further expansion and cooling of the gas, concentrations of this gas in certain regions would be needed in order that local gravitational effects can begin to overcome the overall expansion. Here we run into the unresolved and controversial issue of how galaxies are actually formed, and of what initial irregularities must be present for galaxy formation to be possible. I do not wish to enter into the dispute here. Let us just accept that some kind of irregularities in the initial gas distribution must have been present, and that somehow the right kind of gravitational clumping was initiated so that galaxies could form, with their hundreds of thousands of millions of constituent stars!

We have found where the diffuse gas has come from. It came from the very fireball which was the big bang itself. The fact that this gas has been distributed remarkably uniformly throughout space is what has given us the second law—in the detailed form that this law has come to us—after the entropy-raising procedure of gravitational clumping has become available. How uniformly distributed *is* the material of the actual universe? We have noted that stars are collected together in galaxies. Galaxies, also, are collected together into clusters of galaxies; and clusters into so-called superclusters. There is even some evidence that these superclusters are collected into larger groupings referred to as supercluster complexes. It is important to note, however, that all this irregularity and clustering is 'small beer' by comparison with the impressive uniformity of the structure of the universe as a whole. The further back in time that it has been possible to see, and the larger the portion of the universe that it has been possible to survey, the more uniform the universe appears. The black-body background radiation provides the most impressive evidence for this. It tells us, in particular, that when the universe was a mere million years old, over a range that has now spread to some 10^{23} kilometres—a distance from us that would encompass some 10^{10} galaxies—the universe and all its material contents were *uniform* to one part in one hundred thousand (cf. Davies 1987). The universe, despite its violent origins, was indeed very uniform in its early stages.

Thus, it was the initial fireball that has spread this gas so uniformly throughout space. It is here that our search has led us.

Does the big bang explain the second law?

Has our search come to its end? Is the puzzling fact that the entropy in our universe started out so low—the fact which has given us the second law of thermodynamics—to be 'explained' just by the circumstance that the universe started with a big bang? A little thought suggests that there is something of a paradox involved with this idea. It cannot really be the answer. Recall that the primordial fireball was a *thermal* state—a hot gas in expanding thermal equilibrium. Recall, also, that the term 'thermal equilibrium' refers to a state of *maximum* entropy. (This was how we referred to the maximum entropy state of a gas in a box.) However, the second law demands that in its initial state, the entropy of our universe was at some sort of *minimum*, not a maximum!

What has gone wrong? One 'standard' answer would run roughly as follows:

> True, the fireball was effectively in thermal equilibrium at the beginning, but the universe at that time was very tiny. The fireball represented the state of maximum entropy that could be permitted for a universe of *that* tiny size, but the entropy so permitted would have been minute by comparison with that which is allowed for a universe of the size that we find it to be today. As the universe expanded, the permitted maximum entropy increased with the universe's size, but the actual entropy in the universe lagged well behind this permitted maximum. The second law arises because the actual entropy is always striving to catch up with this permitted maximum.

However, a little consideration tells us that this cannot be the correct explanation. If it were, then, in the case of a (spatially closed) universe model which ultimately recollapses to a big crunch, the argument would ultimately apply again—in the *reverse* direction in time. When the universe finally reaches a tiny size, there would again be a low ceiling on the possible entropy values. The same constraint which served to give us a low entropy in the very early stages of the expanding universe should apply again in the final stages of the contracting universe. It was a low-entropy constraint at 'the beginning of time' which gave us the second law, according to which the entropy of the universe is increasing with time. If this same low-entropy constraint were to apply at 'the end of time', then we should find that there would have to be gross conflict with the second law of thermodynamics!

Of course, it might well be the case that our *actual* universe never recollapses in this way. Perhaps we are living in a universe with zero overall spatial curvature (Euclidean case) or negative curvature (Lobachevsky case). Or perhaps we *are* living in a (positively curved) recollapsing universe, but the recollapse will occur at such a remote time that no violation of the second law would be discernible to us at our present epoch—despite the fact that, on this view, the *entire* entropy of the universe would eventually turn around and decrease to a tiny value—and the second law, as we understand it today, would be grossly violated.

In fact, there are very good reasons for doubting that there could be such a turn-around of entropy in a collapsing universe. Some of the most powerful of these reasons have to do with those mysterious objects known as *black holes*. In a black hole, we have a microcosm of a collapsing universe; so if the entropy were indeed to reverse in a collapsing universe, then observable gross violations of the second law ought also to occur in the neighbourhood of a black hole. However, there is every reason to believe that, with black holes, the second law powerfully holds sway. The theory of black holes will provide a vital input to our discussion of entropy, so it will be necessary for us to consider these strange objects in a little detail.

Black holes

Let us first consider what theory tells us will be the ultimate fate of our sun. The sun has been in existence for some five thousand million years. In another 5–6 thousand million years it will begin to expand in size, swelling inexorably outwards until its surface reaches to about the orbit of the earth. It will then have become a type of star known as a *red giant*. Many red giants are observed elsewhere in the heavens, two of the best known being Aldebaran in Taurus and Betelgeuse in Orion. All the time that its surface is expanding, there will be, at its very core, an exceptionally dense small concentration of matter, which steadily grows. This dense core will have the nature of a *white dwarf* star (Fig. 7.12).

White dwarf stars, when on their own, are actual stars whose material is concentrated to extremely high density, such that a ping-pong ball filled with their material would weigh several hundred tonnes! Such stars are observed in the heavens, in quite considerable numbers: perhaps some ten per cent of the luminous stars in our Milky Way galaxy are white dwarfs. The most famous white dwarf is the companion of Sirius, whose alarmingly high density had provided a great observational puzzle to astronomers in the early part of this century. Later, however, this same star provided a marvellous confirmation of physical theory (originally by R. H. Fowler, in around 1926)—according to which, some stars could indeed have such a large density, and would be held apart by 'electron degeneracy pressure', meaning that it is Pauli's quantum-mechanical exclusion principle (p. 278), as applied to electrons, that is preventing the star from collapsing gravitationally inwards.

Any red giant star will have a white dwarf at its core, and this core will be continually gathering material from the main body of the star. Eventually, the red giant will be completely consumed by this parasitic core, and an actual white dwarf—about the the size of the earth—is all that remains. Our sun will be expected to exist as a red giant for 'only' a few thousand million years. Afterwards, in its last 'visible' incarnation—as a slowly cooling dying ember* of a white dwarf—the sun will persist for a few more thousands of millions of years, finally obtaining total obscurity as an invisible *black dwarf*.

Not all stars would share the sun's fate. For some, their destiny is a considerably more violent one, and their fate is sealed by what is known as the *Chandrasekhar limit*: the maximum possible value for the mass of a white

* In fact, in its final stages, the dwarf will glow dimly as a red star—but what is referred to as a 'red dwarf' is a star of quite a different character!

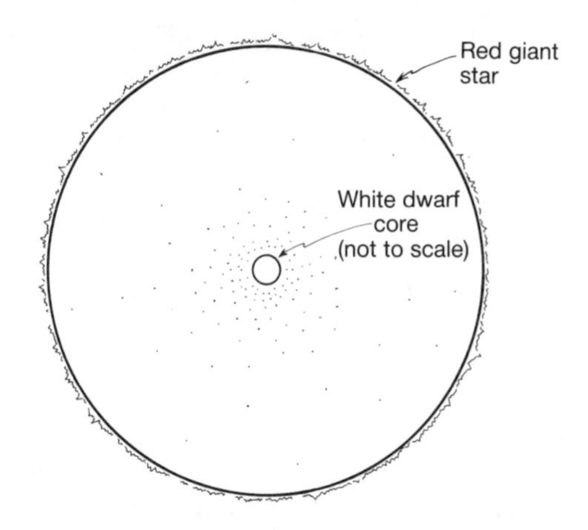

Fig. 7.12. A red giant star with a white dwarf core.

dwarf star. According to a calculation performed in 1929 by Subrahmanyan Chandrasekhar, white dwarfs cannot exist if their masses are more than about one and one-fifth times the mass of the sun. (He was a young Indian research student-to-be, who was travelling on the boat from India to England when he made his calculation.) The calculation was also repeated independently in about 1930 by the Russian Lev Landau. The modern somewhat refined value for Chandrasekhar's limit is about

$$1.4\,M_{\odot}$$

where $M_{\odot}$ is the mass of the sun, i.e. $M_{\odot}$ = one *solar mass*.

Note that the Chandrasekhar limit is not much greater than the sun's mass, whereas many ordinary stars are known whose mass is considerably greater than this value. What would be the ultimate fate of a star of mass $2M_{\odot}$, for example? Again, according to established theory, the star should swell to become a red giant, and its white-dwarf core would slowly acquire mass, just as before. However, at some critical stage the core will reach Chandrasekhar's limit, and Pauli's exclusion principle will be insufficient to hold it apart against the enormous gravitationally induced pressures.[7] At this point, or thereabouts, the core will collapse catastrophically inwards, and hugely increased temperatures and pressures will be encountered. Violent nuclear reactions take place, and an enormous amount of energy is released from the core in the form of neutrinos. These heat up the outer regions of the star, which have been collapsing inwards, and a stupendous explosion ensues. The star has become a supernova!

What then happens to the still-collapsing core? Theory tells us that it reaches enormously greater densities even than those alarming ones already achieved inside a white dwarf. The core can stabilize as a *neutron star*

(p. 321), where now it is *neutron degeneracy pressure*—i.e. the Pauli principle applied to neutrons—that is holding it apart. The density would be such that our ping-pong ball containing neutron star material would weigh as much as the asteroid Hermes (or perhaps Mars's moon Deimos). This is the kind of density found inside the very nucleus itself! (A neutron star is like a huge atomic nucleus, perhaps some ten kilometres in radius, which is, however, extremely tiny by stellar standards!) But there is now a *new* limit, analogous to Chandrasekhar's (referred to as the Landau–Oppenheimer–Volkov limit), whose modern (revised) value is very roughly

$$2.5\, M_{\odot},$$

above which a neutron star cannot hold itself apart.

What happens to this collapsing core if the mass of the original star is great enough that even *this* limit will be exceeded? Many stars are known, of masses ranging between $10 M_{\odot}$ and $100 M_{\odot}$, for example. It would seem highly unlikely that they would invariably throw off so much mass that the resulting core necessarily lies below this neutron star limit. The expectation is that, instead, a *black hole* will result.

What is a black hole? It is a region of space—or of space–time—within which the gravitational field has become so strong that even light cannot escape from it. Recall that it is an implication of the principles of relativity that the velocity of light is the limiting velocity: no material object or signal can exceed the local light speed (pp. 194, 211). Hence, if light cannot escape from a black hole, *nothing* can escape.

Perhaps the reader is familiar with the concept of *escape velocity*. This is the speed which an object must attain in order to escape from some massive body. Suppose that body were the earth; then the escape velocity from it would be approximately 40000 kilometres per hour, which is about 25000 miles per hour. A stone which is hurled from the earth's surface (in any direction away from the ground), with a speed exceeding this value, will escape from the earth completely (assuming that we may ignore the effects of air resistance). If thrown with less than this speed, then it will fall back to the earth. (Thus, it is *not* true that 'everything that goes up must come down'; an object returns only if it is thrown with *less* than the escape velocity!) For Jupiter, the escape velocity is 220000 kilometres per hour i.e. about 140000 miles per hour; and for the sun it is 2200000 kilometres per hour, or about 1400000 miles per hour. Now suppose we imagine that the sun's mass were concentrated in a sphere of just *one quarter* of its present radius, then we should obtain an escape velocity which is *twice* as great as its present value; if the sun were even more concentrated, say in a sphere of *one-hundredth* of its present radius, then the velocity would be *ten times* as great. We can imagine that for a sufficiently massive and concentrated body, the escape velocity could exceed even the velocity of light! When this happens, we have a black hole.[8]

In Fig. 7.13, I have drawn a space–time diagram depicting the collapse of a body to form a black hole (where I am assuming that the collapse proceeds in a way that maintains spherical symmetry reasonably closely, and where I have suppressed one of the spatial dimensions). The light cones have been depicted, and, as we recall from the discussion of general relativity in Chapter 5 (cf. p. 208), these indicate the absolute limitations on the motion of a material object or signal. Note that the cones begin to tip inwards towards the centre, and the tipping gets more and more extreme the more central they are.

There is a critical distance from the centre, referred to as the *Schwarzschild radius*, at which the outer limits of the cones become *vertical* in the diagram. At this distance, light (which must follow the light cones) can simply hover

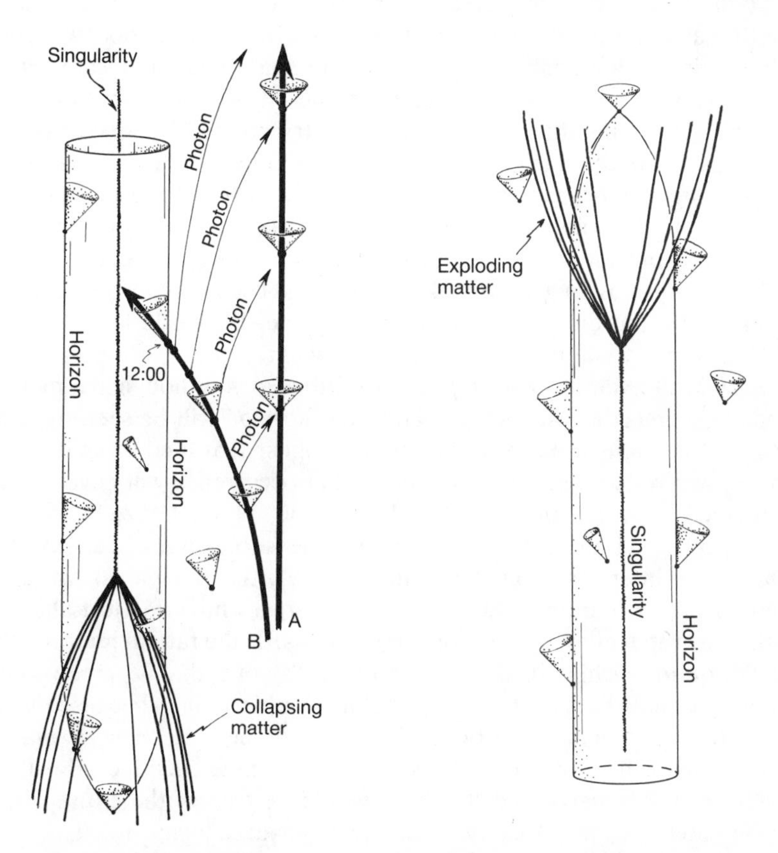

Fig. 7.13. A space–time diagram depicting collapse to a black hole.

Fig. 7.14. A hypothetical space–time configuration: a white hole, ultimately exploding into matter (the time reverse of the space–time of Fig. 7.13).

above the collapsed body, and all the outward velocity that the light can muster is just barely enough to counteract the enormous gravitational pull. The (3-)surface in space–time traced out, at the Schwarzschild radius, by this hovering light (i.e. the light's entire history) is referred to as the (*absolute*) *event horizon* of the black hole. Anything that finds itself within the event horizon is unable to escape or even to communicate with the outside world. This can be seen from the tipping of the cones, and from the fundamental fact that all motions and signals are constrained to propagate within (or on) these cones. For a black hole formed by the collapse of a star of a few solar masses, the radius of the horizon would be a few kilometres. Much larger black holes are expected to reside at galactic centres. Our own Milky Way galaxy may well contain a black hole of about a million solar masses, and the radius of the hole would then be a few million kilometres.

The actual material body which collapses to form the black hole will end up totally within the horizon, and so it is then unable to communicate with the outside. We shall be considering the probable fate of the body shortly. For the moment, it is just the space–time geometry created by its collapse that concerns us—a space–time geometry with profoundly curious implications.

Let us imagine a brave (or foolhardy?) astronaut **B**, who resolves to travel into a large black hole, while his more timid (or cautious?) companion **A** remains safely outside the event horizon. Let us suppose that **A** endeavours to keep **B** in view for as long as possible. What does **A** see? It may be ascertained from Fig. 7.13 that the portion of **B**'s history (i.e. **B**'s world-line) which lies *inside* the horizon will never be seen by **A**, whereas the portion *outside* the horizon will all eventually have become visible to **A**—though **B**'s moments immediately preceding his plunge across the horizon will be seen by **A** only after longer and longer periods of waiting. Suppose that **B** crosses the horizon when his own watch registers 12 o'clock. That occurrence will never actually be witnessed by **A**, but the watch-readings 11:30, 11:45, 11:52, 11:56, 11:58, 11:59, 11:59½, 11:59¾, 11:59⅞, etc. will be successively seen by **A** (at roughly equal intervals, from **A**'s point of view). In principle, **B** will always remain visible to **A** and would appear to be forever hovering just above the horizon, his watch edging ever more slowly towards the fateful hour of 12:00, but never quite reaching it. But, in fact the image of **B** that is perceived by **A** would very rapidly become too dim to be discernible. This is because the light from the tiny portion of **B**'s world-line just outside the horizon has to make do for the whole of the remainder of **A**'s experienced time. In effect, **B** will have vanished from **A**'s view—and the same would be true of the entire original collapsing body. All that **A** can see will indeed be just a 'black hole'!

What about poor **B**? What will be *his* experience? It should first be pointed out that there will be nothing whatever noticeable by **B** at the moment of his crossing the horizon. He glances at his watch as it registers around 12 o'clock and he sees the minutes pass regularly by: 11:57, 11:58, 11:59, 12:00, 12:01,

12:02, 12:03, ... Nothing seems particularly odd about the time 12:00. He can look back at **A**, and will find that **A** remains continuously in view the whole time. He can look at **A**'s own watch, which appears to **B** to be proceeding forwards in an orderly and regular fashion. Unless **B** has *calculated* that he must have crossed the horizon, he will have no means of knowing it.[9] The horizon has been insidious in the extreme. Once crossed, there is no escape for **B**. His local universe will eventually be found to be collapsing about him, and he will be destined shortly to encounter his own private 'big crunch'!

Or perhaps it is not so private. All the matter of the collapsed body that formed the black hole in the first place will, in a sense, be sharing the 'same' crunch with him. In fact, if the universe *outside* the hole is spatially closed, so that the outside matter is also ultimately engulfed in an all-embracing big crunch, then that crunch, also, would be expected to be the 'same' as **B**'s 'private' one.*

Despite **B**'s unpleasant fate, we do not expect that the local physics that he experiences up to that point should be at odds with the physics that we have come to know and understand. In particular, we do not expect that he will experience local violations of the second law of thermodynamics, let alone a complete reversal of the increasing behaviour of entropy. The second law will hold sway just as much inside a black hole as it does elsewhere. The entropy in **B**'s vicinity is still increasing, right up until the time of his final crunch.

To understand how the entropy in a 'big crunch' (either 'private' or 'all-embracing') can indeed be enormously high, whereas the entropy in the big bang had to have been much lower, we shall need to delve a little more deeply into the space–time geometry of a black hole. But before we do so, the reader should catch a glimpse also of Fig. 7.14 which depicts the hypothetical time-reverse of a black hole, namely a *white* hole. White holes probably do *not* exist in nature, but their theoretical possibility will have considerable significance for us.

The structure of space–time singularities

Recall from Chapter 5 (p. 204) how space–time curvature manifests itself as a *tidal effect*. A spherical surface, made up of particles falling freely in the gravitational field of some large body, would be stretched in one direction (along the line towards the gravitating body) and squashed in directions perpendicular to this. This tidal distortion increases as the gravitating body is

* In making this statement I am adopting two assumptions. The first is that the black hole's possible ultimate disappearance—on account of its (extremely slow) 'evaporation' by Hawking radiation that we shall be considering later (cf. p. 324)—would be forestalled by the recollapse of the universe; the second is a (very plausible) assumption known as 'cosmic censorship' (p. 215).

approached (Fig. 7.15), varying inversely with the cube of the distance from it. Such an increasing tidal effect will be felt by the astronaut **B** as he falls towards and into the black hole. For a black hole of a few solar masses, this tidal effect would be huge—far too huge for the astronaut to be able to survive any close approach to the hole, let alone to cross its horizon. For larger holes, the size of the tidal effect at the horizon would actually be smaller. For the million solar-mass black hole that many astronomers believe may reside at the centre of our Milky Way galaxy, the tidal effect would be fairly small as the astronaut crosses the horizon, though it would probably be enough to make him feel a little uncomfortable. This tidal effect would not stay small for long as the astronaut falls in, however, mounting rapidly to infinity in a matter of a few seconds! Not only would the poor astronaut's body be torn to pieces by this rapidly mounting tidal force, but so also, in quick succession, would the very molecules of which he was composed, their constituent atoms, their nuclei, and, finally, even all the subatomic particles! It is thus that the 'crunch' wreaks its ultimate havoc.

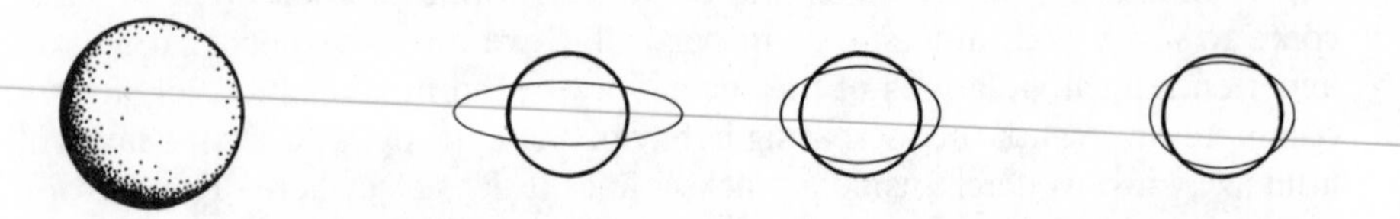

Fig. 7.15. The tidal effect due to a spherical gravitating body increases as the body is approached, according to the inverse cube of the distance from the body's centre.

Not just all matter becomes destroyed in this way, but even the very space–time must find its end! Such an ultimate catastrophe is referred to as a *space–time singularity*. The reader may well be asking how it is that we know such catastrophes must occur, and under what circumstances matter and space–time is destined to suffer this fate. These are conclusions that follow from the classical equations of general relativity, in any circumstance when a black hole is formed. The original black hole model of Oppenheimer and Snyder (1939) exhibited behaviour of this kind. However, for many years astrophysicists had entertained the hope that this singular behaviour was an artefact of the special symmetries that had to be assumed for that model. Perhaps, in realistic (asymmetrical) situations the collapsing matter might swirl around in some complicated way and then escape outwards again. But such hopes were dashed when more general types of mathematical argument became available, providing what are known as *singularity theorems* (cf. Penrose 1965, Hawking and Penrose 1970). These theorems established, within the classical theory of general relativity with reasonable material sources, that space–time singularities are *inevitable* in situations of gravitational collapse.

Likewise, using the reverse direction of time, we again find inevitability for a corresponding *initial* space–time singularity which now represents the big bang, in any (appropriately) expanding universe. Here, rather than representing the ultimate *destruction* of all matter and space–time, the singularity represents the *creation* of space–time and matter. It might appear that there is an exact temporal symmetry between these two types of singularity: the *initial* type, whereby space–time and matter are created; and the *final* type, whereby space–time and matter are destroyed. There is, indeed, an important analogy between these two situations, but when we examine them in detail we find that they are *not* exact time-reverses of one another. The geometric differences are important for us to understand because they contain the key to the origin of the second law of thermodynamics!

Let us return to the experiences of our self-sacrificing astronaut **B**. He encounters tidal forces which mount rapidly to infinity. Since he is travelling in empty space, he experiences the volume-preserving but *distorting* effects that are provided by the kind of space–time curvature tensor that I have denoted by **WEYL** (see Chapter 5, pp. 204, 210). The remaining part of the space–time curvature tensor, the part representing an overall compression and referred to as **RICCI**, is zero in empty space. It might be that **B** does in fact encounter matter at some stage, but even if this is the case (and he is, after all, constituted of matter himself), we still generally find that the measure of **WEYL** is much *larger* than that of **RICCI**. We expect to find, indeed, that the curvature close to a *final* singularity is completely dominated by the tensor **WEYL**. This tensor goes to *infinity*, in general:

$$\mathbf{WEYL} \to \infty,$$

(though it may well do in an oscillatory manner). This appears to be the *generic* situation with a space-time singularity.[10] Such behaviour is associated with a singularity of *high entropy*.

However the situation with the big bang appears to be quite different. The standard models of the big bang are provided by the highly symmetrical Friedmann–Robertson–Walker space–times that we considered earlier. Now the distorting tidal effect provided by the tensor **WEYL** is entirely *absent*. Instead there is a symmetrical inward acceleration acting on any spherical surface of test particles (see Fig. 5.26). This is the effect of the tensor **RICCI**, rather than **WEYL**. In any FRW-model, the tensor equation

$$\mathbf{WEYL} = 0$$

always holds. As we approach the initial singularity more and more closely we find that it is **RICCI** that becomes infinite, instead of **WEYL**, so it is **RICCI** that dominates near the initial singularity, rather than **WEYL**. This provides us with a singularity of *low entropy*.

If we examine the big *crunch* singularity in the *exact* recollapsing FRW-models, we now do find **WEYL** = 0 at the crunch, whereas **RICCI** goes to infinity. However, this is a very special situation and is *not* what we expect for a fully realistic model in which gravitational clumping is taken into account. As time progresses, the material, originally in the form of a diffuse gas, will clump into galaxies of stars. In due course, large numbers of these stars will contract gravitationally: into white dwarfs, neutron stars, and black holes, and there may well be some huge black holes in galactic centres. The clumping—particularly in the case of black holes—represents an enormous increase in the entropy (see Fig. 7.16). It may be puzzling, at first, that the clumpy states represent *high* entropy and the smooth ones low, when we recall that, with a gas in a box, the clumped states (such as the gas being all in one corner of the box) were of *low* entropy, while the *uniform* state of thermal equilibrium was *high*. When gravity is taken into account there is a *reversal* of this, owing to the universally attractive nature of the gravitational field.) The clumping becomes more and more extreme as time goes on, and, at the end, many black holes congeal, and their singularities unite in the very complicated, final, big crunch singularity. The final singularity in no way resembles the idealized big crunch of the recollapsing FRW-model, with its constraint **WEYL** = 0. As more and more clumping takes place, there is, all the time, a tendency for the Weyl tensor to get larger and larger[11], and, in general, **WEYL** $\to \infty$ at any final singularity. See Fig. 7.17, for a space–time picture representing the entire history of a closed universe, in accordance with this general description.

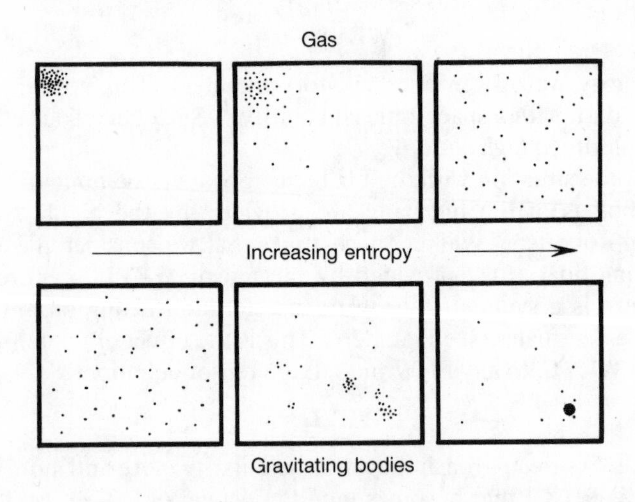

Fig. 7.16. For an ordinary gas, increasing entropy tends to make the distribution more uniform. For a system of gravitating bodies the reverse is true. High entropy is achieved by gravitational clumping—and the highest of all, by collapse to a black hole.

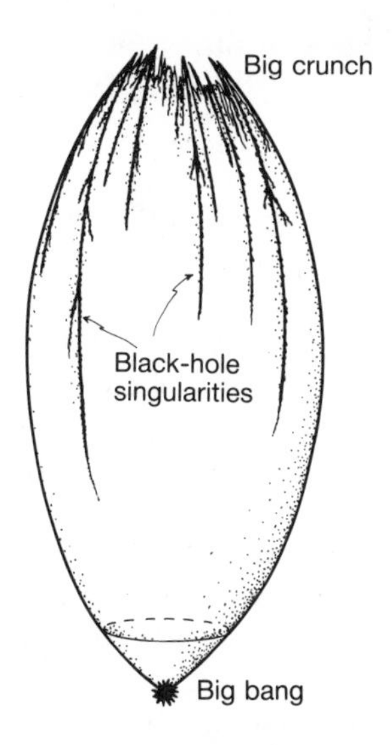

Fig. 7.17. The entire history of a closed universe which starts from a uniform low-entropy big bang, with **WEYL** = 0 and ends with a high-entropy big crunch—representing the congealing of many black holes—with **WEYL** $\rightarrow\infty$.

We see, now, how it is that a recollapsed universe need *not* have a small entropy. The 'lowness' of the entropy at the big bang—which gave us the second law—was thus *not* merely a consequence of the 'smallness' of the universe at the time of the big bang! If we were to time-reverse the picture of the big crunch that we obtained above, then we should obtain a 'big bang' with an enormously *high* entropy, and there would have been no second law! For some reason, the universe was created in a very special (low entropy) state, with something like the **WEYL** = 0 constraint of the FRW-models imposed upon it. If it were not for a constraint of this nature, it would be 'much more probable' to have a situation in which *both* the initial and final singularities were of high-entropy **WEYL** $\rightarrow\infty$ type (see Fig. 7.18). In such a 'probable' universe there would, indeed, be *no* second law of thermodynamics!

How special was the big bang?

Let us try to understand just how much of a constraint a condition such as **WEYL** = 0 at the big bang was. For simplicity (as with the above discussion) we shall suppose that the universe is closed. In order to be able to work out some clear-cut figures, we shall assume, furthermore, that the number B of *baryons*—that is, the number of protons and neutrons, taken together—in the universe is roughly given by

$$B = 10^{80}.$$

(There is no particular reason for this figure, apart from the fact that, observationally B must be *at least* as large as this; Eddington once claimed to have calculated B *exactly*, obtaining a figure which was close to the above value! No-one seems to believe this particular calculation any more, but the value 10^{80} appears to have stuck.) If B were taken to be *larger* than this (and perhaps, in actual fact, $B = \infty$) then the figures that we would obtain would be even *more* striking than the extraordinary figures that we shall be arriving at in a minute!

Try to imagine the phase space (cf. p. 177) of the *entire* universe! Each point in this phase space represents a different possible way that the universe might have started off. We are to picture the Creator, armed with a 'pin'—which is to be placed at some point in the phase space (Fig. 7.19). Each different positioning of the pin provides a different universe. Now the accuracy that is needed for the Creator's aim depends upon the entropy of the universe that is thereby created. It would be relatively 'easy' to produce a high entropy universe, since then there would be a large volume of the phase space available for the pin to hit. (Recall that the entropy is proportional to the logarithm of the volume of the phase space concerned.) But in order to start off the universe in state of low entropy—so that there will indeed be a second law of thermodynamics—the Creator must aim for a much tinier volume of the phase space. How tiny would this region be, in order that a universe closely resembling the one in which we actually live would be the result? In order to answer this question, we must first turn to a very remarkable formula, due to Jacob Bekenstein (1972) and Stephen Hawking (1975), which tells us what the entropy of a *black hole* must be.

Consider a black hole, and suppose that its horizon's surface area is A. The Bekenstein–Hawking formula for the black hole's entropy is then:

$$S_{bh} = \frac{A}{4} \times \left(\frac{kc^3}{G\hbar}\right),$$

where k is Boltzmann's constant, c is the speed of light, G is Newton's gravitational constant, and $\hbar$ is Planck's constant over 2π. The essential part of this formula is the $A/4$. The part in parentheses merely consists of the appropriate physical constants. Thus, the entropy of a black hole is proportional to its surface area. For a spherically symmetrical black hole, this surface area turns out to be proportional to the square of the mass of the hole

$$A = m^2 \times 8\pi(G^2/c^2).$$

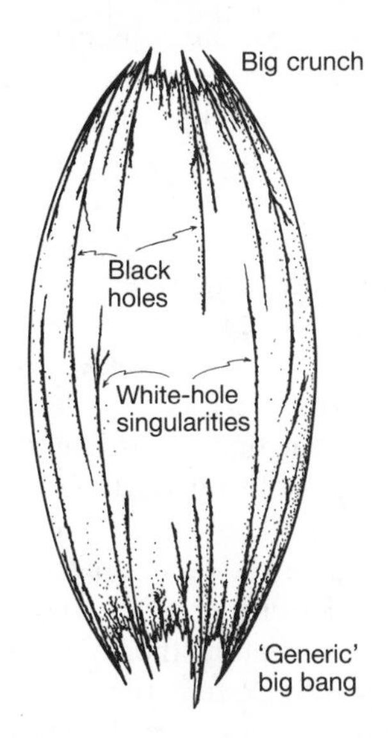

Fig. 7.18. If the constraint **WEYL** = 0 is removed, then we have a high-entropy big bang also, with **WEYL** $\to\infty$ there. Such a universe would be riddled with white holes, and there would be no second law of thermodynamics, in gross contradiction with experience.

Putting this together with the Bekenstein–Hawking formula, we find that the entropy of a black hole is proportional to the square of its mass:

$$S_{bh} = m^2 \times 2\pi(kcG/\hbar).$$

Thus, the *entropy per unit mass* (S_{bh}/m) of a black hole is proportional to its mass, and so gets larger and larger for larger and larger black holes. Hence, for a given amount of mass—or equivalently, by Einstein's $E = mc^2$, for a given amount of *energy*—the greatest entropy is achieved when the material has all collapsed into a black hole! Moreover, two black holes gain (enormously) in entropy when they mutually swallow one another up to produce a single united black hole! Large black holes, such as those likely to be found in galactic centres, will provide absolutely stupendous amounts of entropy—far and away larger than the other kinds of entropy that one encounters in other types of physical situation.

There is actually a slight qualification needed to the statement that the greatest entropy is achieved when all the mass is concentrated in a black hole. Hawking's analysis of the thermodynamics of black holes, shows that there should be a non-zero *temperature* also associated with a black hole. One implication of this is that not quite all of the mass–energy can be contained within the black hole, in the maximum entropy state, the maximum entropy being achieved by a black hole in equilibrium with a 'thermal bath of radiation'. The temperature of this radiation is very tiny indeed for a black hole of any reasonable size. For example, for a black hole of a solar mass, this temperature would be about 10^{-7}K, which is somewhat smaller than the lowest temperature that has been measured in any laboratory to date, and very considerably lower than the 2.7 K temperature of intergalactic space. For larger black holes, the Hawking temperature is even lower!

The Hawking temperature would become significant for our discussion only if either: (i) much tinier black holes, referred to as *mini-black holes*, might exist in our universe; or (ii) the universe does not recollapse before the *Hawking evaporation time*—the time according to which the black hole would evaporate away completely. With regard to (i), mini-black holes could only be produced in a suitably chaotic big bang. Such mini-black holes cannot be very numerous in our actual universe, or else their effects would have already been observed; moreover, according to the viewpoint that I am expounding here, they ought to be absent altogether. As regards (ii), for a solar-mass black hole, the Hawking evaporation time would be some 10^{54} times the present age of the universe, and for larger black holes, it would be considerably longer. It does not seem that these effects should substantially modify the above arguments.

To get some feeling for the hugeness of black-hole entropy, let us consider what was previously thought to supply the largest contribution to the entropy of the universe, namely the 2.7 K black-body background radiation. Astrophysicists had been struck by the enormous amounts of entropy that this radiation contains, which is far in excess of the ordinary entropy figures that one encounters in other processes (e.g. in the sun). The background radiation entropy is something like 10^8 for every baryon (where I am now choosing 'natural units', so that Boltzmann's constant, is unity). (In effect, this means that there are 10^8 photons in the background radiation for every baryon.) Thus, with 10^{80} baryons in all, we should have a total entropy of

$$10^{88}$$

for the entropy in the background radiation in the universe.

Indeed, were it not for the black holes, this figure would represent the *total* entropy of the universe, since the entropy in the background radiation swamps that in all other ordinary processes. The entropy per baryon in the sun, for example, is of order unity. On the other hand, by *black-hole* standards, the background radiation entropy is utter 'chicken feed'. For the Bekenstein–Hawking formula tells us that the entropy per baryon in a solar mass black hole is about 10^{20}, in natural units, so had the universe consisted entirely of solar mass black holes, the total figure would have been very much larger than that given above, namely

$$10^{100}.$$

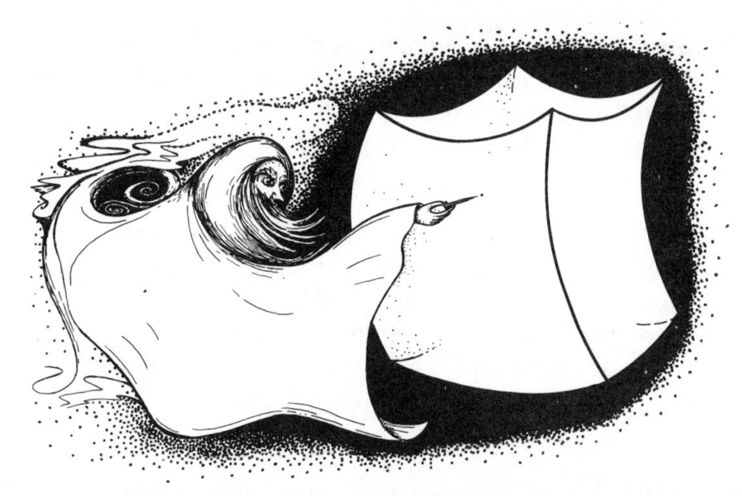

Fig. 7.19. In order to produce a universe resembling the one in which we live, the Creator would have to aim for an absurdly tiny volume of the phase space of possible universes—about $1/10^{10^{123}}$ of the entire volume, for the situation under consideration. (The pin, and the spot aimed for, are not drawn to scale!)

Of course, the universe is not so constructed, but this figure begins to tell us how 'small' the entropy in the background radiation must be considered to be when the relentless effects of gravity begin to be taken into account.

Let us try to be a little more realistic. Rather than populating our galaxies entirely with black holes, let us take them to consist mainly of ordinary stars—some 10^{11} of them—and each to have a million (i.e. 10^6) solar-mass black-hole at its core (as might be reasonable for our own Milky Way galaxy). Calculation shows that the entropy per baryon would now be actually somewhat larger even than the previous huge figure, namely now 10^{21}, giving a total entropy, in natural units, of

$$10^{101}.$$

We may anticipate that, after a very long time, a major fraction of the galaxies' masses will be incorporated into the black holes at their centres. When this happens, the entropy per baryon will be 10^{31}, giving a monstrous total of

$$10^{111}.$$

However, we are considering a closed universe so eventually it should recollapse; and it is not unreasonable to estimate the entropy of the final crunch by using the Bekenstein–Hawking formula as though the whole universe had formed a black hole. This gives an entropy per baryon of 10^{43}, and the absolutely stupendous total, for the entire big crunch would be

$$10^{123}.$$

This figure will give us an estimate of the total phase-space volume **V** available to the Creator, since this entropy should represent the logarithm of the volume of the (easily) largest compartment. Since 10^{123} is the *logarithm* of the volume, the volume must be the *exponential* of 10^{123}, i.e.

$$\mathbf{V} = 10^{10^{123}}.$$

in natural units! (Some perceptive readers may feel that I should have used the figure $e^{10^{123}}$, but for numbers of this size, the e and the 10 are essentially interchangeable!) How big was the original phase-space volume **W** that the Creator had to aim for in order to provide a universe compatible with the second law of thermodynamics and with what we now observe? It does not much matter whether we take the value

$$\mathbf{W} = 10^{10^{101}} \qquad \text{or} \qquad \mathbf{W} = 10^{10^{88}},$$

given by the galactic black holes or by the background radiation, respectively, or a much smaller (and, in fact, more appropriate) figure which would have been the *actual* figure at the big bang. Either way, the ratio of **V** to **W** will be, closely

$$\mathbf{V}/\mathbf{W} = 10^{10^{123}}.$$

(Try it: $10^{10^{123}} \div 10^{10^{101}} = 10^{(10^{123}-10^{101})} = 10^{10^{123}}$ very closely.)

This now tells us how precise the Creator's aim must have been: namely to an accuracy of

$$\text{one part in } 10^{10^{123}}.$$

This is an extraordinary figure. One could not possibly even *write the number down* in full, in the ordinary denary notation: it would be '1' followed by 10^{123} successive '0's! Even if we were to write a '0' on each separate proton and on each separate neutron in the entire universe—and we could throw in all the other particles as well for good measure—we should fall far short of writing down the figure needed. The precision needed to set the universe on its course is seen to be in no way inferior to all that extraordinary precision that we have already become accustomed to in the superb dynamical equations (Newton's, Maxwell's, Einstein's) which govern the behaviour of things from moment to moment.

But *why* was the big bang so precisely organized, whereas the big crunch (or the singularities in black holes) would be expected to be totally chaotic? It would appear that this question can be phrased in terms of the behaviour of the **WEYL** part of the space–time curvature at space–time singularities. What we appear to find is that there is a constraint

$$\mathbf{WEYL} = 0$$

(or something very like this) at *initial* space–time singularities—but not at final singularities—and this seems to be what confines the Creator's choice to this very tiny region of phase space. The assumption that this constraint applies at any initial (but not final) space–time singularity, I have termed *The Weyl Curvature Hypothesis*. Thus, it would seem, we need to understand why such a time-asymmetric hypothesis should apply if we are to comprehend where the second law has come from.[12]

How can we gain any further understanding of the origin of the second law? We seem to have been forced into an impasse. We need to understand why *space-time singularities* have the structures that they appear to have; but space–time singularities are regions where our understanding of physics has reached its limits. The impasse provided by the existence of space–time singularities is sometimes compared with another impasse: that encountered by physicists early in the century, concerning the stability of atoms (cf. p. 228). In each case, the well-established classical theory had come up with the answer 'infinity', and had thereby proved itself inadequate for the task. The singular behaviour of the electromagnetic collapse of atoms was forestalled by *quantum* theory; and likewise it should be quantum theory that yields a finite theory in place of the 'infinite' classical space–time singularities in the gravitational collapse of stars. But it can be no ordinary quantum theory. It must be a quantum theory of the very structure of space and time. Such a theory, if one existed, would be referred to as '*quantum gravity*'. Quantum gravity's lack of existence is not for want of effort, expertise, or ingenuity on the part of the physicists. Many first-rate scientific minds have applied themselves to the construction of such a theory, but without success. This is the impasse to which we have been finally led in our attempts to understand the directionality and the flow of time.

The reader may well be asking what good our journey has done us. In our quest for understanding as to why time seems to flow in just one direction and not in the other, we have had to travel to the very ends of time, and where the very notions of space have dissolved away. What have we learnt from all this? We have learnt that our theories are not yet adequate to provide answers, but what good does this do us in our attempts to understand the mind? Despite the lack of an adequate theory, I believe that there are indeed important lessons that we can learn from our journey. We must now head back for home. Our return trip will be more speculative than was the outward one, but in my opinion, there is no other reasonable route back!

1. Some relativity 'purists' might prefer to use the observers' light cones, rather than their simultaneous spaces. However, this makes no difference at all to the conclusions.
2. Entropy is gained in the combining of light nuclei (e.g. of hydrogen) in stars, into

heavier ones (e.g. of helium or, ultimately, iron). Likewise, there is much 'entropy lowness' in the hydrogen that is present on the earth, some of which we may eventually make use of by converting hydrogen to helium in 'fusion' power stations. The possibility of gaining entropy through this means arises only because gravity has enabled nuclei to be concentrated together, away from the much more numerous photons that have escaped into the vastness of space and now constitute the 2.7K black-body background radiation (cf. p. 322). This radiation contains a vastly greater entropy than is present in the matter in ordinary stars, and if it were all to be concentrated back into the material of the stars, it would serve to disintegrate most of these heavier nuclei back again into their constituent particles! The entropy gain in fusion is thus a 'temporary' one, and is made possible only through the presence of the concentrating effects of gravity. We shall be seeing later that even though the entropy available via the fusion of nuclei is very large in relation to much of that which has so far been directly obtained through gravity—and the entropy in the black-body background is enormously larger—this is a purely local and temporary state of affairs. The entropy resources of gravity are *enormously* greater than of either fusion or the 2.7K radiation (cf. p. 342)!

3. Recent evidence from ultra-deep well drillings in Sweden may be interpreted as support for Gold's theory but the matter is very controversial, there being alternative conventional explanations.
4. I am here assuming that this is what is called a 'type II' supernova. Had it been a supernova of 'type I', we would again be in thinking in terms of the 'temporary' entropy gain provided by fusion (cf. note 2). However, Type I supernovae are unlikely to produce much uranium.
5. I have referred to the models with zero or negative spatial curvature as *infinite* models. There are, however, ways of 'folding up' such models so as to make them spatially finite. This consideration—which is unlikely to be relevant to the actual universe—does not greatly affect the discussion, and I am not proposing to worry about it here.
6. The experimental basis for this confidence arises mainly from two kinds of data. In the first place, the behaviour of particles as they collide with one another at the sorts of speeds that are relevant, and bounce, fragment, and create new particles, are known from high-energy particle accelerators built at various locations on earth, and from the behaviour of cosmic ray particles which strike the earth from outer space. Secondly, it is known that the parameters which govern the way that particles interact have not changed even by one part in 10^6, in 10^{10} years (cf. Barrow 1988) so it is highly likely that they have not changed at all significantly (and probably not at all) since the time of the primordial fireball.
7. Pauli's principle does not actually forbid electrons from being in the same 'place' as one another, but it forbids any two of them from being in the same 'state'—involving also how the electrons are moving and spinning. The actual argument is a little delicate, and it was the subject of much controversy, particularly from Eddington, when it was first put forward.
8. Such reasoning was put forward as early as 1784 by the English astronomer John Michell, and independently by Laplace a little later. They concluded that the most massive and concentrated bodies in the universe might indeed be totally invisi-

ble—like black holes—but their (certainly prophetic) arguments were carried out using Newtonian theory, for which these conclusions are, at best, somewhat debatable. A proper general-relativistic treatment was first given by John Robert Oppenheimer and Hartland Snyder (1939).

9. In fact the *exact* location of the horizon, in the case of a general non-stationary black hole is not something that can be ascertained by direct measurements. It partly depends on a knowledge of all the material that will fall into the hole in its *future*!
10. See the discussions of Belinskii, Khalatnikov, and Lifshitz (1970) and Penrose (1979*b*).
11. It is tempting to identify the gravitational contribution to the entropy of a system with some measure of the total Weyl curvature, but no appropriate such measure has yet come to light. (It would need to have some awkward non-local properties, in general.) Fortunately, such a measure of gravitational entropy is not needed for the present discussion.
12. There is a currently popular viewpoint, referred to as the 'inflationary scenario' which purports to explain why, among other things, the universe is so uniform on a large scale. According to this viewpoint, the universe underwent a vast expansion in its very early stages—of an enormously greater order than the 'ordinary' expansion of the standard model. The idea is that any irregularities would be ironed out by this expansion. However, without some even greater initial constraint, such as is already provided by the Weyl curvature hypothesis, inflation cannot work. It introduces no time-asymmetric ingredient which might explain the difference between the initial and final singularity. (Moreover it is based on unsubstantiated physical theories—the GUT theories—whose status is no better than TENTATIVE, in the terminology of Chapter 5. For a critical assessment of 'inflation', in the context of the ideas of this chapter, see Penrose 1989*b*.)

8
In search of quantum gravity

Why quantum gravity?

What is there that is new to be learnt, concerning brains or minds, from what we have seen in the last chapter? Though we may have glimpsed some of the all-embracing physical principles underlying the directionality of our perceived 'flow of time', we seem, so far, to have gained no insights into the question of why we perceive time to flow or, indeed, why we perceive at all. In my opinion, much more radical ideas are needed. My presentation so far has not been particularly radical, though I have sometimes provided a different emphasis from what is usual. We have made our acquaintance with the second law of thermodynamics, and I have attempted to persuade the reader that the origin of this law—presented to us by Nature in the particular form that she has indeed chosen—can be traced to an enormous geometrical constraint on the big bang origin of the universe: the *Weyl curvature hypothesis*. Some cosmologists might prefer to characterize this initial constraint somewhat differently, but such a restriction on the initial singularity is indeed necessary. The *deductions* that I am about to draw from this hypothesis will be considerably less conventional than is the hypothesis itself. I claim that we shall need a change in the very framework of the quantum theory!

This change is to play its role when quantum mechanics becomes appropriately united with general relativity, i.e. in the sought-for theory of *quantum gravity*. Most physicists do not believe that quantum theory needs to change when it is united with general relativity. Moreover, they would argue that on a scale relevant to our brains the physical effects of *any* quantum gravity must be totally insignificant! They would say (very reasonably) that although such physical effects might indeed be important at the absurdly tiny distance scale known as the *Planck length**—which is 10^{-35} m, some 100000000000000000000 times smaller than the size of the tiniest subatomic particle—these effects should have no direct relevance whatever to pheno-

* This is the distance ($10^{-35}\,\text{m} = \sqrt{(\hbar G c^{-3})}$) at which the so-called 'quantum fluctuations' in the very metric of space–time should be so large that the normal idea of a smooth space–time continuum ceases to apply. (Quantum fluctuations are a consequence of Heisenberg's uncertainty principle—cf. p. 248.)

mena at the far far larger 'ordinary' scales of, say, down only to 10^{-12}m, where the chemical or electrical processes that are important to brain activity hold sway. Indeed, even *classical* (i.e. non-quantum) gravity has almost no influence on these electrical and chemical activities. If classical gravity is of no consequence, then how on earth could any tiny 'quantum correction' to the classical theory make any difference at all? Moreover, since *deviations* from quantum theory have never been observed, it would seem to be even *more* unreasonable to imagine that any tiny putative deviation from standard quantum theory could have any conceivable role to play in mental phenomena!

I shall argue very differently. For I am not concerned so much with the effects that quantum mechanics might have on our theory (Einstein's general relativity) of the structure of space–time, but with the *reverse*: namely the effects that Einstein's space–time theory might have on the very structure of quantum mechanics. I should emphasize that it is an *unconventional* viewpoint that I am putting forward. It is unconventional that general relativity should have any influence at all on the structure of quantum mechanics! Conventional physicists have been very reluctant to believe that the standard structure of quantum mechanics should be tampered with in any way. Although it is true that the direct application of the rules of quantum theory to Einstein's theory has encountered seemingly insurmountable difficulties, the reaction of workers in the field has tended to be to use this as a reason to modify *Einstein's* theory, not quantum theory.[1] My own viewpoint is almost the opposite. I believe that the problems within quantum theory itself are of a fundamental character. Recall the incompatibility between the two basic procedures **U** and **R** of quantum mechanics (**U** obeys the completely deterministic *Schrödinger's equation*—called *unitary* evolution—and **R** was the probabilistic *state-vector reduction* that one must apply whenever an 'observation' is deemed to have been made). In my view, this incompatibility is something which *cannot* be adequately resolved merely by the adoption of a suitable 'interpretation' of quantum mechanics (though the common view seems to be that somehow it must be able to be), but only by some radical new theory, according to which the two procedures **U** and **R** will be seen to be different (and excellent) approximations to a more comprehensive and exact *single* procedure. My view, therefore, is that even the marvellously precise theory of quantum mechanics will have to be changed, and that powerful hints as to the nature of this change will have to come from Einstein's general relativity. I shall go further and say, even, that it is actually the sought-for theory of *quantum gravity* which must contain, as one of its fundamental ingredients, this putative combined **U/R** procedure.

In the *conventional* view, on the other hand, any direct implications of quantum gravity would be of a more esoteric nature. I have mentioned the expectation of a fundamental alteration of the structure of space–time at the ridiculously tiny dimension of the Planck length. There is also the belief (justified, in my opinion) that quantum gravity ought to be fundamentally involved in ultimately determining the nature of the presently observed menagerie of 'elementary particles'. At present, there is, for example, no good theory explaining why the masses of particles should be what they are—whereas 'mass' is a concept intimately bound up with the concept of gravity. (Indeed, mass acts uniquely as the 'source' of gravity.) Also, there is good expectation that (according to an idea put forward in about 1955 by the Swedish physicist Oskar Klein) the correct quantum gravity theory should serve to remove the infinities which plague conventional quantum field theory (cf. p. 289). Physics is a unity, and the *true* quantum gravity theory, when it eventually comes to us, must surely constitute a profound part of our detailed understanding of Nature's universal laws.

We are, however, far from such an understanding. Moreover, any putative quantum gravity theory would surely be very remote from the phenomena governing the behaviour of brains. *Particularly* remote from brain activity would appear to be that (generally accepted) role for quantum gravity that is needed in order to resolve the impasse that we were led to in the last chapter: the problem of *space–time singularities*—the singularities of Einstein's classical theory, arising at the *big bang* and in *black holes*—and also at the *big crunch*, if our universe decides eventually to collapse in upon itself. Yes, this role might well *seem* remote. I shall, however, argue that there is an elusive but important thread of logical connection. Let us try to see what this connection is.

What lies behind the Weyl curvature hypothesis?

As I have remarked, even the conventional viewpoint tells us that it should be quantum gravity that will come to the aid of the classical theory of general relativity and resolve the riddle of space–time singularities. Thus, quantum gravity is to provide us with some coherent physics in place of the nonsensical answer 'infinity' that the classical theory comes up with. I certainly concur with this view: this is indeed a clear place where quantum gravity must leave its mark. However, theorists do not seem to have come much to terms with the striking fact that quantum gravity's mark is blatantly time-asymmetrical! At the big bang—the *past singularity*—quantum gravity must tell us that a condition something like

$$\mathbf{WEYL} = 0$$

must hold, at the moment that it becomes meaningful to speak in terms of the classical concepts of space–time geometry. On the other hand, at the singularities inside black holes, and at the (possible) big crunch—*future singularities*—there is no such restriction, and we expect the Weyl tensor to become infinite:

$$\mathbf{WEYL} \rightarrow \infty,$$

as the singularity is approached. In my opinion, this is a clear indication that the actual theory we seek must be asymmetrical in time:

our sought-for quantum gravity must be a time-asymmetric theory.

The reader is hereby warned that this conclusion, despite its apparently obvious necessity from the way that I have presented things, is *not* accepted wisdom! Most workers in the field appear to be very reluctant to accede to this view. The reason seems to be that there is no clear way in which the conventional and well-understood procedures of quantization (as far as they go) could produce a time-asymmetric quantized theory, when the classical theory to which these procedures are being applied (standard general relativity or one of its popular modifications) is itself time-symmetric. Accordingly (when they consider such issues at all—which is not often!), such gravity quantizers would need to try to look elsewhere for the 'explanation' of the lowness of entropy in the big bang.

Perhaps many physicists would argue that a hypothesis, such as that of vanishing initial Weyl curvature, being a choice of 'boundary condition' and not a dynamical law, is not something that is within the powers of physics to explain. In effect, they are arguing that we have been presented with an 'act of God', and it is not for us to attempt to understand why one boundary condition has been given to us rather than another. However, as we have seen, the constraint that this hypothesis has placed on 'the Creator's pin' is no less extraordinary and no less precise than all the remarkable and delicately organized choreography that constitutes the dynamical laws that we have come to understand through the equations of Newton, Maxwell, Einstein, Schrödinger, Dirac, and others. Though the second law of thermodynamics may seem to have a vague and statistical character, it arises from a geometric constraint of the utmost precision. It seems unreasonable to me that one should despair of obtaining any scientific understanding of the constraints which were operative in the 'boundary condition' which was the big bang, when the scientific approach has proved so valuable for the understanding of dynamical equations. To my way of thinking, the former is just as much a part of science as the latter, albeit a part of science that we do not properly understand, as of now.

The history of science has shown us how valuable has been this idea according to which the *dynamical equations* of physics (Newton's laws, Maxwell's equations, etc.) have been separated from these so-called *boundary conditions*—conditions which need to be imposed in order that the physically appropriate solution(s) of those equations can be singled out from the morass of inappropriate ones. Historically, it has been the dynamical equations that have found simple forms. The motions of particles satisfy simply laws but the *actual arrangements* of particles that we come upon in the universe do not often seem to. Sometimes such arrangements seem at first to be simple—such as with the elliptical orbits of planetary motion, as ascertained by Kepler—but their simplicity is then found to be a *consequence* of the dynamical laws. The deeper understanding has always come through the dynamical laws, and such simple arrangements tend also to be merely approximations to much more complicated ones, such as the perturbed (not-quite-elliptical) planetary motions that are actually observed, these being explained by Newton's dynamical equations. The boundary conditions serve to 'start off' the system in question, and the dynamical equations take over from then on. It is one of the most important realizations in physical science that we can separate the dynamical behaviour from the question of the arrangement of the actual contents of the universe.

I have said that this separation into dynamical equations and boundary conditions has been historically of vital importance. The fact that it is possible to make such a separation at all is a property of the *particular* type of equations (differential equations) that always seem to arise in physics. But I do not believe that this is a division that is here to stay. In my opinion, when we come ultimately to comprehend the laws, or principles, that *actually* govern the behaviour of our universe—rather than the marvellous approximations that we have come to understand, and which constitute our SUPERB theories to date—we shall find that this distinction between dynamical equations and boundary conditions will dissolve away. Instead, there will be just some marvellously consistent comprehensive scheme. Of course, in saying this, I am expressing a very personal view. Many others might not agree with it. But it is a viewpoint such as this that I have vaguely in mind when trying to explore the implications of some unknown theory of quantum gravity. (This viewpoint will also affect some of the more speculative considerations of the final chapter.)

How can we explore the implications of an unknown theory? Things may not be at all as hopeless as they seem. Consistency is the keynote! First, I am asking the reader to accept that our putative theory—which I shall refer to as CQG ('correct quantum gravity'!)—will provide an explanation of the Weyl curvature hypothesis (WCH). This means that *initial* singularities must be constrained so that **WEYL** = 0 in the immediate future of the singularity. This constraint is to be consequent upon the laws of CQG, and so it must apply to *any* 'initial singularity', not just to the particular singularity that we refer to as the 'big bang'. I am not saying that there need *be* any initial singularities in our actual universe other than the big bang, but the point is that *if* there were, then any such singularity would have to be constrained by WCH. An initial singularity would be one out of which, in principle, particles could come. This is the opposite behaviour from the singularities of black holes, those being *final* singularities—into which particles can fall.

A possible type of initial singularity other than that of the big bang would be the singularity in a *white hole*—which, as we recall from Chapter 7, is the time-reverse of a black hole (refer back to Fig. 7.14). But we have seen that the singularities inside black holes satisfy **WEYL** $\rightarrow \infty$, so for a white hole, also, we must have **WEYL** $\rightarrow \infty$. But the singularity is now an *initial* singularity, for which WCH requires **WEYL** $= 0$. Thus WCH *rules out* the occurrence of white holes in our universe! (Fortunately, this is not only desirable on thermodynamic grounds—for white holes would violently disobey the second law of thermodynamics—but it is also consistent with observations! From time to time, various astrophysicists have postulated the existence of white holes in order to attempt to explain certain phenomena, but this always raises many more issues than it solves.) Note that I am not calling the big bang itself a 'white hole'. A white hole would possess a *localized* initial singularity which would not be able to satisfy **WEYL** $= 0$; but the all-embracing big bang *can* have **WEYL** $= 0$ and is allowed to exist by WCH provided that it is so constrained.

There is another type of possibility for an 'initial singularity': namely the very point of *explosion of a black hole* that has finally *disappeared* after (say) 10^{64} years of Hawking evaporation (p. 342; see also p. 361 to follow)! There is much speculation about the precise nature of this (very plausibly argued) presumed phenomenon. I think that it is likely that there is no conflict with WCH here. Such a (localized) explosion could be effectively instantaneous and symmetrical, and I see no conflict with the **WEYL** $= 0$ hypothesis. (In any case, assuming that there are no mini-black holes (cf. p. 342), it is likely that the first such explosion will not take place until after the universe has been in existence for 10^{54} times the length of time T that it has been in existence already! In order to appreciate how long $10^{54} \times T$ is, imagine that T were to be compressed down to the shortest time that can be measured—the tiniest decay time of any unstable particle—then our *actual* present universe age, on this scale, would fall short of $10^{54} \times T$ by a factor of over a million million!

Some would take a different line from the one that I am proposing. They would argue[2] that CQG ought not to be time-asymmetric but that, in effect, it would allow *two* types of singularity structure, one of which requires **WEYL** $= 0$, and the other of which allows **WEYL** $\rightarrow \infty$. There happens to be a singularity of the first kind in our universe, and our perception of the direction of time is (because of the consequent second law) such as to place this singularity in what we call the 'past' rather than what we call the 'future'. However, it seems to me that this argument is not adequate as it stands. It does not explain why there are no *other* initial singularities of the **WEYL** $\rightarrow \infty$ type (nor another of the **WEYL** $= 0$ type). Why, on this view, is the universe not riddled with white holes? Since it presumably *is* riddled with *black* holes, we need an explanation for why there are no white ones.*

Another argument which is sometimes invoked in this context is the so-called *anthropic principle* (cf. Barrow and Tipler 1986). According to this argument, the particular universe that we actually observe ourselves to inhabit is selected from among all the *possible* universes by the fact that *we* (or, at least some kind of sentient creatures) need to be present actually to observe it! (I shall discuss the anthropic principle again in Chapter 10.) It is claimed, by use of this argument, that intelligent beings could only inhabit a universe with a very special type of big bang—and, hence, something like WCH might be a consequence of this principle. However, the argument can get nowhere close to the needed figure of $10^{10^{123}}$, for the 'specialness' of the big bang, as arrived at in Chapter 7 (cf. p. 344). By a very rough calculation, the entire solar system together with all its inhabitants could be created simply from the random collisions of particles much more 'cheaply' than this, namely with an 'improbability' (as measured in terms of phase-space volumes) of 'only' one part in much less than $10^{10^{60}}$. This is all that the anthropic principle can do for us, and we are still enormously short of the required figure. Moreover, as with the viewpoint discussed just previously, this anthropic argument offers no explanation for the absence of white holes.

Time-asymmetry in state-vector reduction

It seems that we are indeed left with the conclusion that CQG must be a time-asymmetric theory, where WCH (or something very like it) is one of the theory's consequences. How is it that we can get a time-asymmetric theory out of two time-symmetric ingredients: quantum theory and general relativity? There are, as it turns out, a number of conceivable technical possibilities for achieving this, none of which have been explored very far (cf. Ashtekar *et al.* 1989). However, I wish to examine a different line. I have indicated that quantum theory is 'time-symmetric', but this really applies only to the **U** part of the theory (Schrödinger equation, etc.). In my discussions of the time-symmetry of physical laws at the beginning of Chapter 7, I deliberately kept away from the **R** part (wavefunction collapse). There seems to be a prevailing view that **R**, also, should be time-symmetric. Perhaps this view arises partly because of a reluctance to take **R** to be an actual 'process' independent of **U**, so the time-symmetry of **U** ought to imply time-symmetry also for **R**. I wish to

* Some might argue (correctly) that observations are not by any means clear enough to support my contention that there are black but not white holes in the universe. But my argument is basically a theoretical one. Black holes are in accordance with the second law of thermodynamics; but white holes are not! (Of course, one could simply *postulate* the second law and the absence of white holes; but we are trying to see more deeply than this, into the second law's origins.)

argue that this is *not so*: **R** is time-*a*symmetric—at least if we simply take '**R**' to mean the procedure that physicists actually adopt when they compute probabilities in quantum mechanics.

Let me first remind the reader of the procedure that is applied in quantum mechanics that is referred to as state-vector reduction (**R**) (recall Fig. 6.23). In Fig. 8.1, I have schematically indicated the strange way that the state-vector $|\psi\rangle$ is taken to evolve in quantum mechanics. For the most part, this evolution is taken to proceed according to *unitary* evolution **U** (Schrödinger's equation), but at various times, when an 'observation' (or 'measurement') is deemed to have taken place, the procedure **R** is adopted, and the state-vector $|\psi\rangle$ *jumps* to another state-vector, say $|\chi\rangle$, where $|\chi\rangle$ is one of two or more orthogonal alternative possibilities $|\chi\rangle$, $|\varphi\rangle$, $|\theta\rangle$, . . . that are determined by the nature of the particular observation O that is being undertaken. Now, the probability p of jumping from $|\psi\rangle$ to $|\chi\rangle$ is given by the amount by which the squared length $|\psi|^2$ of $|\psi\rangle$ is decreased, in the projection of $|\psi\rangle$ in the (Hilbert space) direction of $|\chi\rangle$. (This is mathematically the same as the amount by which $|\chi|^2$ would be decreased, were $|\chi\rangle$ to be projected in the direction of $|\psi\rangle$.) As it stands, this procedure is time-asymmetrical, because immediately *after* the observation O has been made, the state-vector is one of the *given set* $|\chi\rangle$, $|\varphi\rangle$, $|\theta\rangle$, . . . of alternative possibilities *determined by* O, whereas immediately *before* O, the state-vector was $|\psi\rangle$, which need *not* have been one of these given alternatives. However, this asymmetry is only apparent, and it can be remedied by taking a different viewpoint about the evolution of the state-vector. Let us consider a *reversed-time* quantum-mechanical evolution. This eccentric description is illustrated in Fig. 8.2. Now we take the state to be $|\chi\rangle$ immediately *before* O, rather than immediately after it, and we take the unitary evolution to apply *backwards in time* to the time of the *previous* observation O′. Let us suppose that this backwards-evolved state becomes $|\chi'\rangle$ (immediately to the future of the observation O′). In the normal forwards-evolved description of Fig. 8.1, we had some other state $|\psi'\rangle$ just to

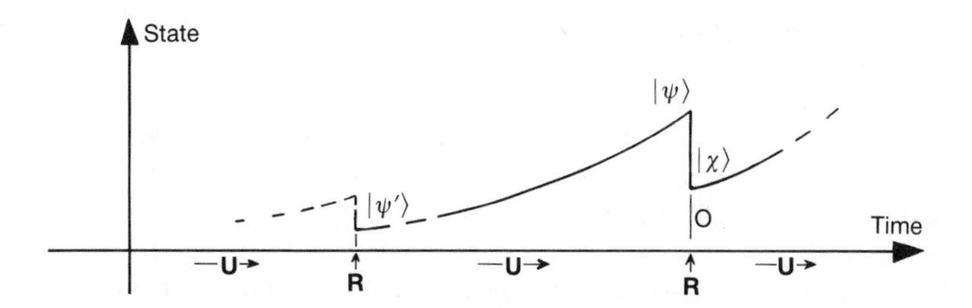

Fig. 8.1. The time-evolution of a state-vector: smooth unitary evolution **U** (by Schrödinger's equation) punctuated by discontinuous state-vector reduction **R**.

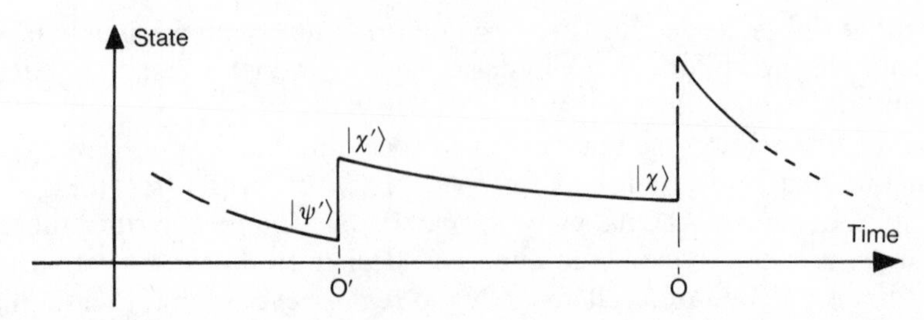

Fig. 8.2. A more eccentric picture of state-vector evolution, where a reversed-time description is used. The calculated probability relating the observation at O to that at O′ would be the same as with Fig. 8.1, but what does this calculated value refer to?

the future of O′ (the result of the observation O′, where $|\psi'\rangle$ would evolve forwards to $|\psi\rangle$ at O in the normal description). Now, in our *reversed* description, the state-vector $|\psi'\rangle$ also has a role: it is to represent the state of the system immediately to the *past* of O′. The state-vector $|\psi'\rangle$ is the state which was actually observed at O′, so in our backwards-evolved viewpoint, we now think of $|\psi'\rangle$ as being the state which is the 'result', in the *reversed-time* sense, of the observation at O′. The calculation of the quantum probability P′ which relates the result of the observation at O′ to that at O, is now given by the amount by which $|\chi'|^2$ is decreased in the projection of $|\chi'\rangle$ in the direction of $|\psi'\rangle$ (this being the same as the amount by which $|\psi'|^2$ is decreased when $|\psi'\rangle$ is projected in the direction of $|\chi'\rangle$). It is a fundamental property of the operation of **U** that in fact this is precisely the same value that we had before.[3]

Thus, it would *seem* that we have established that *quantum theory is time-symmetric*, even when we take into account the discontinuous process described by state-vector reduction **R**, in addition to the ordinary unitary evolution **U**. However, this is *not so*. What the quantum probability p describes—calculated either way—is the probability of finding the result (namely $|\chi\rangle$) at O *given* the result (namely $|\psi'\rangle$) at O′. This is not necessarily the same as the probability of the result at O′ *given* the result at O. The latter[4] would be really what our time-reversed quantum mechanics should be obtaining. It is remarkable how many physicists seem tacitly to assume that these two probabilities are the same. (I, myself, have been guilty of this presumption, cf. Penrose 1979*b*, p. 584.) However, these two probabilities are likely to be wildly different, in fact, and only the former is correctly given by quantum mechanics!

Let us see this in a very simple specific case. Suppose that we have a lamp L and a photo-cell (i.e. a photon detector) P. Between L and P we have a half-silvered mirror M, which is tilted at an angle, say 45°, to the line from L to P (see Fig. 8.3). Suppose that the lamp emits occasional photons from time

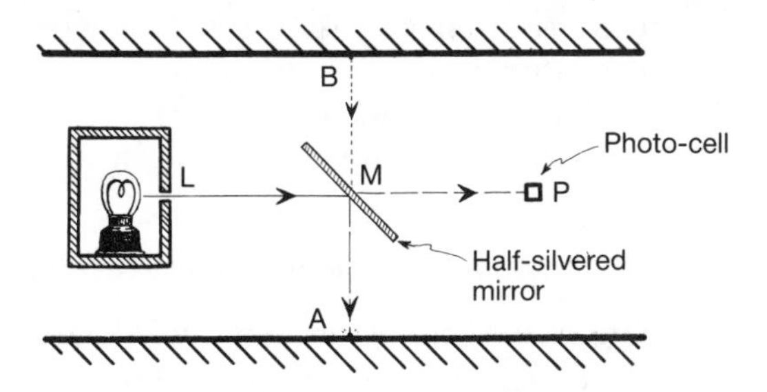

Fig. 8.3. Time-irreversibility of **R** in a simple quantum experiment. The probability that the photo-cell detects a photon *given* that the source emits one is exactly one-half; but the probability that the source has emitted a photon *given* that the photo-cell detects one is certainly *not* one-half.

to time, in some random fashion, and that the construction of the lamp (one could use parabolic mirrors) is such that these photons are always aimed very carefully at P. Whenever the photo-cell receives a photon it registers this fact, and we assume that it is 100 per cent reliable. It may also be assumed that whenever a photon is emitted, this fact is *recorded* at L, again with 100 per cent reliability. (There is no conflict with quantum-mechanical principles in any of these ideal requirements, though there might be difficulties in approaching such efficiency in practice.)

The half-silvered mirror M is such that it reflects exactly one-half of the photons that reach it and transmits the other half. More correctly, we must think of this quantum-mechanically. The photon's wavefunction impinges on the mirror and splits in two. There is an amplitude of $1/\sqrt{2}$ for the reflected part of the wave and $1/\sqrt{2}$ for the transmitted part. Both parts must be considered to 'coexist' (in the normal forward-time description) until the moment that an 'observation' is deemed to have been made. At that point these coexisting alternatives resolve themselves into *actual* alternatives—one *or* the other—with probabilities given by the squares of the (moduli of) these amplitudes, namely $(1/\sqrt{2})^2 = 1/2$, in each case. When the observation is made, the probabilities that the photon was reflected or transmitted turn out both to have been indeed one-half.

Let us see how this applies in our actual experiment. Suppose that L is recorded as having emitted a photon. The photon's wavefunction splits at the mirror, and it reaches P with amplitude $1/\sqrt{2}$, so the photo-cell registers or does not register, each with probability one-half. The other part of the photon's wavefunction reaches a point A on the *laboratory wall* (see Fig. 8.3), again with amplitude $1/\sqrt{2}$. If P does *not* register, then the photon must be considered to have hit the wall at A. For, had we placed another photo-cell at A, then it would always register whenever P does not register—assuming that L has indeed recorded the emission of a photon—and it would not register whenever P does register. In this sense, it is not necessary to place a photo-cell at A. We can infer what the photo-cell at A *would* have done, had it been there, merely by looking at L and P.

It should be clear how the quantum-mechanical calculation proceeds. We ask the question:

'Given that L registers, what is the probability that P registers?'

To answer it, we note that there is an amplitude $1/\sqrt{2}$ for the photon traversing the path LMP and an amplitude $1/\sqrt{2}$ for it traversing the path LMA. Squaring, we find the respective probabilities 1/2 and 1/2 for it reaching P and reaching A. The quantum-mechanical answer to our question is therefore

'one-half.'

Experimentally, this is indeed the answer that one would obtain.

We could equally well use the eccentric 'reverse-time' procedure to obtain the same answer. Suppose that we note that P has registered. We consider a backwards-time wavefunction for the photon, assuming that the photon finally reaches P. As we trace backwards in time, the photon goes back from P until it reaches the mirror M. At this point, the wavefunction bifurcates, and there is a $1/\sqrt{2}$ amplitude for it to reach the lamp at L, and a $1/\sqrt{2}$ amplitude for it to be reflected at M to reach *another* point on the laboratory wall, namely B in Fig. 8.3. Squaring, we again get one-half for the two probabilities. But we must be careful to note what questions these probabilities are the answers to. They are the two questions, 'Given that L registers, what is the probability that P registers?', just as before, and the more eccentric question, 'Given that the photon is ejected from the wall at B, what is the probability that P registers?'

We may consider that both these answers are, in a sense, experimentally 'correct', though the second one (ejection from the wall) would be an inference rather than the result of an *actual* series of experiments! However, neither of these questions is the *time-reverse* question of the one we asked before. That would be:

'Given that P registers, what is the probability that L registers?'

We note that the *correct* experimental answer to this question is not 'one-half' at all, but

'one.'

If the photo-cell indeed registers, then it is virtually certain that the photon came from the *lamp* and not from the laboratory wall! In the case of our time-reversed question, the quantum-mechanical calculation has given us *completely the wrong answer*!

The implication of this is that the rules for the **R** part of quantum mechanics simply cannot be used for such reversed-time questions. If we wish to calculate the probability of a *past* state on the basis of a known *future* state, we get quite the wrong answers if we try to adopt the standard **R** procedure of simply taking the quantum-mechanical amplitude and squaring its modulus. It is only for calculating the probabilities of *future* states on the basis of *past* states that this procedure works—and there it works superbly well! It seems to me to be clear that, on this basis, the procedure **R** *cannot be time-symmetric* (and, incidentally, therefore cannot be a deduction from the time-symmetric procedure **U**).

Many people might take the line that the reason for this discrepancy with time-symmetry is that somehow the second law of thermodynamcis has sneaked into the argument, introducing an additional time-asymmetry not described by the amplitude-squaring procedure. Indeed, it appears to be true that any physical measuring device capable of effecting the **R** procedure must involve a 'thermodynamic irreversibility'—so entropy increases whenever a measurement takes place. I think it is likely that the second law *is* involved in an essential way in the measurement process. Moreover it does not seem to make much physical sense to try to time-reverse the *entire* operation of a quantum-mechanical experiment such as the (idealized) one described above, including the registering of all the measurements involved. I have not been concerned with the question of how far one can go with actually time-reversing an experiment. I have been concerned only with the applicability of that remarkable quantum-mechanical procedure which obtains correct probabilities by squaring the moduli of amplitudes. It is an amazing fact that this simple procedure can be applied in the future direction without any other knowledge of a system needing to be invoked. Indeed, it is part of the theory that one *cannot* influence these probabilities: quantum theoretical probabilities are entirely *stochastic*! However, if one attempts to apply these procedures in the past direction (i.e. for retrodiction rather than prediction) then one goes sadly wrong. Any number of excuses, extenuating circumstances, or other factors may be invoked to explain *why* the amplitude-squaring procedure does not correctly apply in the past-direction, but the fact remains that it does not. Such excuses are simply not needed in the future-direction! The procedure **R**, *as it is actually used*, is just *not* time-symmetric.

Hawking's box: a link with the Weyl curvature hypothesis?

That is as may be, the reader is no doubt thinking, but what has all this to do with WCH or CQG? True, the *second law*, as it operates today, may well be part of the operation of **R**, but where is there any noticeable role for space-time singularities or quantum gravity in these continuing 'everyday' occurrences state-vector reduction? In order to address this question, I wish to describe an outlandish 'thought experiment', originally proposed by Stephen Hawking, though the purpose to which it will be put is not part of what Hawking had originally intended.

Imagine a sealed box of monstrous proportions. Its walls are taken to be totally reflecting and impervious to any influence. No material object may pass through, nor may any electromagnetic signal, or neutrino, or anything. All must be reflected back, whether they impinge from without or from within. Even the effects of gravity are fobidden to pass through. There is no actual substance out of which such walls could be built. No-one could actually *perform* the 'experiment' that I shall describe. (Nor would anyone want to, as we shall see!) That is not the point. In a thought experiment one strives to uncover general principles from the mere mental consideration of experiments that one *might* perform. Technological difficulties are ignored, provided they have no bearing on the general principles under consideration. (Recall the discussion of Schrödinger's cat, in Chapter 6.) In our case, the difficulties in constructing the walls for our box are to be regarded as purely 'technological' for this purpose, so these difficulties will be ignored.

Inside the box is a large amount of material substance, of some kind. It does not matter much what this substance is. We are concerned only with its total mass M, which should be very large, and with the large volume V of the box which contains it. What are we to do with our expensively constructed box and its totally uninteresting contents? The experiment is to be the most boring imaginable. We are to leave it untouched—forever!

The question that concerns us is the ultimate fate of the contents of the box. According to the second law of thermodynamics, its entropy should be increasing. The entropy should increase until the maximum value is attained, the material having now reached 'thermal equilibrium'. Nothing much would happen from then on, were it not for 'fluctuations' where (relatively) brief departures from thermal equilibrium are temporarily attained. In our situation, we assume that M is large enough, and V is something appropriate (*very* large, but not too large), so that when 'thermal equilibrium' is attained most of the material has collapsed into a *black hole*, with just a little matter and radiation running round it—constituting a (very cold!) so-called 'thermal bath', in which the black hole is immersed. To be definite, we could choose M to be the mass of the solar system and V to be the size of the Milky Way galaxy! Then the temperature of the 'bath' would be only about 10^{-7} of a degree above absolute zero!

To understand more clearly the nature of this equilibrium and these fluctuations, let us recall the concept of *phase space* that we encountered in Chapters 5 and 7, particularly in connection with the definition of entropy. Figure 8.4 gives a schematic description of the whole phase space $\mathbb{P}$ of the contents of Hawking's box. As we recall, a phase space is a large-dimensional

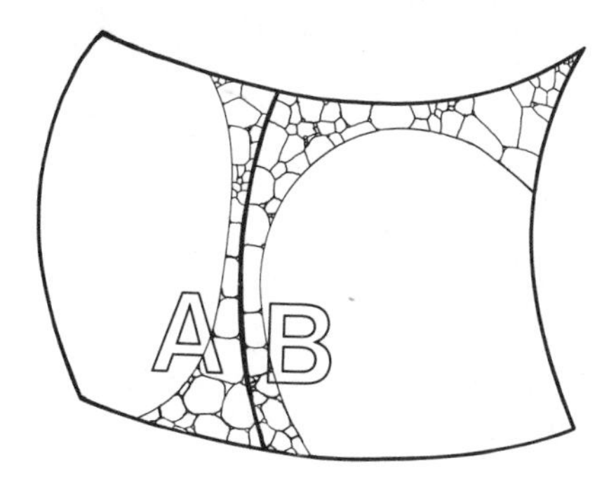

Fig. 8.4. The phase space $\mathbb{P}$ of Hawking's box. The region $\mathbb{A}$ corresponds to the situations where there is *no* black hole in the box and $\mathbb{B}$, to where there *is* a black hole (or more than one) in the box.

space, each single point of which represents an entire possible state of the system under consideration—here the contents of the box. Thus, each point of $\mathbb{P}$ codifies the positions and momenta of all the particles that are present in the box, in addition to all the necessary information about the *space–time geometry* within the box. The subregion $\mathbb{B}$ (of $\mathbb{P}$) on the right of Fig. 8.4 represents the totality of all states in which there is a *black hole* within the box (including all cases in which there is more than one black hole), whilst the subregion $\mathbb{A}$ on the left represents the totality of all states free of black holes. We must suppose that each of the regions $\mathbb{A}$ and $\mathbb{B}$ is further subdivided into smaller compartments according to the 'coarse graining' that is necessary for the precise definition of entropy (cf. Figure 7.3, p. 311), but the details of this will not concern us here. All we need to note at this stage is that the largest of these compartments—representing thermal equilibrium, with a black hole present—is the major portion of $\mathbb{B}$, whereas the (somewhat smaller) major portion of $\mathbb{A}$ is the compartment representing what *appears* to be thermal equilibrium, except that no black hole is present.

Recall that there is a field of arrows (vector field) on any phase space, representing the temporal evolution the physical system (see Chapter 5, p. 177; also Fig. 5.11). Thus, to see what will happen next in our system, we simply follow along arrows in $\mathbb{P}$ (see Fig. 8.5). Some of these arrows will cross over from the region $\mathbb{A}$ into the region $\mathbb{B}$. This occurs when a black hole first forms by the gravitational collapse of matter. Are there arrows crossing back again from region $\mathbb{B}$ into region $\mathbb{A}$? Yes there are, but only if we take into account the phenomenon of *Hawking evaporation* that was alluded to earlier (pp. 342, 353). According to the strict *classical* theory of general relativity, black holes can only swallow things; they cannot emit things. But by taking quantum-mechanical effects into account, Hawking (1975) was able to show that black holes ought, at the quantum level, to be able to emit things after

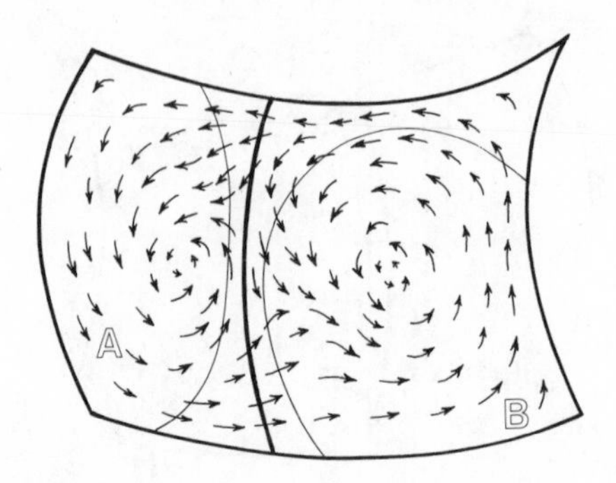

Fig. 8.5. The 'Hamiltonian flow' of the contents of Hawking's box (compare Fig. 5.11). Flow lines crossing from $\mathbb{A}$ to $\mathbb{B}$ represent collapse to a black hole; and those from $\mathbb{B}$ to $\mathbb{A}$, disappearance of a black hole by Hawking evaporation.

all, according to the process of *Hawking radiation*. (This occurs via the quantum process of 'virtual pair creation', whereby particles and anti-particles are continually being created out of the vacuum—momentarily—normally only to annihilate one another immediately afterwards, leaving no trace. When a black hole is present, however, it can 'swallow' one of the particles of such a pair before this annihilation has time to occur, and its partner may escape the hole. These escaping particles constitute Hawking's radiation.) In the normal way of things, this Hawking radiation is very tiny indeed. But in the thermal equilibrium state, the amount of energy that the black hole loses in Hawking radiation exactly balances the energy that it gains in swallowing other 'thermal particles' that happen to be running around in the 'thermal bath' in which the black hole finds itself. Occasionally, by a 'fluctuation', the hole might emit a little too much or swallow too little and thereby lose energy. In losing energy, it loses mass (by Einstein's $E = mc^2$) and, according to the rules governing Hawking radiation, it gets a tiny bit hotter. Very *very* occasionally, when the fluctuation is large enough, it is even possible for the black hole to get into a runaway situation whereby it grows hotter and hotter, losing more and more energy as it goes, getting smaller and smaller until finally it (presumably) disappears completely in a violent explosion! When this happens (and assuming there are no other holes in the box), we have the situation where, in our phase space $\mathbb{P}$, we pass from region $\mathbb{B}$ to region $\mathbb{A}$, so indeed there *are* arrows from $\mathbb{B}$ to $\mathbb{A}$!

At this point, I should make a remark about what is meant by a 'fluctuation'. Recall the coarse-graining compartments that we considered in the last chapter. The phase-space points that belong to a single compartment are to be regarded as (macroscopically) 'indistinguishable' from one another. Entropy increases because by following the arrows we tend to get into huger and huger compartments as time progresses. Ultimately, the phase-space point loses itself in the hugest compartment of all, namely that corresponding to thermal equilibrium (maximum entropy). However, this will be true only up to a point. If one waits for long enough, the phase-space point will *eventually* find a smaller compartment, and the entropy will accordingly go down. This would normally not be for long (comparatively speaking) and the entropy would soon go back up again as the phase-space point re-enters the largest compartment. This is a *fluctuation*, with its momentary lowering of entropy. Usually, the entropy does not fall very much, but very, very occasionally a *huge* fluctuation will occur, and the entropy could be lowered substantially—and perhaps remain lowered for some considerable length of time.

This is the kind of thing that we need in order to get from region $\mathbb{B}$ to region $\mathbb{A}$ via the Hawking evaporation process. A very large fluctuation is needed because a tiny compartment has to be passed through just where the arrows cross over between $\mathbb{B}$ and $\mathbb{A}$. Likewise, when our phase-space point lies inside the major compartment within $\mathbb{A}$ (representing a thermal equilibrium state without black holes), it will actually be a very long while before a gravitational collapse takes place and the point moves into $\mathbb{B}$. Again a large fluctuation is needed. (Thermal radiation does not readily undergo gravitational collapse!)

Are there *more* arrows leading from $\mathbb{A}$ to $\mathbb{B}$, or from $\mathbb{B}$ to $\mathbb{A}$, or is the number of arrows the *same* in each case? This will be an important issue for us. To put the question another way, is it 'easier' for nature to produce a black hole by gravitationally collapsing thermal particles, or to get rid of a black hole by Hawking radiation, or is each as 'difficult' as the other? Strictly speaking it is not the 'number' of arrows that concerns us, but the rate of flow of phase-space volume. Think of the phase space as being filled with some kind of (high-dimensional!) incompressible fluid. The arrows represent the flow of this fluid. Recall *Liouville's theorem*, which was described in Chapter 5 (p. 181). Liouville's theorem asserts that the phase-space volume is preserved by the flow, which is to say that our phase-space fluid is indeed incompressible! Liouville's theorem seems to be telling us that the flow from $\mathbb{A}$ to $\mathbb{B}$ must be *equal* to the flow from $\mathbb{B}$ to $\mathbb{A}$ because the phase-space 'fluid', being incompressible, cannot accumulate either on one side or the other. Thus it would appear that it must be exactly equally 'difficult' to build a black hole from thermal radiation as it is to destroy one!

This, indeed, was Hawking's own conclusion, though he came to this view on the basis of somewhat different considerations. Hawking's main argument was that all the basic physics that is involved in the problem is *time-symmetrical* (general relativity, thermodynamics, the standard unitary procedures of quantum theory), so if we run the clock backwards, we should get the same answer as if we run it forwards. This amounts simply to reversing the directions of all the arrows in $\mathbb{P}$. It would then indeed follow also from *this* argument that there must be exactly as many arrows from $\mathbb{A}$ to $\mathbb{B}$ as from $\mathbb{B}$ to

𝔸 *provided* that it is the case that the time-reverse of the region 𝔹 is the region 𝔹 again (and, equivalently; that the time-reverse of 𝔸 is 𝔸 again). This proviso amounts to Hawking's remarkable suggestion that black holes and their time-reverses, namely white holes, are actually physically identical! His reasoning was that with time-symmetric physics, the thermal equilibrium state ought to be time-symmetric also. I do not wish to enter into a detailed discussion of this striking possibility here. Hawking's idea was that somehow the quantum-mechanical Hawking radiation could be regarded as the time-reverse of the classical 'swallowing' of material by the black hole. Though ingenious, his suggestion involves severe theoretical difficulties, and I do not myself believe that it can be made to work.

In any case, the suggestion is not really compatible with the ideas that I am putting forward here. I have argued that whereas black holes ought to exist, white holes are *forbidden* because of the *Weyl curvature hypothesis!* WCH introduces a *time-asymmetry* into the discussion which was not considered by Hawking. It should be pointed out that since black holes and their space–time singularities are indeed very much part of the discussion of what happens inside Hawking's box, the unknown physics that must govern the behaviour of such singularities is certainly involved. Hawking takes the view that this unknown physics should be a *time-symmetric* quantum gravity theory, whereas I am claiming that it is the time-*a*symmetric CQG! I am claiming that one of the major implications of CQG should be WCH (and, consequently, the second law of thermodynamics in the form that we know it), so we should try to ascertain the implications of WCH for our present problem.

Let us see how the inclusion of WCH affects the discussion of the flow of our 'incompressible fluid' in ℙ. In space-time, the effect of a black-hole singularity is to absorb and destroy all the matter that impinges upon it. More importantly for our present purposes, *it destroys information*! The effect of this, in ℙ, is that some flow lines will now merge together (see Fig. 8.6). Two

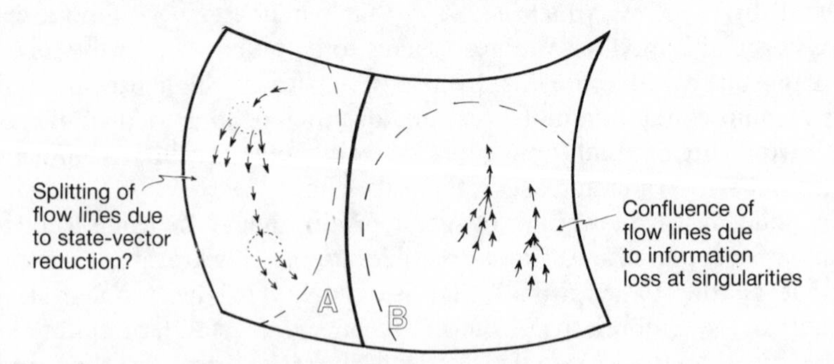

Fig. 8.6. In region 𝔹 flow lines must come together because of information loss at black-hole singularities. Is this balanced by a creation of flow lines due to the quantum procedure **R** (primarily in region 𝔸)?

states which were previously different can become the same as soon as the information that distinguishes between them is destroyed. When flow lines in $\mathbb{P}$ merge together we have an effective *violation* of Liouville's theorem. Our 'fluid' is no longer incompressible, but it is being *continually annihilated* within region $\mathbb{B}$!

Now we seem to be in trouble. If our 'fluid' is being continually destroyed in region $\mathbb{B}$, then there must be *more* flow lines from $\mathbb{A}$ to $\mathbb{B}$ than there are from $\mathbb{B}$ to $\mathbb{A}$—so it is 'easier' to create a black hole than to destroy one after all! This could indeed make sense were it not for the fact that now more 'fluid' flows out of region $\mathbb{A}$ than re-enters it. There are no black holes in region $\mathbb{A}$—and white holes have been excluded by WCH—so surely Liouville's theorem ought to hold perfectly well in region $\mathbb{A}$! However, we now seem to need some means of 'creating fluid' in region $\mathbb{A}$ to make up the loss in region $\mathbb{B}$. What mechanism can there be for increasing the number of flow lines? What we appear to require is that sometimes one and the same state can have more than one possible outcome (i.e. bifurcating flow lines). This kind of uncertainty in the future evolution of a physical system has the 'smell' of quantum theory about it—the **R** part. Can it be that **R** is, in some sense, 'the other side of the coin' to WCH? Whereas WCH serves to cause flow lines to merge within $\mathbb{B}$, the quantum-mechanical procedure **R** causes flow lines to bifurcate. I am indeed claiming that it is an *objective* quantum-mechanical process of state-vector reduction (**R**) which causes flow lines to bifurcate, and so compensate exactly for the merging of flow lines due to WCH (Fig. 8.6)!

In order for such bifurcation to occur, we need **R** to be time-asymmetric, as we have already seen: recall our experiment above, with the lamp, photo-cell, and half-silvered mirror. When a photon is emitted by the lamp, there are two (equally probable) alternatives for the final outcome: either the photon reaches the photo-cell and the photo-cell registers, or the photon reaches the wall at A and the photo-cell does not register. In the phase space for this experiment, we have a flow line representing the emission of the photon and this bifurcates into two: one describing the situation in which the photo-cell fires and the other, the situation in which it does not. This appears to be a genuine bifurcation, because there is only one allowed input and there are two possible outputs. The other input that one might have had to consider was the possibility that the photon could be ejected from the laboratory wall at B, in which case there would be two inputs and two outputs. But this alternative input has been excluded on the grounds of inconsistency with the second law of thermodynamics—i.e. from the point of view expressed here, finally with WCH, when the evolution is traced into the past.

I should re-iterate that the viewpoint that I am expressing is not really a 'conventional' one—though it is not altogether clear to me what a 'conventional' physicist would say in order to resolve all the issues raised. (I suspect that not many of them have given these problems much thought!) I have certainly heard a number of differing points of view. For example, it has from time to time been suggested, by some physicists, that Hawking radiation would never *completely* cause a black hole to disappear, but some small 'nugget' would always remain. (Hence, on this view, there are *no* arrows from $\mathbb{B}$ to $\mathbb{A}$!) This really makes little difference to my argument (and would actually strengthen it). One could evade my conclusions, however, by postulating that the total volume of the phase-space $\mathbb{P}$ is actually *infinite*, but this is at variance with certain rather basic ideas about black-hole entropy and about the nature of the phase space of a bounded (quantum) system; and other ways of technically evading my conclusions that I have heard of seem to me to be no more satisfactory. A considerably more serious objection is that the idealizations involved in the actual construction of Hawking's box are too great, and certain issues of principle are transgressed in assuming that it can be built. I am uncertain about this myself, but I am inclined to believe that the necessary idealizations can indeed be swallowed!

Finally, there is a serious point that I have glossed over. I started the discussion by assuming that we had a *classical* phase space—and Liouville's theorem applies to classical physics. But then the quantum phenomenon of Hawking radiation needed to be considered. (And quantum theory is actually also needed for the *finite-dimensionality* as well as finite volume of $\mathbb{P}$.) As we saw in Chapter 6, the quantum version of phase space is *Hilbert space*, so we should presumably have used Hilbert space rather than phase space for our discussion throughout. In Hilbert space there *is* an analogue of Liouville's theorem. It arises from what is called the '*unitary*' nature of the time-evolution **U**. Perhaps my entire argument could be phrased entirely in terms of Hilbert space, instead of classical phase space, but it is difficult to see how to discuss the classical phenomena involved in the space–time geometry of black holes in this way. My own opinion is that for the *correct* theory neither Hilbert space nor classical phase space would be appropriate, but one would have to use some hitherto undiscovered type of mathematical space which is intermediate between the two. Accordingly, my argument should be taken only at the heuristic level, and it is merely *suggestive* rather than conclusive. Nevertheless, I do believe that it provides a strong case for thinking that WCH and **R** are profoundly linked and that, consequently, **R** *must indeed be a quantum gravity effect.*

To re-iterate my conclusions: I am putting forward the suggestion that quantum-mechanical state-vector reduction is indeed the other side of the coin to WCH. According to this view, the two major implications of our sought-for 'correct quantum gravity' theory (CQG) will be WCH and **R**. The effect of WCH is *confluence* of flow lines in phase space, while the effect of **R** is an exactly compensating *spreading* of flow lines. Both processes are intimately associated with the second law of thermodynamics.

Note that the confluence of flow lines takes place entirely within the region

$\mathbb{B}$, whereas flow-line spreading can take place either within $\mathbb{A}$ *or* within $\mathbb{B}$. Recall that $\mathbb{A}$ represents the *absence* of black holes, so that state-vector reduction can indeed take place when black holes are absent. Clearly it is not necessary to have a black hole in the laboratory in order that **R** be effected (as with our experiment with the photon, just considered). We are concerned here only with a general overall balance between possible things which *might* happen in a situation. On the view that I am expressing, it is merely the *possibility* that black holes might form at some stage (and consequently then destroy information) which must be balanced by the lack of determinism in quantum theory!

When does the state-vector reduce?

Suppose that we accept, on the basis of the preceding arguments, that the reduction of the state-vector might somehow be ultimately a gravitational phenomenon. Can the connections between **R** and gravity be made more explicit? When, on the basis of this view, should a state-vector collapse *actually* take place?

I should first point out that even with the more 'conventional' approaches to a quantum gravity theory there are some serious technical difficulties involved with bringing the principles of general relativity to bear on the rules of quantum theory. These rules (primarily the way in which momenta are re-interpreted as differentiation with respect to position, in the expression for Schrödinger's equation, cf. p. 288) do not fit in at all well with the ideas of curved space–time geometry. My own point of view is that as soon as a 'significant' amount of space–time curvature is introduced, the rules of quantum linear superposition must fail. It is here that the complex-amplitude superpositions of potentially alternative states become replaced by probability weighted actual alternatives—and one of the alternatives indeed *actually* takes place.

What do I mean by a 'significant' amount of curvature? I mean that the level has been reached where the measure of curvature introduced has about the scale of *one graviton*[5] or more. (Recall that, according to the rules of quantum theory, the electromagnetic field is 'quantized' into individual units, called 'photons'. When the field is decomposed into its individual frequencies, the part of frequency ν can come only in whole numbers of photons, each of energy $h\nu$. Similar rules should presumably apply to the gravitational field.) One graviton would be the smallest unit of curvature that would be allowed according to the quantum theory. The idea is that, as soon as this level is reached, the ordinary rules of linear superposition, according to the **U** procedure, become modified when applied to gravitons, and some kind of time-asymmetric 'non-linear instability' sets in. Rather than having complex linear superpositions of 'alternatives' coexisting forever, one of the alternatives begins to win out at this stage, and the system 'flops' into one alternative or the other. Perhaps the choice of alternative is just made by chance, or perhaps there is something deeper underlying this choice. But now, *reality* has become one *or* the other. The procedure **R** has been effected.

Note that, according to this idea, the procedure **R** occurs spontaneously in an entirely objective way, independent of any human intervention. The idea is that the 'one-graviton' level should lie comfortably between the 'quantum level', of atoms, molecules, etc. where the linear rules (**U**) of ordinary quantum theory hold good, and the 'classical level' of our everyday experiences. How 'big' *is* the one-graviton level? It should be emphasized that it is not really a question of physical *size*: it is more a question of mass and energy distribution. We have seen that quantum interference effects can occur over large distances provided that not much energy is involved. (Recall the photon self-interference described on p. 254 and the EPR experiments of Clauser and Aspect, p. 286.) The characteristic quantum-gravitational scale of mass is what is known as the *Planck mass*

$$m_p = 10^{-5} \text{ grams}$$

(approximately). This would seem to be rather larger than one would wish, since objects a good deal *less* massive than this, such as specks of dust, can be directly perceived as behaving in a classical way. (The mass m_p is a little less than that of a flea.) However, I do not think that the one-graviton criterion would apply quite as crudely as this. I shall try to be a little more explicit, but at the time of writing there remain many obscurities and ambiguities as to how this criterion is to be precisely applied.

First let us consider a very direct kind of way in which a particle can be observed, namely by use of a Wilson *cloud chamber*. Here one has a chamber full of vapour which is just on the point of condensing into droplets. When a rapidly moving charged particle enters this chamber, having been produced, say, by the decay of a radioactive atom situated outside the chamber, its passage through the vapour causes some atoms close to its path to become ionized (i.e. charged, owing to electrons being dislodged). These ionized atoms act as centres along which little droplets condense out of the vapour. In this way, we get a track of droplets which the experimenter can directly observe (Fig. 8.7).

Now, what is the quantum-mechanical description of this? At the moment that our radioactive atom decays, it emits a particle. But there are many possible directions in which this emitted particle might travel. There will be an amplitude for this direction, an amplitude for that direction, and an amplitude for each other direction, all these occurring simultaneously in quantum linear superposition. The totality of all these superposed alternatives will constitute a spherical wave emanating from the decayed atom: the

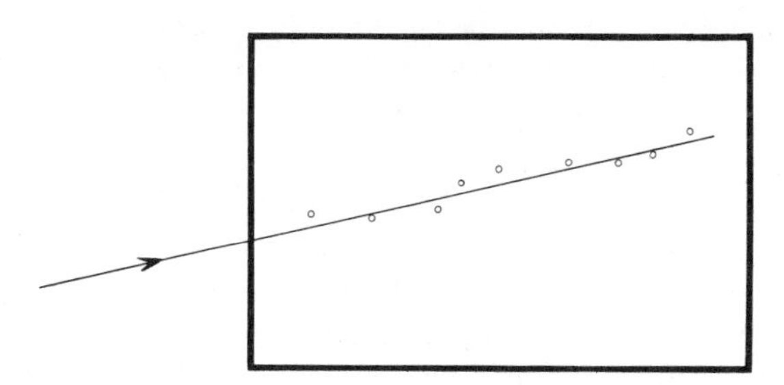

Fig. 8.7. A charged particle entering a Wilson cloud chamber and causing a string of droplets to condense.

wavefunction of the emitted particle. As each possible particle track enters the cloud chamber it becomes associated with a string of ionized atoms each of which begins to act as a centre of condensation for the vapour. All these different possible strings of ionized atoms must also coexist in quantum linear superposition, so we now have a linear superposition of a large number of *different* strings of condensing droplets. At some stage, this complex quantum linear superposition becomes a real probability-weighted collection of *actual* alternatives, as the complex amplitude-weightings have their moduli squared according to the procedure **R**. Only *one* of these is realized in the actual physical world of experience, and this particular alternative is the one observed by the experimenter. According to the viewpoint that I am proposing, that stage occurs as soon as the difference between the gravitational fields of the various alternatives reaches the one-graviton level.

When does this happen? According to a very rough calculation[6], if there had been just *one* completely uniform spherical droplet, then the one-graviton stage would be reached when the droplet grows to about one hundredth of m_P, which is one ten-millionth of a gram. There are many uncertainties in this calculation (including some difficulties of principle), and the size is a bit large for comfort, but the result is not totally unreasonable. It is to be hoped that some more precise results will be forthcoming later, and that it will be possible to treat a whole string of droplets rather than just a single droplet. Also there may be some significant differences when the fact is taken into account that the droplets are composed of a very large number of tiny atoms, rather than being totally uniform. Moreover, the 'one-graviton' criterion itself needs to be made considerably more mathematically precise.

In the above situation, I have considered what might be an actual observation of a quantum process (the decay of a radioactive atom) where quantum effects have been magnified to the point at which the different quantum alternatives produce different, directly observable, macroscopic alternatives. It would be my view that **R** can take place objectively even when such a *manifest* magnification is not present. Suppose that rather than entering a cloud chamber, our particle simply enters a large box of gas (or fluid), of such density that it is practically certain that it will collide with, or otherwise disturb, a large number of the atoms in the gas. Let us consider just two of the alternatives for the particle, as part of the initial complex linear superposition: it might not enter the box at all, or it might enter along a particular path and ricochet off some atom of gas. In the second case, that gas atom will then move off at great speed, in a way that it would not have done otherwise had the particle not run into it, and it will subsequently collide with, and itself ricochet off, some further atom. The two atoms will each then move off in ways that they would not have done otherwise, and soon there will be a cascade of movement of atoms in the gas which would not have happened had the particle not initially entered the box (Fig. 8.8). Before long, in this second case, virtually every atom in the gas will have become disturbed by this motion.

Now think of how we would have to describe this quantum mechanically. Initially it is only the original particle whose different positions must be considered to occur in complex linear superposition—as part of the particle's

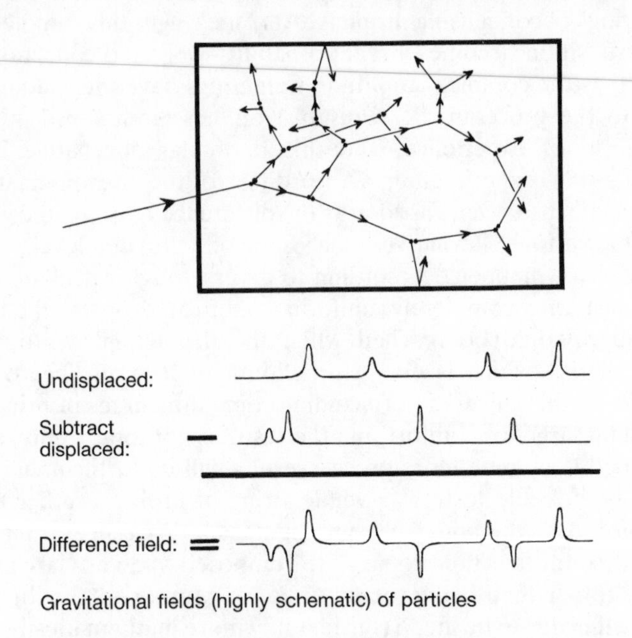

Fig. 8.8. If a particle enters a large box of some gas, then before long virtually every atom of the gas will have been disturbed. A quantum linear superposition of the particle entering, and of the particle not entering, would involve the linear superposition of two different space–time geometries, describing the gravitational fields of the two arrangements of gas particles. When does the *difference* between these geometries reach the one-graviton level?

wavefunction. But after a short while, all the atoms of the gas must be involved. Consider the complex superposition of two paths that might be taken by the particle, one entering the box and the other not. Standard quantum mechanics insists that we extend this superposition to all the atoms in the gas: we must superpose two states, where the gas atoms in one state are all moved from their positions in the other state. Now consider the *difference* in the gravitational fields of the totality of individual atoms. Even though the *overall* distribution of gas is virtually the same in the two states to be superposed (and the overall gravitational fields would be practically identical), if we *subtract* one field from the other, we get a (highly oscillating) *difference* field that might well be 'significant' in the sense that I mean it here—where the one-graviton level for this difference field is easily exceeded. As soon as this level is reached, then state-vector reduction takes place: in the *actual* state of the system, *either* the particle has entered the box of gas *or* it has not. The complex linear superposition has reduced to statistically weighted alternatives, and only *one* of these actually takes place.

In the previous example, I considered a cloud chamber as a way of providing a quantum-mechanical observation. It seems to me to be likely that other types of such observation (photographic plates, spark chambers, etc.) can be treated using the 'one-graviton criterion' by approaching them in the way that I have indicated for the box of gas above. Much work remains to be done in seeing how this procedure might apply in detail.

So far, this is only a germ of an idea for what I believe to be a much-needed new theory.[7] Any totally satisfactory scheme would, I believe, have to involve some very radical new ideas about the nature of space–time geometry, probably involving an essentially non-local description.[8] One of the most compelling reasons for believing this comes from the EPR-type experiments (cf. p.279, 286), where an 'observation' (here, the registering of a photo-cell) at one end of a room can effect the *simultaneous* reduction of the state vector at the other end. The construction of a fully objective theory of state-vector reduction which is consistent with the spirit of relativity is a profound challenge, since 'simultaneity' is a concept which is foreign to relativity, being dependent upon the motion of some observer. It is my opinion that our present picture of physical reality, particularly in relation to the nature of *time*, is due for a grand shake up—even greater, perhaps, than that which has already been provided by present-day relativity and quantum mechanics.

We must come back to our original question. How does all this relate to the physics which governs the actions of our brains? What could it have to do with our thoughts and with our feelings? To attempt some kind of answer, it will be necessary first to examine something of how our brains are actually constructed. I shall return afterwards to what I believe to be the fundamental question: what kind of new *physical* action is likely to be involved when we consciously think or perceive?

1. Popular such modifications are: (i) changing Einstein's actual equations **RICCI** = **ENERGY** (via 'higher-order Lagrangians'); (ii) changing the number of dimensions of space–time from four to some higher number (such as in the so-called 'Kaluza–Klein type theories'); (iii) introducing 'supersymmetry' (an idea borrowed from the quantum behaviour of bosons and fermions, combined into a comprehensive scheme and applied, not altogether logically, to the coordinates of space–time); (iv) string theory (a currently popular radical scheme in which 'world-lines' are replaced by 'string histories'—usually combined with the ideas of (ii) and (iii)). All these proposals, for all their popularity and forceful presentation, are firmly in the category TENTATIVE in the terminology of Chapter 5.
2. As far as I can make out, a viewpoint of this kind is implicit in Hawking's present proposals for a quantum gravity explanation of these matters (Hawking 1987, 1988). A proposal by Hartle and Hawking (1984) of a quantum gravity origin for the initial state is possibly the kind of thing that *could* give theoretical substance to an initial condition of the type **WEYL** = 0, but an (in my opinion) *essential* time-asymmetric input is so far absent from these ideas.
3. These facts are somewhat more transparent in terms of the *scalar product* operation $\langle\psi|\chi\rangle$ given in note 6 of Chapter 6. In the forward-time description we calculate the probability p by

   $$p = |\langle\psi|\chi\rangle|^2 = |\langle\chi|\psi\rangle|^2$$

   and in the backward-time description, by

   $$p = |\langle\chi'|\psi'\rangle|^2 = |\langle\psi'|\chi'\rangle|^2$$

   The fact that these are the same follows from $\langle\psi'|\chi'\rangle = \langle\psi|\chi\rangle$, which is essentially what one means by 'unitary evolution'.
4. Some readers may have trouble understanding what it can mean to ask what is the probability of a past event given a future event! There is no essential problem with this, however. Imagine the entire history of the universe mapped out in space–time. To find the probability of p occurring, given that q happens, we imagine examining all occurrences of q and counting up the fraction of these which are accompanied by p. This is the required probability. It does not matter whether q is the kind of event that would normally occur later or earlier in time than p.
5. These must be allowed to be what are called *longitudinal gravitons*—the 'virtual' gravitons that compose a static gravitational field. Unfortunately, there are theoretical problems involved in defining such things in a clear-cut and 'invariant' mathematical way.
6. My own original rough attempts at calculating this value were very greatly improved upon by Abhay Ashtekar, and I am using his value here (see Penrose 1987*a*). However, he has stressed to me that there is a good deal of arbitrariness in some of the assumptions that one seems compelled to use, so considerable caution must be exercised in adopting the precise mass value obtained.
7. Various other attempts at providing an objective theory of state-vector reduction have appeared from time to time in the literature. Most relevant are Károlyházy (1974), Károlyházy, Frenkel, and Lukács (1986), Komar (1969), Pearle (1985, 1988), Ghirardi, Rimini, and Weber (1986).

8. I have myself been involved, over the years, with trying to develop a non-local theory of space–time, largely motivated from other directions, referred to as 'twistor theory' (see Penrose and Rindler 1986, Huggett and Tod 1985, Ward and Wells 1989). However, this theory is, at best, still lacking in some essential ingredients, and it would not be appropriate to enter into a discussion of it here.

9
Real brains and model brains

What are brains actually like?

Inside our heads is a magnificent structure that controls our actions and somehow evokes an awareness of the world around. Yet, as Alan Turing once put it,[1] it resembles nothing so much as a bowl of cold porridge! It is hard to see how an object of such unpromising appearance can achieve the miracles that we know it to be capable of. Closer examination, however, begins to reveal the brain as having a much more intricate structure and sophisticated organization (Fig. 9.1). The large convoluted (and most porridge-like) portion on top is referred to as the *cerebrum*. It is divided cleanly down the middle into left and right *cerebral hemispheres*, and considerably less cleanly front and back into the frontal lobe and three other lobes: the parietal,

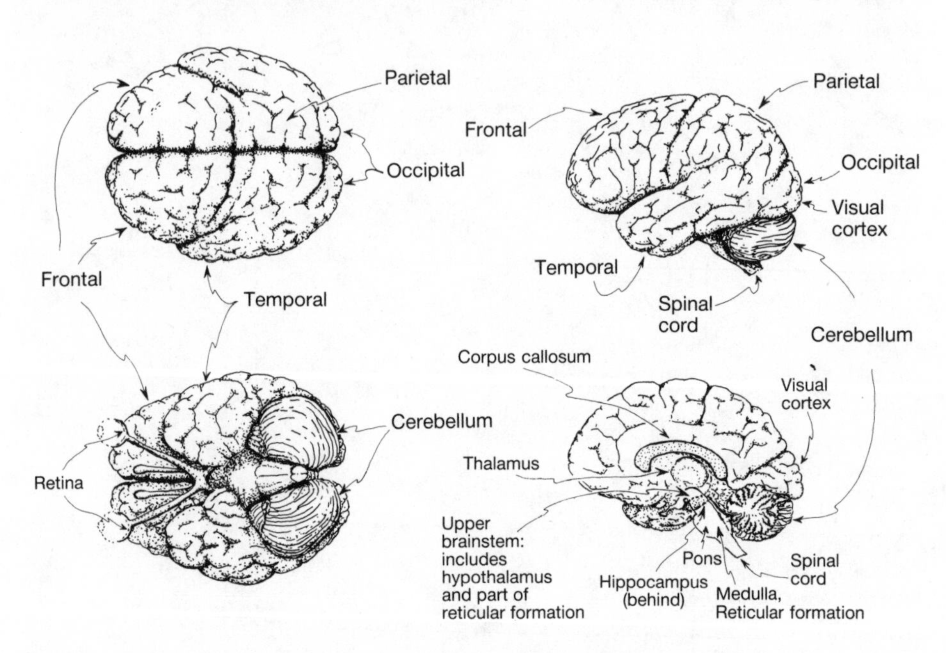

Fig. 9.1. The human brain: above, side, beneath, split.

temporal and occipital. Further down, and at the back lies a rather smaller, somewhat spherical portion of the brain—perhaps resembling two balls of wool—the *cerebellum*. Deep inside, and somewhat hidden under the cerebrum, lie a number of curious and complicated-looking different structures: the pons and medulla (including the reticular formation, a region that will concern us later) which constitute the brain-stem, the thalamus, hypothalamus, hippocampus, corpus callosum, and many other strange and oddly named constructions.

The part that human beings feel that they should be proudest of is the cerebrum—for that is not only the largest part of the human brain, but it is also larger, in its proportion of the brain as a whole, in *man* than in other animals. (The cere*bellum* is also larger in man than in most other animals.) The cerebrum and cerebellum have comparatively thin outer surface layers of *grey matter* and larger inner regions of *white matter*. These regions of grey matter are referred to as, respectively, the *cerebral cortex* and the *cerebellar cortex*. The grey matter is where various kinds of computational task appear to be performed, while the white matter consists of long nerve fibres carrying signals from one part of the brain to another.

Various parts of the cerebral cortex are associated with very specific functions. The *visual cortex* is a region within the occipital lobe, right at the back of the brain, and is concerned with the reception and interpretation of vision. It is curious that Nature should choose this region to interpret signals from the eyes which, at least in man, are situated right at the *front* of the head! But Nature behaves in more curious ways than just this. It is the *right* cerebral hemisphere which is concerned almost exclusively with the *left*-hand side of the body, while the left cerebral hemisphere is concerned with the right-hand side of the body—so that virtually all nerves must cross over from one side to the other as they enter or leave the cerebrum! In the case of the visual cortex, it is not that the right side is associated with the left eye, but with the *left-hand field of vision* of *both eyes*. Similarly the left visual cortex is associated with the right-hand field of vision of both eyes. This means that the nerves from the right-hand side of the retina of each eye must go to the right visual cortex (recall that the image on the retina is inverted) and that the nerves from the left-hand side of the retina of each eye must go to the left visual cortex. (See Fig. 9.2.) In this way a very well-defined map of the left-hand visual field is formed on the right visual cortex and another map is formed of the right-hand visual field on the left visual cortex.

Signals from the ears also tend to cross over to the opposite side of the brain in this curious way. The right auditory cortex (part of the right temporal lobe) deals mainly with sound received on the left and the left auditory cortex, on the whole, with sounds from the right. Smell seems to be an exception to the general rules. The right olfactory cortex, situated at the front of the cerebrum (in the frontal lobe—which is itself exceptional, for a sensory

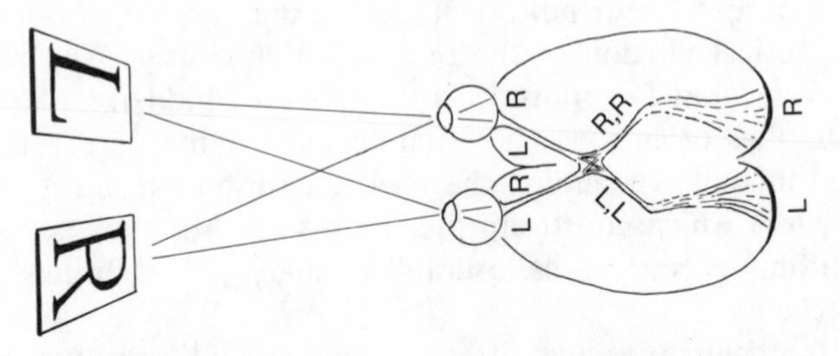

Fig. 9.2. The left-hand field of vision of both eyes is mapped to the right visual cortex and the right-hand field of vision to the left visual cortex. (View from below; note that the images on the retina are inverted.)

area), deals largely with the right nostril and the left with the left nostril.

The sensations of *touch* have to do with the region of the parietal lobe referred to as the *somatosensory* cortex. This region occurs just behind the division between the frontal and parietal lobes. There is a very specific correspondence between the various parts of the surface of the body and the regions of the somatosensory cortex. This correspondence is sometimes graphically illustrated in terms of what is referred to as the 'somatosensory homunculus', which is a distorted human figure pictured as lying along the somatosensory cortex as in Fig. 9.3. The right somatosensory cortex deals with sensations from the left-hand side of the body, and the left, with the right-hand side. There is a corresponding region of the *frontal* lobe, lying just in *front* of the division between the frontal and parietal lobes, known as the *motor* cortex. This is concerned with activating the *movement* of different parts of the body and again there is a very specific correspondence between the various muscles of the body and the regions of the motor cortex. We now have a 'motor homunculus' to depict this correspondence, as given in Fig. 9.4.

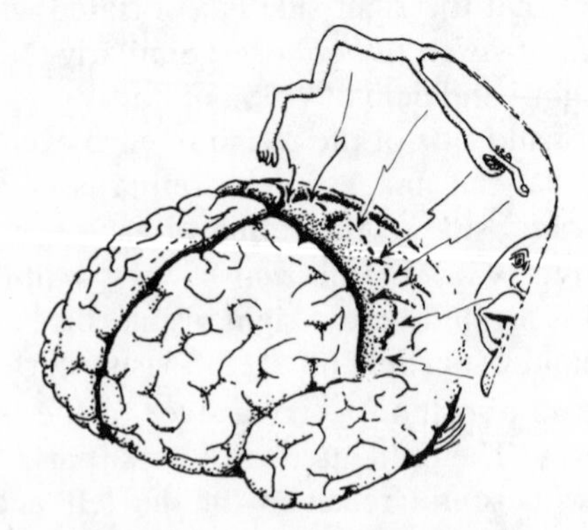

Fig. 9.3. The 'somatosensory homunculus' graphically illustrates the portions of the cerebrum—just behind the division between frontal and parietal lobes—which, for various parts of the body, are most immediately concerned with the sense of touch.

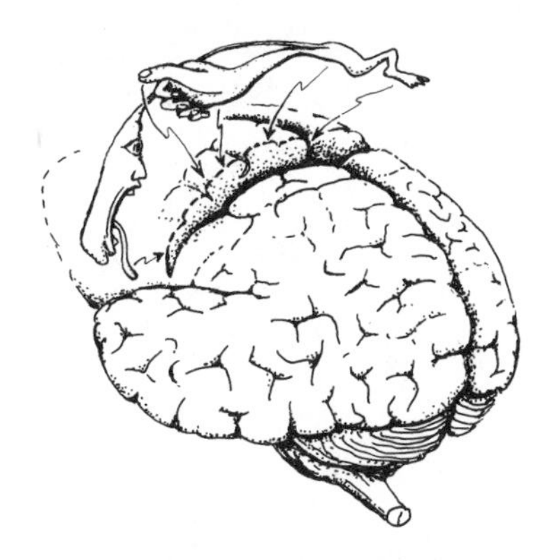

Fig. 9.4. The 'motor homunculus' illustrates the portions of the cerebrum—just in front of the division between frontal and parietal lobes—which, for various parts of the body, most directly activate movement.

The right motor cortex controls the left-hand side of the body, and the left motor cortex, the right-hand side.

The regions of the cerebral cortex just referred to (the visual, auditory, olfactory, somatosensory, and motor) are called *primary*, since they are the ones most directly concerned with the input and output of the brain. Near to these primary regions are the *secondary* regions of the cerebral cortex, which are concerned with a more subtle and complex level of abstraction. (See Fig. 9.5.) The sensory information received by the visual, auditory, and somatosensory cortexes is processed at the associated secondary regions, and the secondary motor region is concerned with conceived plans of motion which get translated into more specific directions for actual muscle movement by the primary motor cortex. (Let us leave aside the olfactory cortex in our considerations, since it behaves differently, and rather little seems to be known about it.) The remaining regions of the cerebral cortex are referred to as *tertiary* (or the *association* cortex). It is largely in these tertiary regions that the most abstract and sophisticated activity of the brain is carried out. It is here—in conjunction, to some extent, with the periphery—that the information from various different sensory regions is intertwined and analysed in a very complex way, memories are laid down, pictures of the outside world are constructed, general plans are conceived and evaluated, and speech is understood or formulated.

Speech is particularly interesting, because it is normally thought of as something very specific to human intelligence. It is curious that (at least in the vast majority of right-handed people and in most left-handed people) the *speech* centres are mainly just on the *left-hand* side of the brain. The essential areas are *Broca's area*, a region in the lower rear part of the frontal lobe, and another called *Wernicke's area*, in and around the upper rear part of the

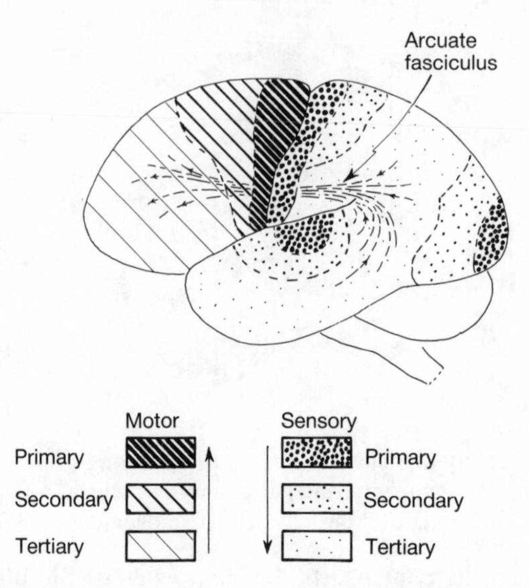

Fig. 9.5. The action of the cerebrum in rough broad outline. External sense-data enter at the primary sensory regions, are processed to successive degrees of sophistication in the secondary and tertiary sensory regions, transferred to the tertiary motor region, and are finally refined into specific instructions for movement at the primary motor regions.

temporal lobe (see Fig. 9.6). Broca's area is concerned with the formulation of sentences, and Wernicke's area with the comprehension of language. Damage to Broca's area impairs speech but leaves comprehension intact, whereas with damage to Wernicke's area, speech is fluent but with little content. A nerve bundle called the *arcuate fasciculus* connects the two areas. When this is damaged, comprehension is not impaired and speech remains fluent, but comprehension cannot be vocalized.

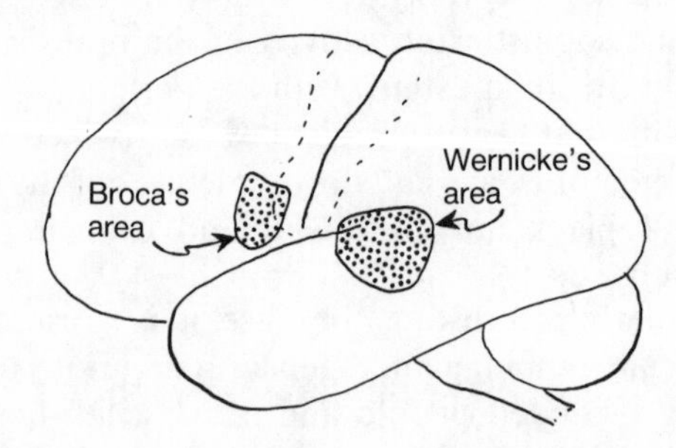

Fig. 9.6. Normally on the left side only: Wernicke's area is concerned with the comprehension and Broca's with the formulation of speech.

We can now form a very crude picture of what the cerebrum does. The brain's *input* comes from visual, auditory, tactile, and other signals which first register in the cerebrum at the *primary* portions of (mainly) the *rear* lobes (parietal, temporal, and occipital). The brain's *output*, in the form of activating movements of the body, is mainly achieved by primary portions of the *frontal* lobes of the cerebrum. In between, some kind of processing takes place. In a general way, there is a movement of the brain's activity starting at the primary portions of the rear lobes, moving on to the secondary portions, as the input data are analysed, and then on to the tertiary portions of the rear lobes as these data become fully comprehended (e.g. as with the comprehension of speech at Wernicke's area). The arcuate fasciculus—the bundle of nerve fibres referred to above, but now on both sides of the brain—then carries this processed information to the frontal lobe, where general plans of action are formulated in its tertiary regions (e.g. as with the formulation of speech in Broca's area). These general plans of action are translated into more specific conceptions about movements of the body in the secondary motor regions and, finally, the brain's activity moves on to the primary motor cortex where signals are sent on, ultimately, to various groups of muscles in the body (and often to several at once).

The picture of a superb computing device seems to be presented to us. The supporters of strong AI (cf. Chapter 1, etc.) would hold that here we have a supreme example of an algorithmic computer—a Turing machine in effect—where there is an input (like the Turing machine's input tape on the left) and an output (like the machine's output tape on the right), and all sorts of complicated computations being carried out in between. Of course the brain's activity can also carry on independently of specific sensory input. This occurs when one simply thinks, calculates, or muses over recollections of the past. To strong AI supporters, these activities of the brain would simply be further algorithmic activity, and they might propose that the phenomenon of 'awareness' arises whenever such internal activity reaches a sufficient level of sophistication.

We must not jump in too fast with our ready explanations, however. The general picture of the cerebrum's activity, as presented above, is only a very rough one. In the first place, even the reception of vision is not as clear-cut as I have presented it. There appear to be several different (though more minor) regions of the cortex where maps of the visual field are made, apparently with various other purposes. (Our *awareness* of vision seems to differ with respect to them.) There also seem to be other subsidiary sensory and motor regions scattered about the cerebral cortex (for example, movements of the eye can be activated by various points in the *rear* lobes).

I have not even mentioned the roles of parts of the brain other than the cerebrum, in my descriptions above. What is the role of the *cerebellum*, for example? Apparently it is responsible for precise co-ordination and control of the body—its timing, balance, and delicacy of movement. Imagine the flowing artistry of a dancer, the easy accuracy of a professional tennis player, the lightning control of a racing driver, and the sure movements of a painter's or musician's hands; imagine also, the graceful bounds of a gazelle and the stealth of a cat. Without the cerebellum, such precision would not be possible, and all movement would become fumbling and clumsy. It seems that, when one is learning a new skill, be it walking or driving a car, initially one must think through each action in detail, and the cerebrum is in control; but when the skill has been mastered—and, has become 'second nature'—it is the cerebellum that takes over. Moreover, it is a familiar experience that if one *thinks about* one's actions in a skill that has been so mastered, then one's easy control may be temporarily lost. *Thinking* about it seems to involve the re-introduction of cerebral control and, although a consequent flexibility of activity is thereby introduced, the flowing and precise cerebellar action is lost. No doubt such descriptions are oversimplified, but they give a reasonable flavour of the cerebellum's role*

It was misleading, also, in my earlier description of the action of the cerebrum, to leave out all account of the other parts of the brain. For example, the *hippocampus* plays a vital role in laying down long-term (permanent) memories, the actual memories being stored somewhere in the cerebral cortex—probably in many places at once. The brain can also retain images on a *short*-term basis in other ways; and it can hold them for some minutes or even hours (perhaps by 'keeping them in mind'). But in order to be able to recall such images after they have left one's attention, it is necessary that they be laid down in a permanent way, and for this the hippocampus is essential. (Damage to the hippocampus causes a fearful condition in which no new memories are retained once they have left the subject's attention.) The *corpus callosum* is the region via which the right and left cerebral hemispheres communicate with one another. (We shall be seeing later some of the striking implications of having the corpus callosum severed.) The *hypothalamus* is the seat of emotion—pleasure, rage, fear, despair, hunger—and it mediates both the mental and the physical manifestations of emotion. There is a continual flow of signals between the hypothalamus and different parts of the cerebrum. The *thalamus* acts as an important processing centre and relay station, and it conveys many of the nerve inputs from the external world to the cerebral cortex. The *reticular formation* is responsible for the general state of alertness or awareness involved in the brain as a whole or in different parts of the brain. There are numerous pathways of nerves connecting these, and many other, vitally important areas.

The description above gives only a sample of some of the more important

* Curiously, the 'crossing-over' behaviour of the cerebrum does not apply to the cerebellum, so that the right-half of the cerebellum largely controls the *right* side of the body and the left-half the *left* side of the body.

parts of the brain. I should end the section by giving a little more about the brain's organization as a whole. Its different parts are classified into three regions which, taken in order, moving away from the spinal column, are called the *hindbrain* (or rhombencephalon), the *midbrain* (or mesencephalon), and the *forebrain* (or prosencephalon). In the developing early embryo one finds these three regions, in this order, as three swellings at the end of the spinal column. The one at the very end, the developing forebrain, sprouts two bulbous swellings, one on either side, which become the cerebral hemispheres. The fully developed forebrain includes many important parts of the brain—not just the cerebrum, but the corpus callosum, thalamus, hypothalamus, hippocampus, and many other parts as well. The cerebellum is part of the hindbrain. The reticular formation has one part in the midbrain and another part in the hindbrain. The forebrain is the 'newest' in the sense of evolutionary development, and the hindbrain is the most 'ancient'.

I hope that this brief sketch, though in numerous ways inadequate, will give the reader some idea of what a human brain is like and what it does in a general way. So far, I have barely touched upon the central issue of *consciousness*. Let us address this issue next.

Where is the seat of consciousness?

Many differing views have been expressed with regard to the relation of the state of the brain to the phenomenon of consciousness. There is remarkably little consensus of opinion for a phenomenon of such obvious importance. It is clear, however, that all parts of the brain are not equally involved in its manifestation. For example, as hinted above, the cerebellum seems to be much more of an 'automaton' than the cerebrum. Actions under cerebellar control seem almost to take place 'by themselves' without one having to 'think about' them. While one may consciously decide to walk from one place to another, one does not often become aware of the elborate plan of detailed muscle movements that would be necessary for controlled motion. The same may be said of unconscious reflex actions, such as the removal of one's hand from a hot stove, which might be mediated not by the brain at all but by the upper part of the spinal column. From this, at least, one may be well inclined to infer that the phenomenon of consciousness is likely to have more to do with the action of the cerebrum than with the cerebellum or the spinal cord.

On the other hand, it is very far from clear that the activity of the cerebrum must itself *always* impinge upon our awareness. For example, as I have described above, in the normal action of walking, where one is *not* conscious of the detailed activity of one's muscles and limbs—the control of this activity being largely cerebellar (helped by other parts of the brain and spinal cord)—primary motor regions of the cerebrum would seem *also* to be involved. Moreover, the same would be true of the primary sensory regions: one might not be aware, at the time, of the varying pressures on the soles of one's feet as one walks, but the corresponding regions of the somatosensory cortex would still be being continually activated.

Indeed, the distinguished Canadian neurosurgeon Wilder Penfield (who, in the 1940s and 1950s, was responsible for much of the detailed mapping of the motor and sensory regions of the human brain) has argued that one's awareness is *not* associated simply with cerebral activity. He suggested, on the basis of his experiences in performing numerous brain operations on conscious subjects, that some region of what he referred to as the *upper brain-stem*, consisting largely of the thalamus and the midbrain (cf. Penfield and Jasper 1947)—though he had mainly in mind the reticular formation—should, in a sense, be regarded as the 'seat of consciousness'. The upper brain-stem is in communication with the cerebrum, and Penfield argued that 'conscious awareness' or 'consciously willed action' would arise whenever this region of brain-stem is in direct communication with the appropriate region of the cerebral cortex, namely the particular region associated with whatever specific sensations, thoughts, memories, or actions are being consciously perceived or evoked at the time. He pointed out that while he might, for example, stimulate the region of the subject's motor cortex which causes the right arm to move (and the right arm would indeed move), this would not cause the subject to *want* to move the right arm. (Indeed, the subject might even reach over with the left arm and stop the right arm's movement—as in Peter Sellers's well-known cinema portrayal of Dr Strangelove!) Penfield suggested that the *desire* for movement might have more to do with the thalamus than the cerebral cortex. His view was that consciousness is a manifestation of activity of the upper brain-stem, but since in addition there needs to be something to be conscious *of*, it is not just the brain-stem that is involved, but also some region in the cerebral cortex which is at that moment in communication with the upper brain-stem and whose activity represents the subject (sense impression or memory) or object (willed action) of that consciousness.

Other neurophysiologists have also argued that, in particular, the reticular formation might be taken to be the 'seat' of consciousness, if such a seat indeed exists. The reticular formation, after all, is responsible for the general state of alertness of the brain (Moruzzi and Magoun 1949). If it is damaged, then unconsciousness will result. Whenever the brain is in a waking conscious state, then the reticular formation is active; when not, then it is not. There does indeed appear to be a clear association between activity of the reticular formation and that state of a person that we normally refer to as 'conscious'. However, the matter is complicated by the fact that in the state of dreaming, where one is indeed 'aware', in the sense of being aware of the dream itself, normally active parts of the reticular formation seem *not* to be active. A thing that also worries people about assigning such an honoured status to the reticular formation is that, in evolutionary terms, it is a very *ancient* part of the brain. If all that one needs to be conscious is an active reticular formation, then frogs, lizards, and even codfish are conscious!

Personally, I do not regard this last argument as a very strong one. What evidence do we have that lizards and codfish do *not* possess some low-level form of consciousness? What right do we have to claim, as some might, that human beings are the only inhabitants of our planet blessed with an actual ability to be 'aware'? Are we alone, among the creatures of earth, as things whom it is possible to 'be'? I doubt it. Although frogs and lizards, and especially codfish, do not inspire me with a great deal of conviction that there is necessarily 'someone there' peering back at me when I look at them, the impression of a 'conscience presence' is indeed very strong with me when I look at a dog or a cat or, especially, when an ape or monkey in the zoo looks at me. I do not demand that they feel as I do, nor even that there is much sophistication about what they feel. I do not ask that they are 'self-aware' in any strong sense (though I would guess that an element of self-awareness can be present*). All I ask is that they sometimes simply *feel*! As for the state of dreaming, I would myself accept that there is some form of awareness present, but presumably at quite a low level. If parts of the reticular formation are, in some way, solely responsible for awareness, then they ought to be active, although at a low level during the dreaming state.

Another viewpoint (O'Keefe 1985) seems to be that it is the action of the *hippocampus* that has more to do with the conscious state. As I remarked earlier, the hippocampus is crucial to the laying down of long-term memories. A case can be made that the laying down of permanent memories is associated with consciousness, and if this is right, the hippocampus would indeed play a central role in the phenomenon of conscious awareness.

Others would hold that it is the cerebral cortex itself which is responsible for awareness. Since the cerebrum is man's pride (though dolphins' cerebrums are as big!) and since the mental activities most closely associated with intelligence appear to be carried out by the cerebrum, then surely it is here that the soul of man resides! That would presumably be the conclusion of the point of view of strong AI, for example. If 'awareness' is merely a feature of the *complexity* of an algorithm—or perhaps with its 'depth' or some 'level of subtlety'—then, according to the strong-AI view, the complicated algorithms being carried out by the cerebral cortex would give that region the strongest claim to be that capable of manifesting consciousness.

Many philosophers and psychologists seem to take the view that human consciousness is very much bound up with human *language*. Accordingly, it is

* There is some persuasive evidence that chimpanzees, at least, are capable of self-awareness, as experiments in which chimpanzees are allowed to play with mirrors seem to show cf. Oakley (1985) Chapters 4 and 5.

only by virtue of our linguistic abilities that we can attain a subtlety of thinking that is the very hallmark of our humanity—and the expression of our very souls. It is language, according to this view, that distinguishes us from other animals, and so provides us with our excuse for depriving them of their freedom and slaughtering them when we feel that such need arises. It is language that allows us to philosophize and to describe how we feel, so we may convince others that *we* possess awareness of the outside world and are aware also of ourselves. From this viewpoint, our language is taken to be the key ingredient of our possession of consciousness.

Now, we must recall that our language centres are (in the vast majority of people) just on the *left*-hand sides of our brains (Broca's and Wernicke's areas). The viewpoint just expressed would seem to imply that consciousness is something to be associated only with the left cerebral cortex and not with the right! Indeed, this appears to be the opinion of a number of neurophysiologists (in particular, John Eccles 1973), though to me, as an outsider, it seems a very odd view indeed, for reasons that I shall explain.

Split-brain experiments

In relation to this, I should mention a remarkable collection of observations concerning human subjects (and animals) who have had their corpus callosums completely severed, so that the left and right hemispheres of the cerebral cortex are unable to communicate with one another. In the case of the humans,[2] the severing of the corpus callosum was performed as a therapeutic operation, it being found that this was an effective treatment for a particularly severe form of epilepsy from which these subjects suffered. Numerous psychological tests were given to them by Roger Sperry and his associates some time after they had had these operations. They would be placed in such a way that the left and right fields of vision would be presented with completely separate stimuli, so the left hemisphere would receive visual information only of what was being displayed on the right-hand side, and the right hemisphere, only of that on the left-hand side. If a picture of a pencil was flashed on the right and a picture of a cup on the left, the subject would vocalize 'That's a *pencil*', because the pencil and not the cup would be perceived by the only side of the brain apparently capable of speech. However, the left hand would be capable of selecting a saucer, rather than a piece of paper, as the appropriate object to associate with the cup. The left hand would be under the control of the right hemisphere, and although incapable of speech, the right hemisphere would perform certain quite complex and characteristically human actions. Indeed, it has been suggested that *geometrical thinking* (particularly in three-dimensions), and also music, may normally be carried out mainly within the *right* hemisphere, to give balance to the verbal and analytical abilities of the left. The right brain can understand common nouns or elementary sentences, and can carry out very simple arithmetic.

What is most striking about these split-brain subjects is that the two sides seem to behave as virtually independent individuals, each of which may be communicated with separately by the experimenter—although communication is more difficult, and on a more primitive level, with the right hemisphere than with the left, owing to the right's lack of verbal ability. One half of the subject's cerebrum may communicate with the other in simple ways, e.g. by watching the motion of the arm controlled by the other side, or perhaps by hearing tell-tale sounds (like the rattling of a saucer). But even this primitive communication between the two sides can be removed by carefully controlled laboratory conditions. Vague emotional feelings can still be passed from one side to the other, however, presumably because structures which are not split, such as the hypothalamus, are still in communication with both sides.

One is tempted to raise the issue: do we have two separately conscious individuals both inhabiting the same body? This question has been the subject of much controversy. Some would maintain that the answer must surely be 'yes', while others claim that neither side is to be regarded as an individual in its own right. Some would argue that the fact that emotional feelings can be common to the two sides is evidence that there is still just a single individual involved. Yet another viewpoint is that only the *left*-hand hemisphere represents a conscious individual, and the right-hand one is an automaton. This view appears to be believed by some who hold that language is an essential ingredient of consciousness. Indeed, only the left hemisphere can convincingly plead 'Yes!' to the verbal question 'Are you conscious?' The right hemisphere, like a dog, a cat or a chimpanzee, might be hard-put even to decipher the words that constitute the question, and it may be unable properly to vocalize its answer.

Yet the issue cannot be dismissed that lightly. In a more recent experiment of considerable interest, Donald Wilson and his coworkers (Wilson *et al.* 1977; Gazzaniga, LeDoux, and Wilson 1977) examined a split-brain subject, referred to as 'P. S.'. After the splitting operation, only the left hemisphere could speak but *both* hemispheres could comprehend speech; later the right hemisphere learned to speak also! Evidently *both* hemispheres were conscious. Moreover, they appeared to be *separately* conscious, because they had different likes and desires. For example, the left hemisphere described that its wish was to be a draughtsman and the right, a racing driver!

I, myself, simply cannot believe the common claim that ordinary human language is necessary for thought or for consciousness. (In the next chapter I shall give some of my reasons.) I therefore side with those who believe, generally, that the two halves of a split-brain subject can be independently conscious. The example of P. S. strongly suggests that, at least in this particular case, both halves indeed can be. In my own opinion the only real difference between P. S. and the others, in this respect, is that his right-hand consciousness could actually convince others of its existence!

If we accept that P. S. indeed has two independent minds, then we are presented with a remarkable situation. Presumably, *before* the operation each split-brain subject possessed but a single consciousness; but afterwards there are two! In some way, the original single consciousness has *bifurcated*. We may recall the hypothetical traveller of Chapter 1 (p. 27), who subjected himself to the teleportation machine, and who (inadvertantly) awoke to find that his allegedly 'actual' self had arrived on Venus. There, the bifurcation of his consciousness would seem to provide a paradox. For we may ask, "Which route did his stream of consciousness 'actually' follow?" If *you* were the traveller, then which one would end up as 'you'? The teleportation machine could be dimissed as science fiction, but in the case of P. S. we appear to have something seemingly analogous, but which has *actually happened*! Which of P. S.'s consciousnesses 'is' the P. S. of before the operation? No doubt many philosophers would dismiss the question as meaningless. For there seems to be no operational way of deciding the issue. Each hemisphere would share memories of a conscious existence before the operation and, no doubt, both would claim to be that person. That may be remarkable, but it is not in itself a paradox. Nevertheless, there is something of a definite puzzle that we are still left with.

The puzzle would be further exacerbated if somehow the two consciousnesses could later be brought together again. Reattaching the individual severed nerves of the corpus callosum would seem out of the question, with present technology, but one might envisage something milder than actual severence of the nerve fibres in the first place. Perhaps these fibres could be temporarily frozen, or paralyzed with some drug. I am not aware that any such experiment has actually been carried out, but I suppose that it could become technically feasible before too long. Presumably, after the corpus callosum has been re-activated, only *one* consciousness would result. Imagine that this consciousness is you! How would it feel to have been two separate people with distinct 'selves' at some time in the past?

Blindsight

The split-brain experiments seem at least to indicate that there need not be a *unique* 'seat of consciousness'. But there are other experiments that appear to suggest that some parts of the cerebral cortex are more to be associated with consciousness than are others. One of these has to do with the phenomenon of *blindsight*. Damage to a region of the visual cortex can cause blindness in the corresponding field of vision. If an object is held in that region of the visual field, then that object will not be perceived. Blindness occurs with respect to that region of vision.

However, some curious findings (cf. Weiskrantz 1987) indicate that things are not so simple as this. A patient, referred to as 'D. B.' had to have some of his visual cortex removed, and this caused him to be unable to see anything in a certain region of the visual field. However, when something was placed in this region and D. B. was asked to *guess* what that something was (usually a mark like a cross or a circle, or a line segment tilted at some angle), he found he could do so with near to 100 per cent accuracy! The accuracy of these 'guesses' came as a surprise to D. B. himself, and he still maintained that he could perceive nothing whatever in that region.*

Images received by the retina are also processed in certain regions of the brain *other* than just the visual cortex, one of the more obscure regions involved lying in the lower temporal lobe. It seems that D. B. may have been basing his 'guesses' on information gained by this lower temporal region. Nothing was directly perceived *consciously* by the activation of these regions, yet the information was there, to be revealed only in the correctness of D. B.'s 'guesses'. In fact, after some training, D. B. was able to obtain a limited amount of actual awareness in respect of these regions.

All this seems to show that some areas of the cerebral cortex (e.g. the visual cortex) are more associated with conscious perception than are other areas, but that with training, some of these other areas can apparently be brought within the scope of direct awareness.

Information processing in the visual cortex

More than any other part of the brain, it is the *visual cortex* that is best understood as to how it deals with the information that it receives; and various models have been put forward to account for this action.[3] In fact, some processing of visual information takes place in the retina itself, *before* the visual cortex is reached. (The retina is actually considered to be part of the brain!) One of the first experiments hinting at how processing is carried out in the visual cortex was that which earned David Hubel and Torsten Wiesel the Nobel Prize in 1981. In their experiments they were able to show that certain cells in the visual cortex of a cat were responsive to lines in the visual field which had a *particular angle of slope*. Other nearby cells would be responsive to lines having some different angle of slope. It would not often matter what it was that had this angle. It could be a line marking the boundary from dark to light or else from light to dark, or just a dark line on a light

* Somewhat complementary to blindsight is a condition known as 'blindness denial' according to which a subject who is in fact totally blind insists on being able to see quite well, seeming to be *visually* conscious of the *inferred* surroundings! (See Churchland 1984, p. 143.)

background. The one feature of 'angle of slope' had been abstracted by the particular cells that were being examined. Yet other cells would respond to particular colours, or to the differences between what is received by each eye, so that perception of depth may be obtained. As one moves away from the primary reception regions, one finds cells that are sensitive to more and more subtle aspects of our perception of what we see. For example, the image of a complete white triangle is perceived when we look at the drawing in Fig. 9.7; yet the lines forming the triangle itself are not actually mainly present in the figure but are *inferred*. Cells in the visual cortex (in what is called the secondary visual cortex) have actually been found which can register the positions of these inferred lines!

Fig. 9.7. Can you see a white triangle, lying above another triangle to which it is held by a ring? The borders of the white triangle are not drawn in everywhere, yet there are cells in the brain which respond to these invisible but perceived lines.

There were claims in the literature,[4] in the early 1970s, of the discovery of a cell in a monkey's visual cortex which responded only when the image of a *face* registered on the retina. On the basis of such information, the 'grandmother-cell hypothesis' was formulated, according to which there would be certain cells in the brain which would respond only when the subject's grandmother entered the room! Indeed, there are recent findings indicating that certain cells are responsive only to particular words. Perhaps this goes some way towards verifying the grandmother-cell hypothesis?

Clearly there is an enormous amount to be learnt about the detailed processing that the brain carries out. Very little is known, as yet, about how the higher brain centres carry out their duties. Let us leave this question now and turn our attention to the actual cells in the brain that enable it to achieve these remarkable feats.

How do nerve signals work?

All the processing that is done by the brain (and also by the spinal column and retina) is achieved by the remarkably versatile cells of the body referred to as *neurons*.[5] Let us try to see what a neuron is like. In Fig. 9.8 I have given a picture of a neuron. There is a central bulb, perhaps somewhat starlike, often shaped rather like a radish, called the *soma*, which contains the nucleus of the cell. Stretching out from the soma at one end is a long nerve fibre—sometimes very long indeed, considering that we are referring to but a single microscopic cell (often up to several centimetres long, in a human)—known as the *axon*. The axon is the 'wire' along which the cell's *output* signal is transmitted. Sprouting off from the axon may be many smaller branches, the axon bifurcating a number of times. Terminating each of these resulting nerve fibres is to be found a little *synaptic knob*. At the other end of the soma and often branching out in all directions from it, are the tree-like *dendrites*, along which *input* data are carried in to the soma. (Occasionally there are synaptic knobs also on dendrites, giving what are called *dendrodendritic* synapses between dendrites. I shall ignore these in my discussion, since the complication they introduce is inessential.)

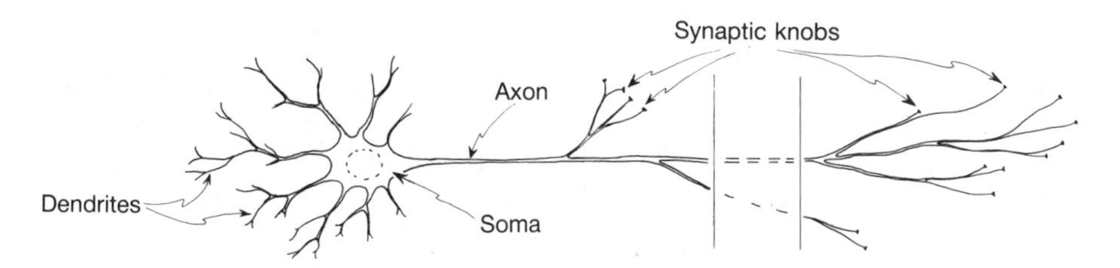

Fig. 9.8. A neuron (often relatively very much longer than indicated). Different types of neuron vary greatly in detailed appearance.

The whole cell, being a self-contained unit, has a cell membrane surrounding soma, axon, synaptic knobs, dendrites, and all. For signals to pass from one neuron to another, it is necessary that they somehow 'leap the barrier' between them. This is done at a junction known as a *synapse* where a synaptic knob of one neuron is attached to some point on another neuron, either on the soma of that neuron itself or else on one of its dendrites, (see Fig. 9.9). Actually there is a very narrow gap between the synaptic knob and the soma or dendrite to which it is attached, called the *synaptic cleft* (see Fig. 9.10). The signal from one neuron to the next has to be propagated across this gap.

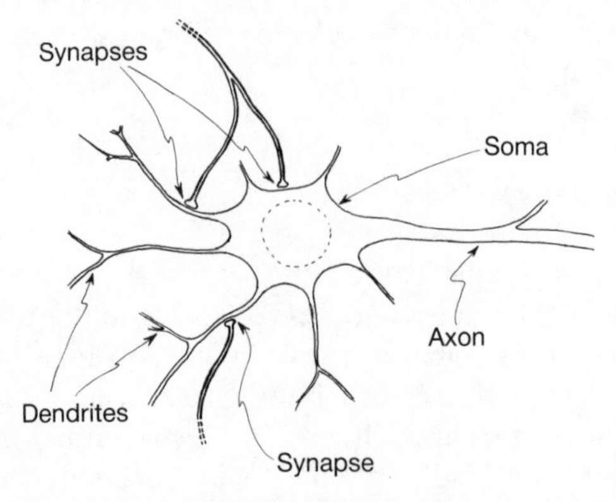

Fig. 9.9. Synapses: the junctions between one neuron and the next.

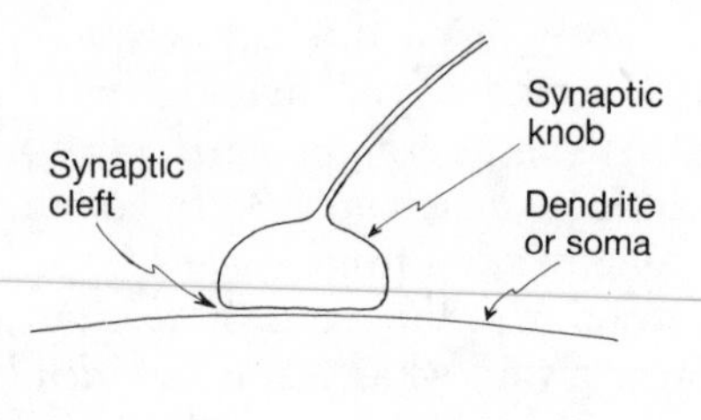

Fig. 9.10. A synapse in closer detail. There is a narrow gap across which neurotransmitter chemicals flow.

What form do the signals take as they propagate along nerve fibres and across the synaptic clefts? What causes the next neuron to emit a signal? To an outsider like myself, the procedures that Nature has actually adopted seem extraordinary—and utterly fascinating! One might have thought that the signals would just be like electric currents travelling along wires, but it is much more complicated than that.

A nerve fibre basically consists of a cylindrical tube containing a mixed solution of ordinary salt (sodium chloride) and potassium chloride, mainly the latter, so there are sodium, potassium, and chloride ions within the tube (Fig. 9.11). These ions are also present outside, but in different proportions, so that outside, there are more sodium ions than potassium ions. In the resting state of the nerve, there is a net negative electric charge inside the tube (i.e. more chloride ions than sodium and potassium together—recall that sodium and potassium ions are positively charged, and chloride ions negatively charged) and a net positive charge outside (i.e. more sodium and potassium than chloride). The cell membrane that constitutes the surface of the cylinder is somewhat 'leaky', so that the ions tend to migrate across it and neutralize

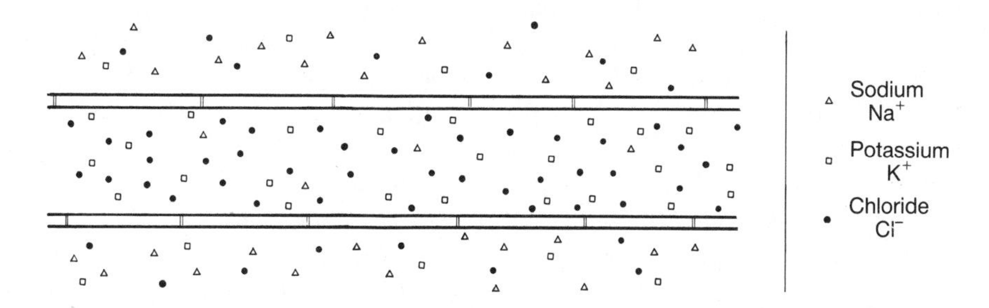

Fig. 9.11. Schematic representation of a nerve fibre. In the resting state there is an excess of chloride over sodium and potassium ions inside, giving a negative charge; and the other way around outside, giving a positive charge. The sodium/potassium balance is also different outside from inside, with more potassium inside and sodium outside.

the charge difference. To compensate for this and keep up the negative charge excess on the inside, there is a 'metabolic pump' which very slowly pumps sodium ions back out through the surrounding membrane. This also partly serves to maintain the excess of potassium over sodium on the inside. There is another metabolic pump which (on a somewhat smaller scale) pumps potassium ions in from the outside, thereby contributing to the potassium excess on the inside (though it works against maintaining the charge imbalance).

A *signal* along the fibre consists of a region in which this charge imbalance is *reversed* (i.e. now positive inside and negative outside) moving along the fibre (Fig. 9.12). Imagine oneself to be situated on the nerve-fibre ahead of

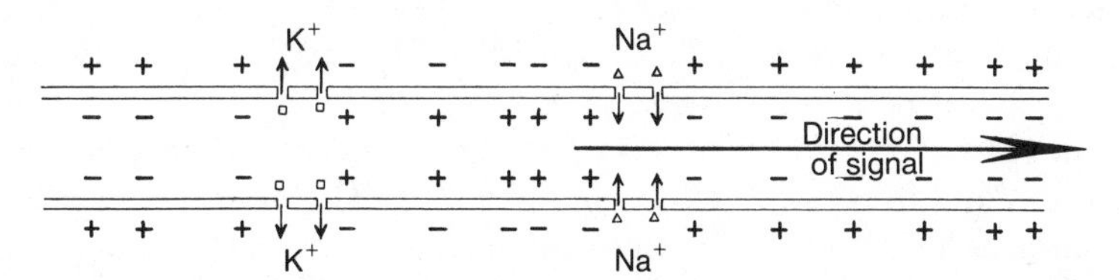

Fig. 9.12. A nerve signal is a region of charge reversal travelling along the fibre. At its head, sodium gates open to allow the sodium to flow inwards; and at its tail, potassium gates open to allow the potassium to flow outwards. Metabolic pumps act to restore the *status quo*.

such a region of charge-reversal. As the region approaches, its electric field causes little 'doors', called *sodium gates*, to open in the cell membrane; this allows sodium ions to flow back through (by a combination of electrical forces and the pressures due to differences in concentration, i.e. 'osmosis') from outside to inside. This results in the charge becoming positive inside and negative outside. When this has happened, the region of charge-reversal which constitutes the signal has reached us. That now causes another set of little 'doors' to open (*potassium gates*) which allow potassium ions to flow back out from the inside, and so begin to restore the excess of negative charge on the inside. The signal has now passed! Finally, as the signal recedes into the distance again, the slow but relentless action of the pumps pushes the sodium ions back out again and potassium ions back in. This restores the resting state of the nerve fibre, and it is ready for another signal.

Note that the signal consists simply of a region of charge reversal moving along the fibre. The actual *material* (i.e. ions) moves very little—just in and out across the cell membrane!

This curiously exotic mechanism appears to work very efficiently. It is employed universally, both by vertebrates and invertebrates. But the vertebrates perfected a further innovation, namely to have the nerve-fibre surrounded by an insulating coating of a whitish fatty substance called *myelin*. (It is this myelin coating which gives the 'white matter' of the brain its colour.) This insulation enables nerve signals to travel unabated (between 'relay stations') at a very respectable speed—up to some 120 metres per second.

When a signal reaches a synaptic knob, it emits a chemical substance known as a neurotransmitter. This substance travels across the synaptic cleft to another neuron—either at a point on one of its dendrites or on the soma itself. Now some neurons have synaptic knobs which emit a neurotransmitter chemical with a tendency to *encourage* the soma of the next neuron to 'fire', i.e. to initiate a new signal out along its axon. These synapses are called *excitatory*. Others tend to *discourage* the next neuron from firing and are called *inhibitory*. The total effect of the excitatory synapses which are active at any moment is added up, and the total of the active inhibitory ones subtracted from this, and if the net result reaches a certain critical threshold, the next neuron is indeed induced to fire. (The excitatory ones cause a positive electrical *potential difference* between the inside and the outside of the next neuron and the inhibitory ones cause a negative potential difference. These potential differences add up appropriately. The neuron will fire when this potential difference reaches a critical level on the attached axon, so that the potassium can't get out fast enough to restore equilibrium.)

Computer models

An important feature of nerve transmission is that the signals are (for the most part) entirely 'all-or-nothing' phenomena. The strength of the signal does not vary: it is either there or it is not. This gives the action of the nervous system a digital computer-like aspect. In fact there are quite a lot of similarities between the action of a great number of interconnected neurons and the internal workings of a digital computer, with its current-carrying wires and logic gates (more about these in a moment). It would not be hard, in principle, to set up a computer simulation of the action of a given such system of neurons. A natural question arises: does this not mean that whatever the detailed wiring of the brain may be, it can always be modelled by the action of a computer?

In order to make this comparison clearer, I should say what a *logic gate* actually is. In a computer, we also have an 'all-or-nothing' situation where either there is a pulse of current down a wire or there is not, the strength of the pulse being always the same when one *is* present. Since everything is timed very accurately, the *absence* of a pulse would be a definite signal, and would be something that would be 'noticed' by the computer. Actually, when we use the term 'logic gate', we are implicitly thinking of the presence or absence of a pulse as denoting 'true' or 'false', respectively. In fact, this has nothing to do with actual truth or falsity; it is just in order to make sense of the terminology that is normally used. Let us also write the digit '1' for '*true*' (presence of pulse) and '0' for '*false*' (absence of pulse) and, as in Chapter 4, we can use '&' for 'and' (which is the 'statement' that both are 'true', i.e. the answer is 1 if and only if both arguments are 1), 'v' for 'or' (which 'means' that one or the other or both are 'true', i.e. 0 if and only if both arguments are 0), '$\Rightarrow$' for 'implies' (i.e. $\mathbf{A} \Rightarrow \mathbf{B}$ means if **A** is true then **B** is true', which is equivalent to 'either **A** is false or **B** is true'), '$\Longleftrightarrow$' for 'if and only if' (both 'true' or both 'false'), and '$\sim$' for 'not' ('true' if 'false'; 'false' if 'true'). One can describe the action of these various logical operations in terms of what are called 'truth tables':

$$\mathbf{A}\,\&\,\mathbf{B}: \begin{pmatrix} 0 & 0 \\ 0 & 1 \end{pmatrix} \qquad \mathbf{A} \vee \mathbf{B}: \begin{pmatrix} 0 & 1 \\ 1 & 1 \end{pmatrix}$$

$$\mathbf{A} \Rightarrow \mathbf{B}: \begin{pmatrix} 1 & 1 \\ 0 & 1 \end{pmatrix} \qquad \mathbf{A} \Longleftrightarrow \mathbf{B}: \begin{pmatrix} 1 & 0 \\ 0 & 1 \end{pmatrix}$$

where, in each case, **A** labels the rows (i.e. $\mathbf{A} = 0$ gives the first row and $\mathbf{A} = 1$, the second) and **B** labels the columns similarly. For example, if $\mathbf{A} = 0$ and $\mathbf{B} = 1$, giving the upper right-hand entry of each table, then in the *third* table we get 1 for the value of $\mathbf{A} \Rightarrow \mathbf{B}$. (For a verbal instance of this in terms of *actual* logic: the assertion 'if I am asleep then I am happy' is certainly borne out—trivially—in a particular case if I happen to be both awake and happy.) Finally, the logic gate 'not', simply has the effect:

$$\sim 0 = 1 \qquad \text{and} \qquad \sim 1 = 0.$$

These are the basic types of logic gates. There are a few others, but all of those can be built up out of the ones just mentioned.[6]

Now, can we in principle build a computer out of *neuron* connections? I shall indicate that, even with just very primitive considerations of neuron firing as have been discussed above, this is indeed possible. Let us see how it would be possible in principle to build logic gates from neuron connections. We need to have some new way of coding the digits, since the *absence* of a signal does not trigger off anything. Let us (quite arbitrarily) take a *double* pulse to denote 1 (or 'true') and a *single* pulse to denote 0 (or 'false'), and take a simplified scheme in which the threshold for the firing of a neuron is always just *two* simultaneous excitatory pulses. It is easy to construct an 'and' gate (i.e. '&'). As is shown in Fig. 9.13, we can take the two input nerve fibres to terminate as the sole pair of synaptic knobs on the output neuron. (If both are double pulses, then both the first pulse and the second pulse will reach the required two-pulse threshold, whereas if either is only a single pulse then only one of the pair will reach this threshold. I am assuming that the pulses are carefully timed and that, for definiteness in the case of a double pulse, it is the *first* of the pair that determines the timing.) The construction of a 'not' gate (i.e. '~') is considerably more complicated, and a way of doing this is given in Fig. 9.14. Here the input signal comes along an axon which divides into two branches. One branch takes a circuitous route, of just such a length as to delay the signal by exactly the time interval between the two pulses of a double pulse, then both bifurcate once more, with one branch from each of them terminating on an inhibitory neuron, but where the one from the delayed branch first splits to give both a direct and a circuitous route. That neuron's output would be *nothing*, in the case of a single-pulse input, and a *double pulse* (in delayed position) in the case of a double-pulse input. The axon carrying this output splits into three branches, all of which terminate with inhibitory synaptic knobs on a final excitatory neuron. The remaining two parts of the original divided axon each divide into two once more and all four branches also terminate on this final neuron now with excitatory synaptic knobs. The reader may care to check that this final excitatory neuron provides the required 'not' output (i.e. a double pulse if the input is a single one and a single pulse if the input is double). (This scheme seems absurdly complicated, but it is the best that I can do!) The reader may care to amuse herself, or himself, in providing direct 'neural' constructions for the other logic gates above.

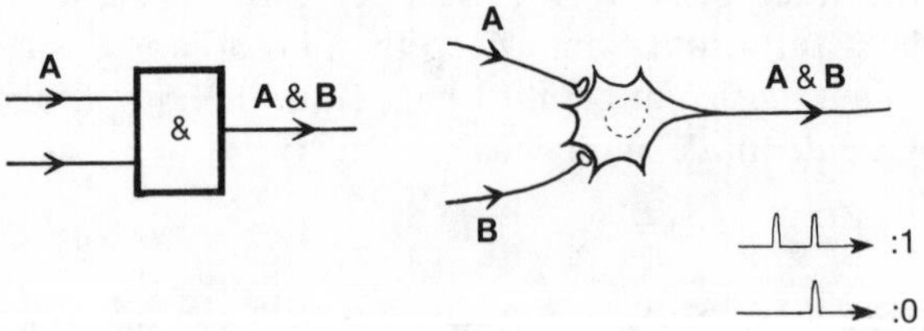

Fig. 9.13. An 'and' gate. In the 'neuron model' on the right, the neuron is taken to fire only when the input reaches twice the strength of a single pulse.

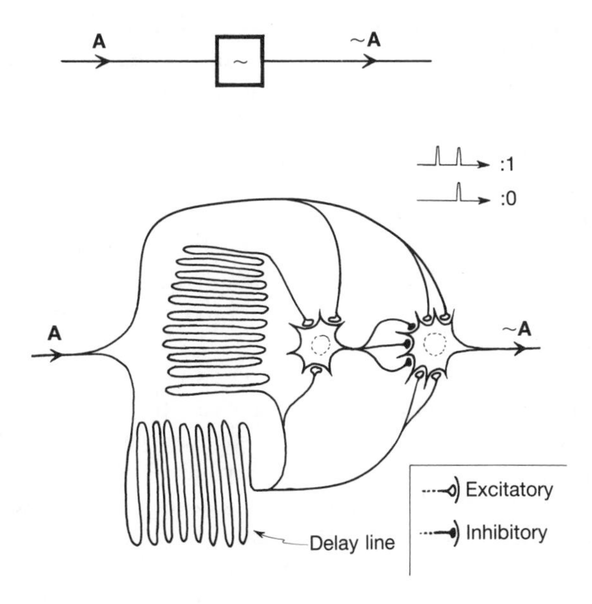

Fig. 9.14. A 'not' gate. In the 'Neuron model' a twofold strength input (at least) is again needed for a neuron to fire.

Of course, these explicit examples are not to be taken as serious models for what the brain actually does in detail. I am only trying to indicate that there is an essential logical equivalence between the model of neuron-firing that I have given above and electronic computer construction. It is easy to see that a computer could simulate any such model of neuron interconnections; whereas the detailed constructions above give an indication of the fact that, conversely, systems of neurons are capable of simulating a computer—and so *could* act as a (universal) Turing machine. Although the discussion of Turing machines given in Chapter 2 did not use logic gates,[7] and in fact we need rather more than just logic gates if we are to simulate a general Turing machine, there is no new issue of principle involved in doing this—*provided* that we allow ourselves to *approximate* the *infinite tape* of a Turing machine by a large but finite bank of neurons. This would seem to argue that brains and computers are essentially equivalent!

But before we jump too hastily to this conclusion, we should consider various differences between brain action and present-day computer action that might possibly be of significance. In the first place, I have oversimplified somewhat in my description of the firing of a neuron as an all-or-nothing phenomenon. That refers to a single pulse travelling along the axon, but in fact when a neuron 'fires' it emits a whole sequence of such pulses in quick succession. Even when a neuron is not activated, it emits pulses, but only at a slow rate. When it fires, it is the *frequency* of these successive pulses which increases enormously. There is also a probabilistic aspect of neuron firing. The same stimulus does not always produce the same result. Moreover, brain action does not have quite the exact timing that is needed for electronic computer currents; and it should be pointed out that the action of neurons—at a maximum rate of about 1000 times per second—is very much slower than that of the fastest electronic circuits, by a factor of about 10^{-6}. Also, unlike the very precise wiring of an electronic computer, there would appear to be a good deal of randomness and redundancy in the detailed way in which neurons are actually connected up—although we now know there is very much more precision in the way that the brain is wired up (at birth) than had been thought some fifty years ago.

Most of the above would seem simply to be to the brain's disadvantage, as compared with the computer. But there are other factors in the brain's favour. With logic gates there are only very few input and output wires (say three or four at most), whereas neurons may have huge numbers of synapses on them. (For an extreme example, those neurons of the cerebellum known as Purkinje cells have about 80000 excitatory synaptic endings.) Also, the total number of neurons in the brain is still in excess of the number of transistors in even the largest computer—probably 10^{11} for the brain and 'only' about 10^9 for the computer! But the figure for the computer, of course, is likely to increase in the future[8]. Moreover, the large brain-cell number arises largely because of the immense number of small *granule cells* that are to be found in the cerebellum—about thirty thousand million (3×10^{10}) of them. If we believe that it is simply the largeness of the neuron number that allows us to have conscious experiences, whilst present-day computers do not seem to, then we have to find some additional explanation of why the action of the *cerebellum* appears to be completely *un*conscious, while consciousness can be associated with the *cerebrum*, which has only about twice as many neurons (about 7×10^{10}), at a much smaller density.

Brain plasticity

There are some other points of difference between brain action and computer action that seem to me to be of much greater importance than the ones so far mentioned, having to do with a phenomenon known as *brain plasticity*. It is actually not legitimate to regard the brain as simply a *fixed* collection of wired-up neurons. The interconnections between neurons are not in fact fixed, as they would be in the above computer model, but are changing all the time. I do not mean that the locations of the axons or dendrites will change. Much of the complicated 'wiring' for these is established in broad outline at birth. I am referring to the synaptic junctions where the communication between different neurons actually takes place. Often these occur at places

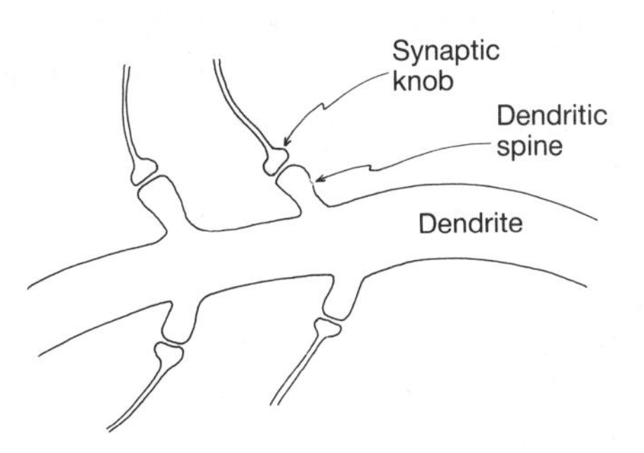

Fig. 9.15. Synaptic junctions involving dendritic spines. The effectiveness of the junction is readily affected by growth or contraction of the spine.

called *dendritic spines*, which are tiny protuberances on dendrites at which contact with synaptic knobs can be made (see Fig. 9.15). Here, 'contact' means just *not* touching, but leaving a narrow gap (synaptic cleft) of just the right distance—about one forty-thousandth of a millimetre. Now under certain conditions, these dendritic spines can shrink away and break contact, or they (or new ones) can grow to make new contact. Thus, if we think of the connections of neurons in the brain as constituting, in effect, a computer, then it is a computer which is capable of changing all the time!

According to one of the leading theories of how long-term memories are laid down, it is such changes in synaptic connections which provide the means of storing the necessary information. If this is so, then we see that brain plasticity is not just an incidental complication, it is an *essential* feature of the activity of the brain.

What is the mechanism underlying these continual changes? How fast can the changes be effected? The answer to the second question seems to be controversial, but there is at least one school of thought which maintains that such changes can occur within seconds. That would be expected if such changes are responsible for the storage of permanent memories, since such memories can indeed be laid down in a matter of seconds (cf. Kandel 1976). This would have significant implications for us later. I shall return to this important issue in the next chapter.

What about the mechanisms underlying brain plasticity? An ingenious theory (due to Donald Hebb 1954) proposes the existence of certain synapses (now called 'Hebb synapses') with the following property: a Hebb synapse between a neuron A and a neuron B would be strenghtened whenever the firing of A is followed by the firing of B and weakened whenever it is not. This is irrespective of whether the Hebb synapse itself is significantly involved in causing B to fire. This gives rise to some form of 'learning'. Various mathematical models have been put forward to try to simulate a learning/problem-solving activity, based on this kind of theory. These are referred to as *neutral networks*. It seems that such models are indeed capable of some kind of rudimentary learning, but so far they are a long way from being realistic models of the brain. In any case, it seems likely that the mechanisms controlling the changes in synaptic connections are likely to be more complicated than the ones considered. More understanding is clearly required.

In relation to this, there is another aspect of the release of neurotransmitters by synaptic knobs. Sometimes these do not occur in synaptic clefts at all, but enter the general intercellular fluid, perhaps to influence other neurons a long way away. Many different neurochemical substances seem to be emitted in this way—and there are different theories of memory from the one that I indicated above, which depend upon the different possible varieties of such chemicals that might be involved. Certainly the state of the brain can be influenced in a general way by the presence of chemicals that are produced by other parts of the brain (e.g. as with hormones). The whole question of neurochemistry is complicated, and it is difficult to see how to provide a reliable detailed computer simulation of everything that might be relevant.

Parallel computers and the 'oneness' of consciousness

Many people appear to be of the opinion that the development of *parallel* computers holds the key to building a machine with the capabilities of a human brain. Let us briefly consider this currently popular idea next. A parallel computer, as opposed to a serial one, has a very great number of separate calculations carried out independently, and the results of these largely autonomous operations are only intermittently combined together to give contributions to the overall calculation. The motivation for this type of computer architecture comes largely from an attempt to imitate the operation of the nervous system, since different parts of the brain indeed seem to carry out separate and independent calculational functions (e.g. with the processing of visual information in the visual cortex).

Two points should be made here. The first is that there is no difference *in principle* between a parallel and a serial computer. Both are in effect *Turing machines* (cf. Chapter 2, p. 48). The differences can only be in the efficiency, or speed, of the calculation as a whole. There are some types of calculation for which a parallel organization is indeed more efficient, but this is by no means always the case. The second point is that, at least in my own opinion, parallel classical computation is very unlikely to hold the key to what is going on with our *conscious* thinking. A characteristic feature of conscious thought (at least when one is in a normal psychological state, and not the subject of a 'split-brain' operation!) is its 'oneness'—as opposed to a great many independent activities going on at once.

Utterances like 'How can you expect me to think of more than one thing at a time?' are commonplace. Is it possible *at all* to keep separate things going on in one's consciousness simultaneously? Perhaps one *can* keep a few things going on at once, but this seems to be more like continual flitting backwards and forwards between the various topics than actually thinking about them simultaneously, consciously, and independently. If one were to think consciously about two things quite independently it would be more like having two *separate consciousnesses*, even if only for a temporary period, while what seems to be experienced (in a normal person, at least) is a *single* consciousness which may be vaguely aware of a number of things, but which is concentrated at any one time on only *one* particular thing.

Of course, what we mean by 'one thing' here is not altogether clear. In the next chapter, we shall be encountering some very remarkable instances of 'single thoughts' in the inspirations of Poincaré and Mozart. But we do not have to go that far in order to recognize that what a person may be conscious of at any one moment can be implicitly very complicated. Imagine deciding what one will have for dinner, for example. There could be a great deal of information involved in one such conscious thought, and a full verbal description might be quite long.

This 'oneness' of conscious perception seems to me to be quite at variance with the picture of a parallel computer. That picture might, on the other hand, be more appropriate as a model for the *unconscious* action of the brain. Various independent movements—walking, fastening a button, breathing, or even talking—can all be carried on simultaneously and more or less autonomously, without one being consciously aware of *any* of them!

On the other hand, it seems to me that there could conceivably be some relation between this 'oneness' of consciousness and *quantum parallelism*. Recall that, according to quantum theory, different alternatives at the quantum level are allowed to coexist in linear superposition! Thus, a *single quantum state* could in principle consist of a large number of different activities, all occurring simultaneously. This is what is meant by quantum parallelism, and we shall shortly be considering the theoretical idea of a 'quantum computer', whereby such quantum parallelism might in principle be made use of in order to perform large numbers of simultaneous calculations. If a conscious 'mental state' might in some way be akin to a quantum state, then some form of 'oneness' or globality of thought might seem more appropriate than would be the case for an ordinary parallel computer. There are some appealing aspects of this idea that I shall return to in the next chapter. But before such an idea can be seriously entertained, we must raise the question as to whether quantum effects are likely to have any relevance at all in the activity of the brain.

Is there a role for quantum mechanics in brain activity?

The above discussions of neural activity have indeed been entirely classical, except in so far as it has been necessary to call upon physical phenomena whose implicit underlying causes are partly quantum-mechanical (e.g. ions, with their unit electrical charges, sodium and potassium gates, the definite chemical potentials that determine the on/off character of nerve signals, the chemistry of neurotransmitters). Is there any more clear-cut role for a genuinely quantum-mechanical control at some key place? This would seem to be necessary if the discussion at the end of the previous chapter is to have any genuine relevance.

There is, in fact, at least one clear place where action at the single quantum level can have importance for neural activity, and this is in the *retina*. (Recall that the retina is technically part of the brain!) Experiments with toads have shown that under suitable conditions, a *single photon* impinging on the dark-adapted retina can be sufficient to trigger a macroscopic nerve signal (Baylor, Lamb, and Yau 1979). The same appears to be true of man (Hecht, Shlaer, and Pirenne 1941), but in this case there is an additional mechanism present which suppresses such weak signals, so that they do not confuse the perceived picture with too much visual 'noise'. A combined signal of about *seven* photons is needed in order that a dark-adapted human subject can actually become aware of their arrival. Nevertheless, cells with single-photon sensitivity do appear to be present in the human retina.

Since there *are* neurons in the human body that can be triggered by single quantum events, is it not reasonable to ask whether cells of this kind might be found somewhere in the main part of the human brain? As far as I am aware, there is no evidence for this. The types of cell that have been examined all require a threshold to be reached, and a very large number of quanta are needed in order that the cell will fire. One might speculate, however, that somewhere deep in the brain, cells are to be found of single quantum sensitivity. If this proves to be the case, then quantum mechanics will be significantly involved in brain activity.

Even this does not yet look very *usefully* quantum-mechanical, since the quantum is being used merely as a means of triggering a signal. No characteristic quantum interference effects have been obtained. It seems that, at best, all that we shall get from this will be an uncertainty as to whether or not a neuron will fire, and it is hard to see how this will be of much use to us.

However, some of the issues involved here are not quite that simple. Let us return to the retina for its consideration. Suppose that a photon arrives at the retina, having been previously reflected off a half-silvered mirror. Its state involves a complex linear superposition of its striking a retinal cell and of its not striking a retinal cell and instead, say, travelling out of the window into space (cf. Fig. 6.17, p. 254). When the time is reached at which it *might* have struck the retina, and so long as the linear rule **U** of quantum theory holds true (i.e. deterministic Schrödinger state-vector evolution, cf. p. 250), we shall have a complex linear superposition of a nerve signal and not a nerve signal. When this impinges upon the subject's consciousness, only *one* of these two alternatives is perceived to take place, and the other quantum procedure **R** (state-vector reduction, cf. p. 250) must have been effected. (In saying this I am ignoring the many-worlds view, cf. p. 295, which has its own multitude of problems!) In line with the considerations touched upon at the end of the previous chapter, we should ask whether enough matter is disturbed by the passage of the signal that the *one-graviton* criterion of that chapter is met. While it is true that an impressively enormous magnification is achieved by the retina in converting the photon's energy into a movement of mass in the actual signal—perhaps by a factor of as much as 10^{20} in mass moved—this mass still falls short of the Planck mass m_p by a very large figure (say about 10^{8}). However, a nerve signal creates a detectable changing *electric field* in its surroundings (a toroidal field, with the nerve as axis, and moving along the nerve). This field could disturb the *surroundings* significantly, and the one-graviton criterion might be easily met within these surroundings. Thus, according to the viewpoint I have been putting forward, the procedure **R** could have been already effected well before we perceive the flash of light, or not, as the case may be. On this viewpoint, our consciousness is not needed in order to reduce the state-vector!

Quantum computers

If we *do* speculate that single-quantum sensitive neurons are playing an important role deep inside the brain, we can ask what effects these might have. I shall first discuss Deutsch's concept of a *quantum computer* (cf. also Chapter 4, p. 146) and then ask whether we can see this as relevant to our considerations here.

As was indicated above, the basic idea is to make use of quantum parallelism, according to which two quite different things must be considered as taking place simultaneously in quantum linear superposition—like the photon simultaneously being reflected and passing through the half-silvered mirror, or perhaps passing through each of two slits. With a quantum computer, these two superposed different things would, instead, be two different *computations*. We are not supposed to be interested in obtaining the answers to *both* computations, but to something which uses partial information extracted from the superposed pair. Finally, a suitable 'observation' would be made on the pair of computations, when both are completed, to obtain the answer that is required.[9] In this way the device may be able to save time by performing two computations simultaneously! So far there would seem to be no significant gain in proceeding in this way, since it would presumably be much more direct to make use of a pair of separate classical computers in parallel (or a single classical parallel computer) than a quantum computer. However, the real gain for the quantum computer might come when a *very large* number of parallel computations are required—perhaps an indefinitely large number—whose individual answers would not interest us, but where some suitable combination of all the results would.

In detail, the construction of a quantum computer would involve a quantum version of a logic gate, where the output would be the result of some 'unitary operation' applied to the input—an instance of the action of **U**—and the entire operation of the computer would be carrying out a **U** process right up until the end, until a final 'act of observation' brings in **R**.

According to Deutsch's analysis, quantum computers cannot be used to perform non-algorithmic operations (i.e. things beyond the power of a Turing machine), but can, in certain very contrived situations, achieve a greater speed, in the sense of *complexity theory* (see p. 140), than a standard Turing machine. So far these results are a little disappointing, for such a striking idea, but these are early days yet.

How might this relate to the action of a brain containing a significant number of single-quantum-sensitive neurons? The main problem with the analogy would be that the quantum effects would very quickly get lost in the 'noise'—the brain is too 'hot' an object to preserve quantum coherence (i.e. behaviour usefully described by the continued action of **U**) for any significant length of time. In my own terms, this would mean that the one-graviton criterion would be continually being met, so that the operation **R** is going on all the time, interspersed with **U**.

So far, this does not look very promising if we expect to get something useful for the brain out of quantum mechanics. Perhaps we are doomed to be computers after all! Personally, I do not believe so. But further considerations are needed if we are to find our way out.

Beyond quantum theory?

I wish to return to an issue that has been an underlying theme of much of this book. Is our picture of a world governed by the rules of classical and quantum theory, as these rules are presently understood, really adequate for the description of brains and minds? There is certainly a puzzle for any 'ordinary' quantum description of our brains, since the action of 'observation' is taken to be an essential ingredient of the valid interpretation of conventional quantum theory. Is the brain to be regarded as 'observing itself' whenever a thought or perception emerges into conscious awareness? The conventional theory leaves us with no clear rule as to how quantum mechanics could take this into account, and thereby apply to the brain as a whole. I have attempted to formulate a criterion for the onset of **R** which is quite independent of consciousness (the 'one-graviton criterion') and if something like this could be developed into fully a coherent theory then there might emerge a way of providing a quantum description of the brain, more clearly than exists at present.

However, I believe that it is not only with our attempts to describe brain action that these fundamental problems arise. The actions of digital computers themselves depend vitally upon quantum effects—effects that are not, in my opinion, entirely free of difficulties inherent in the quantum theory. What is this 'vital' quantum dependence? To understand the role of quantum mechanics in digital computation, we must first ask how we might attempt to make an entirely *classical* object behave like a digital computer. In Chapter 5, we considered the Fredkin–Toffoli classical 'billiard-ball computer' (p. 171); but we also noted that this theoretical 'device' depends upon certain idealizations that sidestep an essential instability problem inherent in classical systems. This instability problem was described as an effective spreading in phase-space, as time evolves (p. 181; Fig. 5.14), leading to an almost inevitable continual loss of accuracy in the operation of a classical device. What halts this degradation of accuracy, ultimately, is quantum mechanics. In modern electronic computers, the existence of *discrete states* is needed (say, coding the digits 0 and 1), so that it becomes a clear-cut matter when the computer is in one of these states and when in another. This is the very essence of the 'digital' nature of computer operation. This discreteness depends ultimately on quantum mechanics. (We recall the quantum discreteness of energy states, of spectral frequencies, of spin, etc., cf. Chapter 6.) Even the old mechanical calculating machines depended upon the *solidity* of their various parts—and solidity, also, actually rests upon the discreteness of quantum theory.

But quantum discreteness is not obtainable solely from the action of **U**. If anything, Schrödinger's equation is *worse* at preventing the undesirable spreading and 'loss of accuracy' than are the equations of classical physics! According to **U**, a single particle's wavefunction, initially localized in space, will spread itself over larger and larger regions as time evolves (p. 251). More complicated systems would also be sometimes subject to such unreasonable lack of localization (recall Schrödinger's cat!), were it not for an action of **R** from time to time. (The *discrete* states of an atom, for example, are those with definite energy, momentum, and total angular momentum. A general state which 'spreads' is a superposition of such discrete states. It is the action of **R**, at some stage, that requires the atom actually to 'be' in one of these discrete states.)

It seems to me that neither classical nor quantum mechanics—the latter without some further fundamental changes which would make **R** into an 'actual' process—can ever explain the way in which we *think*. Perhaps even the digital action of computers needs a deeper understanding of the interrelation between the actions of **U** and **R**. At least with computers we know that this action is *algorithmic* (by our design!), and we do not try to harness any putative *non*-algorithmic behaviour in physical laws. But with brains and minds the situation is, I maintain, very different. A plausible case can be made that there is an essential *non*-algorithmic ingredient to (conscious) thought processes. In the next chapter I shall try to elaborate on the reasons for my belief in such an ingredient, and to speculate on what remarkable actual physical effects might constitute a 'consciousness' influencing the action of the brain.

1. In a BBC radio broadcast; see Hodges (1983), p. 419.
2. The first experiments of this kind were performed on cats (cf. Myers and Sperry 1953). For further information on split-brain experiments, see Sperry (1966), Gazzaniga (1970), MacKay (1987).
3. For a readable account of the workings of the visual cortex, see Hubel (1988).
4. See Hubel (1988) p. 221. Earlier experiments registered cells sensitive only to the image of a hand.
5. The now well-established theory that the nervous system consists of separate individual cells, the neurons, was forcefully put forward by the great Spanish neuroanatomist Ramón y Cajal in around 1900.
6. In fact, *all* logic gates can be constructed from just '~' and '&' (or even just from the *one* operation ~(**A**&**B**) alone).
7. In fact the use of logic gates is closer to the construction of an electronic computer than are the detailed Turing machine considerations of Chapter 2. The emphasis in that chapter was on Turing's approach for theoretical reasons. Actual computer development sprung as much from the work of the outstanding Hungarian/American mathematician John von Neumann as from that of Alan Turing.
8. These comparisons are in many ways misleading. The vast majority of transistors in present-day computers are concerned with 'memory' rather than logical action, and the computer's memory can always be added to externally, virtually indefinitely. With increased parallel operation, more transistors could become directly involved with the logical action than is normally the case at present.
9. Deutsch, in his descriptions, prefers to use the 'many-worlds' viewpoint with regard to quantum theory. However, it is important to realize that this is quite inessential, the quantum-computer concept being equally appropriate whichever viewpoint one takes with regard to standard quantum mechanics.

10
Where lies the physics of mind?

What are minds for?

In discussions of the mind–body problem, there are two separate issues on which attention is commonly focused: 'How is it that a material object (a brain) can actually *evoke* consciousness?'; and, conversely; 'How is it that a consciousness, by the action of its will, actually *influence* the (apparently physically determined) motion of material objects?' These are the passive and active aspects of the mind–body problem. It appears that we have, in 'mind' (or, rather, in 'consciousness'), a non-material 'thing' that is, on the one hand, evoked by the material world and, on the other, can influence it. However, I shall prefer, in my preliminary discussions in this last chapter, to consider a somewhat different and perhaps more scientific question—which has relevance to both the active and passive problems—in the hope that our attempts at an answer may move us a little way towards an improved understanding of these age-old fundamental conundrums of philosophy. My question is; 'What *selective advantage* does a consciousness confer on those who actually possess it?'

There are several implicit assumptions involved in phrasing the question in this way. First, there is the belief that consciousness is actually a *scientifically describable* 'thing'. There is the assumption that this 'thing' actually 'does something'—and, moreover, that what it does is helpful to the creature possessing it, so that an otherwise equivalent creature, but without consciousness, would behave in some less effective way. On the other hand, one might believe that consciousness is merely a passive concomitant of the possession of a sufficiently elaborate control system and does *not*, in itself, actually 'do' *anything*. (This last would presumably be the view of the strong-AI supporters, for example.) Alternatively, perhaps there is some divine or mysterious purpose for the phenomenon of consciousness—possibly a teleological one not yet revealed to us—and any discussion of this phenomenon in terms merely of the ideas of natural selection would miss this 'purpose' completely. Somewhat preferable, to my way of thinking, would be a rather more scientific version of this sort of argument, namely the *anthropic principle*, which asserts that the nature of the universe that we find ourselves in is strongly constrained by the requirement that sentient beings like ourselves must actually be present to observe it. (This principle was briefly alluded to in Chapter 8, p. 354, and I shall be returning to it later.)

I shall address most of these issues in due course, but first we should note that the term 'mind' is perhaps a little misleading when we refer to the 'mind–body' problem. One often speaks, after all, of 'the unconscious mind'. This shows that we do not regard the terms 'mind' and 'consciousness' as synonymous. Perhaps, when we refer to the *un*conscious mind we have a vague image of 'someone back there' who acts behind the scenes but who does not usually (except perhaps in dreams, hallucinations, obsessions, or Freudian slips) directly impinge upon what we perceive. Perhaps, the unconscious mind actually *has* an awareness of its own, but this awareness is normally kept quite separate from the part of the mind that we usually refer to as 'us'.

This may not be quite so far-fetched as it seems at first. There are experiments which seem to indicate that there can be some kind of 'awareness' present even when a patient is being operated upon under general anaesthetic—in the sense that conversations going on at the time can influence the patient 'unconsciously' later, and can be sometimes recalled afterwards under hypnosis as actually having been 'experienced' at the time. Moreover, sensations that seem to have been blocked from consciousness by hypnotic suggestion, can later be recalled under further hypnosis as 'having been experienced', but somehow kept on a 'different track' (cf. Oakley and Eames 1985). The issues are not at all clear to me, though I do not imagine that it would be correct to assign any ordinary 'awareness' to the unconscious mind, and I have no real desire to address such speculations here. Nevertheless, the division between the conscious and the unconscious mind is surely a subtle and complicated issue which we shall need to come back to.

Let us try to be as straightforward as we can about what we mean by 'consciousness' and when we believe that it is present. I do not think that it is wise, at this stage of understanding, to attempt to propose a precise *definition* of consciousness, but we can rely, to good measure, on our subjective impressions and intuitive common sense as to what the term means and when this property of consciousness is likely to be present. I more or less know when I am conscious myself, and I take it that other people experience something corresponding to what I experience. To be conscious, I seem to have to be conscious *of* something, perhaps a sensation such as pain or warmth or a colourful scene or a musical sound; or perhaps I am conscious of a feeling such as puzzlement, despair, or happiness; or I may be conscious of the memory of some past experience, or of coming to an understanding of what someone else is saying, or a new idea of my own; or I may be consciously intending to speak or to take some other action such as get up from my seat. I may also 'step back' and be conscious *of* such intentions or of my feeling of pain or of my experiencing of a memory or of my coming to an understanding; or I may even just be conscious of my own consciousness. I may be asleep and still be conscious to some degree, provided that I am experiencing some dream; or perhaps, as I am beginning to awake, I am consciously influencing the direction of that dream. I am prepared to believe that consciousness is a matter of degree and not simply something that is either there or not there. I take the word 'consciousness' to be essentially synonymous with 'awareness' (although perhaps 'awareness' is just a little more passive than what I mean by 'consciousness'), whereas 'mind' and 'soul' have further connotations which are a good deal *less* clearly definable at present. We shall be having enough trouble with coming to terms with 'consicousness' as it stands, so I hope that the reader will forgive me if I leave the further problems of 'mind' and 'soul' essentially alone!

There is also the question of what one means by the term 'intelligence'. This, after all, is what the AI people are concerned with, rather than the perhaps more nebulous issue of 'consciousness'. Alan Turing (1950), in his famous paper (cf. Chapter 1, p. 6) did not refer, so much directly to 'consciousness', but to 'thinking' and the word 'intelligence' was in the title. In my own way of looking at things, the question of intelligence is a subsidiary one to that of consciousness. I do not think that I would believe that true intelligence could be actually present unless accompanied by consciousness. On the other hand, if it does turns out that the AI people *are* eventually able to simulate intelligence without consciousness being present, then it might be regarded as unsatisfactory not to define the term 'intelligence' to include such simulated intelligence. In that case the issue of 'intelligence' would not be my real concern here. I am primarily concerned with 'consciousness'.

When I assert my own belief that true intelligence requires consciousness, I am implicitly suggesting (since I do not believe the *strong*-AI contention that the mere enaction of an algorithm would evoke consciousness) that intelligence cannot be properly simulated by algorithmic means, i.e. by a computer, in the sense that we use that term today. (See the discussion of the 'Turing test' given in Chapter 1.) For I shall shortly argue strongly (see, particularly, the discussion of mathematical thought given three sections hence, on p. 416) that there must be an essentially *non-algorithmic* ingredient in the action of consciousness.

Next let us address the question of whether there *is* an operational distinction between something that is conscious and an otherwise 'equivalent' thing that is not. Would consciousness in some object always reveal its presence? I should like to think that the answer to this question is necessarily 'yes'. However, my faith in this is hardly encouraged by the total lack of consensus about where in the animal kingdom consciousness is to be found. Some would not allow its possession by any non-human animal at all (and some, even, not in human beings before about 1000 BC cf. Janes, 1980), whilst others would assign consciousness to an insect, a worm, or perhaps even a rock! As for myself, I should be doubtful that a worm or insect—and certainly not a rock—has much, if anything, of this quality, but mammals, in a general way, do give me an impression of some genuine awareness. From this lack of consensus we must infer, at least, that there is no generally accepted criterion for the manifestation of consciousness. It still might be that there *is* a hallmark of conscious behaviour, but one not universally recognized. Even so, it would only be the *active* role of consciousness that this could signify. It is hard to see how the mere presence of awareness, without its active counterpart, could be directly ascertained. This is horrifically borne out in the fact that, for a time in the 1940s, the drug curare was used as an 'anaesthetic' in operations performed on young children—whereas the actual effect that this drug has is to paralyse the action of motor nerves on muscles, so that the agony that *was actually experienced* by these unfortunate children had no way of making its presence known to the surgeon at the time (cf. Dennett 1978, p. 209).

Let us turn to the possible active role that consciousness *can* have. Is it necessarily the case that consciousness can—and indeed sometimes *does*—play an operationally discernible active role? My reasons for believing so are somewhat varied. First, there is the way that by use of our 'common sense', we often feel that we directly perceive that some other person *is* actually conscious. *That* impression is not so likely to be wrong.* Whereas a person who *is* conscious may (like the curare-drugged children) not be obviously so, a person who is *not* conscious is *not* likely to appear to be! Thus there must indeed be some mode of behaviour which is characteristic of consciousness (even though not *always* evidenced by consciousness), which we are sensitive to through our 'common-sense intuitions'.

Second, consider the ruthless process of natural selection. View this process in the light of the fact that, as we have seen in the last chapter, not all of the activity of the brain is directly accessible to consciousness. Indeed, the 'older' cerebellum—with its vast superiority in local density of neurons—seems to carry out very complex actions without consciousness being directly involved at all. Yet Nature has chosen to evolve sentient beings like ourselves, rather than to remain content with creatures that might carry on under the direction of totally unconscious control mechanisms. If consciousness serves no selective purpose, why did Nature go to the trouble to evolve *conscious* brains when non-sentient 'automaton' brains like cerebella would seem to have done just as well?

Moreover, there is a simple 'bottom line' reason for believing that consciousness must have *some* active effect, even if this effect is *not* one of selective advantage. For why is it that beings like ourselves should sometimes

* At least with present-day computer technology (see the discussion of the Turing test given in Chapter 1).

be troubled—especially when probed on the matter—by questions about 'self'? (I could almost say; 'Why are *you* reading this chapter?' or 'Why did *I* feel a strong desire to write a book on this topic in the first place?') It is hard to imagine that an entirely unconscious automaton should waste its time with such matters. Since conscious beings, on the other hand, *do* seem to act in this funny way from time to time, they are thereby behaving in a way that is *different* from the way that they would if *not* conscious—so consciousness has *some* active effect! Of course there would be no problem about deliberately programming a computer to seem to behave in this ridiculous way (e.g. it could be programmed to go around muttering 'Oh dear, what is the meaning of life? Why am I here? What on earth is this "self" that I feel?'). But why should natural selection bother to favour such a race of individuals, when surely the relentless free market of the jungle should have rooted out such useless nonsense long ago!

It seems to me to be clear that the musings and mutterings that we indulge in, when we (perhaps temporarily) become philosophers, are not things that are *in themselves* selected for, but are the necessary 'baggage' (from the point of view of natural selection) that must be carried by beings who indeed *are* conscious, and whose consciousness has been selected by natural selection, but for some quite different and presumably very powerful reason. It is a baggage that is not too detrimental, and is easily (if perhaps grudgingly) borne, I would guess, by the indomitable forces of natural selection. On occasion, perhaps when there is the peace and prosperity that is sometimes enjoyed by our fortunate species, so that we do not always have to fight the elements (or our neighbours) for our survival, the treasures of the baggage's contents can begin to be puzzled and wondered over. It is when one sees others behaving in this strange philosophical way that one becomes *convinced* that one is dealing with individuals, other than oneself, who indeed also have minds.

What does consciousness actually do?

Let us accept that the presence of consciousness in a creature is actually of some selective advantage to that creature. What, specifically, might that advantage be? One view that I have heard expressed is that awareness might be of an advantage to a predator in trying to guess what its prey would be likely to do next by 'putting itself in the place' of that prey. By imagining itself to *be* the prey, it could gain an advantage over it.

It may well be that there is some partial truth in this idea, but I am left very uneasy by it. In the first place, it supposes some pre-existing consciousness on the part of the prey itself, for it would hardly be helpful to imagine oneself to 'be' an automaton, since an automaton—by definition *un*conscious—is not something that is *possible* to 'be' at all! In any case, I could equally well imagine that a totally unconscious automaton predator might itself contain, as part of its program, a subroutine which was the actual program of its automaton prey. It does not seem to me to be *logically* necessary that consciousness need enter into this predator–prey interrelationship at all.

Of course, it is difficult to see how the random procedures of natural selection could have been clever enough to give an *automaton* predator a complete copy of the prey's program. This would sound more like *espionage* than natural selection! And a *partial* program (in the sense of a piece of Turing machine 'tape', or something approximating a Turing machine tape) would hardly be of much selective advantage to the predator. The unlikely possession of the whole tape, or at least some entire self-contained part of it would seem to be necessary. So, as an alternative to this, there could be some partial truth in the idea that some element of consciousness, rather than just a computer program, can be inferred from the predator–prey line of thought. But it does not seem to address the *real* issue as to what the difference between a conscious action and a 'programmed' one actually is.

The idea alluded to above seems to relate to a point of view about consciousness that one often hears put forward, namely that a system would be 'aware' of something if it has a model of that thing within itself, and that it becomes '*self*-aware' when it has a model of *itself* within itself. But a computer program which contains within it (say as a subroutine) some description of another computer program does not give the first program an awareness of the second one; nor would some *self*-referential aspect to a computer program give it *self*-awareness. Despite the claims that seem to be frequently made, the real issues concerning awareness and self-awareness are, in my opinion, hardly being touched by considerations of this kind. A video-camera has no awareness of the scenes it is recording; nor does a video-camera aimed at a mirror possess self-awareness (Fig. 10.1).

Fig. 10.1. A video-camera aimed at a mirror forms a model of itself within itself. Does that make it self-aware?

I wish to pursue a different line. We have seen that not all of the activities carried out by our brains are accompanied by conscious awareness (and, in particular, cerebellar action seems not to be conscious). What is it that we can *do* with conscious thought that cannot be done unconsciously? The problem is made more elusive by the fact that anything that we do seem originally to require consciousness for appears also to be able to be learnt and then later carried out unconsciously (perhaps by the cerebellum). Somehow, consciousness is needed in order to handle situations where we have to form new judgements, and where the rules have not been laid down beforehand. It is hard to be very precise about the distinctions between the kinds of mental activity that seem to require consciousness and those that do not. Perhaps, as the supporters of strong AI (and others) would maintain, our 'forming of new judgements' would again be applying some well-defined algorithmic rules, but some obscure 'high-level' ones, the workings of which we are not aware of. However, I think that the kind of terminology that we tend to use, which distinguishes our conscious from our unconscious mental activity, is at least *suggestive* of a non-algorithmic/algorithmic distinction:

Conscious needed	*Consciousness not needed*
'common sense'	'automatic'
'judgement of truth'	'following rules mindlessly'
'understanding'	'programmed'
'artistic appraisal'	'algorithmic'.

Perhaps these distinctions are not always very clear cut, particularly since many *un*conscious factors enter into our conscious judgements: experience, intuition, prejudice, even our normal use of logic. But the judgements themselve, I would claim, are the manifestations of the action of *consciousness*. I therefore suggest that, whereas unconscious actions of the brain are ones that proceed according to algorithmic processes, the action of consciousness is quite different, and it proceeds in a way that cannot be described by any algorithm.

It is ironic that the views that I am putting forward here represent almost a reversal of some others that I have frequently heard. Often it is argued that it is the *conscious* mind that behaves in the 'rational' way that one can understand, whereas it is the unconscious that is mysterious. People who work with AI often assert that as soon as one can understand some line of thinking consciously, then one can see how to make a computer do it; it is the mysterious *un*conscious processes that one has no idea (yet!) how to deal with. My own line of reasoning has been that unconscious processes could well be algorithmic, but at a very complicated level that is monstrously difficult to disentangle in detail. The fully conscious thinking that can be rationalized as something entirely logical can again (often) be formalized as something algorithmic, but this is at an *entirely different level.* We are not now thinking of the internal workings (firings of neurons, etc.) but of manipulating entire thoughts. Sometimes this thought-manipulation has an algorithmic character (as with early logic: ancient Greek syllogisms as formalized by Aristotle or the symbolic logic of the mathematician George Boole; cf. Gardner 1958), sometimes it has not (as with Gödel's theorem and some of the examples given in Chapter 4). The *judgement-forming* that I am claiming is the hallmark of consciousness is *itself* something that the AI people would have no concept of how to program on a computer.

People sometimes object that the *criteria* for these judgements are not, after all, conscious, so why am I attributing such judgements to consciousness? But this would be to miss the point of the ideas that I am trying to express. I am not asking that we consciously understand *how* we form our conscious impressions and judgements. That would be to make the confusion of levels that I have just been referring to. The underlying *reasons* for our conscious impressions would be things not directly accessible to consciousness. These would have to be considered at a deeper physical level than those of the actual thoughts of which we are aware. (I shall make a stab at a suggestion later!) It is the conscious impressions themselves that *are* the (non-algorithmic) judgements.

It has, indeed, been an underlying theme of the earlier chapters that there seems to be something *non-algorithmic* about our conscious thinking. In particular, a conclusion from the argument in Chapter 4, particularly concerning Gödel's theorem, was that, at least in mathematics, conscious contemplation can sometimes enable one to ascertain the truth of a statement in a way that no algorithm could. (I shall be elaborating on this argument in a moment.) Indeed, algorithms, in themselves, *never* ascertain truth! It would be as easy to make an algorithm produce nothing but falsehoods as it would be to make it produce truths. One needs *external insights* in order to decide the validity or otherwise of an algorithm (more about this later). I am putting forward the argument here that it is this ability to divine (or 'intuit') truth from falsity (and beauty from ugliness!), in appropriate circumstances that is the hallmark of consciousness.

I should make it clear, however, that I do not mean some form of magical 'guessing'. Consiousness is of no help at all in trying to guess the lucky number in a (fairly run) lottery! I am referring to the judgements that one continually makes while one is in a conscious state, bringing together all the facts, sense impressions, remembered experiences that are of relevance, and weighing things against one another—even forming inspired judgements, on occasion. Enough information is in principle available for the relevant judgement to be made, but the process of formulating the appropriate judgement, by extracting what is needed from the morass of data, may be something for which no clear algorithmic process exists—or even where there is one, it may not be a practical one. Perhaps we have a situation where once the judgement *is* made, it may be more of an algorithmic process (or perhaps just an easier one) to *check* that the judgement is an accurate one than it is to form that judgement in the first place. I am guessing that consciousness would, under such circumstances, come into its own as a means of conjuring up the appropriate judgements.

Why do I say that the hallmark of consciousness is a non-algorithmic forming of judgements? Part of the reason comes from my experiences as a mathematician. I simply do not trust my unconscious algorithmic actions when they are inadequately paid attention to by my awareness. Often there is nothing wrong with the algorithm *as* an algorithm, in some calculation that is being performed, but is it the *right* algorithm to choose, for the problem in hand? For a simple example, one will have learnt the algorithmic rules for multiplying two numbers together and also for dividing one number by another (or one may prefer to engage the assistance of an algorithmic pocket calculator), but how does one know whether, for the problem in hand, one should have multiplied or divided the numbers? For that, one needs to *think*, and make a *conscious* judgement. (We shall be seeing shortly why such judgements must, at least sometimes, be *non*-algorithmic!) Of course, once one has done a large number of similar problems, the decision as to whether to multiply or divide the numbers may become second nature and can be carried out algorithmically—presumably by the cerebellum. At that stage, awareness is no longer necessary, and it becomes safe to allow one's conscious mind to wander and to contemplate other matters—although, from time to time one may need to check that the algorithm has not been sidetracked in some (perhaps subtle) way.

The same sort of thing is continually happening at all levels of mathematical thought. One often strives for algorithms, when one does mathematics, but the striving itself does not seem to be an algorithmic procedure. Once an appropriate algorithm is found, the problem is, in a sense, solved. Moreover the mathematical judgement that some algorithm is indeed accurate or appropriate is the sort of thing that requires much conscious attention. Something similar occurred in the discussion of formal systems for mathematics that were described in Chapter 4. One may start with some axioms, from which are to be derived various mathematical propositions. The latter procedure may indeed be algorithmic, but some judgement needs to be made by a conscious mathematician to decide whether the axioms are appropriate. That these judgements are necessarily *not* algorithmic should become clearer from the discussion given in the section after the next one. But before coming to this, let us consider what might be a more prevalent viewpoint as to what our brains are doing and how they have arisen.

Natural selection of algorithms?

If we suppose that the action of the human brain, conscious or otherwise, is merely the acting out of some very complicated algorithm, then we must ask how such an extraordinarily effective algorithm actually came about. The standard answer, of course, would be 'natural selection'. As creatures with brains evolved, those with the more effective algorithms would have a better tendency to survive and therefore, on the whole, had more progeny. These progeny also tended to carry more effective algorithms than their cousins, since they inherited the ingredients of these better algorithms from their parents; so gradually the algorithms improved—not necessarily steadily, since there could have been considerable fits and starts in their evolution—until they reached the remarkable status that we (would apparently) find in the human brain. (Compare Dawkins 1986.)

Even according to my own viewpoint, there would have to be *some* truth in this picture, since I envisage that much of the brain's action is indeed algorithmic, and—as the reader will have inferred from the above discussion—I am a strong believer in the power of natural selection. But I do not see how natural selection, in itself, can evolve algorithms which could have the kind of conscious judgements of the *validity* of other algorithms that we seem to have.

Imagine an ordinary computer program. How would *it* have come into being? Clearly not (directly) by natural selection! Some human computer programmer would have conceived of it and would have ascertained that it correctly carries out the actions that it is supposed to. (Actually, most complicated computer programs contain errors—usually minor, but often subtle ones that do not come to light except under unusual circumstances. The presence of such errors does not substantially affect my argument.) Sometimes a computer program might itself have been 'written' by another, say a 'master' computer program, but then the master program itself would have been the product of human ingenuity and insight; or the program might well be pieced together from ingredients some of which were the products of other computer programs. But in all cases the validity and the very conception of the program would have ultimately been the responsibility of (at least) one human consciousness.

One can imagine, of course, that this need not have been the case, and that, given enough time, the computer programs might somehow have evolved spontaneously by some process of natural selection. If one believes that the actions of the computer programmers' consciousnesses are themselves simply algorithms, then one must, in effect, believe that algorithms *have* evolved in just this way. However, what worries me about this is that the decision as to the validity of an algorithm is *not* itself an algorithmic process! We have seen something of this already in Chapter 2. (The question of whether or not a

Turing machine will actually *stop* is not something that can be decided algorithmically.) In order to decide whether or not an algorithm will actually *work*, one needs *insights*, not just another algorithm.

Nevertheless, one still might imagine some kind of natural selection process being effective for producing *approximately* valid algorithms. Personally, I find this very difficult to believe, however. Any selection process of this kind could act only on the *output* of the algorithms* and not directly on the ideas underlying the actions of the algorithms. This is not simply extremely inefficient; I believe that it would be totally unworkable. In the first place, it is not easy to ascertain what an algorithm actually is, simply by examining its output. (It would be an easy matter to construct two quite different simple Turing machine actions for which the output tapes did not differ until, say, the 2^{65536}th place—and this difference could never be spotted in the entire history of the universe!) Moreover, the slightest 'mutation' of an algorithm (say a slight change in a Turing machine specification, or in its input tape) would tend to render it totally useless, and it is hard to see how actual *improvements* in algorithms could ever arise in this random way. (Even *deliberate* improvements are difficult without 'meanings' being available. This is particularly borne out by the not-infrequent circumstances when an inadequately documented and complicated computer program needs to be altered or corrected; and the original programmer has departed or perhaps died. Rather than try to disentangle all the various meanings and intentions that the program implicitly depended upon, it is probably easier just to scrap it and start all over again!)

Perhaps some much more 'robust' way of specifying algorithms could be devised, which would not be subject to the above criticisms. In a way, this is what I am saying myself. The 'robust' specifications are the *ideas* that underlie the algorithms. But ideas are things that, as far as we know, need conscious minds for their manifestation. We are back with the problem of what consciousness actually is, and what it can actually do that unconscious objects are incapable of—and how on earth natural selection has been clever enough to evolve *that* most remarkable of qualities.

The products of natural selection are indeed astonishing. The little knowledge that I have myself acquired about how the human brain works—and, indeed, any other living thing—leaves me almost dumbfounded with awe and admiration. The working of an individual neuron is extraordinary, but the neurons themselves are organized together in a quite remarkable way, with vast numbers of connections wired up at birth, ready for all the tasks that will be needed later on. It is not just consciousness itself that is remarkable, but all the paraphernalia that appear to be needed in order to support it!

* There is also a somewhat knotty issue of whether two algorithms are to be regarded as equivalent to each other if merely their *outputs* are the same rather than the actual calculations being the same. See Chapter 2, p. 54.

If we ever do discover in detail what quality it is that allows a physical object to become conscious, then, conceivably, we might be able to construct such objects for ourselves—though they might not qualify as 'machines' in the sense of the word that we mean it now. One could imagine that these objects could have a tremendous advantage over us, since they could be designed *specifically* for the task at hand, namely to *achieve consciousness*. They would not have to grow from a single cell. They would not have to carry around the 'baggage' of their ancestry (the old and 'useless' parts of the brain or body that survive in ourselves only because of the 'accidents' of our remote ancestry). One might imagine that, in view of these advantages, such objects could succeed in *actually* superseding human beings, where (in the opinions of such as myself) the algorithmic computers are doomed to subservience.

But there may well be more to the issue of consciousness than this. Perhaps, in some way, our consciousness does depend on our heritage, and the thousands of millions of years of *actual* evolution that lie behind us. To my way of thinking, there is still something mysterious about evolution, with its apparent 'groping' towards some future purpose. Things at least *seem* to organize themselves somewhat better than they 'ought' to, just on the basis of blind-chance evolution and natural selection. It may well be that such appearances are quite deceptive. There seems to be something about the way that the laws of physics work, which allows natural selection to be a much more effective process than it would be with just arbitrary laws. The resulting apparently 'intelligent groping' is an interesting issue, and I shall be returning to it briefly later.

The non-algorithmic nature of mathematical insight

As I have stated earlier, a good part of the reason for believing that consciousness is able to influence truth-judgements in a *non*-algorithmic way stems from consideration of Gödel's theorem. If we can see that the role of consciousness is non-algorithmic when forming *mathematical* judgements, where calculation and rigorous proof constitute such an important factor, then surely we may be persuaded that such a non-algorithmic ingredient could be crucial also for the role of consciousness in more general (non-mathematical) circumstances.

Let us recall the arguments given in Chapter 4 establishing Gödel's theorem and its relation to computability. It was shown there that *whatever* (sufficiently extensive) algorithm a mathematician might use to establish mathematical truth—or, what amounts to the same thing, whatever *formal system* he* might adopt as providing his criterion of truth—there will always

* Of course 'he' means 'she or he'. See footnote on p. 6.

be mathematical propositions, such as the explicit Gödel proposition $P_k(k)$ of the system (cf. p. 107), that his algorithm cannot provide an answer for. If the workings of the mathematician's mind are entirely algorithmic, then the algorithm (or formal system) that he actually uses to form his judgements is not capable of dealing with the proposition $P_k(k)$ constructed from his personal algorithm. Nevertheless, *we* can (in principle) see that $P_k(k)$ is actually *true*! This would seem to provide *him* with a contradiction, since *he* ought to be able to see that also. Perhaps this indicates that the mathematician was *not* using an algorithm at all!

However, in order *actually* to convince ourselves of the truth of $P_k(k)$, we should need to *know* what the mathematician's algorithm really is, and also to be convinced of its validity as a means of arriving at mathematical truth. As strong-AI supporters are quick to point out, if the mathematician is using some very complicated algorithm inside his head, then we should have *no chance* of actually knowing what that algorithm is, and we shall therefore not be able actually to construct its Gödel proposition, let alone be convinced of its validity. That kind of objection is frequently raised against claims like the one that I am making here that Gödel's theorem indicates that human mathematical judgements are non-algorithmic.[1] But I do not myself find the objection convincing. Let us suppose, for the moment, that the ways that human mathematicians form their conscious judgements of mathematical truth *are* indeed algorithmic. We shall try to reduce this, by use of Gödel's theorem, to an absurdity (*reductio ad absurdum*!).

We must first consider the possibility that different mathematicians use *inequivalent* algorithms to decide truth. However, it is one of the most striking features of mathematics (perhaps almost alone among the disciplines) that the truth of propositions can actually be settled by abstract argument! A mathematical argument that convinces one mathematician—provided that it contains no error—will also convince another, as soon as the argument has been fully grasped. This also applies to the Gödel-type propositions. If the first mathematician is prepared to accept all the axioms and rules of procedure of a particular formal system as giving only *true* propositions, then he must also be prepared to accept its Gödel proposition as describing a true proposition. It would be exactly the same for a second mathematician. The point is that the arguments establishing mathematical truth are *communicable*.[2]

Thus we are not talking about various obscure algorithms that might happen to be running around in different particular mathematician's heads. We are talking about *one* universally employed formal system which is *equivalent* to *all* the different mathematicians' algorithms for judging mathematical truth. Now this putative 'universal' system, or algorithm, cannot ever be known as the one that we mathematicians use to decide truth! For if it were, then we *could* construct its Gödel proposition and know that to be a mathematical truth also. Thus, we are driven to the conclusion that the algorithm that mathematicians actually use to decide mathematical truth is so complicated or obscure that its very validity can never be known to us.

But this flies in the face of what mathematics is all about! The whole point of our mathematical heritage and training is that we do *not* bow down to the authority of some obscure rules that we can never hope to understand. We must *see*—at least in principle—that each step in an argument can be reduced to something simple and obvious. Mathematical truth is not a horrendously complicated dogma whose validity is beyond our comprehension. It is something built up from such simple and obvious ingredients—and when we comprehend them, their truth is clear and agreed by all.

To my thinking, this is as blatant a *reductio ad absurdum* as we can hope to achieve, short of an actual mathematical proof! The message should be clear. Mathematical truth is *not* something that we ascertain merely by use of an algorithm. I believe, also, that our *consciousness* is a crucial ingredient in our comprehension of mathematical truth. We must 'see' the truth of a mathematical argument to be convinced of its validity. This 'seeing' is the very essence of consciousness. It must be present *whenever* we directly perceive mathematical truth. When we convince ourselves of the validity of Gödel's theorem we not only 'see' it, but by so doing we reveal the very non-algorithmic nature of the 'seeing' process itself.

Inspiration, insight, and originality

I should attempt to make a few comments concerning those occasional flashes of new insight that we refer to as inspiration. Are these thoughts and images that come mysteriously from the *un*conscious mind, or are they in any important sense the product of consciousness itself. One can quote many instances where great thinkers have documented such experiences. As a mathematician, I am especially concerned with inspirational and original thought in others who are mathematicians, but I imagine that there is a great deal in common between mathematics and the other sciences and arts. For a very fine account, I refer the reader to the slim volume *The Psychology of Invention in the Mathematical Field*, a classic by the very distinguished French mathematician Jacques Hadamard. He cites numerous experiences of inspiration as described by leading mathematicians and other people. One of the best known of these is provided by Henri Poincaré. Poincaré describes, first, how he had intensive periods of deliberate, conscious effort in his search for what he called Fuchsian functions, but he had reached an impasse. Then:

> . . . I left Caen, where I was living, to go on a geologic excursion under the auspices of the School of Mines. The incidents of the travel made me forget my methematical work. Having reached Coutances, we entered an omnibus to go to some place or

other. At the moment when I put my foot on the step, the idea came to me, without anything in my former thoughts seeming to have paved the way for it, that the transformations I had used to define the Fuchsian functions were identical with those of non-Euclidean geometry. I did not verify the idea; I should not have had time, as upon taking my seat in the omnibus, I went on with a conversation already commenced, but I felt a perfect certainty. On my return to Caen, for convenience sake, I verified the result at my leisure.

What is striking about this example (and numerous others cited by Hadamard) is that this complicated and profound idea apparently came to Poincaré in a flash, while his conscious thoughts seemed to be quite elsewhere, and that they were accompanied by this feeling of certainty that they were correct—as, indeed, later calculation proved them to be. It should be made clear that the idea itself would not be something at all easy to explain in words. I imagine that it would have taken him something like an hour-long seminar, given to experts, to get the idea properly across. Clearly it could enter Poincaré's consciousness, fully formed, only because of the many long previous hours of deliberate conscious activity, familiarizing him with many different aspects of the problem at hand. Yet, in a sense, the idea that Poincaré had while boarding the bus was a 'single' idea, able to be fully comprehended in one moment! Even more remarkable was Poincaré's conviction of the truth of the idea, so that his subsequent detailed verification of it seemed almost superfluous.

Perhaps I should try to relate this to experiences of my own which might be in some way comparable. In fact, I cannot recall any occasion when a good idea has come to me so completely out of the blue as in that example of Poincaré's seems to have done (or as in many other quoted examples of genuine inspiration). For myself, it seems to be necessary that I *am* thinking (perhaps vaguely) about the problem at hand—consciously, but maybe at a low level just at the back of my mind. It might well be that I am engaged in some other rather relaxing activity; shaving would be a good example. Probably I would be just beginning to think about a problem that I had set aside for a while. The many hard hours of deliberate conscious activity would certainly be necessary, and sometimes it would take a while to re-acquaint myself with the problem. But the experience of an idea coming 'in a flash', under such circumstances—together with a strong feeling of conviction as to its validity—is not unknown to me.

It is perhaps worth relating a particular example of this, which has an additional curious point of interest. In the autumn of 1964, I had been worrying about the problem of black-hole singularities. Oppenheimer and Snyder had shown, in 1939, that an exactly *spherical* collapse of a massive star could lead to a central space–time singularity—at which the classical theory of general relativity is stretched beyond its limits (see Chapter 7, p. 332). Many people felt that this unpleasant conclusion could be avoided if their (unreasonable) assumption of *exact* spherical symmetry were removed. In the spherical case all the collapsing matter is aimed at a single central point where, perhaps not unexpectedly on account of this symmetry, a singularity of infinite density occurs. It seemed to be not unreasonable to suppose that *without* such symmetry, the matter would arrive at the central region in a more higgledy-piggledy way, and no singularity of infinite density would arise. Perhaps, even, the matter would all swirl out again to give a behaviour quite different from Oppenheimer and Snyder's idealized black hole.[3]

My own thoughts had been stimulated by the renewed interest in the black-hole problem that had stemmed from the fairly recent discovery of quasars (in the early 1960s). The physical nature of these remarkably bright and distant astronomical objects had led some people to speculate that something like Oppenheimer–Snyder black holes might reside at their centres. On the other hand, many thought that the Oppenheimer–Snyder assumption of spherical symmetry might provide a totally misleading picture. However, it had occurred to me (from the experience of work that I had done in another context) that there might be a precise mathematical theorem to be proved, showing that space–time singularities would be *inevitable* (according to standard general relativity theory), and hence vindicating the black-hole picture—provided that the collapse had reached some kind of 'point of no return'. I did not know of any mathematically definable criterion for a 'point of no return' (not using spherical symmetry) let alone any statement or proof of an appropriate theorem. A colleague (Ivor Robinson) had been visiting from the USA and he was engaging me in voluble conversation on a quite different topic as we walked down the street approaching my office in Birkbeck college in London. The conversation stopped momentarily as we crossed a side road, and resumed again at the other side. Evidently, during those few moments, an idea occurred to me, but then the ensuing conversation blotted it from my mind!

Later in the day, after my colleague had left, I returned to my office. I remember having an odd feeling of elation that I could not account for. I began going through in my mind all the various things that had happened to me during the day, in an attempt to find what it was that had caused this elation. After eliminating numerous inadequate possibilities, I finally brought to mind the thought that I had had while crossing the street—a thought which had momentarily elated me by providing the solution to the problem that had been milling around at the back of my head! Apparently, it was the needed criterion—that I subsequently called a 'trapped surface'—and then it did not take me long to form the outline of a proof of the theorem that I had been looking for (Penrose 1965). Even so, it was some while before the proof was formulated in a completely rigorous way, but the idea that I had had while crossing the street had been the key. (I sometimes wonder what would have happened if some unimportant *other* elating experience had happened to me during that day. Perhaps I should never have recalled the trapped-surface idea at all!)

The above anecdote brings me to another issue concerning inspiration and insight, namely that *aesthetic* criteria are enormously valuable in forming our judgements. In the arts, one might say that it is aesthetic criteria that are paramount. Aesthetics in the arts is a sophisticated subject, and philosophers have devoted lifetimes to its study. It could be argued that in mathematics and the sciences, such criteria are merely incidental, the criterion of *truth* being paramount. However, it seems to be impossible to separate one from the other when one considers the issues of inspiration and insight. My impression is that the strong conviction of the *validity* of a flash of inspiration (not 100 per cent reliable, I should add, but at least far more reliable than just chance) is very closely bound up with its aesthetic qualities. A beautiful idea has a much greater chance of being a correct idea than an ugly one. At least that has been my own experience, and similar sentiments have been expressed by others (cf. Chandrasekhar 1987). For example, Hadamard (1945, p. 31) writes:

> . . . it is clear that no significant discovery or invention can take place without the *will* of finding. But with Poincaré, we see something else, the intervention of the sense of beauty playing its part as an indispensable *means* of finding. We have reached the double conclusion:
>
> that invention is choice
>
> that this choice is imperatively governed by the sense of scientific beauty.

Moreover Dirac (1982), for example, is unabashed in his claim that it was his *keen sense of beauty* that enabled him to divine his equation for the electron (the 'Dirac equation' alluded to on p. 289), while others had searched in vain.

I can certainly vouch for the importance of aesthetic qualities in my own thinking, both in relation to the 'conviction' that would be felt with ideas that might possibly qualify as 'inspirational' and with the more 'routine' guesses that would continually have to be made, as one feels one's way towards some hoped-for goal. I have written on this matter elsewhere in relation, in particular, to the discovery of the aperiodic tiles described in Figs. 10.3 and 4.11. Undoubtedly it was the aesthetic qualities of the first of these tiling patterns—not just its visual appearance, but also its intriguing mathematical properties—that had allowed the intuition to come to me (probably in a 'flash', but with only about 60 per cent certainty!) that its arrangement could be forced by appropriate matching rules (i.e. jig-saw assembly). We shall be seeing more about these tiling patterns shortly. (Cf. Penrose 1974.)

It seems clear to me that the importance of aesthetic criteria applies not only to the instantaneous judgements of inspiration, but also to the much more frequent judgements that we make all the time in mathematical (or scientific) work. Rigorous argument is usually the *last* step! Before that, one has to make many guesses, and for these, aesthetic convictions are enormously important—always contrained by logical argument and known facts.

It is these judgements that I consider to be the hallmark of conscious thinking. My guess is that even with the sudden flash of insight, apparently produced ready-made by the unconscious mind, it is *consciousness* that is the arbiter, and the idea would be quickly rejected and forgotten if it did not 'ring true'. (Curiously, I *did* actually forget my trapped surface, but that is not at the level that I mean. The idea broke through into consciousness for long enough for it to leave a lasting impression.) The 'aesthetic' rejection that I am referring to might, I am supposing, be such as to forbid unappealing ideas to reach any very appreciably permanent level of consciousness at all.

What, then is my view as to the role of the *unconscious* in inspirational thought? I admit that the issues are not so clear as I would like them to be. This is an area where the unconscious seems indeed to be playing a vital role, and I must concur with the view that unconscious processes are important. I must agree, also, that it cannot be that the unconscious mind is simply throwing up ideas at random. There must be a powerfully impressive selection process that allows the conscious mind to be disturbed only by ideas that 'have a chance'. I would suggest that these criteria for selection—largely 'aesthetic' ones, of some sort—have been already strongly influenced by conscious desiderata (like the feeling of ugliness that would accompany mathematical thoughts that are inconsistent with already established general principles).

In relation to this, the question of what constitutes genuine *originality* should be raised. It seems to me that there are two factors involved, namely a 'putting-up' and a 'shooting-down' process. I imagine that the putting-up could be largely unconscious and the shooting-down largely conscious. Without an effective putting-up process, one would have no new ideas at all. But, just by itself, this procedure would have little value. One needs an effective procedure for forming judgements, so that only those ideas with a reasonable chance of success will survive. In dreams, for example, unusual ideas may easily come to mind, but only very rarely do they survive the critical judgements of the wakeful consciousness. (For my own part, I have never had a successful scientific idea in a dreaming state, while others, such as the chemist Kekulé in his discovery of the structure of benzene, may have been more fortunate.) In my opinion, it is the conscious shooting-down (judgement) process that is central to the issue of originality, rather than the unconscious putting-up process; but I am aware that many others might hold to a contrary view.

Before leaving things in this rather unsatisfactory state, I should mention another striking feature of inspirational thought, namely its *global* character. Poincaré's anecdote above was a striking example, since the idea that came into his mind in a fleeting moment would have encompassed a huge area of mathematical thought. Perhaps more immediately accessible to the non-mathematical reader (though no more comprehensible, no doubt) is the way that (some) artists can keep the totality of their creations in mind all at once. A striking example is given vividly by Mozart (as quoted in Hadamard, 1945, p. 16)

When I feel well and in a good humour, or when I am taking a drive or walking after a good meal, or in the night when I cannot sleep, thoughts crowd into my mind as easily as you could wish. Whence and how do they come? I do not know and I have nothing to do with it. Those which please me I keep in my head and hum them; at least others have told me that I do so. Once I have my theme, another melody comes, linking itself with the first one, in accordance with the needs of the composition as a whole: the counterpoint, the part of each instruments and all the melodic fragments at last produce the complete work. Then my soul is on fire with inspiration. The work grows; I keep expanding it, conceiving it more and more clearly until I have the entire composition finished in my head though it may be long. Then my mind seizes it as a glance of my eye a beautiful picture or a handsome youth. It does not come to me successively, with various parts worked out in detail, as they will later on, but in its entirety that my imagination lets me hear it.

It seems to me that this accords with a putting-up/shooting-down scheme of things. The putting-up seems to be unconscious ('I have nothing to do with it') though, no doubt, highly selective, while the shooting-down is the conscious arbiter of taste ('those which please me I keep . . .'). The globality of inspirational thought is particularly remarkable in Mozart's quotation ('It does not come to me successively . . . but in its entirety') and also in Poincaré's ('I did not verify the idea; I should not have had time'). Moreover, I would maintain that a remarkable globality is already present in our conscious thinking generally. I shall return to this issue shortly.

Non-verbality of thought

One of the major points that Hadamard makes in his study of creative thinking is an impressive refutation of the thesis, so often still expressed, that verbalization is necessary for thought. One could hardly do better than repeat a quotation from a letter he received from Albert Einstein on the matter:

The words or the language, as they are written or spoken, do not seem to play any role in my mechanism of thought. The psychical entities which seem to serve as elements of thought are certain signs and more or less clear images which can be "voluntarily" reproduced and combined. . . . The above mentioned elements are, in my case, of visual and some muscular type. Conventional words or other signs have to be sought for laboriously only in a second stage, when the mentioned associative play is sufficiently established and can be reproduced at will.

The eminent geneticist Francis Galton is also worth quoting:

It is a serious drawback to me in writing, and still more in explaining myself, that I do not think as easily in words as otherwise. It often happens that after being hard at work, and having arrived at results that are perfectly clear and satisfactory to myself, when I try to express them in language I feel that I must begin by putting myself upon quite another intellectual plane. I have to translate my thoughts into a language that does not run very evenly with them. I therefore waste a vast deal of time in seeking appropriate words and phrases, and am conscious, when required to speak on a sudden, of being often very obscure through mere verbal maladroitness, and not through want of clearness of perception. That is one of the small annoyances of my life.

Also Hadamard himself writes:

I insist that words are totally absent from my mind when I really think and I shall completely align my case with Galton's in the sense that even after reading or hearing a question, every word disappears the very moment that I am beginning to think it over; and I fully agree with Schopenhauer when he writes, 'thoughts die the moment they are embodied by words'.

I quote these examples because they very much accord with my own thought-modes. Almost all my mathematical thinking is done visually and in terms of non-verbal concepts, although the thoughts are quite often accompanied by inane and almost useless verbal commentary, such as 'that thing goes with that thing and that thing goes with that thing'. (I might use words sometimes for simple logical inferences.) Also, the difficulties that these thinkers have had with translating their thoughts into words is something that I frequently experience myself. Often the reason is that there simply are not the words available to express the concepts that are required. In fact I often calculate using specially designed diagrams which constitute a shorthand for certain types of algebraic expression (cf. Penrose and Rindler 1984, pp. 424–434). It would be a very cumbersome process indeed to have to translate such diagrams into words, and this is something that I would do only at a last resort if it becomes necessary to give a detailed explanation to others. As a related observation, I had noticed, on occasion, that if I have been concentrating hard for some while on mathematics and someone would engage me suddenly in conversation, then I would find myself almost unable to speak for several seconds.

This is not to say that I do not sometimes think in words, it is just that I find words almost useless for *mathematical* thinking. Other kinds of thinking, perhaps such as *philosophizing*, seem to be much better suited to verbal expression. Perhaps this is why so many philosophers seem to be of the opinion that language is essential for intelligent or conscious thought! No doubt different people think in very different ways—as has certainly been my own experience, even just amongst mathematicians. The main polarity in mathematical thinking seems to be analytical/geometrical. It is interesting that Hadamard considered himself to be on the analytical side, even though he used visual rather than verbal images for his mathematical thinking. As for myself, I am very much on the geometrical end of things, but the spectrum amongst mathematicians generally is a very broad one.

Once it is accepted that much of conscious thinking can indeed be of a non-verbal character—and, to my own mind, this conclusion is an inescapable conclusion of considerations like the above—then perhaps the reader may not find it so hard to believe that such thinking might also have a non-algorithmic ingredient!

Recall that in Chapter 9 (p. 384) I referred to the frequently expressed viewpoint that only the half of the brain which is capable of speech (the left half, in the vast majority) would also be capable of consciousness. It should be clear to the reader, in view of the above discussion, why I find that viewpoint totally unacceptable. I do not know whether, on the whole, mathematicians would tend to use one side of their brains more, or the other; but there can be no doubt of the high level of consciousness that is needed for genuine mathematical thought. Whereas analytical thinking seems to be mainly the province of the left side of the brain, geometrical thinking is often argued to be the *right* side's, so it is a very reasonable guess that a good deal of *conscious* mathematical activity actually *does* actually take place on the right!

Animal consciousness?

Before leaving the topic of the importance of verbalization to consciousness, I should address the question, raised briefly earlier, of whether non-human animals can be conscious. It seems to me that people sometimes rely on animals' inability to speak as an argument against their having any appreciable consciousness—and, by implication, against their having any 'rights'. The reader may well perceive that I regard this as an untenable line of argument, since much sophisticated conscious thinking (e.g. mathematical) can be carried out without verbalization. The right side of the brain, also, is sometimes argued as having as 'little' consciousness as a chimpanzee, also because of its lack of verbal ability (cf. LeDoux, 1985, pp. 197–216).

There is a considerable controversy about whether chimpanzees and gorillas are, in fact, capable of genuine verbalization when they are allowed to use *sign language* rather than speak in the normal human way (which they cannot do owing to the lack of suitable vocal chords). (See various articles in Blakemore and Greenfield 1987.) It seems clear, despite the controversy, that they are able to communicate at least to a certain elementary degree by such means. In my own opinion, it is a little churlish of some people not to allow this to be called 'verbalization'. Perhaps, by denying apes entry to the verbalizer's club, some would hope to exclude them from the club of conscious beings!

Leaving aside the question of speech, there is good evidence that chimpanzees are capable of genuine *inspiration*. Konrad Lorenz (1972) describes a chimpanzee in a room which contains a banana suspended from the ceiling just out of reach, and a box elsewhere in the room:

> The matter gave him no peace, and he returned to it again. Then, suddenly—and there is no other way to describe it—his previously gloomy face 'lit up'. His eyes now moved from the banana to the empty space beneath it on the ground, from this to the box, then back to the space, and from there to the banana. The next moment he gave a cry of joy, and somersaulted over to the box in sheer high spirits. Completely assured of his success, he pushed the box below the banana. No man watching him could doubt the existence of a genuine 'Aha' experience in anthropoid apes.

Note that, just as with Poincaré's experience as he boarded the omnibus, the chimpanzee was 'completely assured of his success' before he had verified his idea. If I am right that such judgements require consciousness, then there is evidence here, also, that non-human animals can indeed be conscious.

An interesting question arises in connection with dolphins (and whales). It may be noted that the cerebra of dolphins are as large (or larger) than our own, and dolphins can also send extremely complex sound signals to one another. It may well be that their large cerebra are needed for some purpose other than 'intelligence' on a human or near-human scale. Moreover, owing to their lack of prehensile hands, they are not able to construct a 'civilization' of the kind that we can appreciate—and, although they cannot write books for the same reason, they might sometimes be philosophers, and ponder over the meaning of life and why they are 'there'! Might they sometimes transmit their feelings of 'awareness' via their complex underwater sound signals? I am not cognizant of any research indicating whether they use one particular side of their brains to 'verbalize' and communicate with one another. In relation to the 'split-brain' operations that have been performed on humans, with their puzzling implications of the continuity of 'self', it should be remarked that dolphins do not go to sleep[4] with their whole brains simultaneously, but one side of the brain goes to sleep at a time. It would be instructive for us if we could ask them how *they* 'feel' about the continuity of consciousness!

Contact with Plato's world

I have mentioned that there seem to be many different ways in which different people think—and even in which different mathematicians think about their mathematics. I recall that when I was about to enter university to study that subject I had expected to find that the others, who would be my mathematical colleagues, would think more-or-less as I did. It had been my experience at school that my classmates seemed to think rather differently from myself, which I had found somewhat disconcerting. 'Now', I had thought to myself excitedly, 'I shall find colleagues with whom I can much more readily communicate! Some will think more effectively than I, and some less; but all will share my particular wavelength of thought.' How wrong I was! I believe that I encountered more differences in modes of thinking than I had experienced ever before! My own thinking was much more geometrical and less analytical than the others, but there were many other differences between my various colleagues' thought-modes. I always had particular trouble with comprehending a verbal description of a formula, while many of my colleagues seemed to experience no such difficulty.

A common experience, when some colleague would try to explain some piece of mathematics to me, would be that I should listen attentively, but almost totally uncomprehending of the logical connections between one set of words and the next. However, some guessed image would form in my mind as to the ideas that he was trying to convey—formed entirely on my own terms and seemingly with very little connection with the mental images that had been the basis of my colleague's own understanding—and I would reply. Rather to my astonishment, my own remarks would usually be accepted as appropriate, and the conversation would proceed to and fro in this way. It would be clear, at the end of it, that some genuine and positive communication had taken place. Yet the actual sentences that each one of us would utter seemed only very infrequently to be actually understood! In my subsequent years as a professional mathematician (or mathematical physicist) I have found this phenomenon to be no less true than it had been when I was an undergraduate. Perhaps, as my mathematical experience has increased, I have got a little better at guessing what others are meaning by their explanations, and perhaps I am a little better at making allowances for other modes of thinking when I am explaining things myself. But, in essence, nothing has changed.

It had often been a puzzle to me how communication is possible at all according to this strange procedure, but I should like, now, to venture an explanation of a sort, because I think that it could possibly have some deep relevance to the other issues that I have been addressing. The point is that with the conveying of mathematics, one is *not* simply communicating *facts*. For a string of (contingent) facts to be communicated from one person to another, it is necessary that the facts be carefully enunciated by the first, and that the second should take them in individually. But with mathematics, the *factual* content is small. Mathematical statements are necessary truths (or else necessary falsehoods!) and even if the first mathematician's statement represents merely a groping for such a necessary truth, it will be that truth itself which gets conveyed to the second mathematician, provided that the second has adequately understood. The second's mental images may differ in detail from those of the first, and their verbal descriptions may differ, but the relevant mathematical idea will have passed between them.

Now this type of communication would not be possible at all were it not for the fact that *interesting* or *profound* mathematical truths are somewhat sparcely distributed amongst mathematical truths in general. If the truth to be conveyed were, say, the *un*interesting statement $4897 \times 512 = 2507264$, then the second would indeed have to be comprehending of the first, for the precise statement to be conveyed. But for a mathematically interesting statement, one can often latch on to the concept intended, even if its description has been very imprecisely provided.

There may seem to be a paradox in this, since mathematics is a subject where precision is paramount. Indeed, in written accounts, much care is taken in order to make sure that the various statements are both precise and complete. However, in order to convey a mathematical idea (usually in verbal descriptions), such precision may sometimes have an inhibiting effect at first, and a more vague and descriptive form of communication may be needed. Once the idea has been grasped in essence, then the details may be examined afterwards.

How is it that mathematical ideas can be communicated in this way? I imagine that whenever the mind perceives a mathematical idea, it makes contact with Plato's world of mathematical concepts. (Recall that according to the Platonic viewpoint, mathematical ideas have an existence of their own, and inhabit an ideal Platonic world, which is accessible via the intellect only; cf. pp. 97, 158.) When one 'sees' a mathematical truth, one's consciousness breaks through into this world of ideas, and makes direct contact with it ('accessible via the intellect'). I have described this 'seeing' in relation to Gödel's theorem, but it is the essence of mathematical understanding. When mathematicians communicate, this is made possible by each one having a *direct route to truth*, the consciousness of each being in a position to perceive mathematical truths directly, through this process of 'seeing'. (Indeed, often this act of perception is accompanied by words like 'Oh, I see'!) Since each can make contact with Plato's world directly, they can more readily communicate with each other than one might have expected. The mental images that each one has, when making this Platonic contact, might be rather different in each case, but communication is possible because each is directly in contact with the *same* externally existing Platonic world!

According to this view, the mind is always capable of this direct contact. But only a little may come through at a time. Mathematical discovery consists of broadening the area of contact. Because of the fact that mathematical truths are necessary truths, no actual 'information', in the technical sense, passes to the discoverer. All the information was there all the time. It was just a matter of putting things together and 'seeing' the answer! This is very much in accordance with Plato's own idea that (say mathematical) discovery is just a form of *remembering*! Indeed, I have often been struck by the similarity between just not being able to remember someone's name, and just not being able to find the right mathematical concept. In each case, the sought-for concept is, in a sense *already present* in the mind, though this is a less usual form of words in the case of an undiscovered mathematical idea.

In order for this way of viewing things to be helpful, in the case of mathematical communication, one must imagine that the interesting and profound mathematical ideas somehow have a stronger existence than the uninteresting or trival ones. This will have significance in relation to the speculative considerations of the next section.

A view of physical reality

Any viewpoint as to how consciousness can arise, within the universe of physical reality, must address, at least implicitly, the question of physical reality itself.

The viewpoint of strong AI, for example, maintains that a 'mind' finds its existence through the embodiment of a sufficiently complex algorithm, as this algorithm is acted out by some objects of the physical world. It is not supposed to matter what actual objects these are. Nerve signals, electric currents along wires, cogs, pulleys, or water pipes would do equally well. The algorithm itself is considered to be all-important. But for an algorithm to 'exist' independently of any particular physical embodiment, a Platonic viewpoint of mathematics would seem to be essential. It would be difficult for a strong-AI supporter to take the alternative line that 'mathematical concepts exist only in minds', since this would be circular, requiring pre-existing minds for the existence of algorithms and pre-existing algorithms for minds! They might try to take the line that algorithms can exist as marks on a piece of paper, or directions of magnetization in a block of iron, or charge displacements in a computer memory. But such arrangements of material do not in themselves actually constitute an algorithm. In order to become algorithms, they need to have an *interpretation*, i.e. it must be possible to *decode* the arrangements; and that will depend upon the 'language' in which the algorithms are written. Again a pre-existing mind seems to be required, in order to 'understand' the language, and we are back where we were. Accepting, then, that algorithms inhabit Plato's world, and hence *that* world, according to the strong-AI view, is where minds are to be found, we would now have to face the question of how the physical world and Plato's world can relate to one another. This, it seems to me, is the strong-AI version of the mind–body problem!

My own viewpoint is different from this, since I believe that (conscious) minds are *not* algorithmic entities. But I am somewhat disconcerted to find that there are a good many points in common between the strong-AI viewpoint and my own. I have indicated that I believe consciousness to be closely associated with the sensing of necessary truths—and thereby achieving a direct contact with Plato's world of mathematical concepts. This is not an algorithmic procedure—and it is not the algorithms that might inhabit that world which have special concern for us—but again the mind–body problem is seen, according to this view, to be intimately related to the question of how Plato's world relates to the 'real' world of actual physical objects.

We have seen in Chapters 5 and 6 how the real physical world seems to accord in a remarkable way with some very precise mathematical schemes (the SUPERB theories cf. p. 152). It has often been remarked how extraordinary this precision actually is (cf. especially Wigner 1960) It is hard for me to believe, as some have tried to maintain, that such SUPERB theories could have arisen merely by some random natural selection of ideas leaving only the good ones as survivors. The good ones are simply much *too* good simply to be the survivors of ideas that have arisen in that random way. There must, instead, be some deep underlying reason for the accord between mathematics and physics, i.e. between Plato's world and the physical world.

To speak of 'Plato's world' at all, one is assigning some kind of reality to it which is in some way comparable to the reality of the physical world. On the other hand, the reality of the physical world itself seems more nebulous than it had seemed to be before the advent of the SUPERB theories of relativity and quantum mechanics (see the remarks on pp. 152, 153, and 286, especially). The very precision of these theories has provided an almost abstract mathematical existence for actual physical reality. Is this in any way a paradox? How can concrete reality become abstract and mathematical? This is perhaps the other side of the coin to the question of how abstract mathematical concepts can achieve an almost concrete reality in Plato's world. Perhaps, in some sense, the two worlds are actually the *same*? (Cf. Wigner 1960; Penrose 1979*a*; Barrow 1988; also Atkins 1987.)

Though I have some strong sympathy with this idea of actually identifying these two worlds, there must be more to the issue than just that. As I mentioned in Chapter 3, and earlier in this chapter, some mathematical truths seem to have a stronger ('deeper', 'more interesting', 'more fruitful'?) Platonic reality than others. These would be the ones that would be more strongly to be identified with the workings of physical reality. (The system of complex numbers (cf. Chapter 3) would be a case in point, these being the fundamental ingredients of quantum mechanics, the probability amplitudes.) With such an identification, it might be more comprehensible how 'minds' could seem to manifest some mysterious connection between the physical world and Plato's world of mathematics. Recall also that, as described in Chapter 4, there are many parts of the mathematical world—some of its deepest and most interesting parts, moreover—that have a non-algorithmic character. It would seem to be likely, therefore, on the basis of the viewpoint that I have been attempting to expound, that non-algorithmic action ought to have a role within the physical world of very considerable importance. I am suggesting that this role is intimately bound up with the very concept of 'mind'.

Determinism and strong determinism

So far I have said little about the question of 'free will' which is normally taken to be the fundamental issue of the *active* part of the mind–body problem. Instead, I have concentrated on my suggestion that there is an essential *non-algorithmic* aspect to the role of conscious action. Normally, the issue of free will is discussed in relation to determinism in physics. Recall that in most of our SUPERB theories there is a clear-cut determinism, in the sense that if the state of the system is known at any one time,[5] then it is completely fixed at all later (or indeed earlier) times by the equations of the theory. In this way there seems to be no room for 'free will' since the future behaviour of a system seems to be totally determined by the physical laws. Even the **U** part of quantum mechanics has this completely deterministic character. However, the **R** 'quantum-jump' part is not deterministic, and it introduces a completely random element into the time-evolution. Early on, various people leapt at the possibility that here might be a role for free will, the action of consciousness perhaps having some direct effect on the way that an individual quantum system might jump. But if **R** is *really* random, then it is not a great deal of help either, if we wish to do something positive with our free wills.

My own point of view, although it is not very well formulated in this respect, would be that some new procedure (CQG; cf. Chapter 8) takes over at the quantum–classical borderline which interpolates between **U** and **R** (each of which are now regarded as approximations), and that this new procedure would contain *an essentially non-algorithmic* element. This would imply that the future would *not be computable* from the present, even though it *might* be *determined* by it. I have tried to be clear in distinguishing the issue of computability from that of determinism, in my discussions in Chapter 5. It seems to me to be quite plausible that CQG might be a deterministic but non-computable theory.* (Recall the non-computable 'toy model' that I described in Chapter 5, p. 170.)

Sometimes people take the view that even with classical (or **U**-quantum) determinism there is no *effective* determinism, because the initial conditions cannot ever be well-enough known that the future could *actually* be computed. Sometimes very small changes in the initial conditions can lead to very large differences in the final outcome. This is what happens, for example, in the phenomenon known as 'chaos' in a (classical) deterministic system—an example being the uncertainty of weather prediction. However, it is very hard

* It may be pointed out that there is at least one approach to a quantum gravity theory that seems to involve an element of non-computability (Geroch and Hartle 1987).

to believe that this kind of classical uncertainty can be what allows us our (illusion of?) free will. The future behaviour would still be *determined*, right from the big bang, even though we would be unable to compute it (cf. p. 173).

The same objection might be raised against my suggestion that a lack of *computability* might be intrinsic to the dynamical laws—now assumed to be non-algorithmic in character—rather than to our lack of information concerning initial conditions. Even though not computable, the future would, on this view, still be completely *fixed* by the past—all the way back to the big bang. In fact, I am not being so dogmatic as to insist that CQG ought to be deterministic but non-computable. My guess would be that the sought-for theory would have a more subtle description than that. I am only asking that it should contain non-algorithmic elements of some essential kind.

To close this section, I should like to remark on an even more extreme view that one might hold towards the issue of determinism. This is what I have referred to (Penrose 1987*a*) as *strong determinism*. According to strong determinism, it is not just a matter of the future being determined by the past; the *entire history of the universe is fixed*, according to some precise mathematical scheme, *for all time*. Such a viewpoint might have some appeal if one is inclined to identify the Platonic world with the physical world in some way, since Plato's world is fixed once and for all, with no 'alternative possibilities' for the universe! (I sometimes wonder whether Einstein might have had such a scheme in mind when he wrote 'What I'm really interested in is whether God could have made the world in a different way; that is, whether the necessity of logical simplicity leaves any freedom at all!' (letter to Ernst Strauss; see Kuznetsov 1977, p. 285).

As a variant of strong determinism, one might consider the *many-worlds* view of quantum mechanics (cf. Chapter 6, p. 295). According to this, it would not be a *single* individual universe-history that would be fixed by a precise mathematical scheme, but the totality of myriads upon myriads of 'possible' universe-histories that would be so determined. Despite the unpleasant nature (at least to me) of such a scheme and the multitude of problems and inadequacies that it presents us with, it cannot be ruled out as a possibility.

It seems to me that if one has strong determinism, but *without* many worlds, then the mathematical scheme which governs the structure of the universe would probably *have* to be non-algorithmic.[6] For otherwise one could in principle calculate what one was going to do next, and then one could 'decide' to do something quite different, which would be an effective contradiction between 'free will' and the strong determinism of the theory. By introducing non-computability into the theory one can evade this contradiction—though I have to confess that I feel somewhat uneasy about this type of resolution, and I anticipate something much more subtle for the *actual* (non-algorithmic!) rules that govern the way that the world works!

The anthropic principle

How important is consciousness for the universe as a whole? Could a universe exist without any conscious inhabitants whatever? Are the laws of physics specially designed in order to allow the existence of conscious life? Is there something special about our particular location in the universe, either in space or in time? These are the kinds of question that are addressed by what has become known as the *anthropic principle*.

This principle has many forms. (See Barrow and Tipler 1986.) The most clearly acceptable of these addresses merely the spatio-temporal location of conscious (or 'intelligent') life in the universe. This is the *weak* anthropic principle. The argument can be used to explain why the conditions happen to be just right for the existence of (intelligent) life on the earth at the present time. For if they were not just right, then we should not have found ourselves to be here now, but somewhere else, at some other appropriate time. This principle was used very effectively by Brandon Carter and Robert Dicke to resolve an issue that had puzzled physicists for a good many years. The issue concerned various striking numerical relations that are observed to hold between the physical constants (the gravitational constant, the mass of the proton, the age of the universe, etc.). A puzzling aspect of this was that some of the relations hold only at the present epoch in the earth's history, so we appear, coincidentally, to be living at a very special time (give or take a few million years!). This was later explained, by Carter and Dicke, by the fact that this epoch coincided with the lifetime of what are called main-sequence stars, such as the sun. At any other epoch, so the argument ran, there would be no intelligent life around in order to measure the physical constants in question—so the coincidence *had* to hold, simply because there would be intelligent life around only at the particular time that the coincidence *did* hold!

The *strong* anthropic principle goes further. In this case, we are concerned not just with our spatio-temporal location within the universe, but within the infinitude of *possible* universes. Now we can suggest answers to questions as to why the physical constants, or the laws of physics generally, are specially designed in order that intelligent life can exist at all. The argument would be that if the constants or the laws were any different, then we should not be in this particular universe, but we should be in some other one! In my opinion, the strong anthropic principle has a somewhat dubious character, and it tends to be invoked by theorists whenever they do not have a good enough theory to explain the observed facts (i.e. in theories of particle physics, where the masses of particles are unexplained and it is argued that if they had different values from the ones observed, then life would presumably be impossible, etc.). The weak anthropic principle, on the other hand, seems to me to be unexceptionable, provided that one is very careful about how it is used.

By the use of the anthropic principle—either in the strong or weak form—one might try to show that consciousness was *inevitable* by virtue of the fact that sentient beings, that is 'we', have to be around in order to observe the world, so one need *not* assume, as I have done, that sentience has any selective advantage! In my opinion, this argument is technically correct, and the weak anthropic argument (at least) *could* provide a reason that consciousness is here without it having to be favoured by natural selection. On the other hand, I cannot believe that the anthropic argument is the *real* reason (or the only reason) for the evolution of consciousness. There is enough evidence from other directions to convince me that consciousness *is* of powerful selective advantage, and I do not think that the anthropic argument is needed.

Tilings and quasicrystals

I now turn away from the sweeping speculations of the last few sections, and consider instead an issue that, though still somewhat speculative, is much more scientific and 'tangible'. At first this issue will appear to be an irrelevant digression. But the significance of it for us will become apparent in the next section.

Recall the tiling patterns depicted in Fig. 4.12, p. 137. These patterns are somewhat remarkable in that they 'almost' violate a standard mathematical theorem concerning crystal lattices. The theorem states that the only rotational symmetries that are allowed for a crystalline pattern are twofold, threefold, fourfold, and sixfold. By a crystalline pattern, I mean a discrete system of points which has a *translational symmetry*; that is, there is a way of sliding the pattern over itself without rotating it, so that the pattern goes over into itself (i.e. is unchanged by that particular motion) and therefore has a *period parallelogram* (cf. Fig. 4.8). Examples of tiling patterns with these allowed rotational symmetries are given in Fig. 10.2. Now the patterns of Fig. 4.12, like that depicted in Fig. 10.3 (which is essentially the tiling produced by fitting together the tiles of Fig. 4.11), on the other hand, *almost* have translational symmetry and they *almost* have *fivefold* symmetry—where 'almost' means that one can find such motions of the pattern (translational and rotational, respectively) for which the pattern goes into itself to any desired pre-assigned degree of agreement just short of 100 per cent. There is no need to worry about the precise meaning of this here. The only point that will be of relevance for us is that if one had a substance in which the atoms were arranged at the vertices of the pattern, then it would appear to be crystalline, yet it would exhibit a forbidden fivefold symmetry!

In December 1984, the Israeli physicist Dany Shechtman who had been working with colleagues at the National Bureau of Standards in Washington DC, USA, announced the discovery of a phase of an aluminium–manganese

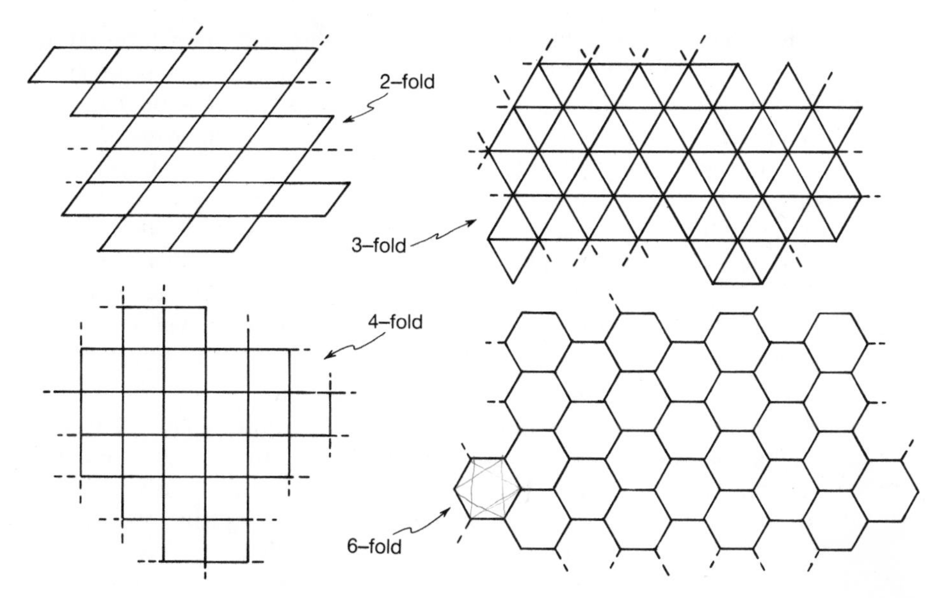

Fig. 10.2. Periodic tilings with various symmetries (where the centre of symmetry, in each case, is taken to be the centre of a tile).

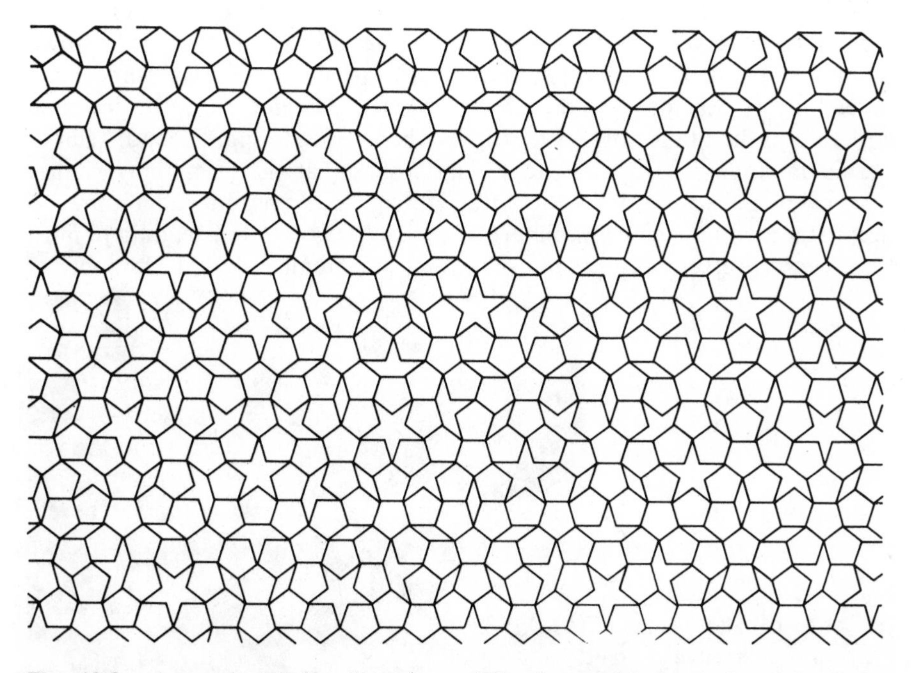

Fig. 10.3. A quasi-periodic tiling (essentially that which is produced by fitting together the tiles of Fig. 4.11) with a crystallographically 'impossible' fivefold quasi-symmetry.

alloy which seemed indeed to be a crystalline-like substance—now referred to as a *quasicrystal*—with fivefold symmetry. In fact this quasicrystalline substance also exhibited a symmetry in *three*-dimensions, and not only in the plane—giving a forbidden *icosahedral* symmetry in all (Shechtman *et al* 1984). (Three-dimensional 'icosahedral' analogues of my fivefold plane tilings had been found by Robert Ammann in 1975, see Gardner 1989.) Shechtman's alloys formed only tiny microscopic quasicrystals, about 10^{-3} of a millimetre across, but later other quasicrystalline substances were found and, in particular, an aluminium–lithium–copper alloy for which the icosahedrally symmetric units can grow to about a millimetre in size, and are quite visible to the naked eye (see Fig. 10.4).

Now, a remarkable feature of the quasicrystalline tiling patterns that I have been describing is that their assembly is necessarily *non-local*. That is to say, in assembling the patterns, it is necessary, from time to time, to examine the state of the pattern many, many 'atoms' away from the point of assembly, if one is to be sure of not making a serious error when putting the pieces together. (This is perhaps akin to the apparently 'intelligent groping' that I referred to in relation to natural selection.) This kind of feature is an ingredient of a considerable controversy that surrounds the question of quasicrystal structure and growth at the present time, and it would be unwise to attempt to draw definitive conclusions until some of the outstanding issues are resolved. Nevertheless, one may speculate; and I shall venture my own opinion. First, I believe that some of these quasicrystalline substances are indeed highly organized, and their atomic arrangements are rather close in structure to the tiling patterns that I have been considering. Second, I am of the (more tentative) opinion that this implies that their assembly cannot reasonably be achieved by the local adding of atoms one at a time, in accordance with the *classical* picture of crystal growth, but instead there must be a *non-local* essentially quantum-mechanical ingredient to their assembly.[7]

Fig. 10.4. A quasicrystal (an Al–Li–Cu alloy) with an apparently impossible crystal symmetry. (From Gayle 1987.)

The way that I picture this growth as taking place is that, instead of having atoms coming individually and attaching themselves at a continually moving growth line (classical crystal growth), one must consider an evolving quantum linear superposition of many different alternative arrangements of attaching atoms (by the quantum procedure **U**). Indeed, this is what quantum mechanics tells us *must* (almost always) be occurring! There is not just one thing that happens; many alternative atomic arrangements must coexist in complex linear superposition. A few of these superposed alternatives will grow to very much bigger conglomerations and, at a certain point, the difference between the gravitational fields of some of the alternatives will reach the one-graviton level (or whatever is appropriate; see Chapter 8, p. 367). At this stage one of the alternative arrangements—or, more likely, still a superposition, but a somewhat reduced superposition—will become singled out as the 'actual' arrangement (quantum procedure **R**). This superposed assembly, together with reductions to more definite arrangements, would continue at a larger and larger scale until a reasonable-sized quasicrystal is formed.

Normally, when Nature seeks out a crystalline configuration, she is searching for a configuration of *lowest energy* (taking the background temperature to be zero). I am envisaging a similar thing with quasicrystal growth, the difference being that this state of lowest energy is much more difficult to find, and the 'best' arrangement of the atoms cannot be discovered simply by adding on atoms one at a time in the hope that each individual atom can get away with just solving its *own* minimizing problem. Instead, we have a *global* problem to solve. It must be a co-operative effort among a large number of atoms all at once. Such co-operation, I am maintaining, must be achieved quantum mechanically; and the way that this is done is by many different combined arrangements of atoms being 'tried' simultaneously in linear superposition (perhaps a little like the quantum computer considered at the end of Chapter 9). The selection of an appropriate (though probably not the best) solution to the minimizing problem must be achieved as the one-graviton criterion (or appropriate alternative) is reached—which would presumably only occur when the physical conditions are just right.

Possible relevance to brain plasticity

Let me now carry these speculations further, and ask whether they might have any relevance to the question of brain functioning. The most likely possibility, as far as I can see, is in the phenomenon of brain plasticity. Recall that the brain is not really quite like a computer, but it is more like a computer which is continually changing. These changes can apparently come about by synapses becoming activated or de-activated through the growth or contraction of dendritic spines (see Chapter 9, p. 392; Fig. 9.15). I am putting my neck out here and speculating that this growth or contraction could be governed by something like the processes involved in quasicrystal growth. Thus, not just one of the possible alternative arrangements is tried out, but vast numbers, all superposed in complex linear superposition. So long as the effects of these alternatives are kept below the one-graviton level (or whatever) then they will indeed coexist (and almost invariably *must* coexist, according to the rules of **U**-quantum mechanics). If kept below this level, simultaneous superposed calculations can begin to be performed, very much in accordance with the principles of a quantum computer. However, it seems unlikely that these superpositions can be maintained for long, since nerve signals produce electric fields which would significantly disturb the surrounding material (though their myelin sheaths might help to insulate them). Let us speculate that such superpositions of calculations can actually be maintained for at least a sufficient length of time that something of significance *is* actually calculated before the one-graviton level (or whatever) is reached. The successful result of such a calculation would be the 'goal' that takes the place of the simple minimizing energy 'goal' in quasicrystal growth. Thus the achieving of that goal is like the successful growth of the quasicrystal!

There is obviously much vagueness and doubtfulness about these speculations, but I believe they represent a genuinely plausible analogy. The growth of a crystal or quasicrystal is strongly influenced by the concentrations of the appropriate atoms and ions that are in its neighbourhood. Similarly, one might envisage that the growth or contraction of families of dendritic spines could well be as much influenced by the concentrations of the various neurotransmitter substances that might be around (such as might be affected by emotions). Whichever atomic arrangements finally get resolved (or 'reduced') as the *actuality* of the quasicrystal involve the solution of an energy-minimizing problem. In a similar way, so I am speculating, the actual thought that surfaces in the brain is again the solution of some problem, but now not just an energy-minimizing problem. It would generally involve a goal of a much more complicated nature, involving desires and intentions that themselves are related to the computational aspects and capabilities of the brain. I am speculating that the action of conscious thinking is very much tied up with the resolving out of alternatives that were previously in linear superposition. This is all concerned with the unknown physics that governs the borderline between **U** and **R** and which, I am claiming, depends upon a yet-to-be discovered theory of quantum gravity—CQG!

Could such a physical action be non-algorithmic in nature? We recall that the general tiling problem, as described in Chapter 4, is one without an algorithmic solution. One might envisage that assembly problems for atoms might share this non-algorithmic property. If these problems can in principle be 'solved' by the kind of means that I have been hinting at, then there is indeed some possibility for a non-algorithmic ingredient in the type of brain-action that I have in mind. For this to be so, however, we need something non-algorithmic in CQG. Clearly there is considerable speculation here. But *something* of a non-algorithmic character seems to me to be definitely needed, in view of the arguments put forward above.

How fast can these changes in brain connection take place? The matter seems to be somewhat controversial among neurophysiologists, but since permanent memories can be laid down within a few fractions of a second, it is plausible that such changes in connections can be effected in that sort of time. For my own ideas to have a chance, this kind of speed would indeed be necessary.

The time-delays of consciousness

I wish, next, to describe two experiments (described in Harth 1982) that have been performed on human subjects, and which appear to have rather remarkable implications for our considerations here. These have to do with the time that consciousness takes to act and to be enacted. The first of these is concerned with the active role of consciousness, and the second, its passive role. Taken together, the implications are even more striking.

The first was performed by H. H. Kornhuber and associates in Germany in 1976. (Deecke, Grötzinger, and Kornhuber 1976.) A number of human subjects volunteered to have electrical signals recorded at a point on their heads (electroencephalograms, i.e. EEGs), and they were asked to flex the index finger of their right hands suddenly at various times *entirely of their own choosing*. The idea is that the EEG recordings would indicate something of the mental activity that is taking place within the skull, and which is involved in the actual conscious decision to flex the finger. In order to obtain a significant signal from the EEG traces, it is necessary to average the traces from several different runs, and the resulting signal is not very specific. However, what is found is remarkable, namely that there is a gradual build-up of recorded electric potential for a *full second*, or perhaps even up to a second and a half, *before* the finger is actually flexed. This seems to indicate that the conscious decision process takes over a second in order to act! This may be contrasted with the much shorter time that it takes to respond to an external signal if the mode of response has been laid down beforehand. For example, instead of it being 'freely willed', the finger flexing might be in response to the flash of a light signal. In that case a reaction time of about one-fifth of a second is normal, which is about five times faster than the 'willed' action that is tested in Kornhuber's data (see Fig. 10.5).

In the second experiment, Benjamin Libet, of the University of California, in collaboration with Bertram Feinstein of the Mount Zion Neurological Institute in San Francisco (Libet *et al.* 1979), tested subjects who had to have

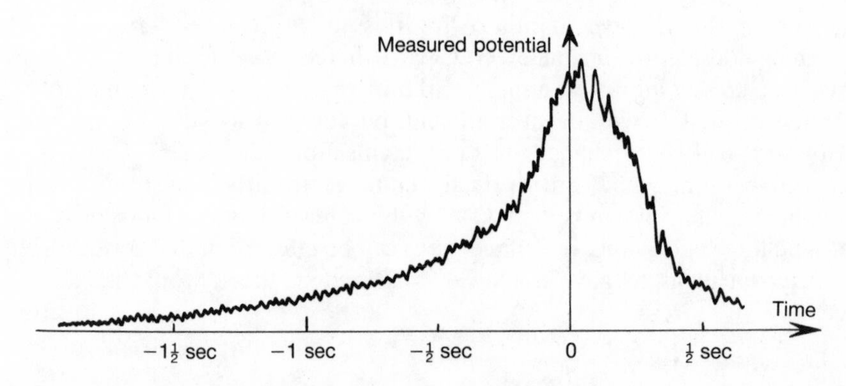

Fig. 10.5. Kornhuber's experiment. The decision to flex the finger appears to be made at time 0, yet the precursor signal (averaged over many trials) suggest a 'foreknowledge' of the intention to flex.

brain surgery for some reason unconnected with the experiment and who consented to having electrodes placed at points in the brain, in the somatosensory cortex. The upshot of Libet's experiment was that, when a stimulus was applied to the skin of these patients, it took about half a second before they were consciously aware of that stimulus, despite the fact that the brain itself would have received the signal of the stimulus in only about a hundredth of a second, and a pre-programmed 'reflex' response to such a stimulus (cf. above) could be achieved by the brain in about a tenth of a second (Fig. 10.6). Moreover, despite the delay of half a second before the stimulus reaches awareness, there would be the subjective impression by the patients themselves that no delay had taken place at all in their becoming aware of the stimulus! (Some of Libet's experiments involved stimulation of the thalamus, cf. p. 380, with similar results as for the somatosensory cortex.)

Recall that the somatosensory cortex is the region of the cerebrum at which sensory signals enter. Thus, the electrical stimulation of a point of the somatosensory cortex, corresponding to some particular point on the skin, would appear to the subject just as though something had actually touched the skin at that corresponding point. However, it turns out that if this electrical stimulation is too brief—for less than about half a second—then the subject does not become aware of any sensation at all. This is to be contrasted with a direct stimulation of the point on the skin itself, since a momentary touching of the skin can be felt.

Now, suppose that the skin is first touched, and then the point in the somatosensory cortex is electrically stimulated. What does the patient feel? If the electrical stimulation is initiated about a quarter of a second after the touching of the skin, then the skin touching is not felt at all! This is an effect referred to as *backwards masking*. Somehow the stimulation of the cortex

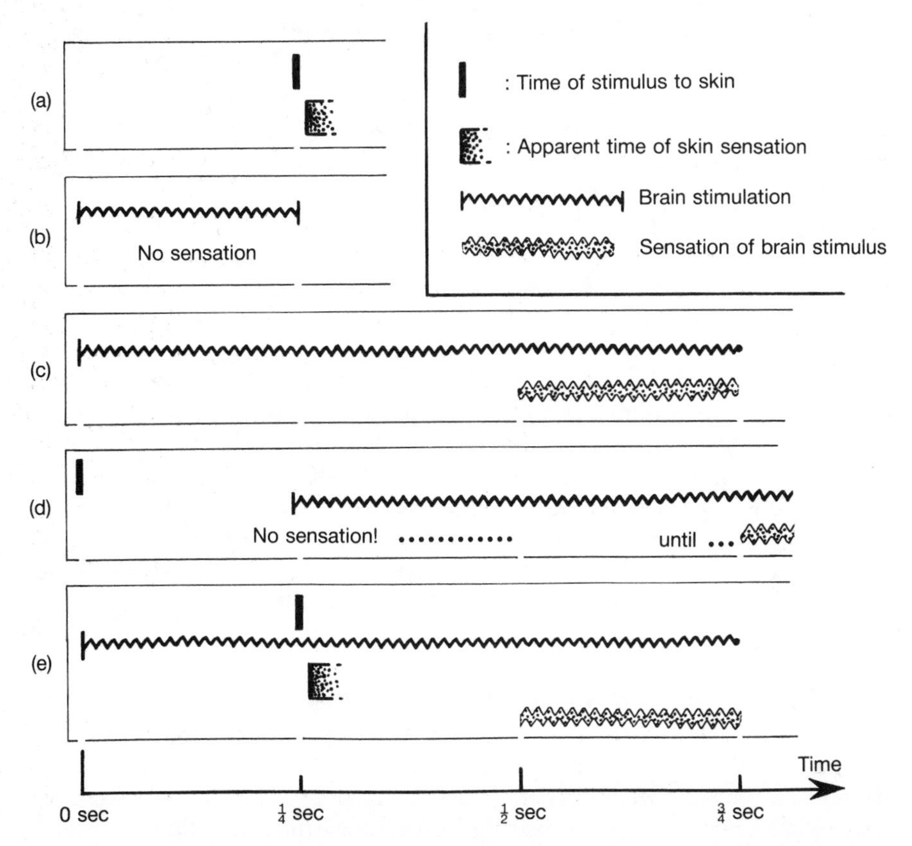

Fig. 10.6. Libet's experiment. (a) The stimulus to the skin 'seems' to be perceived at about the actual time of the stimulus. (b) A cortical stimulus of less than half a second is not perceived. (c) A cortical stimulus of over half a second is perceived from half a second onwards. (d) Such a cortical stimulus can 'backwards mask' an earlier skin stimulus, indicating that awareness of the skin stimulus had actually *not yet taken place* by the time of the cortical stimulus. (e) If a skin stimulus is applied shortly *after* such a cortical stimulus, then skin awareness is 'referred back' but the cortical awareness is not.

serves to prevent the normal skin-touching sensation from being consciously felt. The conscious perception can be prevented ('masked') by a later event, provided that that event occurs within about half a second. This in itself tells us that the conscious awareness of such a sensation occurs at something like half a second after the actual event producing that sensation!

However, one does not seem to be 'aware' of such a long time-delay in one's perceptions. One way of making sense of this curious finding might be to imagine that the 'time' of all one's 'perceptions' is actually delayed by about half a second from the 'actual time'—as though one's internal clock is simply 'wrong', by half a second or so. The time at which one perceives an event to take place would then always be half a second *after* the actual occurrence of that event. This would present a consistent, albeit disturbingly delayed, picture of sense impressions.

Perhaps something of this nature is borne out in a second part of Libet's experiment, where he initiated an electrical stimulation of the cortex *first*, continuing this stimulation for a good while longer than half a second, and also touching the skin while this stimulation was still going on, but less than half a second after its initiation. Both the cortical stimulation and the touching of the skin were separately perceived, and it was clear to the subject which was which. When asked which stimulus occurred *first*, however, the subject would reply that it was the skin-touching, despite the fact that the cortical stimulation was in fact initiated first! Thus, the subject does appear to refer the perception of the skin-touching *backwards in time* by about half a second (see Fig. 10.6.) However, this seems not to be simply an overall 'error' in the internally perceived time, but a more subtle rearranging of the temporal perception of events. For the cortical stimulation, assuming that this is actually perceived no later than half a second after its initiation, seems *not* to be referred back in this way.

From the first of the above experiments, we seem to deduce that conscious action takes something like a second to a second and a half before it can be effected, while according to the second experiment, consciousness of an external event does not seem to occur until half a second after that event has taken place. Imagine what happens when one responds to some unanticipated external occurrence. Suppose that the response is something that requires a moment's conscious contemplation. It would appear, on the basis of Libet's findings, that half a second must elapse before consciousness is called into play; and then, as Kornhuber's data seem to imply, well over a second is needed before one's 'willed' response can take effect. The whole process, from sensory input to motor output, would seem to require something like two seconds! The apparent implication of these two experiments taken together is that consciousness cannot even be called into play *at all* in response to an external event, if that response is to take place within a couple of seconds or so!

The strange role of time in conscious perception

Can we take these experiments at their face value? If so, we appear to be driven to the conclusion that we act entirely as 'automatons' when we carry out any action that would take less than a second or two in which to modify a response. No doubt consciousness is slow-acting, as compared with other mechanisms of the nervous system. I have, myself, noticed occasions such as watching helplessly as my hand closes the car door a moment after I have noticed something within the car that I had wished to retrieve, and my willed command to stop the motion of my hand acts disturbingly slowly—too slowly to stop the closing of the door. But does this take a whole second or two? It seems unlikely to me that such a long timescale was involved. Of course, my *conscious* awareness of the object within the car together with my imagined 'free willing' of the command to stop my hand *could* all have occurred well after both events. Perhaps consciousness is, after all, merely a spectator who experiences nothing but an 'action replay' of the whole drama. Similarly, on the face of it, there would, on the basis of the above findings, be no time for consciousness to be playing any role at all when, for example, one plays a shot at tennis—and certainly not at ping-pong! No doubt the experts at these pursuits would have all the essentials of their responses superbly pre-programmed in cerebellar control. But that consciousness should be playing *no* role at all in the decision as to what shot should be played at the time is something that I find a little hard to credit. No doubt there is a lot in the anticipation of what one's opponent might do, and many pre-programmed responses might be available to each possible action of the opponent, but this seems to me to be inefficient, and a *total* absence of conscious involvement at the time is something that I would find difficult to accept. Such comments would be even more pertinent in relation to ordinary conversation. Again, although one might partly anticipate what the other would be saying, there must often be something unexpected in the other's remarks, or conversation would be entirely unnecessary! It certainly does not take as much as a couple of seconds to respond to someone in the normal way of conversation!

There is perhaps some reason to doubt that Kornhuber's experiments demonstrate that consciousness 'actually' takes a second and a half to act. While it is true that the *average* of all the EEG-traces intentions to flex the finger had a signal emerging that early, it could be that only in *some* cases is there an intention to flex the finger that far ahead—where often this conscious intention might *not* actually materialize—whereas in many other cases the conscious action came about much nearer to the finger flexing than this.

Let us, for the moment, accept that both experimental conclusions *are* actually valid. I should like to make an alarming suggestion in relation to this. I suggest that we may actually be going badly wrong when we apply the usual physical rules for *time* when we consider consciousness! There is, indeed, something very odd about the way that time actually enters into our conscious perceptions in any case, and I think that it is possible that a very different conception may be required when we try to place conscious perceptions into a conventionally time-ordered framework. Consciousness is, after all, the one phenomenon that we know of, according to which time needs to 'flow' at all! The way in which time is treated in modern physics is not essentially different from the way in which *space* is treated* and the 'time' of physical descriptions does not really 'flow' at all; we just have a static-looking fixed 'space–time' in which the events of our universe are laid out! Yet, according to our perceptions, time *does* flow (see Chapter 7). My guess is that there is something illusory here too, and the time of our perceptions does not 'really' flow in quite the linear forward-moving way that we perceive it to flow (whatever that might mean!). The temporal ordering that we 'appear' to perceive is, I am claiming, something that we impose upon our perceptions in order to make sense of them in relation to the uniform forward time-progression of an external physical reality.

Some people might detect a good deal of philosophical 'unsoundness' in the above remarks—and no doubt they would be correct in such accusations. How can one be 'wrong' about what one actually perceives? Surely, one's actual perceptions are just the things of which one is directly aware, by *definition*; and so one cannot be 'wrong' about it. Nevertheless, I think it is indeed likely that we *are* 'wrong' about our perceptions of temporal progression (despite my inadequacies in the use of ordinary language to describe this) and that there is evidence to support such a belief (see Churchland 1984).

An extreme example (p. 423) is Mozart's ability to 'seize at a glance' an entire musical composition 'though it may be long'. One must assume, from Mozart's description, that this 'glance' contained the essentials of the entire composition, yet that the actual external time-span, in ordinary physical terms, of this conscious act of perception, could be in no way comparable with the time that the composition would take to perform. One might imagine that Mozart's perception would have taken a different form altogether, perhaps spatially distributed like a visual scene or an entire musical score laid out. But even a musical score would take a considerable time to peruse—and I would very much doubt that Mozart's perception of his compositions could have initially taken this form (or surely he would have said so!). The visual scene would seem closer to his descriptions, but (as with the most common of mathematical imagery that is more familiar to me personally) I would greatly doubt that there would be anything like a direct translation of music into visual terms. It seems to me to be much more likely that the best interpretation of Mozart's 'glance' must indeed be taken purely *musically*, with the distinctly temporal connotations that the hearing (or performing) of a piece of music would have. Music consists of sounds that take a definite time to perform, the *time* that in Mozart's actual description allows '. . . that my imagination lets me hear it.'

* This symmetry between time and space would be even more striking for a *two*-dimensional space–time. The equations of two-dimensional space–time physics would be essentially symmetrical with respect to the interchange of space with time—yet nobody would take space to 'flow' in two-dimensional physics. It is hard to believe that what makes time 'actually flow' in our experiences of the physical world we know is merely the asymmetry between the number of space dimensions (3) and time dimensions (1) that our space–time happens to have.

Listen to the quadruple fugue in the final part of J. S. Bach's *Art of Fugue*. No one with a feeling for Bach's music can help being moved as the music stops after ten minutes of performance, just after the third theme enters. The composition as a whole still seems somehow to be 'there', but now it has faded from us in an instant. Bach died before he was able to complete the work, and his musical score simply stops at that point, with no written indication as to how he intended it to continue. Yet it starts with such an assurance and total mastery that one cannot imagine that Bach did not hold the essentials of the entire composition in his head at the time. Would he have needed to play it over to himself in its entirety in his mind, at the normal pace of a performance, trying it again and again, and yet again, as various different improvements came to him? I cannot imagine that it was done this way. Like Mozart, he must somehow have been able to conceive the work in its entirety, with the intricate complication and artistry that fugal writing demands, all conjured up together. Yet, the temporal quality of music is one of its essential ingredients. How is it that music can remain music if it is not being performed in 'real time'?

The conceiving of a novel or of history might present a comparable (though seemingly less puzzling) problem. In the comprehending of an individual's entire life, one would need to contemplate various events whose proper appreciation would seem to require their mental enaction in 'real time'. Yet this seems not to be necessary. Even the impressions of memories of one's own time-consuming experiences seem somehow to be so 'compressed' that one can virtually 're-live' them in an instant of recollection!

There is perhaps some strong similarity between musical composition and mathematical thinking. People might suppose that a mathematical proof is conceived as a logical progression, where each step follows upon the ones that have preceded it. Yet the conception of a new argument is hardly likely actually to proceed in this way. There is a globality and seemingly vague conceptual content that is necessary in the construction of a mathematical argument; and this can bear little relation to the time that it would seem to take in order fully to appreciate a serially presented proof.

Suppose, then, that we accept that the timing and temporal progression of consciousness is not in accord with that of external physical reality. Are we not in some danger of leading into a paradox? Suppose that there is even something vaguely teleological about the effects of consciousness, so that a future impression might affect a past action. Surely *this* would lead us into a contradiction, like the paradoxical implications of faster-than-light signalling that we considered—and justly ruled out—in our discussions towards the end of Chapter 5 (cf. p. 212)? I wish to suggest that there need be *no* paradox—by the very nature of what I am contending that consciousness actually achieves. Recall my proposal that consciousness, in essence, is the 'seeing' of a necessary truth; and that it may represent some kind of actual contact with

Plato's world of ideal mathematical concepts. Recall that Plato's world is itself timeless. The perception of Platonic truth carries no actual information—in the technical sense of the 'information' that can be transmitted by a message—and there would be no actual contradiction involved if such a conscious perception were even to be propagated backwards in time!

But even if we accept that consciousness itself has such a curious relation to time—and that it represents, in some sense, contact between the external physical world and something timeless—how can this fit in with a physically determined and time-ordered action of the material brain? Again, we seem to be left with a mere 'spectator' role for consciousness if we are not to monkey with the normal progression of physical laws. Yet, I *am* arguing for some kind of active role for consciousness, and indeed for a powerful one, with a strong selective advantage. The answer to this dilemma, I believe, lies with the strange way that CQG must act, in its resolution of the conflict between the two quantum-mechanical processes **U** and **R**.

Recall the problems with time that the process **R** encounters when we try to make it consistent with (special) relativity (Chapters 6, 8, pp. 286, 371). The process does not seem to make sense at all when described in ordinary space–time terms. Consider a quantum state for a pair of particles. Such a state would normally be a *correlated* state (i.e. *not* of the simple form $|\psi\rangle|\chi\rangle$, where $|\psi\rangle$ and $|\chi\rangle$ each describes just one of the particles, but as a sum, like $|\psi\rangle|\chi\rangle + |\alpha\rangle|\beta\rangle + \ldots + |\rho\rangle|\sigma\rangle$). Then an observation on one of the particles will affect the other one in a non-local way which cannot be described in ordinary space–time terms consistently with special relativity (EPR; the Einstein–Podolsky–Rosen effect). Such non-local effects would be implicitly involved in my suggested 'quasicrystal' analogy to dendritic spine growth and contraction.

Here, I am interpreting 'observation' in the sense of an amplification of the action of each observed particle until something like the 'one-graviton' criterion of CQG is met. In more 'conventional' terms, an 'observation' is a much more obscure thing and it is hard to see how one could begin to develop a quantum-theoretical description of brain action when one might well have to regard the brain as 'observing itself' all the time!

My own idea is that CQG would, on the other hand, provide an *objective* physical theory of state-vector reduction (**R**) which would *not* have to depend on any ideas of consciousness. We do not yet have such a theory, but at least the finding of it will not be hampered by the profound problems of deciding what consciousness actually 'is'!

I envisage that once CQG has been actually found, it may *then* become possible to elucidate the phenomenon of consciousness in terms of it. In fact, I believe that the required properties of CQG, when that theory arrives, will be even further from having a conventional space–time description than the puzzling two-particle EPR phenomena alluded to above. If, as I am suggesting, the phenomenon of consciousness depends upon this putative CQG, then consciousness itself will fit only very uncomfortably into our present conventional space–time descriptions!

Conclusion: a child's view

In this book I have presented many arguments intending to show the untenability of the viewpoint—apparently rather prevalent in current philosophizing—that our thinking is basically the same as the action of some very complicated computer. When the explicit assumption is made that the mere enaction of an algorithm can evoke *conscious awareness*, Searle's terminology 'strong AI' has been adopted here. Other terms such as 'functionalism' are sometimes used in a somewhat less specific way.

Some readers may, from the start, have regarded the 'strong-AI supporter' as perhaps largely a straw man! Is it not 'obvious' that mere computation cannot evoke pleasure or pain; that it cannot perceive poetry or the beauty of an evening sky or the magic of sounds; that it cannot hope or love or despair; that it cannot have a genuine autonomous purpose? Yet science seems to have driven us to accept that we are all merely small parts of a world governed in full detail (even if perhaps ultimately just probabilistically) by very precise mathematical laws. Our brains themselves, which seem to control all our actions, are also ruled by these same precise laws. The picture has emerged that all this precise physical activity is, in effect, nothing more than the acting out of some vast (perhaps probabilistic) computation—and, hence our brains and our minds are to be understood solely in terms of such computations. Perhaps when computations become extraordinarily complicated they can begin to take on the more poetic or subjective qualities that we associate with the term 'mind'. Yet it is hard to avoid an uncomfortable feeling that there must always be something missing from such a picture.

In my own arguments I have tried to support this view that there must indeed be something essential that is missing from any purely computational picture. Yet I hold also to the hope that it is through science and mathematics that some profound advances in the understanding of mind must eventually come to light. There is an apparent dilemma here, but I have tried to show that there *is* a genuine way out. Computability is not at all the same thing as being mathematically precise. There is as much mystery and beauty as one might wish in the precise Platonic mathematical world, and most of this mystery resides with concepts that lie outside the comparatively limited part of it where algorithms and computation reside.

Consciousness seems to me to be such an important phenomenon that I simply cannot believe that it is something just 'accidentally' conjured up by a complicated computation. It is the phenomenon whereby the universe's very existence is made known. One can argue that a universe governed by laws that do not allow consciousness is no universe at all. I would even say that all the mathematical descriptions of a universe that have been given so far must fail this criterion. It is only the phenomenon of consciousness that can conjure a putative 'theoretical' universe into actual existence!

Some of the arguments that I have given in these chapters may seem tortuous and complicated. Some are admittedly speculative, whereas I believe that there is no real escape from some of the others. Yet beneath all this technicality is the feeling that it is indeed 'obvious' that the *conscious* mind cannot work like a computer, even though much of what is actually involved in mental activity might do so.

This is the kind of obviousness that a child can see—though that child may, in later life, become browbeaten into believing that the obvious problems are 'non-problems', to be argued into non-existence by careful reasoning and clever choices of definition. Children sometimes see things clearly that are indeed obscured in later life. We often forget the wonder that we felt as children when the cares of the activities of the 'real world' have begun to settle upon our shoulders. Children are not afraid to pose basic questions that may embarrass us, as adults, to ask. What happens to each of our streams of consciousness after we die; where was it before each was born; might we become, or have been, someone else; why do we perceive at all; why are we here; why is there a universe here at all in which we can actually be? These are puzzles that tend to come with the awakenings of awareness in any one of us—and, no doubt, with the awakening of genuine self-awareness, within whichever creature or other entity it first came.

I remember, myself, being troubled by many such puzzles as a child. Perhaps my own consciousness might suddenly get exchanged with someone else's. How would I ever know whether such a thing might not have happened to me earlier—assuming that each person carries only the memories pertinent to that particular person? How could I explain such an 'exchange' experience to someone else? Does it really mean anything? Perhaps I am simply living the same ten minutes' experiences over and over again, each time with exactly the same perceptions. Perhaps only the present instant 'exists' for me. Perhaps the 'me' of tomorrow, or of yesterday, is really a quite different person with an independent consciousness. Perhaps I am actually living backwards in time, with my stream of consciousness heading into the past, so my memory really tells me what is *going* to happen to me rather than what *has* happened to me—so that unpleasant experience at school is really something that is in store for me and I shall, unfortunately, shortly actually encounter. Does the distinction between *that* and the normally experienced time-progression actually 'mean' something, so that the one is 'wrong' and the other 'right'? For the answers to such questions to be resolvable in principle, a theory of consciousness would be needed. But how could one even *begin* to explain the substance of such problems to an entity that was not itself conscious. . . ?

1. See Lucas (1961) for an argument that Gödel's theorem implies non-computability and Good (1969), Benacerraf (1967), Bowie (1982) for various counter-arguments.
2. Some readers may be troubled by the fact that there are indeed different points of view among mathematicians. Recall the discussion given in Chapter 4. However the differences, where they exist, need not greatly concern us here. They refer only to esoteric questions concerning very large sets, whereas we can restrict our attention to propositions in arithmetic (with a finite number of existential and universal quantifiers) and the foregoing discussion will apply. (Perhaps this overstates the case somewhat, since a reflection principle referring to infinite sets can sometimes be used to derive propositions in arithmetic.) As to the very dogmatic Gödel-immune formalist who claims not even to recognize that there *is* such a thing as mathematical truth, I shall simply ignore him, since he apparently does not possess the truth-divining quality that the discussion is all about!
3. The term 'black hole' became common usage only much later, in about 1968 (largely via the prophetic ideas of the American physicist John A. Wheeler).
4. It seems to me that the fact that animals require sleep in which they appear sometimes to *dream* (as is often noticeable with dogs) is evidence that they can possess consciousness. For an element of consciousness seems to be an important ingredient of the distinction between dreaming and non-dreaming sleep.
5. In the case of special or general relativity, read 'simultaneous spaces' or 'spacelike surfaces' (pp. 200, 214) in place of 'times'.
6. There is a let-out in the case of a spatially infinite universe, however, since then (rather like in the case of many worlds) it turns out that there would be infinitely many copies of oneself and one's immediate environment! The future behaviour for each copy might be slightly different, and one would never be quite sure which of the approximate copies of oneself modelled in the mathematics one might actually 'be'!
7. Even the growth of some actual crystals could involve similar problems, for example, where the basic unit cell involves several hundred atoms, as in what are called Frank–Casper phases. It should be mentioned, on the other hand, that a theoretical 'almost local' (though still not local) growth procedure for five-fold symmetric quasicrystals has been suggested by Onoda, DiVincenzo, Steinhardt and Socolar (1988).

Epilogue

'. . . . . feel like? Oh, . . . a most interesting question, my lad . . . er . . . rather like to know the answer to that myself', said the Chief Designer. 'Let's see what our friend has to say about . . . that's odd . . . er . . . Ultronic says it doesn't see what . . . it can't even understand what you're getting at!' The ripples of laughter about the room burst into a roar.

Adam felt acutely embarrassed. Whatever they should have done, they should not have laughed.

References

Aharonov, Y. and Albert, D. Z. (1981). Can we make sense out of the measurement process in relativistic quantum mechanics? *Phys. Rev.*, **D24**, 359–70.

Aharonov, Y., Bergmann, P., and Lebowitz, J. L. (1964). Time symmetry in the quantum process of measurement. In *Quantum theory and measurement* (ed. J. A. Wheeler and W. H. Zurek), Princeton University Press, 1983; originally in *Phys. Rev.*, **134B**, 1410–16.

Ashtekar, A., Balachandran, A. P., and Sang Jo (1989). The CP problem in quantum gravity. *Int. J. Mod. Phys.*, **A6**, 1493–514.

Aspect, A. and Grangier, P. (1986). Experiments on Einstein–Podolsky–Rosen-type correlations with pairs of visible photons. In *Quantum concepts in space and time* (ed. R. Penrose and C. J. Isham), Oxford University Press.

Atkins, P. W. (1987). *Why mathematics works.* Oxford University Extension Lecture in series: Philosophy and the New Physics (13 March).

Barrow, J. D. (1988). *The world within the world.* Oxford University Press.

Barrow, J. D. and Tipler, F. J. (1986). *The anthropic cosmological principle.* Oxford University Press.

Baylor, D. A., Lamb, T. D., and Yau, K.-W. (1979). Responses of retinal rods to single photons. *J. Physiol.*, **288**, 613–34.

Bekenstein, J. (1972). Black holes and entropy. *Phys. Rev.*, **D7**, 2333–46.

Belinfante, F. J. (1975). *Measurement and time reversal in objective quantum theory.* Pergamon Press, New York.

Belinskii, V. A., Khalatnikov, I. M., and Lifshitz, E. M. (1970). Oscillatory approach to a singular point in the relativistic cosmology. *Adv. Phys.* **19**, 525–573.

Bell, J. S. (1987). *Speakable and unspeakable in quantum mechanics.* Cambridge University Press.

Benacerraf, P. (1967). God, the Devil and Gödel. *The Monist*, **51**, 9–32.

Blakemore, C. and Greenfield, S., (ed.) (1987). *Mindwaves: thoughts on intelligence, identity and consciousness.* Basil Blackwell, Oxford.

Blum, L., Shub, M., and Smale, S. (1989). On a theory of computation and complexity over the real numbers: NP completeness, recursive functions and universal machines. *Bull. Amer. Math. Soc.* (In press.)

Bohm, D. (1951). The Paradox of Einstein, Rosen and Podolsky. In *Quantum theory and measurement* (ed., J. A. Wheeler and W. H. Zurek), Princeton University Press, 1983; originally in *Quantum theory*, D. Bohm, Ch. 22, sect. 15–19. Prentice-Hall, Englewood-Cliffs.

Bohm, D. (1952). A suggested interpretation of the quantum theory in terms of 'hidden' variables, I and II, in *Quantum theory and measurement.* (ed., J. A. Wheeler and W. H. Zurek), Princeton University Press, 1983; originally in *Phys. Rev.*, **85**, 166–93.

Bondi, H. (1960). Gravitational waves in general relativity. *Nature* (*London*), **186**, 535.

Bowie, G. L. (1982). Lucas' number is finally up. *J. of Philosophical Logic,* **11**, 279–85.

Cartan, É. (1923). Sur les variétés a connexion affine et la théorie de la relativité generalisée. *Ann. Sci. Ec. Norm. Sup.*, **40**, 325–412.

Chandrasekhar, S. (1987). *Truth and beauty: aesthetics and motivations in science*, Univeristy of Chicago Press.

Church, A. (1941). *The calculi of lambda–conversion*. Annals of Mathematics Studies, no. 6. Princeton University Press.

Churchland, P. M. (1984). *Matter and consciousness*. Bradford Books, MIT Press, Cambridge, Mass.

Clauser, J. F., Horne, A. H., Shimony, A., and Holt, R. A. (1969). Proposed experiment to test local hidden-variable theories. In *Quantum theory and measurement* (ed. J. A. Wheeler and W. H. Zurek), Princeton University Press, 1983; originally in *Phys. Rev. Lett.*, **23**, 880–84.

Close, F. (1983). *The cosmic orion: quarks and the nature of the universe.* Heinemann, London.

Cohen, P. C. (1966). *Set theory and the continuum hypothesis*, Benjamin, Menlo Park, CA.

Cutland, N. J. (1980). *Computability: an introduction to recursive function theory*. Cambridge University Press.

Davies, P. C. W. (1974). *The physics of time-asymmetry*. Surrey University Press.

Davies, P. C. W. and Brown, J. (1988). Superstrings: a theory of everything? Cambridge University Press.

Davies, R. D., Lasenby, A. N., Watson, R. A., Daintree, E. J., Hopkins, J., Beckman, J., Sanchez-Almeida, J., and Rebolo, R. (1987). Sensitive measurement of fluctuations in the cosmic microwave background. *Nature*, **326**, 462–5.

Davis, M. (1988). Mathematical logic and the origin of modern computers. In *The universal Turing machine: a half-century survey* (ed. R. Herken), Kammerer & Unverzagt, Hamburg.

Dawkins, R. (1986). *The blind watchmaker.* Longman, London.

de Broglie, L. (1956). *Tentative d'interpretation causale et nonlineaire de la mechanique ondulatoire*. Gauthier-Villars, Paris.

Deeke, L., Grötzinger, B., and Kornhuber, H. H. (1976). Voluntary finger movements in man: cerebral potentials and theory. *Biol. Cybernetics*, **23**, 99.

Delbrück, M. (1986). *Mind from matter?* Blackwell Scientific Publishing, Oxford.

Dennett, D. C. (1978). *Brainstorms*. Philosophical Essays on Mind and Psychology, Harvester Press, Hassocks, Sussex.

Deutch, D. (1985). Quantum theory, the Church–Turing principle and the universal quantum computer. *Proc. Roy. Soc. (Lond.),* **A400**, 97–117.

Devlin, K. (1988). *Mathematics: the new golden age.* Penguin Books, London.

De Witt, B. S. and Graham, R. D. (ed.) (1973). *The many-worlds interpretation of quantum mechanics.* Princeton University Press.

Dirac, P. A. M. (1928). The quantum theory of the electron. *Proc. Roy. Soc. (Lond.),* **A117**, 610–24; *ditto*, part II, *ibid*, **A118**, 361.

Dirac, P. A. M. (1938). Classical theory of radiating electrons. *Proc. Roy. Soc. (Lond.)*, **A167**, 148.

Dirac, P. A. M. (1939). The relations between mathematics and physics. *Proc. Roy.*

Soc., Edinburgh, **59**, 122.

Dirac, P. A. M. (1947). *The principles of quantum mechanics* (3rd edn), Oxford University Press.

Dirac, P. A. M. (1982). Pretty mathematics. *Int. J. Theor. Phys.*, **21**, 603–5.

Drake, S. (trans.) (1953). *Galileo Galilei: dialogue concerning the two chief world systems — Ptolemaic and Copernican*. University of California, Berkeley, 1953.

Drake, S. (1957). *Discoveries and opinions of Galileo.* Doubleday, New York.

Eccles, J. C. (1973). *The understanding of the brain*. McGraw-Hill, New York.

Einstein, A., Podolsky, P., and Rosen, N. (1935). Can quantum-mechanical description of physical reality be considered complete? In *Quantum theory and measurement* (ed. J. A. Wheeler and W. H. Zurek), Princeton University Press, 1983; originally in *Phy. Rev.*, **47**, 777–80.

Everett, H. (1957). 'Relative State' formulation of quantum mechanics. In *Quantum theory and measurement* (ed. J. A. Wheeler and W. H. Zurek), Princeton University Press, 1983; originally in *Rev. of Mod. Phys.*, **29**, 454–62.

Feferman, S. (1988). Turing in the Land of O(z). In *The universal Turing machine: a half-century survey* (ed. R. Herken), Kammerer & Unverzagt, Hamburg.

Feynman, R. P., Leighton and Sands (1965). *The Feynman Lectures* . Addison-Wesley.

Feynman, R. P. (1985). *QED: the strange theory of light and matter*. Princeton University Press.

Fodor, J. A. (1983). *The modularity of mind.* MIT Press, Cambridge, Mass.

Fredkin, E. and Toffoli, T. (1982). Conservative Logic. *Int. J. Theor. Phys.*, **21**, 219–53.

Freedman, S. J. and Clauser, J. F. (1972). Experimental test of local hidden-variable theories. In *Quantum theory and measurement* (ed. J. A. Wheeler and W. H. Zurek) Princeton University Press, 1983; originally in *Phys. Rev. Lett.*, **28**, 938–41.

Galilei, G. (1638). *Dialogues concerning two new sciences*. Macmillan edn. 1914; Dover Inc.

Gandy, R. (1988). The Confluence of Ideas in 1936. In *The universal Turing machine: a half-century survey* (ed. R. Herken), Kammerer & Unverzagt, Hamburg.

Gardner, M. (1958). *Logic machines and diagrams*. University of Chicago Press.

Gardner, M. (1983). *The whys of a philosophical scrivener.* William Morrow and Co., Inc., New York.

Gardner, M. (1989). *Penrose tiles to trapdoor ciphers.* W. H. Freeman and Company, New York.

Gayle, F. W. (1987). Free-surface solidification habit and point group symmetry of a faceted icosahederal Al-Li-Cu phase. *J. Mater. Sci.*, **2**, 1–4.

Gazzaniga, M. S. (1970). *The bisected brain.* Appleton-Century-Crofts, New York.

Gazzaniga, M. S., LeDoux, J. E., and Wilson, D. H. (1977). Language, praxis, and the right hemisphere: clues to some mechanisms of consciousness. *Neurology*, **27**, 1144–7.

Geroch, R and Hartle, J. B. (1986). Computability and physical theories. *Found. Phys.*, **16**, 533.

Ghirardi, G. C., Rimini, A., and Weber, T. (1980). A general argument against superluminal transmission through the quantum mechanical measurement

process. *Lett. Nuovo. Chim.*, **27**, 293–8.
Ghirardi, G. C., Rimini, A., and Weber, T. (1986). Unified dynamics for microscopic and macroscopic systems. *Phys. Rev.*, **D34**, 470.
Gödel, K. (1931). Über formal unentscheidbare Sätze der Principia Mathematica und verwandter Systeme I. *Monatshefte für Mathematik und Physik*, **38**, 173–98.
Good, I. J. (1969). Gödel's theorem is a red herring. *Brit. J. Philos. Sci.*, **18**, 359–73.
Gregory, R. L. (1981). *Mind in science; A history of explanations in psychology and physics.* Weidenfeld and Nicholson Ltd.
Grey Walter, W. (1953). *The living brain.* Gerald Duckworth and Co. Ltd.
Grünbaum, B. and Shephard, G. C. (1981). Some problems on plane tilings. In *The mathematical Gardner* (ed. D. A. Klarner), Prindle, Weber and Schmidt, Boston.
Grünbaum, B. and Shephard, G. C. (1987). *Tilings and patterns.* W. H. Freeman.
Hadamard, J. (1945). *The psychology of invention in the mathematical field.* Princeton University Press.
Hanf, W. (1974). Nonrecursive tilings of the plane, I. *J. Symbolic Logic*, **39**, 283–5.
Harth, E. (1982). *Windows on the Mind.* Harvester Press, Hassocks, Sussex.
Hartle, J. B. and Hawking, S. W. (1983). Wave function of the universe. *Phys. Rev.*, **D31**, 1777.
Hawking, S. W. (1975). Particle creation by black holes. *Commun. Math. Phys.*, **43**, 199–220.
Hawking, S. W. (1987). Quantum cosmology. In *300 years of gravitation* (ed. S. W. Hawking and W. Israel), Cambridge University Press.
Hawking, S. W. (1988). *A brief history of time.* Bantam Press. London.
Hawking, S. W. and Penrose, R. (1970). The singularities of gravitational collapse and cosmology. *Proc. Roy. Soc. (London),* **A314**, 529–48.
Hebb, D. O. (1954). The problem of consciousness and introspection. In *Brain mechanisms and consciousness* (ed. J. F. Delafresnaye), Blackwell, Oxford.
Hecht, S., Shlaer, S., and Pirenne, M. H. (1941). Energy, quanta and vision. *J. of Gen. Physiol.*, **25**, 891–40.
Herken, R. ed., (1988). *The universal Turing machine: a half-century survey,* Kammerer & Unverzagt, Hamburg.
Hiley, B. J. and Peat, F. D., eds. (1987). *Quantum implications. Essays in honour of David Bohm.* Routledge and Kegan Paul, London & New York.
Hodges, A. P. (1983). *Alan Turing: the enigma.* Burnett Books and Hutchinson, London; Simon and Schuster, New York.
Hofstadter, D. R. (1979). *Gödel, Esher, Bach: an eternal golden braid.* Harvester Press, Hassocks, Sussex.
Hofstadter, D. R. (1981). A conversation with Einstein's brain. In *The mind's I* (ed. D. R. Hofstadter and D. C. Dennett), Basic Books, Inc.; Penguin Books, Ltd; Harmondsworth, Middx.
Hofstadter, D. R. and Dennett, D. C. ed. (1981). *The mind's I*, Basic Books, Inc.; Penguin Books, Ltd; Harmondsworth, Middx.
Hubel, D. H. (1988). *Eye, brain and vision*, Scientific American Library Series #22.
Huggett, S. A. and Tod, K. P. (1985). *An introduction to twistor theory.* London Math. Soc. student texts, Cambridge University Press.
Janes, J., (1980). *The origin of consciousness in the breakdown of the bicameral mind.* Pengiun Books Ltd. Harmondsworth, Middx.

Kandel, E. R. (1976). *The cellular basis of behaviour*. Freeman, San Francisco.
Károlyházy, F. (1974). Gravitation and quantum mechanics of macroscopic bodies. *Magyar Fizikai Polyoirat*, **12**, 24.
Károlyházy, F., Frenkel, A., and Lukács, B. (1986). On the possible role of gravity on the reduction of the wave function. In *Quantum concepts in space and time* (ed. R. Penrose and C. J. Isham), Oxford University Press.
Keene, R. (1988). Chess: Henceforward. *The Spectator*, **261**, (no. 8371), 52.
Knuth, D. M. (1981). *The art of computer programming*, Vol. 2 (2nd edn), Addison-Wesley, Reading, MA.
Komar, A. B. (1964). Undecidability of macroscopically distinguishable states in quantum field theory. *Phys. Rev.*, **133B**, 542–4.
Komar, A. B. (1969). Qualitative features of quantized gravitation. *Int. J. Theor. Phys.* **2**, 157–60.
Kuznetsov, B. G. (1977). *Einstein: Leben, Tod, Unslerblichkeit* (trans. into German by H. Fuchs). Birkhauser, Basel.
LeDoux, J. E. (1985). Brain, mind and language. In *Brain and mind* (ed. D. A. Oakley), Methuen, London and New York.
Levy, D. W. L. (1984). *Chess computer handbook*, Batsford.
Libet, B., Wright, E. W. Jr., Feinstein, B., and Pearl, D. K. (1979). Subjective referral of the timing for a conscious sensory experience. *Brain*, **102**, 193–224.
Lorenz, K. (1972). Quoted in: *From ape to Adam* by H. Wendt, Bobbs Merrill, Indianapolis.
Lucas, J. R. (1961). Minds, machines and Gödel. *Philosophy* **36**, 120–4; reprinted in Alan Ross Anderson (1964). *Minds and machines*, Englewood Cliffs.
MacKay, D. (1987). Divided brains–divided minds? In *Mindwaves* (ed. C. Blakemore and S. Greenfield), Basil Blackwell, Oxford.
Majorana, E. (1932). Atomi orientati in campo magnetico variabile. *Nuovo Cimento*, **9**, 43–50.
Mandelbrot, B. B. (1986). Fractals and the rebirth of iteration theory. In *The beauty of fractals: images of complex dynamical systems*, H.-O. Peitgen and P. H. Richter, Springer-Verlag, Berlin. pp. 151–60.
Maxwell, J. C. (1865). A dynamical theory of the electromagnetic field. *Philos. Trans. Roy. Soc. (Lond.),* **155**, 459–512.
Mermin, D. (1985). Is the moon there when nobody looks? Reality and the quantum theory. *Physics Today*, **38** (no. 4), 38–47.
Michie, D. (1988). The fifth generation's unbridged gap. In *The universal Turing machine: a half-century survey* (ed. R. Herken), Kammerer & Unverzagt, Hamburg.
Minsky, M. L. (1968), Matter, mind, and models. In Semantic information processing, (ed. M. L. Minsky), M.I.T. Press, Cambridge, Mass.
Misner, C. W. (1969). Mixmaster universe. *Phys. Rev. Lett.*, **22**, 1071–4.
Moruzzi, G. and Magoun, H. W. (1949). Brainstem reticular formation and activation of the EEG. *Electroencephalography and Clinical Neurophysiology*, **1**, 455–73.
Mott, N. F. (1929). The wave mechanics of α-ray tracks. In *Quantum theory and measurement* (ed. J. A. Wheeler and W. H. Zurek), Princeton University Press, 1983; originally in *Proc. Roy. Soc. (Lond.)*, **A126**, 79–84.
Mott, N. F. and Massey, H. S. W. (1965). Magnetic moment of the electron. In *Quantum theory and measurement* (ed. J. A. Wheeler and W. H. Zurek), Prince-

ton University Press, 1983; originally in *The theory of atomic collisions* by N. F. Mott and H. S. W. Massey (Clarendon Press, Oxford; 1965).

Myers, D. (1974). Nonrecursive tilings of the plane, II. *J. Symbolic Logic*, **39**, 286–94.

Myers, R. E. and Sperry, R. W. (1953). Interocular transfer of a visual form discrimination habit in cats after section of the optic chiasm and corpus callosum. *Anatomical Record*, **175**, 351–2.

Nagel, E. and Newman, J. R. (1958). *Gödel's proof.* Routledge & Kegan Paul Ltd.

Nelson, D. R. and Halperin, B. I. (1985). Pentagonal and icosahedral order in rapidly cooled metals. *Science*, **229**, 233.

Newton, I. (1730). *Opticks*, 1952 Dover, Inc.

Newton, I. (1687). *Principia*, Cambridge University Press.

Oakley, D. A., (ed.) (1985). *Brain and mind*, Methuen, London and New York.

Oakley, D. A. and Eames, L. C. (1985). The plurality of consciousness. In *Brain and mind* (ed. D. A. Oakley), Methuen, London and New York.

O'Connell, K. (1988). Computer chess. *Chess*, **15**.

O'Keefe, J. (1985). Is consciousness the gateway to the hippocampal cognitive map? A speculative essay on the neural basis of mind. In *Brain and mind* (ed. D. A. Oakley), Methuen, London and New York.

Onoda, G. Y., Steinhardt, P. J., DiVincenzo, D. P., and Socolar, J. E. S. (1988). Growing perfect quasicrystals. *Phys. Rev. Lett.*, **60**, 2688.

Oppenheimer, J. R. and Snyder, H. (1939). On continued gravitational contraction. *Phys. Rev.* **56**, 455–9.

Pais, A. (1982). *'Subtle is the Lord . . .': the science and the life of Albert Einstein.* Clarendon Press, Oxford.

Paris, J. and Harrington, L. (1977). A mathematical incompleteness in Peano arithmetic. In *Handbook of mathematical logic* (ed. J. Barwise), North-Holland, Amsterdam.

Pearle, P. (1985). 'Models for reduction'. In *Quantum concepts in space and time* (ed. C. J. Isham and R. Penrose), Oxford University Press.

Pearle, P. (1989). Combining stochastic dynamical state-vector reduction with spontaneous localization. *Phys. Rev. A*, **39**, 2277–89.

Peitgen, H.-O. and Richter, P. H. (1986). *The beauty of fractals*, Springer-Verlag, Berlin and Heidelberg.

Peitgen, H.-O. and Saupe, D. (1988). *The science of fractal images*, Springer-Verlag, Berlin.

Penfield, W. and Jasper, H. (1947). *Highest Level Seizures*, Research Publications of the Association for Research in Nervous and Mental Diseases (New York) **26**, 252–71.

Penrose, R. (1965) Gravitational collapse and space-time singularities. *Phys. Rev. Lett.*, **14**, 57–9.

Penrose, R. (1974). The rôle of aesthetics in pure and applied mathematical research, *Bull. Inst. Math. Applications* **10**, no 7/8, 266–71.

Penrose, R. (1979*a*). Einstein's vision and the mathematics of the natural world. *The Sciences* (March) 6–9.

Penrose, R. (1979*b*). Singularities and time-asymmetry. In *General relativity: an Einstein centenary* (ed. S. W. Hawking and W. Israel), Cambridge University Press.

Penrose, R. (1987*b*). Newton, quantum theory and reality. In *300 years of gravity* (ed.

S. W. Hawking and W. Israel), Cambridge University Press.

Penrose, R. (1987*c*). *Quantum implications. Essays in honour of David Bohm.* Routledge and Kegan Paul, London & New York.

Penrose, R. (1989*a*) Tilings and quasi-crystals; a non-local growth problem? In *Aperiodicity and order 2* (ed. M. Jarič), Academic Press, New York.

Penrose, R. (1989*b*). Difficulties with inflationary cosmology, to appear in the *Proceeding of the 14th Texas Symposium on Relativistic Astrophysics* (ed. E. Fenves), NY Acad. Sci., New York.

Penrose, R. and Rindler, W. (1984). *Spinors and space–time*, Vol. 1: *Two-spinor calculus and relativistic fields*, Cambridge University Press.

Penrose, R. and Rindler, W. (1986). *Spinors and space–time*, Vol. 2: *Spinor and twistor methods in space–time geometry*, Cambridge University Press.

Pour-El, M. B. and Richards, I. (1979). A computable ordinary differential equation which possesses no computable solution. *Ann. Math. Logic*, **17**, 61–90.

Pour-El, M. B. and Richards, I. (1981). The wave equation with computable initial data such that its unique solution is not computable. *Adv. in Math.*, **39**, 215–39.

Pour-El, M. B. and Richards, I. (1982). Noncomputability in models of physical phenomena. *Int. J. Theor. Phys.*, **21**, 553–5.

Rae, A. (1986). *Quantum physics: illusion or reality?* Cambridge University Press.

Resnikoff, H. L. and Wells, R. O. Jr. (1973). *Mathematics and civilization*, Holt, Rinehart and Winston, Inc., New York, reprinted with additions 1984 Dover Publications, Inc., Mineola, NY.

Rindler, W. (1977). *Essential relativity*, Springer-Verlag, New York.

Rindler, W. (1982). *Introduction to special relativity*, Clarendon Press, Oxford.

Robinson, R. M. (1971). Undecidability and nonperiodicity for tilings of the plane. *Invent. Math.*, **12**, 177–209.

Rouse Ball, W. W. (1892). Calculating prodigies. In *Mathematical recreations and essays*.

Rucker, R. (1984). *Infinity and the mind: the science and philosophy of the infinite.* Paladin Books, Granada Publishing Ltd., London (first published by Harvester Press Ltd, 1982).

Sachs, R. K. (1962). Gravitational waves in general relativity. VIII. Waves in asymptotically flat space-time. *Proc. Roy. Soc. London*, **A270**, 103–26.

Schank, R. C. and Abelson, R. P. (1977). *Scripts, plans, goals and understanding.* Erlbaum, Hillsdale, NJ.

Schechtman, D., Blech, I., Gratias, D., and Cahn J. W. (1984). Metallic phase with long-range orientational order and no translational symmetry. *Phys. Rev. Lett.*, **53**, 1951.

Schrödinger, E. (1935). Die gegenwärtige Situation in der Quantenmechanik, *Naturwissenschaftenp*, **23**, 807–12, 823–8, 844–9. (Translation by J. T. Trimmer (1980). In *Proc. Amer. Phil. Soc.*, **124**, 323–38.) In *Quantum theory and measurement* (ed. J. A. Wheeler and W. H. Zurek), Princeton University Press, 1983.

Schrödinger, E. (1967). *'What is life?' and 'Mind and matter'*, Cambridge University Press.

Searle, J. (1980). Minds, brains and programs, in *The behavioral and brain sciences, Vol. 3.* Cambridge University Press, reprinted in *The mind's I* (ed. D. R. Hofstadter and D. C. Dennett), Basic Books, Inc., Penguin Books Ltd., Harmondsworth, Middx. 1981.

Searle, J. R. (1987). Minds and brains without programs. In *Mindwaves* (ed. C. Blakemore and S. Greenfield), Basil Blackwell, Oxford.

Smith, S. B. (1983). *The great mental calculators*. Columbia University Press.

Smorynski, C. (1983). "Big" news from Archimedes to Friedman. *Notices Amer. Math. Soc.*, **30**, 251–6.

Sperry, R. W. (1966). Brain bisection and consciousness. In *Brain and conscious experience* (ed. J. C. Eccles), Springer, New York.

Squires, E. (1985). *To acknowledge the wonder*, Adam Hilger Ltd., Bristol.

Squires, E. (1986). *The mystery of the quantum world*, Adam Hilger Ltd., Bristol.

Tipler, F. J., Clarke, C. J. S., and Ellis, G. F. R. (1980). Singularities and horizons – a review article. In *General relativity and gravitation* (ed. A. Held), Vol. 2, pp. 97–206. Plenum Press, New York.

Turing, A. M. (1937). On computable numbers, with an application to the Entscheidungsproblem. *Proc. Lond. Math. Soc. (ser. 2)*, **42**, 230–65; a correction **43**, 544–6.

Turing, A. M. (1939). Systems of logic based on ordinals. *P. Lond. Math. Soc.*, **45**, 161–228.

Turing, A. M. (1950). Computing machinery and intelligence. *Mind* **59** no. 236; reprinted in *The mind's I* (ed. D. R. Hofstadter and D. C. Dennett), Basic Books, Inc.; Penguin Books, Ltd; Harmondsworth, Middx. 1981.

von Neumann, J. (1955). *Mathematical foundations of quantum mechanics*. Princeton University Press.

Waltz, D. L. (1982). Artificial intelligence. *Scientific American*, **247** (4), 101–22.

Ward, R. S. and Wells, R. O. Jr (1989). *Twistor geometry*, Cambridge University Press.

Weinberg, S. (1977). *The first three minutes: A modern view of the origin of the universe*, Andre Deutsch, London.

Weiskrantz, L. (1987). Neuropsychology and the nature of consciousness. In *Mindwaves* (ed. C. Blakemore and S. Greenfield), Blackwell, Oxford.

Westfall, R. S. (1980). *Never at rest*, Cambridge University Press.

Wheeler, J. A. (1983). Law without law. In *Quantum theory and measurement* (ed. J. A. Wheeler, W. H. Zurek), Princeton University Press. pp. 182–213.

Wheeler, J. A. and Feynman, R. P. (1945). Interaction with the absorber as the mechanism of radiation. *Revs. Mod. Phys.*, **17**, 157–81.

Wheeler, J. A. and Zurek, W. H., (ed.) (1983). *Quantum theory and measurement*. Princeton University Press.

Whittaker, E. T. (1910). The history of the theories of aether and electricity, Longman, London.

Wigner, E. P. (1960). The unreasonable effectiveness of mathematics. *Commun. Pure Appl. Math.*, **13**, 1–14.

Wigner, E. P. (1961). Remarks on the mind-body question. In *The scientist speculates* (ed. I. J. Good), Heinemann, London. (Reprinted in E. Wigner (1967). *Symmetries and reflections*, Indiana University Press, Bloomington) In *Quantum theory and measurement* (ed. J. A. Wheeler and W. H. Zurek), Princeton University Press, 1983.

Will, C. M. (1987). Experimental gravitation from Newton's *Principia* to Einstein's general relativity. In *300 years of gravitation* (ed. S. W. Hawking and W. Israel), Cambridge University Press.

Wilson, D. H., Reeves, A. G., Gazzaniga, M. S., and Culver, C. (1977). Cerebral commissurotomy for the control of intractable seizures. *Neurology*, **27**, 708–15.

Winograd, T. (1972). Understanding natural language. *Cognitive Psychology*, **3**, 1–191.

Wootters, W. K. and Zurek, W. H. (1982). A single quantum cannot be cloned. *Nature*, **299**, 802–3.

Index

Note: index entries to reference notes are indicated by suffix 'n'; Figures and Tables are indicated by *italic page numbers*.